Intersubband Transitions in Quantum Structures

McGraw-Hill Nanoscience and Technology Series
Series Editor: Omar Manasreh

GILLEO • *MEMS/MOEMS Packaging: Concepts, Designs, Materials, and Processes*

KOO • *Polymer Nanocomposites: Processing, Characterization, and Applications*

LIOU and FANG • *Microfluid Mechanics: Principles and Modeling*

LOBONTIU • *Mechanical Design of Microresonators: Modeling and Applications*

PAIELLA (ed.) • *Intersubband Transitions in Quantum Structures*

PETERS • *Molecular Thermodynamics and Transport Phenomena*

TOMM and JIMÉNEZ • *Quantum-Well Laser Array Packaging*

ZHANG • *Nano/Microscale Heat Transfer*

Intersubband Transitions in Quantum Structures

Roberto Paiella
Editor
Boston University
Boston, Massachusetts

McGraw-Hill
New York Chicago San Francisco Lisbon London Madrid
Mexico City Milan New Delhi San Juan Seoul
Singapore Sydney Toronto

The McGraw-Hill Companies

CIP Data is on file with the Library of Congress

1 2 3 4 5 6 7 8 9 0 DOC/DOC 0 1 2 1 0 9 8 7 6

ISBN 0-07-145792-5

The sponsoring editor for this book was Kenneth P. McCombs, the editing supervisors were Caroline Levine and Stephen M. Smith, and the production supervisor was Pamela A. Pelton. It was set in Century Schoolbook by Digital Publishing Solutions. The art director for the cover was Handel Low.

Printed and bound by RR Donnelley.

This book is printed on acid-free paper.

This book is dedicated to my wife, Lei Choon.

Contents

Contributors

Alexey Belyanin *Department of Physics, Texas A&M University, College Station, Texas* (Chaps. 5, 6)

Pallab Bhattacharya *Department of Electrical Engineering and Computer Science, University of Michigan, Ann Arbor, Michigan* (Chap. 8)

Federico Capasso *Division of Engineering and Applied Sciences, Harvard University, Cambridge, Massachusetts* (Chaps. 3, 6)

Subhananda Chakrabarti *Department of Electrical Engineering and Computer Science, University of Michigan, Ann Arbor, Michigan* (Chap. 8)

Thomas Elsaesser *Max-Born-Institut für Nichtlineare Optik und Kurzzeitspektroskopie, Berlin, Germany* (Chap. 4)

Carl H. Fischer *M.I.T. Lincoln Laboratory, Lexington, Massachusetts* (Chap. 8)

Claire Gmachl *Department of Electrical Engineering and PRISM, Princeton University, Princeton, New Jersey* (Chap. 5)

Rüdeger Köhler *NEST CNR-INFM and Scuola Normale Superiore, Pisa, Italy* (Chap. 2)

H. C. Liu *Institute for Microstructural Sciences, National Research Council, Ottawa, Ontario, Canada* (Chaps. 3, 7)

Oana Malis *Bell Laboratories, Lucent Technologies, Murray Hill, New Jersey* (Chap. 5)

Rainer Martini *Department of Physics and Engineering Physics, Stevens Institute of Technology, Hoboken, New Jersey* (Chap. 3)

Arup Neogi *Department of Physics and Materials Engineering, University of North Texas, Denton, Texas* (Chap. 10)

Roberto Paiella *Department of Electrical and Computer Engineering and Photonics Center, Boston University, Boston, Massachusetts* (Chap. 3)

Harald Schneider *Fraunhofer-Institute for Applied Solid-State Physics, Freiburg, Germany* (Chap. 7)

Hans Sigg *Paul Scherrer Institut, Villigen PSI, Switzerland* (Chap. 9)

Carlo Sirtori *Matériaux et Phénomènes Quantique, Université Paris 7, Paris, France; THALES Research & Technology, Palaiseau cede, France* (Chap. 1)

Alexander Soibel *Jet Propulsion Laboratory, California Institute of Technology, Pasadena, California* (Chap. 3)

Adrienne D. Stiff-Roberts *Department of Electrical and Computer Engineering, Duke University, Durham, North Carolina* (Chap. 8)

Xiaohua Su *Department of Electrical Engineering and Computer Science, University of Michigan, Ann Arbor, Michigan* (Chap. 8)

Roland Teissier *CEM2, Université de Montpellier 2, Montpellier, France* (Chap. 1)

Alessandro Tredicucci *NEST CNR-INFM and Scuola Normale Superiore, Pisa, Italy* (Chap. 2)

Mariano Troccoli *Division of Engineering and Applied Sciences, Harvard University, Cambridge, Massachusetts* (Chap. 6)

Preface

Semiconductor quantum structures, such as quantum wells, quantum wires, quantum dots, tunneling barriers, and superlattices, have been the subject of extensive research over the past three decades. Their basic appeal lies in the ability they provide to observe and engineer a wide variety of quantum mechanical effects, both for fundamental studies and for device applications in electronics and photonics. A particularly interesting class of semiconductor quantum devices is based on intersubband transitions, i.e., optical transitions between quantized states within the conduction band (or within the valence bands) in lowdimensional systems. These transitions offer several remarkable features for device applications, including the ability to tailor their center wavelengths, dipole moments, and relaxation times through bandgap engineering, ultrafast carrier lifetimes, and giant optical nonlinearities.

The past several years have witnessed the demonstration and development of many practical devices based on intersubband transitions, most notably the quantum well infrared photodetector (QWIP) and the quantum cascade (QC) laser. These devices have so far been primarily realized with InGaAs/InAlAs or GaAs/AlGaAs quantum wells grown by molecular beam epitaxy, which offer excellent material quality in terms of interface abruptness and thickness control and uniformity. Following extensive development efforts, both QWIPs and QC lasers operating at mid-infrared wavelengths have reached a remarkable level of performance, to the point that they have now become commercially available. In particular, QC lasers are the only mid-infrared semiconductor lasers capable of continuous-wave operation at room temperature, and they are playing an ever-increasing role in many sensing applications. Similarly, QWIPs offer excellent performance for infrared thermal imaging and are finding new applications utilizing their distinctive high-speed response.

At the same time, considerable research efforts are ongoing to extend the spectral range covered by these devices as well as to explore novel functionalities of intersubband transitions. For example, a major milestone of the past few years has been the development of QC lasers emitting at wavelengths beyond the *reststrahlen* band of the underlying material, i.e., in the terahertz region. The integration of bandgap-engineered intersubband optical nonlinearities within the QC active layer has also led to promising new device concepts. Other areas of research that are attracting increasing attention include the investigation of the high-speed properties of both QC lasers and QWIPs as well as the development of intersubband devices based on novel material systems. Most notably, these include quantum dots (which have advantageous features, especially for detector applications), Si/SiGe heterostructures (which are compatible with the silicon microelectronics platform), and wide-conduction-band-offset quantum wells such as the InGaAs/AlAsSb or GaN/AlGaN systems (which can accommodate intersubband transitions at near-infrared wavelengths).

This book is intended to provide a comprehensive review of the fundamental properties of intersubband optoelectronic devices, and of the more promising and exciting new developments in this rich field of research. QC lasers are the subject of Chaps. 1 through 3, which are devoted to high-performance mid-infrared devices, terahertz lasers, and QC sources of ultrafast mid-infrared pulses, respectively. The (sub)picosecond carrier dynamics of intersubband systems, which form the basis of the unique high-speed properties of QC lasers and QWIPs, are reviewed in Chap. 4, together with the ultrafast spectroscopic techniques used to investigate them. Chapters 5 and 6 provide a comprehensive description of the nonlinear optical properties of intersubband transitions, and of novel devices based on such nonlinearities in QC lasers. The principle of operation and fundamental properties of QWIPs are described in Chap. 7, together with new developments involving their high-speed properties and nonlinear response. Chapters 8 and 9 are devoted to the investigation of intersubband photodetectors and light emitters based on two "novel" material systems, In(Ga)As quantum dots and Si-based quantum structures, respectively. Finally, Chap. 10 describes ultrafast all-optical switching at communication wavelengths based on intersubband transitions in wide-conduction-band-offset InGaAs/AlAsSb quantum wells—another exciting new application enabled by the uniquely fast dynamics of these systems.

It is a pleasure to thank all the contributors for the time and energy they have devoted in preparing their chapters and in delivering them in a timely fashion. It is my hope that the resulting collection will be of interest and benefit to a broad audience of scientists, both conducting research on the device physics of intersubband transitions and using these devices for a large and increasing number of applications.

Roberto Paiella

Intersubband Transitions in Quantum Structures

Chapter

1

Quantum Cascade Lasers: Overview of Basic Principles of Operation and State of the Art

Carlo Sirtori
Matériaux et Phénomènes Quantique
Université Paris 7, 75251 Paris, France;
THALES Research & Technology
91767 Palaiseau cede, France

Roland Teissier
CEM2, Université de Montpellier 2
34095 Montpellier, France

1.1 Introduction

1.1.1 Historical introduction

Before the invention of the diode laser, a semiconductor laser based on transitions between Landau levels in a strong magnetic field was proposed by Lax in 1960. This is the first proposal of a semiconductor laser in which the optical transition occurs between low dimensional states of the same band (conduction or valence) rather than by the recombination of electron-hole pairs across the semiconductor bandgap. The idea of a unipolar laser was then ignored for many years since only 2 years after the proposal of Lax the first diode laser was demonstrated (Hall et al. 1962). This exploit drew all the attention of the semiconductor community on bandgap lasers and began the race for the first diode laser *continuous-wave* (cw) operation at room temperature. The race ended in 1970 when the first AlGaAs/GaAs heterostructure laser in cw operation at room temperature was achieved at nearly the same time by Alferov's group at the Ioffe Institute in St. Petersburg (Alferov et al. 1970) and by Panish et al. (1970) at Bell Labs (Murray Hill, NJ).

The effort to improve the performance of diode lasers and the development of the transistor technology in III-V compounds had an enormous impact on the epitaxial techniques for the growth of thin semiconductor layers and gave rise to the concept of two-dimensional structures such as quantum wells and inversion layers (Cho 1994). In fact, also in 1970, while the diode lasers reached technological maturity, Esaki and Tsu published their seminal paper presenting the concepts of a superlattice (Esaki and Tsu 1970). One year after, Kazarinov and Suris (1971), two scientists also from the Ioffe Institute, suggested that optical gain could be obtained by using transitions between two-dimensional states in a superlattice biased by an external electric field. This structure, very innovative with respect to the other semiconductor lasers, introduced the concept of a unipolar device in which the optical transitions could be completely engineered by the judicious choice of the thickness of well and barrier materials, regardless of their energy gap. Years later, after the demonstration of the first *quantum cascade* (QC) lasers, at a conference Capasso proudly declared, "The QC lasers make us finally free from the bandgap slavery."

It is hard to say if the Kazarinov and Suris structure could really show optical gain, and most likely it is not so for two main reasons. The first is related to the lack of reservoirs of electrons that inject carriers in each active region of the cascade. Without the reservoir it is necessary to bring electrons from the contacts, which makes the structure electrically unstable due to the formation of space charge domains. The second reason, more subtle, comes from the lack of a region where high-energy electrons can be accumulated without backfilling, by thermal effect, the ground state of the laser transition. In other words, the structure proposed by Kazarinov and Suris was lacking the *injector*, which today is thought to be an essential part of a quantum cascade laser.

For more than 15 years after the first proposal no real progress was made toward the realization of a unipolar laser. At the end of the 1980s, and beginning of the 1990s, researchers working on resonant tunneling reawakened the subject (Liu 1988, Henderson 1993). In the span of a few years, several proposals appeared on how to achieve population inversion by using intersubband transitions in superlattices or in coupled quantum wells, but none was implemented into a real laser structure.

In 1988, Capasso at Bell Labs was also working on resonant tunneling, and that year he published a review paper in which he proposed a unipolar laser based on a superlattice structure (Capasso et al. 1986). As a matter of fact, he really began to work on light interaction with intersubband transitions only in the 1990s, when Sirtori joined his group to study the linear and nonlinear optical properties of these transitions (Sirtori et al. 2000). The laser project really gained momentum one year later when Faist joined the group of Capasso. A little more than 2 years after his arrival, on January 14, 1994, for the first time Jerome Faist observed laser action from an electrically injected intersubband laser at 4.3-μm wavelength: The quantum cascade laser was born. The main results were published a couple of months later in *Science* (Faist et al. 1994). It is interesting to note that in the mid-infrared regime (3 μm < λ < 15 μm) the demonstration of a laser practically anticipated the observation of electroluminescence. This is related to the intrinsic difficulty of observing spontaneous emission from an electron on an excited subband which has a radiative efficiency close to 10^{-5}!

In a few years after the first demonstration, the performance of the quantum cascade laser dramatically improved: In 1996, above-room-temperature operation in pulsed mode was achieved, and the long-wavelength range was extended already to 11 μm (Faist et al. 1996, Sirtori et al. 1996). Today quantum cascade lasers in the range of 4- to 10-μm wavelength operate routinely continuous-wave at room temperature with hundreds of milliwatts of optical power (Bewley 2005). At low temperature, the concepts of quantum cascade lasers have been extended into the terahertz region and the whole wavelength range where lasers have demonstrated spans at present from 3.5 to 150 μm (Faist et al. 1998).

Between the first QC laser demonstration and the present, several contributions were made by many different groups. Before concluding this historical introduction, we mention some of what we believe are the breakthroughs that played a major role in this field. In chronological order, in 1998 a quantum cascade laser in a GaAs-based heterostructure was demonstrated at the Thomson (today Thales) Laboratories (Sirtori et al. 1998); in 2002, terahertz QC lasers were realized at the Scuola Normale Superiore of Pisa, and during the same year the University of Neuchâtel demonstrated the first laser operating in cw at room temperature (Köhler et al. 2002, Beck 2002); and in 2004 room-temperature high-power devices were developed (Evans et al. 2004a, b).

At present there remain several scientific challenges in the field of QC lasers: devices with short wavelength (λ < 3 μm), where the development of less conventional heterostructures is needed (GaN- or Sb-based materials); the use of one-dimensional or zero-dimensional structures for the realization of QC lasers based on quantum wires or dots; the improvement of the wall plug efficiency and the comprehension of the gain saturation that limits the peak output power.

That said, we are also convinced that the main challenge for this field lies in the development of real-world applications and the establishment of a commercial market.

1.1.2 Quantum engineering

In the last century, the work of engineers dealing with condensed matter, such as for electronics, for optoelectronics, and in general for material science, was in great part related to the manipulation of the material properties directly descendent from the chemistry of the principal constituents. Quantum engineering (Fig. 1.1) does not play with the chemical bonds between atoms as has been done until recently to discover and produce new and advanced materials. Rather, quantum engineering explores the possibility of controlling the material properties by defining the size and the spatial distribution of the constituents at the nanolevel independently of their chemical nature. The final characteristics of this new class of materials come directly from the redefinition of topological properties at the atomic level and can be applied with the same results on different combinations of materials. In this respect nanotechnologies are a direct emanation of the quantum engineering. The quantum cascade laser is an excellent example of how quantum engineering can be used to conceive efficient devices and emitters in the mid-infrared (mid-ir). In these devices the princi-

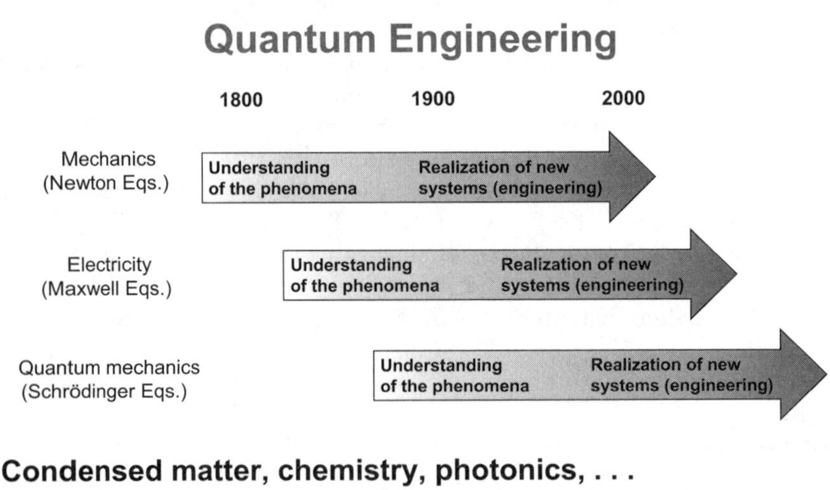

Figure 1.1 Quantum engineering is a consequence of our time. The rules of quantum mechanics are today sufficiently well known and verified so as to start a new engineering that develops devices based on Schrödinger equation. A similar scheme can be observed, by looking back through the centuries, in mechanics and electronics.

ples of operation are not based on the physical properties of the constituent materials, but arise from the layer sequence forming the heterostructure.

Quantum engineering associated with the recent progress of the epitaxial growth techniques allows one to ascribe within a semiconductor crystal artificial potentials with the desired electronic energy levels and wave functions. This approach is the basis for modifying, in a unique way, the optical and transport properties of semiconductors, opening avenues to artificial materials and the creation of useful devices. A remarkable illustration of this concept is the QC terahertz laser, in which by judiciously introducing less than 2% of Al atoms into a GaAs crystal we transform a piece of bulk semiconductor into a performing far-infrared laser!

Another crucial aspect of QC lasers, related to the quantum engineering, is that the fundamental principles of this device are essentially independent of the specific semiconductor system used. As of today, QC lasers have been demonstrated using basically three material systems: GaInAs/AlInAs//InP, which is the original system and the one that still gives the best performance for lasers in the mid-infrared range; GaAs/AlGaAs//GaAs, which is the material system for the terahertz laser; and AlSb/InAs//InAs (or //GaSb), which is the most recently exploited and whose very high conduction band discontinuity gives hope for short-wavelength QC lasers. Interestingly, in these three material systems, QC lasers have been fabricated, for instance, at 10 μm, confirming that the emission wavelength is totally independent of a particular transition intrinsic to the compounds material. This is unique, and no other laser, semiconductor or not, has this property.

1.1.3 Organization of the chapter

After the introduction in Sec 1.1 of the history of the quantum cascade laser and presentation of the scientific context in which this research is situated, the chapter has three main parts and a brief conclusion.

In Sec. 1.2 we recall the fundamentals of quantum cascade lasers. First we describe the main properties of a QC laser and its differences from diode lasers. Then in Sec. 1.2.2 we describe the rules that have to be respected to obtain population inversion between two subbands of the same band, and in Sec. 1.2.3 we give a description of the optical gain based on a simple rate equations model.

In Sec. 1.3 we review the state-of-the-art of QC lasers and emphasize some of the best results. We then present lasers at short and very long wavelengths of the spectral range and finally give an overview of the material systems in which QC lasers have been fabricated up to now.

The physical parameters relevant for the design and control of perfomance, such as conduction band discontinuity and electronic effective mass, are discussed in Sec. 1.4. In Sec. 1.4.5 we also give a basic description of the waveguides used for QC lasers.

1.2 Fundamental Principles of Quantum Cascade Lasers

1.2.1 Unipolarity and cascading

Semiconductor diode lasers, including quantum well lasers, rely on transitions between energy bands in which conduction electrons and valence band holes are injected into the active layer through a forward-biased *p-n* junction and radiatively recombine across the material bandgap (Yariv 1989). The latter essentially determines the emission wavelength. In addition, because the electron and hole populations are broadly distributed in the conduction and valence bands according to Fermi's statistics, the resulting gain spectrum is quite broad and its width is on the order of the thermal energy.

The unipolar intersubband laser or QC laser differs in many fundamental ways from diode lasers. All the differences are consequences of two main features which are unique to quantum cascade lasers and distinguish them from conventional semiconductor light emitters: *unipolarity* (electrons only) and a *cascading scheme* (electron recycling). These two features, shown schematically in Fig. 1.2, are

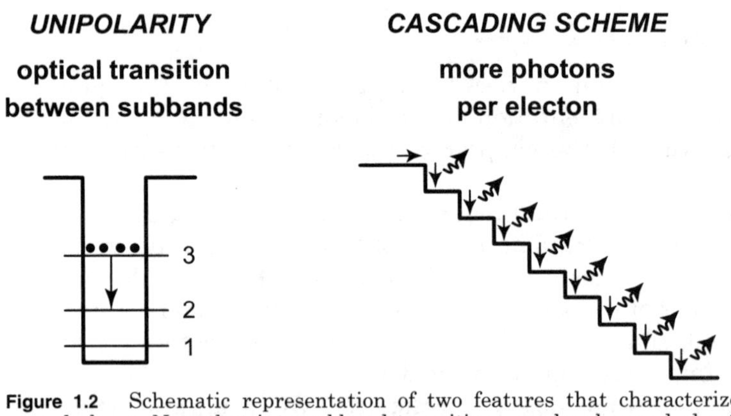

Figure 1.2 Schematic representation of two features that characterize a quantum cascade laser. Note that intersubband transitions can be observed also for holes, and therefore QC lasers are also conceivable in the valence band, for instance, using SiGe quantum wells.

independent and can be used separately, as has been demonstrated experimentally. In 1997 Garcia et al. demonstrated a cascade interband laser at ~830 nm, whereas Gmachl et al. in 1998 demonstrated an intersubband laser without a cascade scheme.

The unipolarity in QC lasers is a consequence of the optical transitions that occur between conduction band states (subbands) arising from size quantization in quantum wells. These transitions are commonly denoted as intersubband transitions. Their initial and final states are in the conduction band and therefore have the same curvature in the reciprocal space. If one neglects nonparabolicity, the joint density of states is very sharp, similar to the case of atomic transitions. In contrast to interband transitions, the gain linewidth is now only indirectly dependent on temperature through collision processes and many body effects. Moreover, for these devices the emission wavelength is not dependent on the bandgap of constituent materials, but can be tuned by tailoring the layer thickness. The highest achievable photon energy is ultimately set by the constituent conduction band discontinuity, while on the long-wavelength side there are no fundamental limits preventing the fabrication of QC devices emitting in the far infrared.

The other fundamental feature of QC lasers is the multistage cascade scheme, whereby electrons are recycled from period to period, contributing each time to the gain and the photon emission. Thus each electron injected above threshold can generate, in principle, N_p laser photons, where N_p is the number of stages. This leads to a quantum efficiency $\eta_q \gg 1$ and therefore very high optical output power, since both quantities are proportional to N_p.

1.2.2 Rules to get intersubband population inversion

In a QC laser, one period of the active zone can be approximated to a three-level system, as shown in Fig. 1.3. The role of the injector region is to transport electrons from level e_1 to level e_3 of the next period. Intersubband gain arises from a population inversion between levels e_3 and e_2. An important parameter for QC laser operation is the injection efficiency η_i. It is defined as the ratio of the current injected in level e_3 to the total current

$$\eta_i = \frac{J_3}{J} \tag{1.1}$$

In the ideal case $\eta_i = 1$. Deviations from this value can be due either to thermal activation from the injector to continuum or highly excited

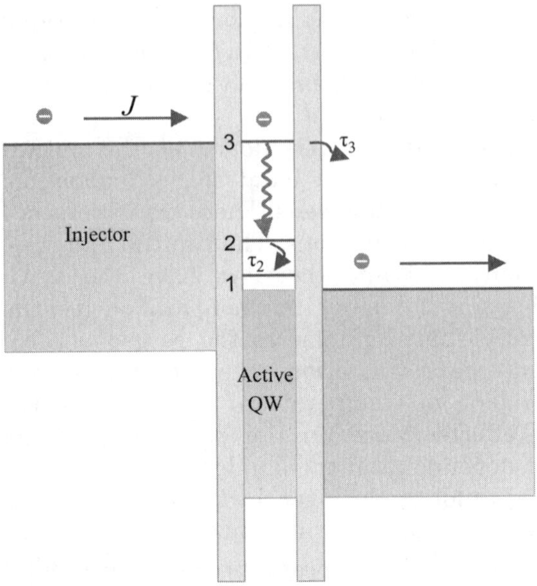

Figure 1.3 Schematic representation of one active region of a QC laser.

states, or to direct injection of electrons from the injector to the lower states e_2 and e_1. The first mechanism can be avoided by a proper design of the active zone and by the use of higher energy barriers. The second mechanism can be reduced by the use of a narrow quantum well after the injection barrier that enhances the injector coupling with e_3, but lowers the coupling with e_2 and e_1.

If τ_3 is the total lifetime of electrons in level e_3, the steady-state electron sheet density in this level is given by

$$n_3 = \frac{\eta_i J}{e}\tau_3 \tag{1.2}$$

If one assumes that level e_2 is populated only through the direct scattering of electrons from level e_3, the electron density in e_2 is simply given by

$$n_2 = n_3 \frac{\tau_2}{\tau_{32}} \tag{1.3}$$

where τ_{32} is the mean scattering time from e_3 to e_2 and τ_2 the lifetime of electrons in e_2. In this picture, the population inversion reads

$$n_3 - n_2 = \frac{\eta_i J}{e}\tau_3\left(1 - \frac{\tau_2}{\tau_{32}}\right) \tag{1.4}$$

In the most unfavorable case in which all the electrons flow sequentially from e_3 to e_2 (equivalent to saying that $\tau_3 = \tau_{32}$), n_2 has its maximum value

$$n_2 = \frac{\tau_2 \eta_i J}{e} \tag{1.5}$$

and

$$n_3 - n_2 = \frac{\eta_i J}{e}(\tau_3 - \tau_2) \tag{1.6}$$

If the condition $\tau_3 > \tau_2$ is satisfied, population inversion and optical gain are achieved for any current density. A typical way to achieve shorter τ_2 is to use the three-level scheme with the e_2 to e_1 transition resonant with the LO phonon energy. This is a remarkable feature of *QC lasers* (QCLs): There is no threshold for population inversion, and gain is present from the first flowing electron. As a result, the laser threshold current density is directly proportional to the optical cavity losses and not related to a transparency condition.

Hence, a main issue for achieving a low-threshold QCL is to reduce optical losses, both by fabricating low-loss waveguides and by using long devices (2 to 4 mm) to reduce the contribution of mirror losses.

1.2.3 Rate equations

The exact determination of the steady-state populations in the two states e_2 and e_3 involved in the intersubband optical transition is, in general, not straightforward. It results from the quasi-equilibrium distribution of the total electron population in the active zone, involving different intersubband scattering mechanisms. These include direct scattering from the previous injector across the injection barrier, or backscattering of hot or low-energy electrons from the next injector. The contribution of thermalized electrons to the population of e_2 is known as the thermal backfilling effect.

Backfilling can be avoided by proper design of the injector, giving a large enough energy Δ (Fig. 1.4) between level e_2 and the lower state of the injector. This energy is strongly dependent on the applied electric field, and in that respect, the field necessary to achieve injector alignment to e_3 is a critical parameter.

Beyond the simple description presented above, the phenomena that govern the population inversion in a more realistic QC laser structure are complex. Different approaches based on the rate-equation approximation that include a description of various scattering mechanisms

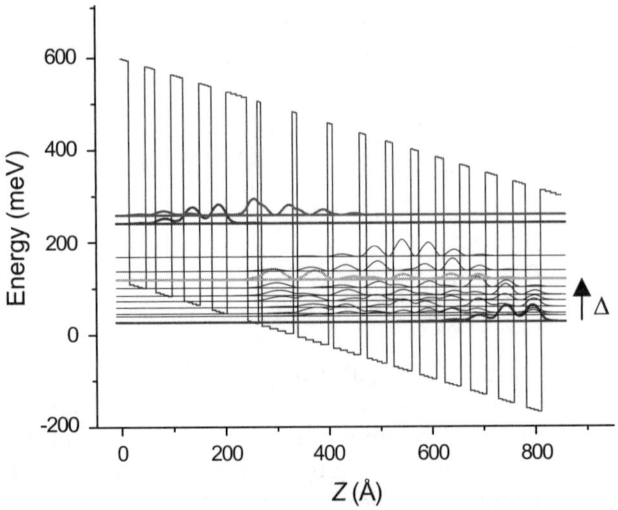

Figure 1.4 Schematic conduction band diagram of a portion of the laser heterostructure at threshold bias. The wavy lines indicate the moduli squared of the calculated wave functions. Indicated also is Δ, the energy that separates the ground state of the laser transition from the lower state of the injector.

between all quantum levels of the active zone have shown fair agreement with experimental characteristics, such as current-voltage curves (Donovan et al. 2001). However, not all scattering mechanisms are included in these calculations, and typically electron-electron interaction and interface roughness scattering are not taken into account.

Another approach is a nonequilibrium Green's function theory (Lee and Wacker 2002). It allows all the important scattering mechanisms to be included and accesses the current-voltage characteristics and gain spectra of QCLs. Its advantage over semiclassical rate equation approaches is clear for long-wavelength QCLs (terahertz), where the extension of the wave functions is easily beyond the coherence length of a quantum state in a semiconductor.

1.2.4 Gain derivation

To derive the gain of a quantum cascade active region, first we focus on the determination of the intersubband transition probabilities induced by the presence of an incident electromagnetic wave.

We consider a linearly polarized electromagnetic plane wave with an electric field $\mathbf{E} = E_0 \boldsymbol{\varepsilon} \cos(\omega t - \mathbf{q} \cdot \mathbf{r})$ of polarization $\boldsymbol{\varepsilon}$, pulsation ω, and propagation vector $\mathbf{q}$. In a semiconductor material of refractive index n, we have $q = n\omega/c$. The vector potential $\mathbf{A}$ associated with this incident electromagnetic wave is given by the relation $\mathbf{E} = -\partial \mathbf{A}/\partial t$ and reads

$$\mathbf{A} = -\frac{E_0}{2\omega}\,\boldsymbol{\varepsilon}\left(e^{i(\omega t - \mathbf{q}\cdot\mathbf{r})} - e^{-i(\omega t - \mathbf{q}\cdot\mathbf{r})}\right) \tag{1.7}$$

The action of this incident wave on the electron eigenstates of the heterostructure is given by the time-dependent hamiltonian H, written by using the dipolar approximation and the effective mass description (Bastard 1988) as

$$H = -\frac{e}{m^*}\mathbf{A}\cdot\mathbf{p} = V\left(e^{i\omega t} - e^{-i\omega t}\right) \tag{1.8}$$

where e is the electron charge, m^* is the effective mass in the quantum well material, and

$$V = \frac{eE_0}{2m^*\omega}\boldsymbol{\varepsilon}\cdot\mathbf{p} \tag{1.9}$$

Under the action of this perturbation hamiltonian, the probability per unit time that an electron makes a transition from the initial state $|\,i\rangle$ of energy E_i to the final state $|\,f\rangle$ of energy E_f is given by the Fermi golden rule:

$$W_{if}(\hbar\omega) = \frac{2\pi}{\hbar}\,|\langle f\,|\,V\,|\,i\rangle|^2\,\delta\left(E_f - E_i \pm \hbar\omega\right) \tag{1.10}$$

Using the momentum matrix element, we have

$$W_{if}(\hbar\omega) = \frac{2\pi}{\hbar}\frac{e^2E_0^2}{4m^{*2}\omega^2}\,|\langle f\,|\,\boldsymbol{\varepsilon}\cdot\mathbf{p}\,|\,i\rangle|^2\,\delta\left(E_f - E_i \pm \hbar\omega\right) \tag{1.11}$$

The $-\hbar\omega$ term is associated to the first component of **A** and corresponds to the absorption of an incident photon $\left(E_f = E_i + \hbar\omega\right)$, whereas the $+\hbar\omega$ term is associated to the second component of **A** and corresponds to the stimulated emission of a photon $\left(E_f = E_i - \hbar\omega\right)$.

In the effective mass approximation, the wave functions of the heterostructure states are in the form

$$\begin{aligned}\psi_i(\mathbf{r}) &= u_{\upsilon_i} f_i(\mathbf{r})\\ f_i(\mathbf{r}) &= \frac{1}{\sqrt{S}}e^{i\mathbf{k}_{\perp i}\cdot\mathbf{r}_\perp}\chi_i(z)\end{aligned} \tag{1.12}$$

where u_{υ_i} is the periodic part of the Bloch function, $\mathbf{k}_{\perp i}$ and $\mathbf{r}_\perp$ are the two-dimensional wave and position vectors in the plane of the layers of area S, and $\chi_i(z)$ is the envelope function, which describes the extension of the electron state in the direction perpendicular to the heterostructure layers.

For intersubband transitions, the initial and final states originate from the same band $\left(u_{\upsilon_i} = u_{\upsilon_f}\right)$. In this case, the matrix element of Eq. 1.11 simplifies to the matrix element of the envelope functions

$$\begin{aligned}\langle f \mid \varepsilon \cdot \mathbf{p} \mid i \rangle &= \langle f_f \mid \varepsilon \cdot \mathbf{p} \mid f_i \rangle \\ &= \langle \chi_f \mid \chi_i \rangle \frac{1}{S} \int dx\, dy\, e^{-i\mathbf{k}_{\perp f} \cdot \mathbf{r}_\perp} \left(\varepsilon_x p_x + \varepsilon_y p_y\right) e^{i\mathbf{k}_{\perp i} \cdot \mathbf{r}_\perp} \\ &\quad + \delta\left(\mathbf{k}_{\perp f} - \mathbf{k}_{\perp i}\right) \varepsilon_z \langle \chi_f \mid p_z \mid \chi_i \rangle\end{aligned} \tag{1.13}$$

Since the envelope functions are orthogonal, the term $\langle \chi_f \mid \chi_i \rangle$ is null if the final subband is different from the initial one. Consequently, the transition rate from state $\mid i \rangle$ to state $\mid f \rangle$ is given by

$$\begin{aligned}W_{if}(\hbar\omega) = \frac{2\pi}{\hbar} \frac{e^2 E_0^2}{4 m^{*2} \omega^2} \varepsilon_z^2 \mid \langle \chi_f \mid p_z \mid \chi_i \mid \rangle \mid^2 \\ \times \delta(\mathbf{k}_{\perp f} - \mathbf{k}_{\perp i}) \delta(E_f - E_i \pm \hbar\omega)\end{aligned} \tag{1.14}$$

Alternatively, we can choose to express the matrix element in the **r** representation, which is a much more common notation. In this case, the transition rate is written using the z dipole moment matrix element and reads

$$\begin{aligned}W_{if}(\hbar\omega) &= \frac{2\pi}{\hbar} \frac{e^2 E_0^2}{4} \varepsilon_z^2 \mid \langle \chi_f \mid z \mid \chi_i \rangle \mid^2 \\ &\quad \times \delta(\mathbf{k}_{\perp f} - \mathbf{k}_{\perp i}) \delta(E_f - E_i \pm \hbar\omega)\end{aligned} \tag{1.15}$$

Equations 1.14 and 1.15 contain several important pieces of physical information about intersubband transitions:

- The optical transitions occur only when the electric field is parallel to the direction of growth z. In fact, only the ε_z component of the polarization is left in the expressions of the transition rates. This has the immediate consequence that no transitions are possible if the light propagates perpendicular to the sample surface. This is the well-known intersubband polarization selection rule, which is very well observed experimentally.

- The optical transitions are vertical in **k**-space ($\mathbf{k}_{\perp f} = \mathbf{k}_{\perp i}$) and, in the parabolic band approximation, the transition energy $E_{if} = E_i - E_f$ and the transition rate W_{if} do not depend on the in-plane wave vector.
- The matrix element depends exclusively on the envelope functions and can be tailored by designing the shape of the wave functions in coupled well structures.

In summary, W_{if} is the general transition rate of an electron from subband i to subband f independent of its two-dimensional nature.

In reality, the subbands have a finite width that can be accounted for by replacing the delta function for energy conservation by a lorentzian function of half width at half maximum γ:

$$W_{if}\ (\hbar\omega) = \frac{\pi e^2 E_0^2\ |\langle z_{if}\rangle|^2}{2\hbar} \frac{\gamma / \pi}{(E_{if} - \hbar\omega)^2 + \gamma^2} \tag{1.16}$$

As far as QC lasers are concerned, the relevant quantity is the maximum stimulated emission rate, obtained for $\hbar\omega = E_{if}$, of

$$W_{if}^{\ \max} = \frac{e^2 E_0^2 |\langle z_{if}\rangle|^2}{2\gamma\hbar} \tag{1.17}$$

Equation 1.17 is the probability of stimulated emission or absorption of a photon per unit time and per electron present in the initial subband, in the presence of electromagnetic radiation of pulsation $\omega = E_{if} / \hbar$ and of electric field amplitude E_0 in the direction perpendicular to the layers.

To derive the propagation gain of a QCL active region, one must consider the geometry of the device (Fig. 1.5). Let us consider an electromagnetic plane wave propagating over a width w in the plane of a heterostructure of thickness L_p containing one QC active period consisting of the three levels of Fig. 1.3.

The power density carried by the plane wave of amplitude E_0 is given by

$$P = \frac{1}{2}\varepsilon_0 n c E_0^2 \tag{1.18}$$

and the number of photons of energy $\hbar\omega$ crossing the structure per unit time is

$$\Phi = \frac{1}{2}\frac{\varepsilon_0 n c E_0^2}{\hbar\omega} w L_p \tag{1.19}$$

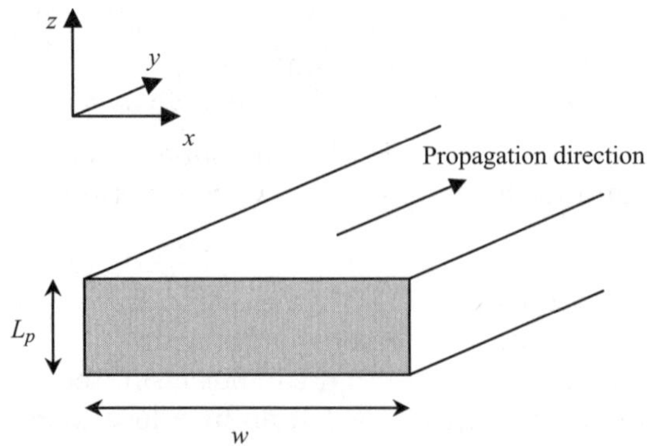

Figure 1.5 Schematic geometry of the device used to derive the material gain. The growth direction is z, and y is the direction of light propagation.

If $\hbar\omega = E_{32}$, the variation of the photon flux over a distance dy is

$$d\Phi = W_{32}^{\max} n_3 w\, dy - W_{32}^{\max} n_2 w\, dy \tag{1.20}$$

where $n_3 w\, dy$ (and $n_2 w\, dy$) is the total number of electrons of level e_3 (and e_2) in the slice of length dy. The first term corresponds to the stimulated emission of photons due to the presence of electrons in e_3, while the second corresponds to the absorption of photons due to the presence of electrons in e_2. The *propagation gain* (often also referred to as *material gain*) is defined as the variation of the photon flux divided by the number of photons (definition equivalent to the absorption coefficient)

$$G = \frac{d\Phi / dy}{\Phi} \tag{1.21}$$

By using Eqs. 1.19 and 1.20 and the intersubband transition rate (Eq. 1.17), one gets

$$G = \frac{2e^2 \langle z_{32} \rangle^2 \omega}{\varepsilon_0 n c 2\gamma L_p}(n_3 - n_2) \tag{1.22}$$

With the expression of the population inversion in Eq. 1.4 and by using the wavelength $\lambda = 2\pi c / \omega$ (in vacuum) of the propagating light, one finally gets the usual expression of the gain of a QC laser, proportional to the current density,

$$G = gJ \tag{1.23}$$

where g is the *gain coefficient*, defined as

$$g = \frac{4\pi e\langle z_{32}\rangle^2}{\varepsilon_0 n\lambda 2\gamma L_p}\eta_i\tau_3\left(1 - \frac{\tau_2}{\tau_{32}}\right) \tag{1.24}$$

A quantity commonly used in these expressions is the oscillator strength f_{32} of the intersubband transition

$$f_{32} = \frac{2m_0\omega}{\hbar}\langle Z_{32}\rangle^2 \tag{1.25}$$

The gain coefficient can then be written as

$$g = \eta_i \frac{e\hbar}{\varepsilon_0 c m_0}\frac{1}{2\gamma n L_p} f_{32}\tau_3\left[1 - \frac{\tau_2}{\tau_{32}}\right] \tag{1.26}$$

In a QC laser, normally the guided optical mode extends in the z direction outside of the active region. The gain is then reduced in the proportion of the spatial overlap Γ of the guided mode with the active zone

$$G = \Gamma g J \tag{1.27}$$

Up to now, we have considered only one period. If we have an active region composed of N_p periods, each having a comparable overlap with the optical mode, we can write $\Gamma = \Gamma_p N_p$, and therefore

$$G = \Gamma_p N_p g J \tag{1.28}$$

which makes it evident that the gain is proportional to the number of periods. This consideration is very well observed when $\Gamma < 50\%$ and the active region is centered in a flat part of the mode. To conclude, it is interesting to note that the gain coefficient g is not proportional to the number of periods of the active region. The gain multiplication, when using N_p periods, is canceled in Eq. 1.24 by the increase of the thickness L_p of the active region with the same factor. As mentioned before, the benefit of a large number of active periods comes from the corresponding increase of the overlap factor Γ with the guided optical mode.

The gain is quite small compared to that of interband semiconductor lasers. Threshold currents for the QC laser are higher, but much less sensitive to T since the excited state lifetime is by nature intrinsically short.

1.3 Quantum Cascade Lasers

1.3.1 Summary of present results

Today's best performance. Since their invention in 1994, QC lasers have reached maturity. They now are a well-established class of optoelectronic devices that can produce coherent radiation in a wide part of the infrared spectrum, although the best performances are obtained in the 5- to 13-μm wavelength range (Evans et al. 2004a). A remarkable feature of QCLs is their low sensitivity to temperature. Characteristic temperatures T_0 of 150 to 200 K are the standard, enabling high-temperature operation of the lasers; in pulsed mode a maximum operating temperature of 450 K (180°C) is currently achieved. However, wall plug efficiencies are in the range of 10% to 15% for the best devices, and thermal power management remains an important issue to be solved.

Room-temperature pulsed operation was rapidly obtained (Faist et al. 1996), as well as cw operation (Sirtori et al. 1996). However, the first continuous-wave operation of a QC laser at room temperature was reported only 8 years after its first demonstration (Beck et al. 2002).

Today, there is no doubt that in the wavelength range of 3.5 to 150 μm the QC laser outperforms all other semiconductor laser technologies based on current injection. Room temperature operation in pulsed mode has been achieved on a very wide spectral range, from 4 to 16 μm, and peak power on the order of 1 W is routinely obtained (Sirtori et al. 2002, Faist et al. 2002). In the last couple of years, continuous-wave operation above room temperature has been demonstrated between 4.8 and 9 μm (Beck et al. 2002, Evans et al. 2004*b*). Record cw optical power of 0.5 W at 6 μm (Fig. 1.6) has been recently reported by Evans et al., and the same authors have also reported average power close to 1 W for devices operating at 290 K (Evans et al. 2004a). This is the most important technological result that has been demonstrated for QC lasers, and it opens the route for new important applications, such as wireless communications and high-sensitivity optical sensors, based on mid-infrared radiation. However, room temperature cw operation is still one of the major challenges for QC lasers, due to the very high threshold power densities that generate strong self-heating of the devices. If we look at the numbers, we see that at 300 K the best lasers reach threshold at current densities on the order of 2 kA/cm^2. This would represent a reasonable injected power for laser diodes, where the voltage is typically on the order of the bandgap. However, in QC lasers, due to the cascade

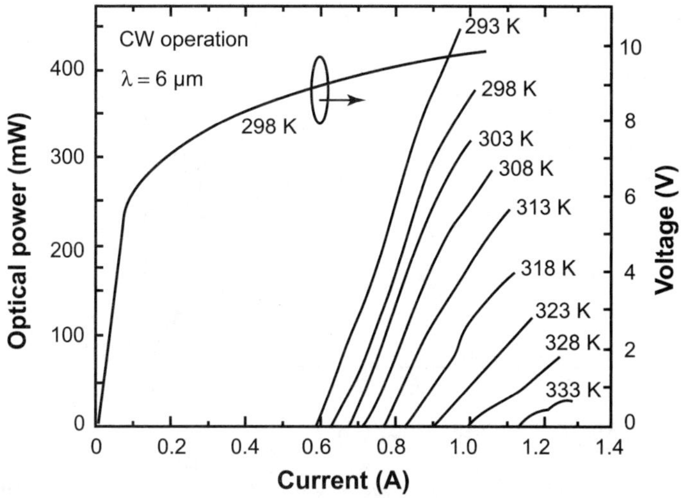

Figure 1.6 Continuous-wave light versus current ($L - I$) curve of an HR-coated, 9-μm-wide and 3-mm-long buried heterostructure laser at various heat sink temperatures. The V–I curve at 298 K is also shown. (Reprinted with permission from Evans et al. 2004a. Copyright © 2004, American Institute of Physics.)

scheme, the voltage is a function of the number of periods, and it can easily reach 10 V, as shown in Fig. 1.6.

High-speed modulation of quantum cascade lasers and mode locking have been the subject of intense investigations by a group of researchers at Bell Labs led by Capasso in recent years (Paiella et al. 2000, 2001). Among all these brilliant experiments, we would like to draw attention to the small-signal analysis of a laser under high-frequency modulation. The data in Fig. 1.7 are a clear demonstration of the absence of relaxation oscillation resonance for QC lasers. This is direct experimental evidence that the bandwidth of these lasers will be ultimately determined by the photon lifetime in the cavity rather than the resonant coupling between the photon field and the optical gain, as in conventional diode lasers.

In parallel, a new type of QCL emerged in 2002 (Köhler et al. 2002)—the terahertz laser emitting in the far infrared at a wavelength of around 100 μm (ω = 3 THz). These devices are discussed in the next section.

This rapid progress is the result of intense research, complex quantum engineering, and technological developments in active region and waveguide designs, in the growth of optimized heterostructure materials, and in device processing. We try to summarize hereafter the contributions that we feel are more important in each of the fields just mentioned.

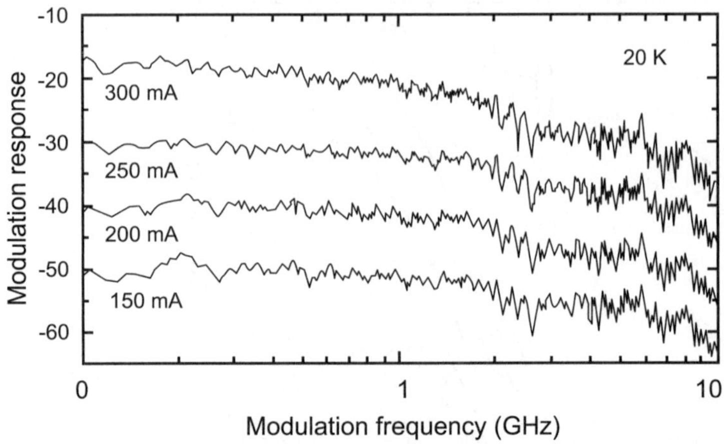

Figure 1.7 High-frequency modulation response traces of an 8-μm QC laser at 20 K for different values of the drive current ranging from very near threshold (150 mA), up to one order of magnitude higher photon density (300 mA). These traces were normalized to the experimental frequency response of the receiver and reflect only the modulation response of the QC laser. (After Paiella et al. 2001.)

Active region. The active region of the first QCL was based on a simple three-quantum-well (QW) scheme, where the laser transition was diagonal between two adjacent QWs. It emitted at a wavelength of 4.3 μm at cryogenic temperatures (Faist et al. 1994). Then more vertical transitions were explored, in order to improve the intersubband gain (Sirtori et al. 1998). The most efficient design was commonly considered to be a vertical intersubband transition in two coupled QWs, with a third very thin QW in the injection barrier. The role of the latter QW is to selectively enhance the amplitude of the excited state of the laser in the injection barrier, to increase resonant tunneling injection while preventing direct injection into the lower states.

For high-temperature and high-power operation of QC lasers, the efficiency of electron extraction from the lower state of the active quantum wells is an important issue. To overcome the bottleneck in electron extraction from the active region, new designs have been proposed. Superlattice-based active regions (Scamarcio 1997) have produced high-efficiency QCLs, thanks to a very rapid carrier extraction in the superlattice miniband. This was, however, to the detriment of electron injection into the upper state e_3. The most efficient scheme, called *bound-to-continuum* (Faist et al. 2001), combines efficient electron injection into a bound state e_3 and rapid extraction from a delocalized lower state.

Waveguide. Waveguide design is another fundamental issue. As already discussed, QCL performances are directly related to the low optical losses of the laser cavity. The undisputable champion for high performance in the mid-infrared range is the InP material used as optical cladding layers. It is a binary compound that provides a large dielectric refractive index contrast with the active region. In addition, InP has high electrical and thermal conductivity, without the need for a high doping level and therefore allows very low free-carrier optical losses. For this reason, QCLs made of materials grown on InP have considerable advantages. Alternative materials can, however, have their interest in specific cases. This material issue for QCL is discussed in greater detail in the following sections.

Growth. QCLs are very demanding devices in terms of heterostructure growth. Molecular beam epitaxy, with atomic monolayer control, is the technique of choice for their fabrication. The evolution of laser performances was directly related to the progress in growth quality. The question of growth rate stability is very important for the control of the emission wavelength. Short-term stability (hour scale) impacts the homogeneity of all periods of the active region and consequently the optical gain.

The question of interface roughness, due to atomic intermixing or segregation, is of great relevance (Offermans et al. 2003). First, it induces inhomogeneous broadening of the intersubband transition and consequently of the gain curve. In addition, interface fluctuations create a scattering potential in the plane of the epilayers. This is the cause of elastic scattering of electrons in the active region of the QCLs, which affects the electron dynamics and possibly the lifetime of electrons in the excited state (Leulliet et al. 2005).

Most important is certainly the decrease of epitaxial defects and subsequent reduction of optical losses. The more advanced realizations are the monolithic growth of complete QCL structures including active region and high-purity cladding layers (Evans et al. 2004a) which reduced optical losses to few cm^{-1}. This appears to be the key for the fabrication of very high-performance devices.

Processing. Beyond the conventional ridge geometry, solutions have been proposed to manage the dissipation of the high thermal power generated in the active part of QCLs. The reduction of the total area of QC devices, without the addition of extra waveguide losses, is one of the most important issues presently under investigation. To this end, two processing technologies are under development: (1) the conventional buried heterostructure used by Beck and coworkers (Beck et al. 2002) and (2) the selective current channeling by ion implantation recently demonstrated at the corporate laboratory of THALES (Sirtori et al. 2002). By exploiting this second way of realizing devices Page et al. were able to demonstrate cw operation of GaAs-based quantum cascade lasers up to 150 K (Page et al. 2004). Additional thick electroplated gold on the top contact is another way to spread the heat away from the active region (Evans et al. 2004a, Forget et al. 2005).

Monomode emission from distributed feedback (DFB) QCLs has been obtained on InP (Faist et al. 1997) and on GaAs (Schrenk 2000). With such technology, QCLs are a new class of single-mode sources available for applications to chemical sensing or pollution monitoring, of particular interest in the 3- to 5-μm and 8- to 12-μm ranges. The implementation of photonic crystals is another field explored toward more compact or surface-emitting QC devices (Colombelli et al. 2003). QC lasers are a very attractive light source for molecular spectroscopy when they are processed into distributed feedback lasers for wavelength control and stabilization. In Fig. 1.8 the spectra of GaAs-based QC lasers, mounted on a Peltier cooler, are shown. Notice that the emission wavelength varies as function of the temperature, due to the change of the material's refractive index. This is a very important parameter that allows one to fine-tune the emission frequency with the molecular resonance. The linewidth of free-running DFB lasers has been measured in different experiments and gives a value in the 2 to 5 MHz (Kosterev and Tittel 2002, Blaser et al. 2001). When stabilized by means of an electronic feedback loop, for high-stability

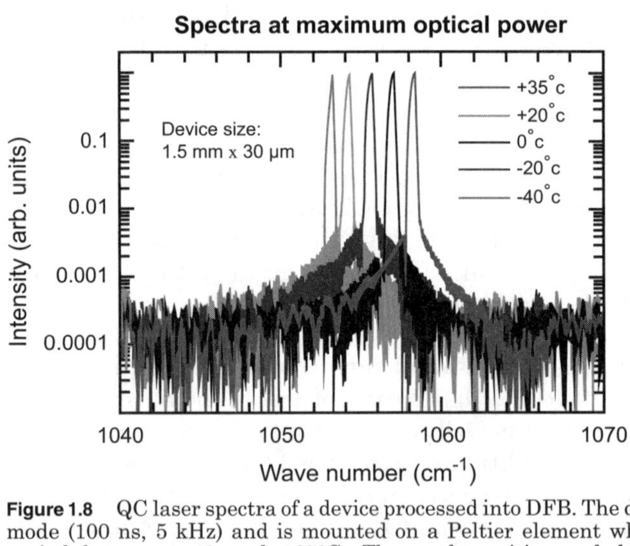

Figure 1.8 QC laser spectra of a device processed into DFB. The device operates in pulsed mode (100 ns, 5 kHz) and is mounted on a Peltier element where the temperature is varied between –40 and +35°C. The peak position red-shifts with temperature; therefore the far left curve corresponds to +35°C and that on the far right to –40°C. The peak optical power is in excess of 100 mW at all temperatures. Note that in this temperature range the device can be tuned over 5 cm^{-1}. (After Sirtori and Nagle 2003.)

operation, these devices have shown lines with intrinsic width well bellow 1 kHz (Myers et al. 2002).

Applications. As they produce compact light sources of coherent radiation working cw at room temperature in the mid-infrared range, from wavelengths of 4 to 10 μm, QCLs become attractive for many applications in this spectral range such as gas-sensing, free-space optical communications, or optical countermeasures. In the terahertz range, molecular spectroscopy or imaging applications promise the potential for long-wavelength QCLs.

Gas-sensing applications include, for example, trace gas detection, pollution or process monitoring, isotope separation, etc. Monomode lasers with narrow spectral width and high side mode suppression ratio are needed for spectroscopic applications. The tunability of the emission wavelength is an important property, obtained through the current-induced temperature shift of the laser line. High-T operation is also required for most applications. Continuous-wave operation is indispensable for high-resolution spectroscopic applications. However, cw operation is not absolutely necessary, as demonstrated by the company Cascade Technology. It developed a QCL-based spectroscopic system able to acquire the entire spectrum over one single current pulse (Stevenson 2004).

Free-space communications can also benefit from QCL sources emitting in the 3- to 5-μm and 8- to 12-μm atmospheric windows. Longer wavelengths are theoretically better because they are less sensitive to Raleigh scattering or bad atmospheric conditions.

QCLs are also proposed as candidates for optical countermeasures. However, for such applications consisting of deceiving or blinding adverse infrared sensors, emitted power is still an issue. The versatility of QCL technology and its ability to produce multiwavelength sources could be an advantage in this domain.

As far as commercial applications are concerned, new perspectives have appeared with the first MOCVD-grown QCLs (Green et al. 2003). This technique provides a more reliable and industrial-friendly approach for QCL fabrication. Despite its presumed impossibility to produce sharp interfaces, rapidly improving performances of MOCVD-grown QC lasers are reported (Troccoli et al. 2005a).

1.3.2 Frontiers

Beyond the conventional mid-infrared QC lasers of increasing maturity, new frontiers are being explored. Thanks to the flexibility of the active region and the cascading scheme, new functionalities or phenomena can be explored inside the laser cavity, such as nonlinear optics or Raman lasing (Troccoli et al. 2005b). However, one of the most attractive research areas is at the frontiers of the emission spectrum, to extend the spectral width of QC laser operation. The long-wavelength side is the emerging field of terahertz lasers, which opens new perspectives in terms of imaging or molecular spectroscopy. On the other end of the spectrum, short-wavelength QCLs ($\lambda < 4$ μm) could be the candidate to fulfill the lack of semiconductor laser sources operating at room temperature in the 3 to 4 μm range.

New physics. The high intensity of optical field in the cavity of QCLs allows generation of intracavity nonlinear effects. Nonlinearity can originate either from the semiconductor crystal itself or from specifically designed structures inserted in series with the active region. Frequency doubling has been demonstrated (Owschimikow et al. 2003) as well as sideband generation (Dihlon 2005).

The coherent light generated in the QCL cavity can also be used to pump intersubband transitions and generate emissions at different wavelengths. This principle gave birth to the intracavity Raman laser (Troccoli 2005b).

Terahertz domain. In only a couple of years, since the first demonstration in 2002 (Köhler et al. 2002), the progress on QC terahertz lasers has been astonishing. Not only has the wavelength range been extended by more than a factor of 2 from 67 to 150 μm (Worral et al. 2005), but also the operating temperatures have noticeably improved: In continuous-wave, terahertz devices operate a few degrees above liquid nitrogen, and in pulsed-mode they reach 140 K (Williams et al. 2005). In Fig. 1.9 we report *V-I* and *L-I* characteristics at temperature $T = 10$ K of a device with an emission wavelength of 103 μm (2.9 THz) (Alton et al. 2005).

The principles of operation of terahertz QC lasers relie on rather different concepts than do devices in the mid-infrared, which take advantage of electron-LO-phonon scattering processes above the material *reststrahlenband* to achieve large population inversions. Also for terahertz lasers, efficient depletion of the lower level is essential, while long lifetimes of the upper level are highly desirable. To this end, terahertz active regions are designed to have the optical transition, with a large dipole matrix element, across a minigap 10 to 15 meV wide. The lower laser state is strongly coupled to a wide injector miniband (comprising a high number of subbands), which provides a large phase space where electrons scatter, thus ensuring a fast depletion of the lower state of the laser transition. Moreover, the minibands allow efficient electrical transport, even at high current densities, and suppress thermal backfilling.

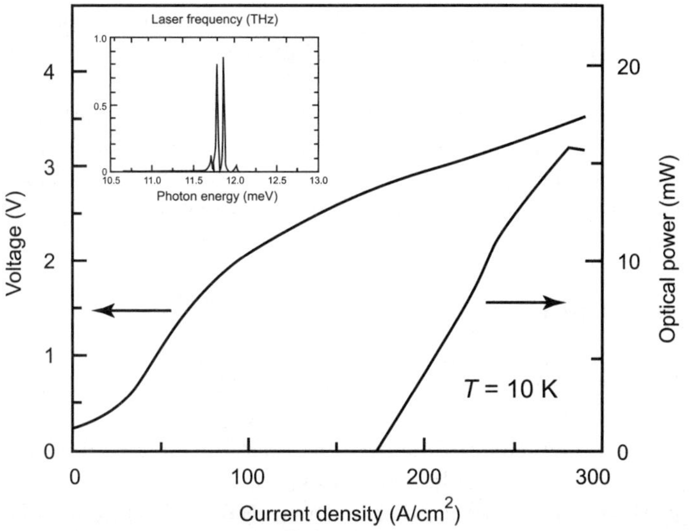

Figure 1.9 Electrical and optical characteristics (continuous-wave) of a 100-μm-wide ridge terahertz laser at 10 K. The length of the device is 2 mm. In the insert is the high-resolution emission spectrum.

An additional relevant issue for the terahertz range is the fact that conventional laser waveguides are not suitable, owing to large free-carrier absorption losses and practical limitations on the thickness of epilayer growth. For these long-wavelength lasers, the optical confinement is never achieved by using dielectric claddings, but by metallic layers very much as in the case of microwave strips. Even if operating at low temperature, terahertz QC lasers are very promising candidates to become compact sources for imaging systems, or local oscillators for heterodyne detection (Gao et al. 2005).

Short wavelength. The first demonstration of a QC laser (Faist et al. 1994) was a device emitting at a wavelength of 4.2 μm. The shortest wavelength demonstrated today is 3.5 μm, not too far from the first QC laser (Faist et al. 1998). As a matter of fact, the standard wavelength for a QCL is restrained in a wavelength range of around 5 to 13 μm. The latter corresponds to photon energies between 100 and 200 meV, easily accessible for intersubband transitions in GaInAs/AlInAs quantum wells.

The realization of QC lasers emitting at short wavelength is more difficult, since they are limited by the finite depth of the quantum wells (ΔE_c in Fig. 1.14). This parameter sets a limit for the energy level e_3, the excited state of the laser transition. In typical QCL designs the energy level e_3, measured from the conduction band edge in the QW, has a value of twice the photon energy. Thus, a QCL made of

GaAs/AlGaAs materials with $\Delta E_c = 0.36$ eV will not be able to operate with photons of energy greater than 0.18 eV, that is, below a wavelength of 7 μm.

This limitation on the short wavelength is fundamental and dependent on the materials. Alternative materials presenting higher conduction band discontinuities are required to achieve QC laser emission below 4 μm. This is the main motivation of most of the research on new material systems for QCLs presented below.

1.3.3 Results on different materials

By nature, the concept of a QC laser does not depend on the choice of the well and barrier materials that constitute the heterostructure. However, material-dependent properties will play a significant role in the device characteristics, through their influence on barrier heights, scattering mechanisms, waveguide properties, etc.

The different III-V materials that have been investigated for QC lasers are presented in Fig. 1.10. Among all the possible combinations two material systems have been already amply exploited to produce QC lasers: InGaAs/InAlAs grown lattice matched on indium phosphide (InP) substrate and GaAs/AlGaAs grown on gallium arsenide (GaAs) substrate.

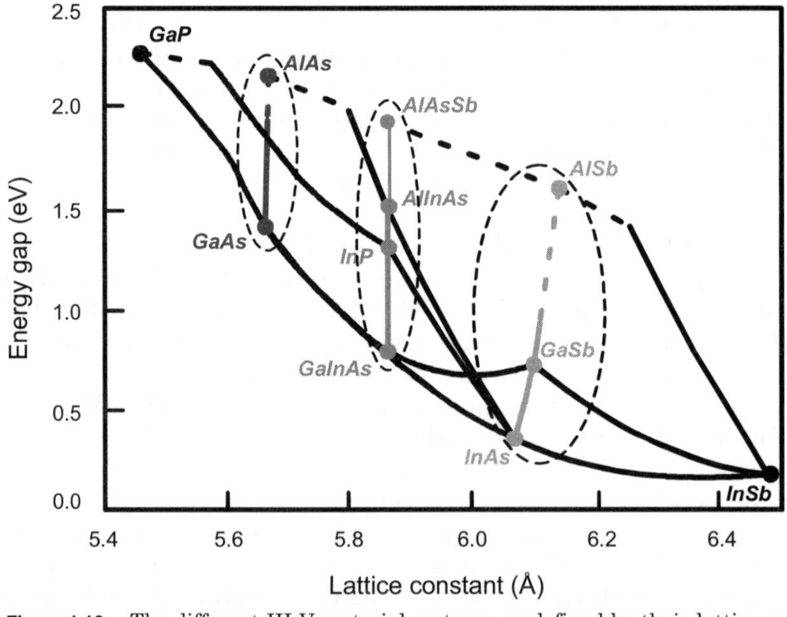

Figure 1.10 The different III-V material systems, as defined by their lattice constants. The heterostructures that have been already used to realize quantum cascade lasers are indicated by the dashed ovals.

InGaAs/InAlAs on InP. This is the historical material system in which the laser was first demonstrated (Faist et al. 1994). In the mid-infrared, InGaAs/AlInAs quantum wells lattice matched on InP substrates are the best-performing solution for QC lasers. Today, they cover the spectrum from 3.6 to 24 μm. CW operation at room temperature is obtained from $\lambda = 4.3$ μm (Yu et al. 2005) to $\lambda = 9.6$ μm (Beck et al. 2002).

With a ΔE_c of 0.52 eV, the standard InP system fails to produce high-performance QCLs below 6 μm. However, its spectral range can be extended toward short wavelengths by using strained compensated materials, with increased indium content in the InGaAs wells and increased aluminium content in the InAlAs barriers. This provides a larger band discontinuity of 650 to 750 meV as a function of the In and Al content. As a matter of fact, QCL devices made of these strained materials have shown performances of 300-mW continuous-wave emission at a wavelength of 4.8 μm at room temperature (Evans et al. 2004b) and 160 mW at $\lambda = 4.3$ μm (Yu et al. 2005). DFB lasers operating cw at room temperature have also been realized with InP-based QCLs, first at a wavelength of 5.4 μm (Blaser et al. 2005).

GaAs. A few years later, GaAs-based QCLs were demonstrated (Sirtori et al. 1998a). They now range from 8 μm to the terahertz domain at wavelengths up to 160 μm. In the mid-infrared, this material system has less interesting performance that of InP-based materials. The major drawback is the relatively higher threshold at room temperature ($J_{\mathrm{th}}^{\mathrm{GaAs}} \sim 5 J_{\mathrm{th}}^{\mathrm{InP}}$) which makes cw operation almost impossible for these devices above 200 K. On the long-wavelength side, GaAs becomes better performing than InP fundamentally for reasons of material purity, which dramatically increase the mobility and the conductivity of the semiconductor at low temperature and guarantee lower waveguide losses. This has allowed the birth and development of the terahertz QC lasers (Köhler et al. 2002).

Antimonides at 6.1 Å. Antimonide compound semiconductors are the family of III-V materials with a crystal lattice constant of about 6.1 Å (Fig. 1.10), which can be grown lattice-matched on gallium antimonide (GaSb) or indium arsenide (InAs) substrates. This family includes the three binary compounds InAs, GaSb, and AlSb and their alloys.

The most interesting property of antimonides for the design of QCL is the remarkably large depth of InAs QW with AlSb barriers $\Delta E_c = 2.1$ eV. In principle, it should enable the design of very short-wavelength QCLs down to the near-infrared range (1.5 μm). The available QW depth is, however, limited by the position of the lateral valleys (X or L) of the Brillouin zone. The L valley is 800 meV above the conduction band edge in InAs. Therefore, the emission range for this novel material system is probably restricted to wavelengths greater than 3 μm. Other specific issues for the realization of short-wavelength QCLs are presented in Sec. 1.4.

A second specificity of antimonides is the very small effective mass in InAs. This leads to significantly stronger intrinsic intersubband optical gain, compared to other material systems (see Fig 1.13 below). Hence, from a material point of view, it is clear that antimonides are a very attractive system for mid-infrared QCLs.

In spite of the very high conduction band discontinuity, the first QC laser demonstrated in this material system was at 10 μm (Ohtani and Ohno 2003). More recently the potential of this heterostructure has been better exploited and lasers emitting at ~4 μm have been demonstrated up to room temperature (Teissier et al. 2004). Even though the performances of these devices are very encouraging, they have not yet reached those of the QC laser based on InP. We cannot address any fundamental reasons for this, but just remind that these very recently born devices need more work and optimization.

InGaAs/AlAsSb on InP. An intermediate solution has been proposed and developed from 2003—to use the antimonide alloy AlAsSb lattice-matched on InP (Fig. 1.10) as a barrier material with InGaAs as the well material both lattice-matched on InP. This heterostructure benefits both from the advantages of the InP system and from a high ΔE_c provided by the AlAsSb barriers. It is, in our opinion, a very attractive material choice for the short wavelengths, for the most part for technological reasons related to the maturity of the processing and InP waveguide claddings. In principle, however, GaInAs has a higher mass and lower lateral valley energy than InAs, which should give lower gain and reduced effective QW depth. QCLs made of these InGaAs/AlAsSb materials were realized for the first time in 2004 (Revin et al. 2004). The devices emitted at $\lambda = 4.3$ μm in pulsed mode up to a temperature of 240 K. Recently, very high-temperature operation of similar devices around $\lambda = 4.5$ μm has been reported (Yang et al. 2005).

Si/SiGe. For Si/SiGe the steps toward a unipolar laser are less advanced. Nevertheless, intersubband electroluminescence has been recently observed from QC active regions realized on metamorphic substrate $Si_{0.5}Ge_{0.5}$ (Diehl et al. 2002). The realization of QC lasers in Si/SiGe would represent a major breakthrough simply because it would represent the first laser based on Si. In this material system, intersubband transitions occur in the valence band, which is a serious complication from the point of view of the theoretical description.

Moreover, the presence of the different hole dispersions (heavy, light, and split off) increases enormously the number of subbands and makes it more difficult to control their energy separation. Even if preliminary, these recent results on electroluminescence represent, in our opinion, the closest yet to the demonstration of a Si-based semiconductor laser.

1.4 Material Issues

1.4.1 Fundamental Parameters m^* and ΔE_c

In Sec. 1.1 and in particular in the paragraph where we describe our definition of quantum engineering, we insist on the minor dependence that the quantum cascade laser concept has on the material parameters of the heterostructure. Indeed, there are some physical parameters, related to the material system, that have a direct influence on the laser characteristics. At the first order, there are two parameters originating from the heterostructure that have to be taken into account: the conduction band discontinuity ΔE_c and the effective mass of the well material m^*.

1.4.2 Role of the effective mass m^*

The role of m^* is complex, but can be easily illustrated by analyzing the formula of the gain (Eq. 1.24). We take the expression of the gain coefficient g and compare its value for two material systems: GaAs and InAs. Moreover, to reduce the influence of the active region design and have a fair comparison of the two materials, we apply this formula to the first two states (E_1 and E_2) of a quantum well with very high barriers (Fig. 1.11). In the limit of $\tau_2 << \tau_{32}$ (or $\tau_1 << \tau_{21}$ in the present two-level picture), normally very well satisfied at low temperature, and with unity injection g can be written as

$$\begin{aligned} g &= \frac{4\pi e}{\varepsilon_0 n \lambda\, 2\gamma L_p} \,|\, z_{21} \,|^2\, \tau_2 = \frac{2e}{\varepsilon_0 n c \hbar} \frac{1}{2\gamma L_p} E_{21} \tau_2 \,|\, z_{21} \,|^2 \\ &= \frac{e\hbar}{\varepsilon_0 c m_0} \frac{1}{2\gamma n L_p} f_{21} \tau_2 \end{aligned} \tag{1.29}$$

In this expression the effective mass has a direct influence on the three terms on the right, namely, the transition energy E_{21}, the squared dipole matrix element, $|z_{21}|^2$, and the lifetime of the excited state of the laser transition τ_2. If we set the transition energy to a fixed value for both materials, for instance, 124 meV ($\lambda = 10$ μm), then the quantity

$1/m^*L^2$ is identical in the two materials, where L is the width of the quantum well.

Therefore the lighter the mass, the wider is the quantum well, as schematically illustrated in Fig. 1.11. It can be easily demonstrated that in a quantum well the absolute value of the dipole matrix element $|z_{21}|$ is proportional to L. Thus it can be concluded that for a given material the square of the dipole matrix element is inversely proportional to the effective mass. If we now concentrate on τ_2, the excited state lifetime, we can safely assume that it is merely controlled by the electron optical-phonon interaction. In this case it can be shown that τ_2 depends on the inverse of the square root of the mass. In summary, the gain coefficient g is

$$g \propto |z_{32}|^2 \tau_2 \propto \left(m^*\right)^{-3/2} \tag{1.30}$$

which means that the gain coefficient is proportional to $(m^*)^{-3/2}$ and that the ratio of the gain coefficients of the two materials InAs and GaAs can be evaluated to

$$\frac{g^{\text{InAs}}}{g^{\text{GaAs}}} = \left[\frac{m^*_{\text{GaAs}}}{m^*_{\text{GaInAs}}}\right]^{3/2} = \left[\frac{0.067}{0.023}\right]^{3/2} \cong 5 \tag{1.31}$$

This example evidently illustrates the strong dependence of the gain on the effective mass. However, the full picture is more complicated due to the effect of the conduction-band nonparabolicity that makes the electronic effective mass a function of the energy $m^* = m^*(E)$. This effect arises from the coupling of the electronic wave functions with those of the valence bands, and it is more pronounced for materials with small gaps, such as InAs. To calculate the energy-dependent effective mass, we used a simplified 3×3 Kane hamiltonian which describes the conduction, light-hole, and split-off bands, since near the Γ point the heavy-hole band is decoupled from the others (Sirtori et al. 1994). The value of m^* as a function of the energy for different material systems is

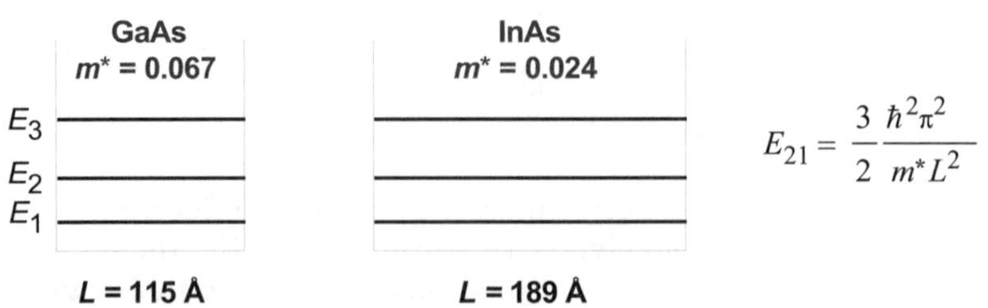

Figure 1.11 Quantum wells made of different materials. Notice that if we fix the energy separation E_{21} the width of the quantum well is inversely proportional to $\sqrt{m^*}$.

plotted in Fig. 1.12. As shown, the mass is almost linear as a function of energy. We remark that the slope is almost the same for all materials. This is so because the slope is directly proportional to the Kane energy, which is constant within less than 10% for all the III-V materials (Bastard 1988). This is not in contradiction to the fact that for narrow-gap materials nonparabolicities are stronger, since the latter have to be seen as a relative increase of the mass with the energy. As an example, we compare the relative increase of the effective mass at 200 meV above the bottom of the conduction band (the zero of the energy in our calculations) for GaAs and InAs. We can see that m^*_{GaAs} (200 meV) has increased only 10%, with respect to its zero energy value m^*_{GaAs}. If we now take $\left[m^*_{\mathrm{InAs}}(200\,\mathrm{meV}) - m^*_{\mathrm{InAs}}\right] / m^*_{\mathrm{InAs}}$ we can see that relative increase corresponds to more than 40%.

We are now in a position to comment on the energy-dependent gain that takes into account the masses of the initial and final states. In this case the definition of the dipole matrix element gets more complicated since the conduction and valence components of the wave function have to be taken into account (Sirtori et al. 1994). It is intuitive that the higher the laser transition energy, the heavier are the effective masses of the states. This has as a final consequence that the gain as a function of the energy can saturate, as in the case of InAs. An illustration of this behavior of the gain in our simplified model is given in Fig. 1.13. In this figure we plot the product of the oscillator strength f and the upper state lifetime τ as a function of the transition energy in a single quantum well for four different material systems. On the left panel we have calculated the product $f^*\tau$ for the 2–1 transition and on the right that of the 3–2 transition. Note that the value of the gain in the right panel is almost

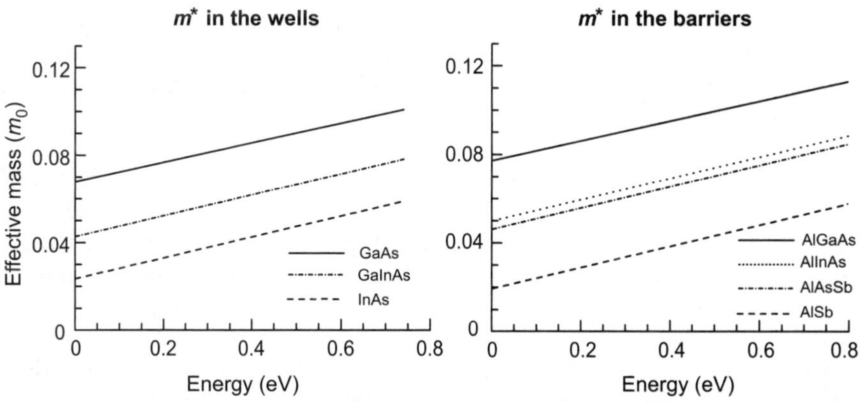

Figure 1.12 Energy dependence of the effective mass in the wells (left) and in the barriers (right) for the different material systems. For the barrier materials the zero of the energy is set at the bottom of the conduction band of the associated well materials, so that one can read the value of the masses on the same energy scale on both panels.

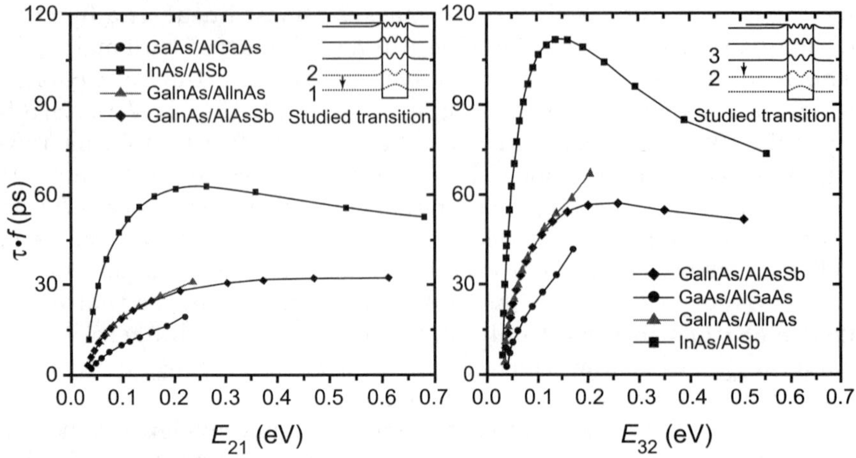

Figure 1.13 The product $f{\cdot}\tau$ between levels 2 and 1 is plotted on the left, and between levels 3 and 2 on the right. The quantity $f{\cdot}\tau$ directly proportional to the gain.

twice that on the left, showing the importance of designing lasers between excited states.

1.4.3 Role of the conduction band discontinuity ΔE_c

At first glance, the role of ΔE_c is self-evident and defines the highest photon energy achievable in a given material system, as mentioned earlier. Therefore, the range in which the emission wavelength of the QC laser can be tuned by adjusting the width of the layers is finite and has an upper limit set by the depth of the quantum wells. Nonetheless, on the long-wavelength side, there are no fundamental limits preventing the fabrication of QC lasers emitting in the far-infrared. Apart from the wavelength range, there are less evident effects controlled by ΔE_c that need to be mentioned.

Escape to the continuum. When a confined level gets closer to the top of the barriers, the probability of thermal activation of electrons to a delocalized state increases dramatically (Fig. 1.14). The scattering rate of an electron out of the confined state e_3 is controlled by the occupation probability in the subband at an energy greater than that of the top of the barrier, and by the typical scattering time τ_{scatt} from this subband. The model to describe this effect is identical to the one used for calculating dark current in quantum well infrared photodetectors (QWIPs) (Liu et al. 1993).

The two-dimensional density of electrons localized in the quantum well and with energy greater than ΔE_c is given by

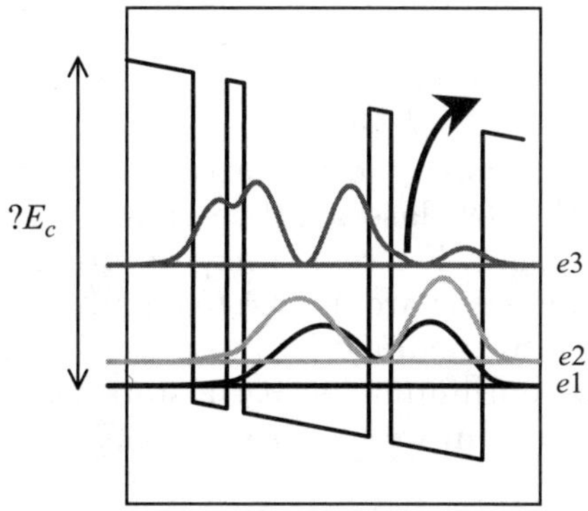

Figure 1.14 Schematic representation of a QC laser active region. The arrow indicates the path of electrons that, due to thermal effects, are activated into the continuum states.

$$n_{2D} = n_3 \exp\left(-\frac{\Delta E_c - e_3}{kT_e}\right) \tag{1.32}$$

where n_3 is the total sheet density in level e_3, and T_e is the electron temperature. It has been measured (Spagnolo et al. 2004) to be sensibly greater than the lattice temperature by a value on the order of 100 K.

The leakage current density to the continuum is then simply given by

$$J_{esc} = \frac{qn_{2D}}{\tau_{scatt}} \tag{1.33}$$

The important issue is the estimation of the scattering time. It can be derived from the drift velocity and mean transit time of unbound electrons over the quantum well, as usually assumed in QWIP theory. In the case of QCLs, however, the band structure of the active region is much more complex with a large number of confined states, and the applied electric field is far greater than the one in QWIPs. For these reasons, we can neglect the probability of electron recapture to the initial state. The time τ_{scatt} is then the mean scattering time of the confined state to a continuum of states. We can safely assume that in InGaAs/InAlAs structures τ_{scatt} is dominated by LO phonon scattering and is on the order of 0.2 ps.

This escape current has to be compared to the cascade current J_c which is governed by LO phonon scattering to the lower subbands. The electron lifetime in the excited state of the active region τ_3 is on the order of 1 ps, and its temperature dependence is given by the variation of the phonon population

$$\tau_3 = \frac{\tau_{3(0)}}{1 + 2/\left[\exp(\hbar\omega_{LO}/kT) - 1\right]} \tag{1.34}$$

The calculated ratio of leakage current density to the cascade current is shown in Fig. 1.15 for InGaAs/InAlAs QC lasers lattice-matched on InP, as a function of emission wavelength. For room temperature operation, the limit set by the escape current is around 4.5 to 5 μm.

The effective activation energy can be modified by the presence of minigap or quasi-bound states in the continuum density of states, due to quantum interference with the quantum wells. These effects have been reviewed in Faist et al. (2000).

When a high electric field is applied, a lowering of the barrier occurs. Moreover, due to tunneling the continuum states penetrate into the active region, thus also modifying the effective activation energy. All these effects contribute to make the design of short-wavelength QC lasers more and more difficult.

Injector barrier transparency. The basis of QC laser engineering is the control of the coupling between adjacent QWs. In that respect the band discontinuity ΔE_c is a critical parameter, since it sets the transparency of the barriers. The wave function decay rate in the barrier for a level of energy E is given by

$$k_B = \sqrt{\frac{2m_B^*(E)}{\hbar^2}(\Delta E_c - E)} \tag{1.35}$$

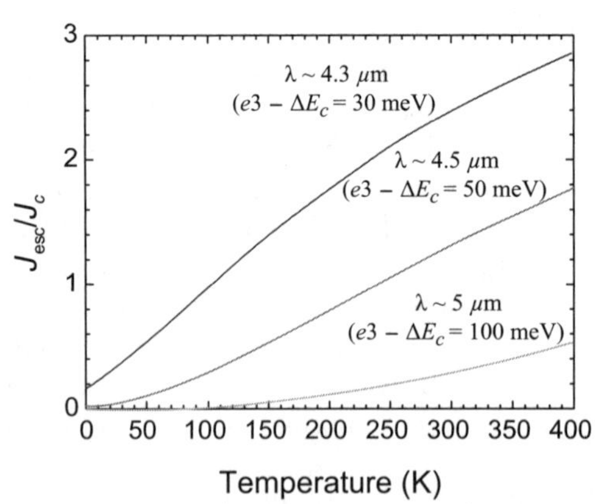

Figure 1.15 Ratio between the current in the active region and the leakage current due to the thermal leakage in the continuum. It is clear that for activation energy smaller than 50 meV, room temperature operations are completely hindered by the thermal activated current.

where $m_B^*(E)$ is the energy-dependent effective mass in the barrier material. The transmission of the barrier is proportional to the attenuation of the squared wave function over the thickness of the barrier L_B.

$$T_B \approx \exp(-2k_B L_B) \tag{1.36}$$

The exponent leads to very rapid variation of the transparency as a function of the barrier parameters. Note that not only the band discontinuity, but also the effective mass in the barrier plays an important role in this effect. For materials with large ΔE_c, such as antimonides, much thinner barriers are needed for a given coupling strength. This is illustrated in the Fig. 1.16 where the coupling is observed through the splitting energy of anticrossed levels in symmetric double quantum wells.

A typical value for the injection barrier in InGaAs/InAlAs QCLs is 40 to 50 Å. In GaAs/AlGaAs QCLs, equivalent transmission is obtained for 60- to 70-Å-thick barriers, while in InAs/AlSb QCLs much thinner barriers (about 25 Å) are needed. As a consequence, greater precision on layer thicknesses is required in material systems of larger ΔE_c. The problem is more severe for thinner barriers. It is current practice to use 10- to 20-Å-thick coupling barriers in the active quantum well or in the injector of InP-based QCLs. From Fig. 1.16, their transposition in the InAs/AlSb system leads to 4- to 6-Å-thick AlSb layers, i.e. one to two atomic monolayers. This constitutes a limitation on the use of high

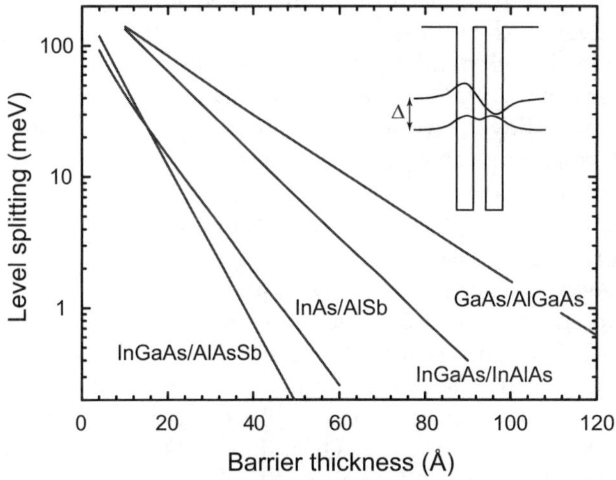

Figure 1.16 Calculated splitting of the levels of two coupled quantum wells, giving a picture of barrier transparency for different material systems. The confinement energy is kept constant at a value of 250 meV from the bottom of the QW.

ΔE_c materials. In that respect, optimum design schemes may be significantly different in these alternative material systems.

The absolute barrier thicknesses also influence the inhomogeneous broadening of confined levels and the associated interface roughness scattering mechanisms. A fluctuation of one atomic monolayer will lead to a larger shift of intersubband transitions when thinner barriers are used. In a similar way, interface scattering is expected to be more important for higher barriers, since a given thickness roughness will produce a larger scattering potential. However, these mechanisms are deeply linked to the microscopic structure of the interfaces, which is very dependent on the material system and on the growth conditions. In practice, intersubband emission linewidths comparable to those of InP- or GaAs-based structures have been reported in InAs/AlSb structures (Barate et al. 2005).

Terahertz QCLs require very thick quantum wells. For this reason the electron wave functions have a very small amplitude in the barriers, and adjacent QWs are much less coupled than in mid-infrared QC lasers. In this case, too, thinners barriers are necessary. One advantage of the GaAs system for terahertz lasers is the possibility of reducing the height of AlGaAs barriers by reducing the Al content of the ternary alloy. Typical values of 15% Al content are used to avoid the problems inherent in too thin barriers.

1.4.4 Lateral valleys

Effective barrier height. The presence of lateral valleys can be another limitation to short-wavelength operation of QC lasers. Heterostructure quantum states are built independently from each minimum of the conduction band in the $k \cdot p$ theory (Bastard 1988). However, if the states associated with lateral valleys have an energy close to that of the Γ states of the active region, a possible intervalley transfer can occur and deteriorate the gain of the laser.

The positions of the relevant lateral valley minima are shown in Fig. 1.17 for the different material systems. The large conduction band discontinuity of GaAs/AlAs cannot be fully exploited because of its indirect bandgap and the presence of X valley confined states in the AlAs barriers. Tunneling channels introduced by X states have been observed experimentally and extensively studied (Finley et al. 1998). They can produce significant leakage currents for electrons of energy greater than 200 meV. For this reason, GaAs-based QCLs use only direct gap barrier with Al content of 30% to 40%.

In an InGaAs/AlInAs system, lateral valley minima L and X are close to the top of the barriers and are not a main limitation for wavelenghs

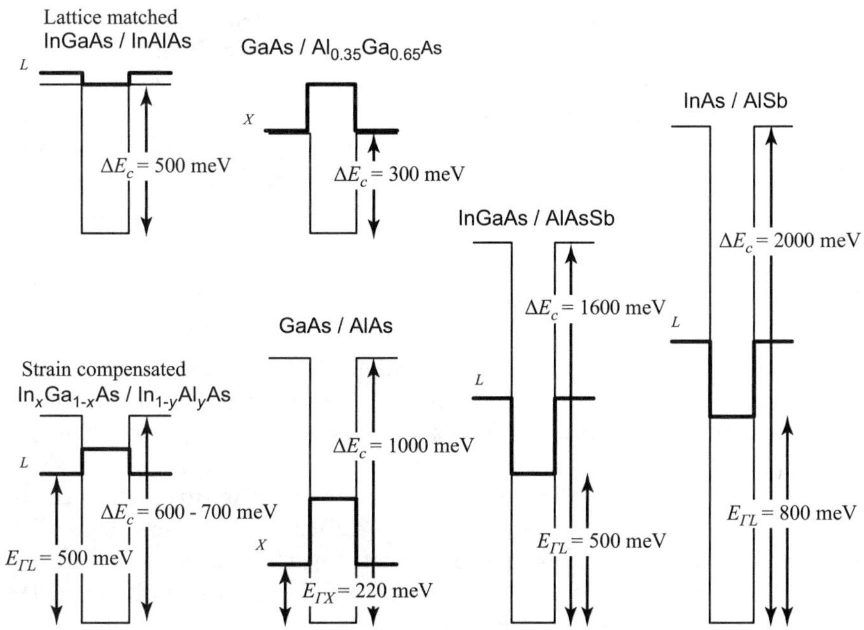

Figure 1.17 Band offsets of the different material systems used for QC lasers. Both Γ and relevant lateral valleys (X or L) are shown for the wells and the barriers.

up to 6 μm. When one is using strained compensated InGaAs and AlInAs, the situation is different. The strained ternary alloy $Al_xIn_{1-x}As$ ($x > 0.48$) has indirect bandgap due to both contributions of composition change and strain. Its L valley minimum energy decreases compared to the lattice-matched alloy. Altogether, the Γ(InGaAs) to L(AlInAs) energy separation remains unchanged in the strained balanced pair of materials (see Fig. 1.17). In the same way, the use of AlAsSb alloy lattice-matched on InP does not change the L point energy in InGaAs ($\Gamma - L = 500$ meV), despite the strong increase of ΔE_c. In these cases, the L valley could be a limitation to the realization of high-performance, short-wavelength QCLs.

In InAs the L valley is about 0.8 eV above the conduction band minimum. Hence, the system InAs/AlSb provides the larger conduction band energy range, free of parasitic intervalley transfer channels. As a matter of fact, the shortest-wavelength intersubband emission ($\lambda = 2.5$ μm; $h\nu = 500$ meV) has been obtained in electroluminescence devices made in these materials systems (Barate et al. 2005).

Intervalley transfer. The real impact on the laser performance of intervalley scattering is not clear at present. The tunneling currents associated to Γ-X transfer have been studied for AlAs barriers coupled to a continuum of states (Finley et al. 1998). It has been shown to be dominant for thicker barriers and higher applied electric fields.

In the case of QC lasers, the lateral valley states are not coupled to a continuum. When L or X states come below the Γ point conducting levels, the more likely transfer mechanism is a sequential hopping transport between X or L localized states. Intervalley scattering times have been demonstrated (Teissier et al. 1996) to be in the form

$$\tau_{ij} = \frac{\tau_0}{|\langle \Gamma_i \mid X_j \rangle|^2} \tag{1.37}$$

where the wave function overlap is calculated between the two localized states Γ_i and X_j and τ_0 is a constant scattering time, characteristic of the transfer mechanism. Typical values are $\tau_0 = 0.15$ ps for elastic scattering to X_z valleys (in the growth direction) and $\tau_0 = 0.33$ ps for phonon-assisted scattering to X_{xy} valleys (in the direction parallel to the layers). The situation should be significantly different depending on whether these states are localized in the barriers or in the wells. In the first case, the overlap between Γ state localized in the wells and X state localized in the barrier is small, leading to a slow intervalley transfer. In addition, coupling between X states in different barriers is also weak. The final result is a charge buildup in the barriers. On the contrary, in the case of X or L states localized in the wells, the overlap factor expressed in Eq. 1.37 is closer to unity, and more efficient Γ-$X(L)$ transfer is expected. Therefore when the X or L valley is localized in the well, the scattering in the lateral valley can give rise to an important parasitic leakage current, such as depicted in Fig. 1.18.

The exploration of the short-wavelength limit of antimonide QCLs (InGaAs/AlAsSb or InAs/AlSb) could be a way of studying the real impact of intervalley transfer, in a case where the escape of carriers to the continuum is negligible because of the huge barrier height.

1.4.5 Waveguide

The waveguide design and its realization are crucial steps to obtain high-performance QC lasers. The vertical confinement of the optical mode propagating in the cavity of the laser is obtained, as in most edge-emitting semiconductor lasers, by a multilayer structure: a core waveguide made of the active region and spacer layers, surrounded by optical cladding layers made of low-refractive-index material. This is another aspect for which the material system impacts QCL performances.

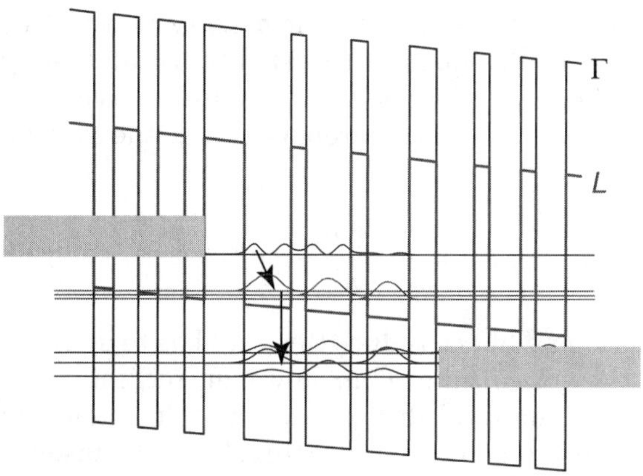

Figure 1.18 Schematic diagram of the Γ and L conduction band profile in a short-wavelength InAs/AlSb QC structure, showing the relevant states issued from both valleys. The arrows show the possible intervalley leakage current mechanism.

The typical structure of a waveguide for a QC laser is presented in Fig. 1.19. The intersubband transitions are only coupled to transverse magnetic (TM) polarized light (electric field prependicular to the layers). The electric field intensity of the fundamental TM mode of the waveguide is shown in the same figure. An important parameter to define is the overlap factor Γ, which is the percentage of the mode interacting with the active region and can be written as

$$\Gamma = \frac{\int_{AR} |E(z)|^2 \, dz}{\int_{-\infty}^{+\infty} |E(z)|^2 \, dz} \tag{1.38}$$

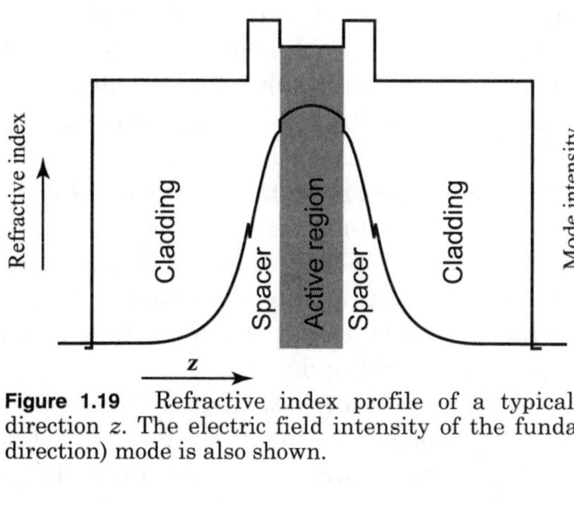

Figure 1.19 Refractive index profile of a typical QCL waveguide, in the growth direction z. The electric field intensity of the fundamental TM (electric field in the z direction) mode is also shown.

The gain of a QCL is proportional to the current density and to this overlap factor (Eq. 1.27). As a result, the laser threshold current density is directly proportional to the cavity losses, which include waveguide losses α_w and mirror losses α_m, and inversely proportional to the overlap factor

$$\Gamma g J_{th} = \alpha_w + \alpha_m \Rightarrow J_{th} = \frac{\alpha_w + \alpha_m}{\Gamma g} \tag{1.39}$$

A typical way to increase Γ, and hence reduce threshold, is to increase the number of periods in the active zone. This, however, increases the applied voltage and therefore the total dissipated power injected into the device. The main criterion to achieve a low-threshold QC laser is to find the waveguide that minimizes the optical losses and simultaneously maximizes the overlap factor.

Due to the long wavelength a major contribution to optical losses is the free-carrier absorption in the doped region of the waveguide, manly in the cladding layers. These losses increase rapidly for longer wavelength and are the dominant factor for terahertz QC lasers. The optimum waveguide design depends both on the material system and on the wavelength range.

The InP-based system provides, with no doubt, the best waveguide properties for mid-infrared QC lasers. The main advantages of this material system stem from the possibility of using InP as cladding material. In fact, due to its low refractive index and good electrical properties, InP provides optical confinement without the need of high doping concentration and therefore allow very low free-carrier optical losses. Moreover, this is the only binary material that can be used as cladding, with very important consequences to the thermal resistance of the devices. The thermal conductivity of InP (0.7 W/cm/K) is quite high, almost twice that of GaAs and at least 10 times higher than the ternary or quaternary alloys, typically used for the optical confinement in waveguides grown on GaAs, InAs, or GaSb substrates. Nowadays, practically all the results relevant for technological applications are referred to InP-based lasers.

In GaAs-based QCLs, the substrate index is higher than the index of the active region. Alternative materials must be used as cladding layers. Different solutions have been proposed. The most straightforward are the transpositions to the mid-ir of the waveguides used in the near-ir GaAs-based diode lasers, using the alloys AlGaAs (Sirtori et al. 1998) or GaInP (Green 2002) as cladding layers.

An alternative to dielectric waveguides (using low-dielectric-constant material as cladding layers) was proposed in the early days of QCL history (Sirtori et al. 1995): a plasmon-enhanced waveguide using highly doped cladding layers. The low refractive index is obtained thanks to the strong dispersion when the laser frequency gets close to the plasma frequency of the doped material. This type of waveguide has been efficiently applied to mid-infrared GaAs QCLs, either as a pure GaAs waveguide (Sirtori et al. 1999) or as AlGaAs waveguide enhanced by n^+ GaAs plasmons (Sirtori et al. 1998). Its drawback is obviously its intrinsic high optical losses, due to the highly doped layers. Thick, low-doped spacers are needed to separate the active region from the highly doped layers, thus reducing the overlap factor.

Plasmon-enhanced waveguide is the solution for InAs systems too. Today, all InAs/AlSb QCLs have been made of n^+-doped InAs cladding layers (Ohtani and Ohno, 2003). For short-wavelength lasers, the free-carrier losses (proportional to λ^2) are reduced to a reasonable level, and high-performance waveguides with low loss (α_w = 7 cm^{-1}) and high confinement factor (Γ = 47%) have been demonstrated at λ = 4.5 μm (Teissier et al. 2004). The main limitation of the InAs-based waveguide is the small bandgap energy (E_g = 0.36 eV) of InAs, which prevents its use for wavelengths shorter than 4 μm.

Waveguides for terahertz QCLs require different solutions. Dielectric waveguides cannot be used due to the very long wavelength that would impose the use of very thick layers. On the other hand, at terahertz frequencies the very high free-carrier absorption prevents the use of doped layers. The solution is to use optical confinement through surface plasmons at metal semiconductor interfaces (Köhler et al. 2002). The ultimate confinement, with Γ close to 100%, has been obtained in a terahertz QCL using double metal waveguides (Williams et al. 2003), analog to a mm-wave microstrip line.

References

Alferov, Zh. I., V. M. Andreev, E. L. Portnoi, M. K. Trukan, *Fiz. Tekhn. Polupr.* **3**, 132 (1969); *Sov. Phys. Semicond.* **3**, 1107 (1970).

Alton, J., S. Dhillon, A. de Rossi, M. Calligaro, H. E. Beere, S. Barbieri, E. Linfield, D. Ritchie, and C. Sirtori, "Buried waveguides via two dimensional surface plasmon modes in THz quantum cascade lasers," *Appl. Phys. Lett.* **86**, 071109 (2005).

Barate, D., R. Teissier, Y. Wang, and A. N. Baranov, "Short wavelength intersubband emission from InAs/AlSb quantum cascade structures," *Appl. Phys. Lett.* **87,** 051103 (2005).

Bastard, G., *Wave Mechanics Applied to Semiconductor Heterostructures*, Les Editions de Physique, Les Ulis, France, 1988.

Beck, M., D. Hofstetter, T. Aellen, J. Faist, U. Oesterle, M. Ilegems, E. Gini, and H. Melchior, "Continuous wave operation of a mid-infrared semiconductor laser at room temperature," *Science* **295,** 301 (2002).

Bewley, W. W., J. R. Lindle, C. S. Kim, I. Vurgaftman, J. R. Meyer, A. J. Evans, J. S. Yu, S. Slivken, and M. Razeghi, "Beam steering in high-power cw quantum cascade lasers," *IEEE J. Quantum Electron.* **41,** 833 (2005).

Blaser, S., L. Diehl, M. Beck, J. Faist, U. Oesterle, J. Xu, S. Barbieri, and F. Beltram, "Characterization and modeling of quantum cascade lasers based on a photon-assisted tunneling transition," *IEEE J. Quantum Electron.* **37,** 448 (2001).

Blaser, Stéphane, Dmitri A. Yarekha, Lubos Hvozdara, Yargo Bonetti, Antoine Muller, Marcella Giovannini, and Jérôme Faist, "Room-temperature, continuous-wave, single-mode quantum-cascade lasers at lambda ≅ 5.4 μm," *Appl. Phys. Lett.* **86,** 041109 (2005).

Capasso, F., K. Mohamed, and A. Y. Cho, "Resonant tunneling trough double barriers, perpendicular quantum transport phenomena in superlattices, and their device applications," *IEEE J. Quantum Electron.* **22,** 1853–1869 (1986).

Cho, A., *Molecular Beam Epitaxy*, AIP Press, Woodbury, NW, 1994.

Colombelli, R., K. Srinivasan, M. Troccoli, O. Painter, C. Gmachl, D. M. Tennant, A. M. Sergent, D. L. Sivco, A. Y. Cho, and F. Capasso, "Quantum cascade surface-emitting photonic crystal laser," *Science* **302,** 1374 (2003).

Diehl, L., S. Mentese, E. Müller, D. Grützmacher, H. Sigg, U. Gennser, I. Sagnes, Y. Campidelli, O. Kermarrec, D. Bensahel, and J. Faist, "Electroluminescence from strain-compensated $Si_{0.2}Ge_{0.8}$/Si quantum-cascade structures based on a bound-to-continuum transition," *Appl. Phys. Lett.* **81,** 4700 (2002).

Dhillon, S. S., C. Sirtori, S. Barbieri, A. de Rossi, M. Calligaro, H. E. Beere, and D. A. Ritchie, "THz sideband generation at telecom wavelengths in a GaAs-based quantum cascade laser," *Appl. Phys. Lett.* **87,** 071101 (2005).

Donovan K., P. Harrison, and R. W. Kelsall, "Self-consistent solutions to the intersubband rate equations in quantum cascade lasers: Analysis of a GaAs/ $Al_xGa_{1-x}As$ device," *J. Appl. Phys.* **89,** 3084 (2001).

Esaki, L., and R. Tsu, "Supperlattice and negative differential conductivity in semiconductors," *IBM J. Res. Devel.* **14**, 61–65 (1970).

Evans, A., J. S.Yu, J. David, L. Doris, K. Mi, S. Slivken, and M. Razeghi, "High-temperature, high-power, continuous-wave operation of buried heterostructure quantum-cascade lasers," *Appl. Phys. Lett.* **84,** 314 (2004a).

Evans, A., J. S. Yu, S. Slivken, and M. Razeghi, "Continuous-wave operation of $\lambda \sim 4.8$ μm quantum cascade lasers at room temperature," *Appl. Phys. Lett.* **85,** 2166 (2004b).

Faist, J., F. Capasso, D. L. Sivco, C. Sirtori, A. L. Hutchinson, and A. Y. Cho, "Quantum cascade laser," *Science* **264,** 553 (1994).

Faist, J., F. Capasso, C. Sirtori, D. L. Sivco, J. N. Baillargeon, A. L. Hutchinson, S.-N. G. Chu, and A. Y. Cho, "High power mid-infrared ($\lambda \sim 5$ μm) quantum cascade lasers operating above room temperature," *Appl. Phys. Lett.* **68,** 3680 (1996).

Faist J., C. Gmachl, F. Capasso, C. Sirtori, D. L. Sivco, J. N. Baillargeon, and A. Y. Cho, "Distributed feedback quantum cascade lasers," *Appl. Phys. Lett.* **70**, 2670 (1997).

Faist, J., F. Capasso, D. L. Sivco, A. L. Hutchinson, S.-N. G. Chu, and A. Y. Cho, "Short wavelength ($\lambda \approx 3.4$ μm) QCL based on strained compensated InGaAs/AlInAs," *Appl. Phys. Lett.* **72,** 680 (1998).

Faist, J., C. Sirtori, F. Capasso, L. N. Pfeiffer, K. W. West, D. L. Sivco, and A. Y. Cho, "Quantum interference effects in intersubband transitions," *Intersubband Transitions in Quantum Wells: Physics and Device Applications I*, H. C. Liu and F. Capasso (eds.), *Semiconductor and Semimetals*, vol. 62, chap. 2, pp. 101–128, Academic Press, 2000.

Faist, J., M. Beck, T. Aellen, and E. Gini, "Quantum-cascade lasers based on a bound-to-continuum transition," *Appl. Phys. Lett.* **78,** 147 (2001).

Faist, J., D. Hofstetter, M. Beck, T. Allen, M. Rochat, and S. Blaser, "Bound-to-continuum and two-phonon resonance quantum cascade lasers for high duty cycle, high temperature operation," *IEEE J. Quantum Electron* **38,** 533–546 (2002).

Finley, J. J., R. Teissier, M. S. Skolnick, J. W. Cockburn, G. A. Roberts, R. Grey, G. Hill, M. A. Pate, and R. Planel, "Role of the X minimum in transport through AlAs single-barrier structures," *Phys. Rev.* **B58,** 10619 (1998).

Forget, S., C. Faugeras, J.-Y. Bengloan, M. Calligaro, O. Parillaud, M. Giovannini, J. Faist, and C. Sirtori, "High-power spatial single-mode quantum cascade lasers at 8.9 μm," *Electron. Lett.* **41,** 418–419 (2005).

Gao, J. R. J. N. Hovenier, Z. Q. Yang, J. J. A. Baselmans, A. Baryshev, M. Hajenius, T. M. Klapwijk, A. J. L. Adam, T. O. Klaassen, B. S. Williams, S. Kumar, Q. Hu, and J. L. Reno, "Terahertz heterodyne receiver based on a quantum cascade laser and a superconducting bolometer," *Appl. Phys. Lett.* **86,** 244104 (2005).

Garcia, J.-C., E. Rosencher, Ph. Collot, N. Laurent, J. L. Guyaux, B. Vinter, and J. Nagle, "Epitaxially stacked lasers with Esaki junction: A bipolar cascade laser," *Appl. Phys. Lett.* **71,** 3752–3754 (1997).

Gmachl, C., F. Capasso, A. Tredicucci, D. L. Sivco, A. L. Hutchinson, S. N. G. Chu, and A. Y. Cho, "Noncascaded intersubband injection lasers at $\lambda \sim 7.7$ μm," *Appl. Phys. Lett.* **73,** 3830–3832 (1998).

Green, R. P., L. R. Wilson, D. A. Carder, J. W. Cockburn, M. Hopkinson, M. J. Steer, R. J. Airey, and G. Hill, "Room temperature GaAs-based quantum cascade laser with GaInP waveguide cladding," *Electron. Lett.* **38,** 1539 (2002).

Green, R. P., A. Krysa, J. S. Roberts, D. G. Revin, L. R. Wilson, E. A. Zibik, W. H. Ng, and J. W. Cockburn, "Room-temperature operation of InGaAs/AlInAs quantum cascade lasers grown by metal organic vapor phase epitaxy," *Appl. Phys. Lett.* **83,** 1921 (2003).

Hall, R. N., G. E. Fenner, J. D. Kingsley, T. J. Soltys, and R. O. Carlson, "Coherent light emission from GaAs junctions," *Phys. Rev. Lett.* **9,** 366 (1962).

Hayashi, I., M. B. Panish, P. W. Foy, and S. Sumski, "A semiconductor junction lasers which operate continuously at room temperature," *Appl. Phys. Lett.* **17,** 109 (1970).

Henderson, G. N., L. C. West, T. K. Gaylord, C. W. Roberts, E. N. Glytsis, and M. T. Asom, "Optical transitions to above-barrier quasi-bound states in asymmetric semiconductor heterostructures," *Appl. Phys. Lett.* **62,** 1432 (1993).

Kazarinov, R. F., and R. A. Suris, "Possibility of amplification of electromagnetic waves in a semiconductor with a superlattice," *Fiz. Tekh.* Poluprov., **5,** 797–800 (1971); transl. in *Sov. Phys. Semicond.,* **5,** 707–709 (1971).

Köhler, R., A. Tredicucci, F. Beltram, H. E. Beere, E. H. Linfeld, A. G. Davies, D. A. Ritchie, R. C. Iotti, and F. Rossi, "Terahertz semiconductor heterostructure laser," *Nature* **417,** 156 (2002).

Kosterev, A. A., and F. K. Tittel, "Chemical sensors based on quantum cascade lasers," *IEEE J. Quantum Electron.* **38,** 582 (2002).

Lax, B., *Proceedings of the International Symposium on Quantum Electronics,* C. H. Townes (ed.), Columbia Univ. Press, New York, 1960, p.428.

Lee, S.-C., and A. Wacker, "Nonequilibrium Green's function theory for transport and gain properties of quantum cascade structures," *Phys. Rev.* **B66**, 245314 (2002).

Leuliet, A., A. Vasanelli, C. Sirtori, A. Wade, G. Fedorov, D. Smirnov, and G. Bastard, "Electron scattering spectroscopy by high magnetic field in quantum cascade lasers," submitted for publication to *Phys. Rev. B* (2005).

Liu, H. C., "A novel superlattice infrared source," *J. Appl. Phys.* **63,** 2856 (1988).

Liu, H. C., A. G. Steele, M. Buchanan, and Z. R. Wasilewski, "Dark current in quantum well infrared photodetectors," *J. Appl. Phys.* **73** (4), 2029–2031 (1993).

Myers, T. L., R. M. Williams, M. S. Taubman, C. Gmachl, F. Capasso, D. L. Sivco, and A. Y. Cho, "Free-running frequency stability of mid-infrared quantum cascade lasers," *Opt. Lett.* **27,** 170 (2002).

Offermans, P., P. M. Koenraad, J. H. Wolter, M. Beck, T. Aellen, and J. Faist, "Digital alloy interface grading of an InAlAs/InGaAs quantum cascade laser structure studied by cross-sectional scanning tunneling microscopy," *Appl. Phys. Lett.* **83**(20), 4131–4133, (2003).

Ohtani, K., and H. Ohno, "InAs/AlSb quantum cascade lasers operating at 10 μm," *Appl. Phys. Lett.* **82,** 1003 (2003).

Owschimikow, N., C. Gmachl, A. Belyanin, V. Kocharovsky, D. L. Sivco, R. Colombelli, F. Capasso, and A. Y. Cho, "Resonant second-order nonlinear optical processes in quantum cascade lasers," *Phys. Rev. Lett.* **90**, 043902 (2003).

Page, H., S. Dhillon, M. Calligaro, C. Becker, V. Ortiz, and C. Sirtori, "Improved cw operation of GaAs-based QC lasers: T_{max} = 150 K," *IEEE J. Quantum Electron.* **40,** 665 (2004).

Paiella, R., F. Capasso, C. Gmachl, D. L. Sivco, J. N. Baillargeon, A. L. Hutchinson, A. Y. Cho, and H. C. Liu, "Self-mode-locking of quantum cascade lasers with giant ultrafast optical nonlinearities," *Science* **290,** 1793 (2000).

Paiella, R., R. Martini, F. Capasso, C. Gmachl, H. Y. Hwang, D. L. Sivco, J. N. Baillargeon, A. Y. Cho, E. A. Wittaker, and H. C. Liu, "High-frequency modulation without the relaxation oscillation resonance in quantum cascade lasers," *Appl. Phys. Lett.* **79,** 2526 (2001).

Revin, D. G., L. R.Wilson, E. A. Zibik, R. P. Green, J. W. Cockburn, M. J. Steer, R. J. Airey, and M. Hopkinson, "InGaAs/AlAsSb quantum cascade lasers," *Appl. Phys. Lett.* **85**, 3992–3994 (2004).

Scamarcio, G., F. Capasso, C. Sirtori, J. Faist, A. L. Hutchinson, D. L. Sivco, and A. Y. Cho, "High-power infrared (8-micrometer wavelength) superlattice lasers," *Science* **276,** 773 (1997).

Schrenk, W., N. Finger, S. Gianordoli, L. Hvozdara, G. Strasser, and E. Gornik, "GaAs/AlGaAs distributed feedback quantum cascade lasers," *Appl. Phys. Lett.* **76,** 253 (2000).

Sirtori, C., F. Capasso, J. Faist, and S. Scandolo, "Nonparabolicity and a sum rule associated with bound-to-bound and bound-to-continuum intersubband transitions in quantum wells," *Phys. Rev.* **B50,** 8663 (1994).

Sirtori, C., J. Faist, F. Capasso, D. L. Sivco, A. L. Hutchinson, and A. Y. Cho, "Quantum cascade laser with plasmon-enhanced waveguide operating at 8.4 μm wavelength," *Appl. Phys. Lett.* **66,** 3242 (1995).

Sirtori, C., J. Faist, F. Capasso, D. L. Sivco, A. L. Hutchinson, S.-N. G. Chu, and A. Y. Cho, "Continuous wave operation of midinfrared (7.4–8.6 μm) quantum cascade lasers up to 110 K temperature," *Appl. Phys. Lett.* **68,** p1745 (1996).

Sirtori, C., P. Kruck, S. Barbieri, P. Collot, J. Nagle, M. Beck, J. Faist, and U. Oesterle, "GaAs/$Al_xGa_{1-x}As$ quantum cascade lasers," *Appl. Phys. Lett.* **73,** 3486 (1998a).

Sirtori, C., F. Capasso, J. Faist, D. L. Sivco, A. L. Hutchinson, and A. Y. Cho, "Resonant tunneling effect in quantum cascade lasers," *IEEE J. Quantum Electron.* **34,** 1722 (1998b).

Sirtori, C., P. Kruck, S. Barbieri, H. Page, J. Nagle, M. Beck, J. Faist, and U. Oesterle, "Low-loss Al-free waveguides for unipolar semiconductor lasers," *Appl. Phys. Lett.* **75,** 3911 (1999).

Sirtori, C., F. Capasso, D. L. Sivco, and A. Y. Cho, "Nonlinear optics in coupled-quantum-wells," *Intersubband Transitions in Quantum Wells: Physics and Device Appliations II,* H. C. Liu and F. Capasso (eds.), *Semiconductor and Semimetals,* vol. **66,** chap. 2, pp. 85–123, Academic Press, 2000.

Sirtori, C., H. Page, C. Becker, and V. Ortiz, "GaAs/AlGaAs quantum cascade lasers: physics technology and prospects," *IEEE J. Quantum Electron.* **38,** 547 (2002).

Sirtori, C., and J. Nagle, "Quantum cascade lasers: The quantum technology for semiconductor lasers in the mid-far-infrared," *C. R. Physique* **4** (2003).

Spagnolo, V., G. Scamarcio, H. Page, and C. Sirtori, "Simultaneous measurement of the electronic and lattice temperatures in GaAs/$Al_{0.45}Ga_{0.55}As$ quantum-cascade lasers: Influence on the optical performance," *Appl. Phys. Lett.* **84,** 3690 (2004).

Stevenson, R., "QCL's: Mass market devices or just interesting physics?" *Compound Semiconductor*, September 2004, p. 28.

Teissier, R., J. J. Finley, M. S. Skolnick, J. W. Cockburn, J. L. Pelouard, R. Grey, G. Hill, M. A. Pate, and R. Planel, "Experimental determination of Γ-X intervalley transfer mechanisms in GaAs/AlAs heterostructures," *Phys. Rev.* **B54,** 8329 (1996).

Teissier, R., D. Barate, A. Vicet, C. Alibert, A. N. Baranov, X. Marcadet, C. Renard, M. Garcia, C. Sirtori, D. Revin, and J. Cockburn, "Room temperature operation of InAs/AlSb quantum cascade lasers," *Appl. Phys. Lett.* **85,** 167 (2004).

Troccoli, M., S. Corzine, D. Bour, J. Zhu, O. Assayag, L. Diehl, B. G. Lee, G. Höfler, and F. Capasso, "Room temperature continuous-wave operation of quantum-cascade lasers grown by metal organic vapour phase epitaxy," *Electron. Lett.* **41,** 1059 (2005a).

Troccoli, M., A. Belyanin, F. Capasso, E. Cubukcu, D. L. Sivco and A. Y. Cho, "Raman injection laser," *Nature* **433,** 845 (2005b).

Williams, B. S., S. Kumar, H. Callebaut, Q. Hu, and J. L. Reno, "Terahertz quantum-cascade laser at $\lambda \sim 100$ μm using metal waveguide for mode confinement," *Appl. Phys. Lett.* **83,** 2124 (2003).

Williams, B. S., S. Kohen, S. Kumar, Q. Hu, and J. L. Reno, "Terahertz quantum cascade lasers with metal-metal waveguides," *Proceedings of SPIE,* vol. 5738, *Novel In-Plane Semiconductor Lasers IV*, Carmen Mermelstein and David P. Bour (eds.), April 2005, pp. 170–179.

Worral, Chris, Jesse Alton, Mark Houghton, Stefano Barbieri, Carlo Sirtori, Harvey E. Beere, and David Ritchie, "Continuous wave operation of a quantum cascade laser emitting at 2 THz" submitted for publication in *Optics Express,* 2005.

Yang, Q., C. Manz, W. Bronner, Ch. Mann, L. Kirste, K. Köhler, and J. Wagner, "GaInAs/AlAsSb quantum-cascade lasers operating up to 400 K," *Appl. Phys. Lett.* **86,** 131107 (2005).

Yariv, A., *Quantum Electronics*, 3d ed., Wiley, New York, 1989.

Yu, J. S., A. Evans, S. Slivken, S. R. Darvish, and M. Razeghi, "Short wavelength ($\lambda \approx 4.3$ μm) high performance continuous wave quantum cascade lasers," *IEEE Phot. Technol. Lett.* **17,** 1154 (2005).

Chapter

2

Terahertz Quantum Cascade Lasers

Alessandro Tredicucci and Rüdeger Köhler

NEST CNR-INFM and Scuola Normale Superiore
Piazza dei Cavalieri 7, I-56126 Pisa, Italy

2.1 Introduction

The existence of electromagnetic waves was proved in 1880 by Hertz, who demonstrated that two circuits exchanged energy with no direct connection. Soon after, it was recognized that the same principle could be used to transmit information, leading to the invention of the radio in 1895 by Marconi. Later, the development of radar in the 1930s sparked interest in the generation and detection of radiation at ever-higher frequencies. The study of the interaction of electromagnetic waves with matter contributed substantially to the birth of modern physics. The observation of distinct and characteristic spectral lines of atoms and molecules at ultraviolet to infrared wavelengths led to the Bohr-Sommerfeld model of the atom. The discovery of X-ray diffraction by von Laue revealed for the first time the crystal structure of solids. Ever since, the use of electromagnetic waves of various frequencies has been a powerful tool both for scientific investigation and for everyday applications. Now the generation and detection of radiation with frequencies from kilohertz up to that of ultraviolet light are a vital part of modern life. This revolution is largely a consequence of the tremendous progress in semiconductor technology.

Since the initial demonstration of microwave devices from 1958 to 1965 [1–7] and of the first semiconductor laser in 1962 [8], both electronics and photonics have experienced a rapid evolution. The progress in solid-state oscillators that now reach frequencies as high as 1 THz was driven by the need of ever-shorter wavelength in precision radar systems and the demand for ever-higher bandwidth in wireless communication links. Likewise, the extension of the emission wavelength in semiconductor lasers from the near-infrared of the first GaAs laser diodes to the ultraviolet was stimulated by the need of small, compact visible sources for optical data storage and retrieval, displays, optical sensing, and medical treatment. Equally, the extension to mid- and far-infrared wavelengths was pursued for the paramount opportunities in infrared gas spectroscopy and sensing, industrial process control, military and law enforcement applications, and free-space point-to-point optical communication links. Today, laser diodes cover the entire range from ultraviolet to far-infrared light (~10,000–12 THz). For the terahertz region (1–10 THz, $\lambda \sim 300$–$30\ \mu m$), on the contrary, the first compelling applications are just now being identified. Even from a purely scientific point of view, perhaps except for particular spectroscopic applications in astrophysics and atmospheric studies, the interaction between terahertz waves and matter was not considered for a long time to be of any specific use. A noteworthy statement in this respect is the opening remark of the 1970 *Proceedings of the Symposium on Submillimeter Waves* by Senitzky and Oliver [9]: "It is appropriate to view the submillimeter wave region as a transition region lying between the millimeter wave and the infrared portions of the electromagnetic spectrum and possessing no hallmark of its own." Already in the 1970s, however, this began to change when Yang et al. succeeded in generating terahertz radiation by nonlinear frequency conversion [10]. The continuing progress in the technology to produce and detect terahertz waves then enabled initial studies in terahertz spectroscopy. Significant milestones were the design of ultrafast switches by Mourou, Auston, and others [11–14]; optical rectification [15, 16] demonstrated by Hu et al. [17] and Xu et al. [18]; and the use of photoconductive antennas [19, 20]. Even more important was the realization of free-space electrooptical sensing by Zhang and others [21–25], which allows an extremely wide detection bandwidth from 100 GHz up to 50 THz.

The development of the above-mentioned components and techniques in turn allowed the invention of terahertz *time-domain spectroscopy* (TDS) [26–28] in the middle to late 1980s and the first use of terahertz TDS for imaging in 1995 [29–32]. Subsequently, even tomographic images were obtained [33–35], a technique particularly suitable in industrial process control [30], medical diagnosis [36, 37], analysis of biological tissue in general, and security screening, since the most common packaging materials such as plastic or cardboard are transparent at terahertz wavelengths [38, 39]. Recently, a novel technique for the label-free characterization of DNA fragments has been demonstrated [40, 41]. Terahertz pulses are also a powerful tool to reveal the dynamics of carrier-carrier interactions on ultrafast time scales [42].

Despite these advances, in all these sources the generation of terahertz radiation is of an indirect nature. The only fundamental terahertz oscillators are optically pumped gas lasers and *p*-doped germanium lasers [43–48], both of which have shortcomings that prevent their widespread use. The former deliver rather high output powers (up to 200 mW), but their emission is not tunable owing to the molecular nature of the transition. Tunable emission can be obtained from *p*-doped germanium lasers, but their operation is limited so far to low temperatures and pulsed excitation. In fact, the rapid stream of advancement in solid-state emitters has left the region of 1–10 THz essentially untouched, so much so that the term *terahertz gap* was coined. The reason is inherent to the spectral position of the terahertz gap itself. Located between electronics and photonics, it requires ideas and parts of both technologies, which, however, are quite distinct. Oscillators in electronics are based on transport characteristics that exhibit some kind of negative differential resistance, which is used as an amplifier or nonlinear element for frequency multiplication. Consequently, design and analysis of these devices rely on quasi-classical concepts for electronic transport. Another aspect is the fact that the wavelength of the emitted radiation is comparable to or larger than the characteristic device size. Photonics, on the contrary, utilizes optical transitions between well-defined energy levels or bands, whose population is changed by electrical injection or optical pumping. Here, the underlying theory focuses on the quantum-mechanical description of the transition process. In most photonic devices, the emission wavelength is smaller than the device size, and consequently the emission contains many optical modes.

Historically, major steps in shrinking the terahertz gap to its present span (1–10 THz) came from electronics, where the operation frequencies of microwave devices such as IMPATT, TUNNETT, or Gunn diodes have been pushed up to several hundred gigahertz [49], and resonant tunneling diodes have even shown operation at 712 GHz [50]. With the help of frequency multipliers, emission up to 2 THz has been demonstrated [51]. The power levels of these devices, however, follow approximately a $1/\nu^2$ law [52] leading to output powers at 1 THz limited to about 0.1 mW. Nevertheless, according to recent theoretical studies, better performance in terms of both power and operation frequency can be expected from GaN-based devices or quantum-engineered structures [49, 53, 54].

In the 1980s semiconductor lasers emitting at wavelengths as long as $\lambda \sim 46$ μm (6.5 THz) were demonstrated [55]. These devices rely on narrow-gap materials such as quaternary PbSnTeSe solid solutions [55, 56] or ternary $Pb_{1-x} Sn_x Te$ heterostructures [57, 58]. The performance of these lasers is limited for both technological and fundamental reasons. First, growth and fabrication of lead-salt materials are still less reliable than those of common III–V or group IV materials. In addition, once fabricated, such devices experience rapid aging and suffer in particular from thermal cycling. Second, narrow-gap materials are particularly sensitive to Auger processes, in which the energy and momentum of a recombining electron-hole pair are transferred to a third carrier, rather than to a photon.

The search for materials with a fundamental band gap in the terahertz energy range can be abandoned when turning to light sources that rely on intersubband transitions rather than interband transitions. In semiconductor quantum well structures, such subbands are formed as a consequence of the confinement in the direction perpendicular to the quantum well. Quite like in the textbook case of a "particle in a box," the energy of these subbands can be controlled by the width of the well. Quantum wells can be fabricated by employing modern semiconductor growth techniques such as metal-organic chemical vapor deposition or molecular beam epitaxy. In particular, the latter has reached a level of control that allows the repeatable, reliable, and homogeneous deposition of layers with a thickness down to that of a single monolayer. The possibility of using transitions between subbands in a quantum well for the generation and amplification of light was first discussed by Kazarinov and Suris in 1971 and 1972 [59, 60]. The energy separation between the subbands is at first instance independent of the fundamental band gaps of the constituent materials. Therefore, operation at arbitrarily long wavelengths can be imagined. The subsequent theoretical and experimental work focused indeed on the realization of such an emitter in the far-infrared [61–64], at energies below the reststrahlen band of the materials. The reason for this choice was the extremely fast nonradiative intersubband relaxation of electrons via the emission of optical phonons, if the subband separation exceeds the phonon energy [65, 66]. It was thus thought that population inversion would be achieved more easily below the reststrahlen band than above. Quite the contrary, however, turned out to be true: Precisely this fast scattering process can be used to conveniently control nonradiative lifetimes and thus achieve population inversion. Consequently, the first intersubband laser [67] demonstrated in 1994 by Faist et al. emitted at mid-infrared wavelengths ($\lambda \sim 4.6$ μm). As witnessed by the present book, the *quantum cascade* (QC) laser, so named for the cascade-like electronic potential, has made remarkable progress since then. The first devices operated only in pulsed mode and at temperatures up to 80 K. Today, *continuous-wave* (cw) operation at room temperature is reached with output powers in excess of 0.5 W [68]. By using distributed feedback resonators [69, 70], continuously tunable single-mode emission with high side-mode suppression ratios better than 30 dB was achieved [71–74]. By using InP-based materials, emission wavelengths from as short as $\lambda \sim 3.4$ μm [75] to as long as $\lambda \sim 24$ μm [76] were obtained. The QC principle was then later extended to the GaAs/AlGaAs [77] and InAs/AlAsSb [78, 79] material systems. Despite this rapid success, the emission wavelength was limited for long time to the range above the reststrahlen band (8–9 THz, $\lambda \sim 33$–37 μm, in GaAs/AlGaAs).

The first encouraging step toward the accomplishment of a QC laser emitting at frequencies below 8 THz was made in 1998 by Rochat et al. [80] through the observation of terahertz electroluminescence from a QC structure, which triggered a series of experimental investigations [81–88]. However, in none of these structures was population inversion observed, and electroluminescence output powers were typically not more than tens of picowatts. This can be mainly attributed to difficulties in attaining precise control of carrier dynamics at such small energy separation between subbands, and the fact that the concept of exploiting electron-phonon scattering to achieve population inversion could not be extended in a straightforward manner. The fabrication of a suitable resonator capable of confining the terahertz radiation to a thickness of the order of 10 μm without high propagation losses represented another challenge.

The first terahertz QC lasers were finally realized in 2002 [89]; the key ingredients to this accomplishment were the invention of a novel type of partially metallic waveguide and its combination with an active gain medium based on the chirped superlattice concept [90]. The devices emitted at 4.4 THz ($\lambda \sim 67$ μm) with output powers of a few milliwatts and a maximum operating temperature of 50 K under pulsed excitation. Since then, their development has proceeded at a very rapid pace. Most of the research has focused on new concepts for the quantum design of the gain medium as well as on improving the device structure by optimizing fabrication and waveguide technology [91–93]. These efforts by various groups have led to maximum operating temperatures of about 140 K in pulsed mode, output powers of up to 80 mW, and lasing in continuous wave up to 93 K [94–96]. Emission frequencies from 2.1 to 4.8 THz have been demonstrated [97, 98]. Such advances are making terahertz QC lasers more and more appealing for applications in various fields such as chemical sensing, astronomy, spectroscopy, and imaging.

This chapter discusses the fundamental aspects of terahertz QC lasers, from the analysis of the innovations that made their realization possible, to the description of the latest advances and open challenges. After a brief review of the status and performance of common terahertz sources and detectors in Sec. 2.2, Sec. 2.3 will address the progress of terahertz intersubband emission up to the first lasing demonstration. In Sec. 2.4, the developments of these last 3 years toward higher operating temperatures and longer emission wavelengths are presented. New aspects, such as single-mode devices and operation in high magnetic fields, are dealt with in Sec. 2.5.

2.2 Terahertz Photonics

2.2.1 Sources of terahertz radiation

The first widely used sources of terahertz radiation rely on *harmonic generation* or frequency multiplication. The fundamental oscillation frequency is typically around 200 GHz [49] and is generated by conventional microwave electronics such as IMPATT/Gunn diodes or tunnel injection transit-time devices. Resonant-tunneling diodes emitting at frequencies up to 700 GHz were also reported [50]. By using frequency doublers or triplers and their combinations, emission frequencies above 1 THz can be generated [99]. The output power, however, scales rapidly down with increasing frequency, and in the best cases only a few microwatts are attained around 2 THz [51]. Better performance and higher frequencies can be expected in principle from GaN devices, thanks to the more favorable material parameters of GaN [53, 54, 99].

Another method of generating terahertz emission that had already been recognized in the early 1970s is *optical heterodyne conversion*, also termed *photomixing*. Two optical sources, typically frequency-stabilized lasers, with frequencies ν_1 and ν_2 are focused onto an antenna patterned on top of a layer of *low-temperature grown* (LTG) GaAs [19, 20]. The central part of this antenna consists of interlaced electrode fingers, which are dc-biased by an external power source. For average laser powers P_1 and P_2 and laser frequencies ν_1 and ν_2, the instantaneous optical power on the antenna is given by

$$P_i = P_1 + P_2 + 2\sqrt{P_1 P_2}\{\cos[2\pi(\nu_1 - \nu_2)t] + \cos[2\pi(\nu_1 + \nu_2)t]\} \qquad (2.1)$$

The photoconductance between the electrodes is modulated by the difference-frequency term $\alpha \cos\left[2\pi\left(\nu_1 - \nu_2\right)t\right]$ of the mixed signal of the two lasers. The sum frequency term is not coupled well into the antenna and is therefore negligible. The antenna then radiates only the electrical signal at the difference frequency (see Fig. 2.1). Photomixers have the advantage of room temperature operation and a wide tunability from ~25 GHz to ~3 THz, but their output power is limited to about 1 μW [100, 101].

A similar approach to generate terahertz radiation is the *Auston switch* [13, 14]. In this case two electrodes, separated by a distance of about 10 μm, are patterned onto a semiconductor such as LTG GaAs or silicon on sapphire. The electrodes are biased at very high voltage (usually 100–300 V), and a laser pulse of femtosecond duration is focused onto the region of maximum electric field close to the anode. A schematic view is shown in Fig. 2.2. The laser pulse generates electron-hole pairs, a process that almost instantaneously follows the temporal shape of the laser pulse, and carriers are accelerated by the high electric field generating a rapidly varying current transient. The Fourier transform of the latter is a broadband spectrum that extends well into the terahertz region. Typical spectra are shown in Fig. 2.3. Owing to

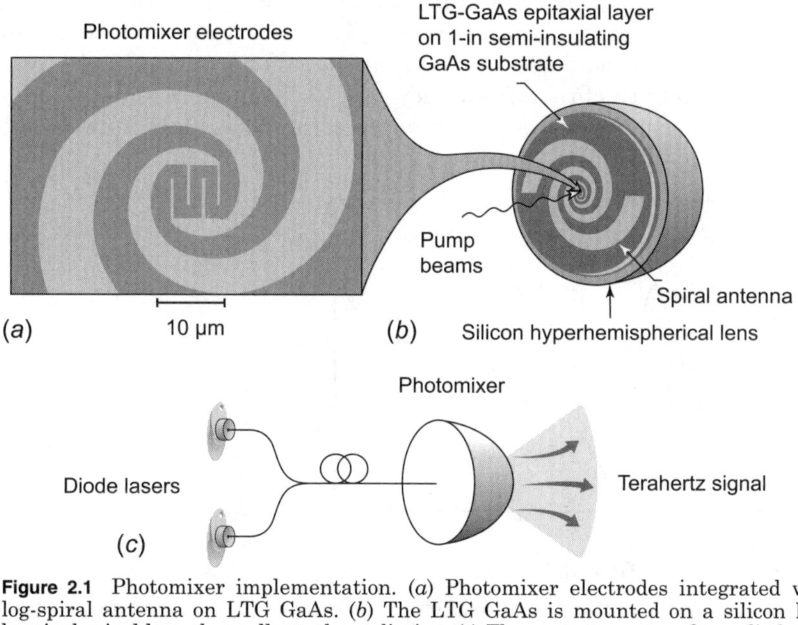

Figure 2.1 Photomixer implementation. (*a*) Photomixer electrodes integrated with a log-spiral antenna on LTG GaAs. (*b*) The LTG GaAs is mounted on a silicon hyperhemispherical lens that collects the radiation. (*c*) The output powers of two diode lasers are combined in a fiber and illuminate the electrodes so as to generate a terahertz signal. (Reprinted from Fig. 1 of [100], with kind permission of Springer Science and Business Media.)

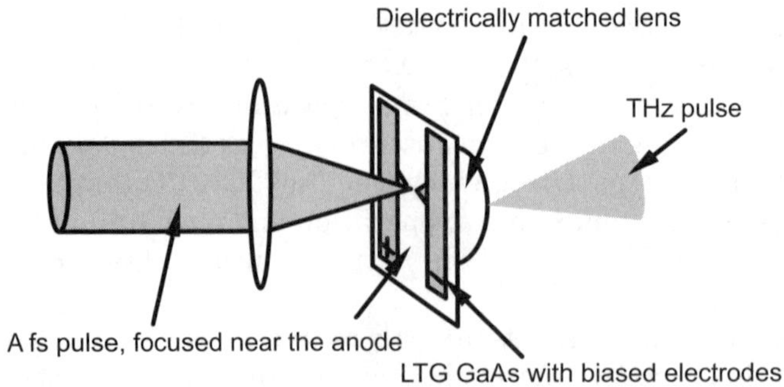

Figure 2.2 Schematic view of an Auston switch. A femtosecond laser pulse is focused onto the biased region of a photoconductor. The current transient contains a broadband of frequencies far into the terahertz range (see Fig. 2.3), and the radiation is coupled out via an appropriate lens.

the short carrier lifetimes in LTG GaAs [103], the carrier population relaxes again on a subpicosecond time scale. Besides Auston switches also cold plasma oscillations [104] and surface electric fields are used to generate terahertz radiation with similar mechanisms [105–107].

An all-optical technique to generate terahertz radiation is *optical rectification*, a second-order nonlinear effect. It was first found in 1962 by Bass et al. [15] and independently proposed by A'skaryan [16]. In essence, optical rectification is a difference-frequency generation where the frequency difference is close to zero. In recent years, femtosecond laser pulses have been used to generate terahertz radiation by this process. Owing to the broad frequency range contained in femtosecond-long laser pulses, many frequency components can contribute to the

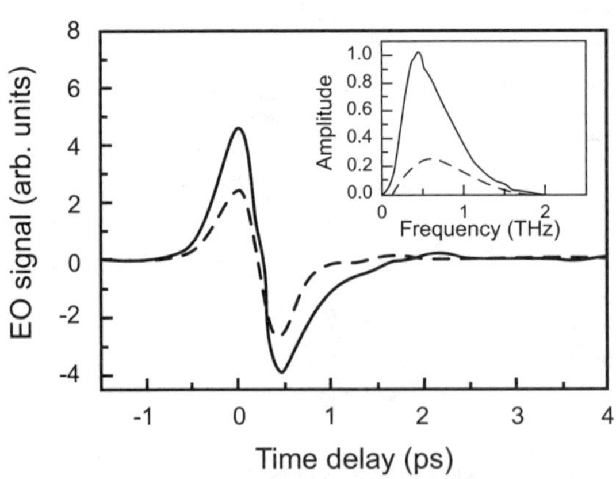

Figure 2.3 Terahertz transients emitted from an Auston switch after excitation with an 800-nm Ti:Sa laser emitting a pulse of 10-fs duration and 2.5-nJ energy. The two curves correspond to two different switches: The solid line is recorded from a structure with an enhanced absorption in the LTG GaAs layer. The inset shows the corresponding power spectra. (Reprinted with permission from [102]; copyright © 2002 IEEE.)

difference-frequency generation, and therefore a broadband spectrum is obtained [108, 109]. Peak output powers in this technique are generally lower than those obtained from photoconductive antennas. Narrow-bandwidth terahertz radiation can be generated with this approach by using periodically poled lithium niobate (PPLN) [110].

There also exist types of laser sources emitting at terahertz frequencies. The first ones were gas lasers using organic molecules such as methanol, ethanol, or chloroform as the active medium and were optically pumped by a continuous-wave (cw) CO_2 laser. They deliver cw powers up to 200 mW at far-infrared wavelengths, but owing to the molecular nature of their active medium they are not tunable. However, they can be very well frequency-stabilized and are therefore suitable as local oscillators in heterodyne detection setups. The second type of laser is the *p*-doped Ge laser that was invented about 20 years ago by Andronov et al. [43]. Here, radiation is generated by achieving population inversion between the heavy-hole and light-hole valence bands in crossed magnetic and electric fields [43]. In a certain range of electric and magnetic fields, light holes follow closed orbitals in the $k_x k_y$ plane while heavy holes, owing to their larger effective mass, are accelerated up to the energy of the optical phonon. Emission of an optical phonon then brings them back to the origin, where they can also enter the light hole band, eventually establishing population inversion between the two hole bands. This, however, requires a very pure material to prevent the light holes from being scattered out of the closed trajectories. The foremost advantages of the *p*-doped Ge laser are its tunability by magnetic field and the high peak output powers of up to 10 W. These are, however, obtained at the expense of very high input powers, and the poor efficiency causes a large excess heating of the crystal that hampers continuous-wave operation. Progress has been made toward this goal by reducing the physical dimension [111, 112] of the crystal, naturally reducing output powers. Moreover, cryogenic temperatures are necessary to minimize scattering and high magnetic fields are required for tunability. Until now, such difficulties have prevented the use of this laser in many applications. On the other hand, operation on a single longitudinal mode [113] and mode locking of the emission [114] have been recently demonstrated. Besides the classical *p*-doped Ge laser that follows the above-described concept, there are lasers relying on transitions between light hole Landau levels [44, 45] and lasers based on resonances between the hole band and impurity levels [46, 115].

Terahertz radiation can also be generated by *optical parametric oscillation* (OPO) [116]. This is a second-order nonlinear effect leading to the transfer of energy from a pump wave oscillating at frequency ω_3 to the idler and signal waves at frequencies ω_2 and

ω_1, respectively. From the requirement of conservation of energy, the condition $\omega_3 = \omega_2 + \omega_1$ must be satisfied. Moreover, efficient transfer of energy to the signal and idler waves takes place only when the three electromagnetic waves are phase-matched. If the nonlinear crystal used for parametric oscillation is anisotropic, i.e., if the refractive index depends on the direction of propagation with respect to the crystal orientation, then the condition for phase matching can be written as

$$\begin{aligned} K_3 &= K_2 + K_1 \\ n_3\omega_3 &= n_2\omega_2 + n_1\omega_1 \end{aligned} \tag{2.2}$$

where n_1, n_2, n_3 denote the refractive indices in the direction of propagation for the waves at frequencies ω_1, ω_2, ω_3, respectively. In an experimental setup like the one shown in Fig. 2.4, the frequency thus can be tuned by simply rotating the crystal. By using a $LiNbO_3$ crystal and a Q-switched Nd:YAG laser, high pulse energies up to ~1 nW are obtained for pulse lengths of 15 ns (full width at half maximum) [117–119]. By changing the pump frequency and/or the angle of the crystal, emission frequencies from 0.6 to 2.4 THz are generated. By replacing the $LiNbO_3$ by GaP, frequencies up to 7 THz have been generated [120, 121].

Recently, lasing between impurity levels [122–124] of antimonide, phosphorous, or bismuth in silicon was observed. These lasers are optically pumped by CO_2 lasers or free-electron lasers and so far operate

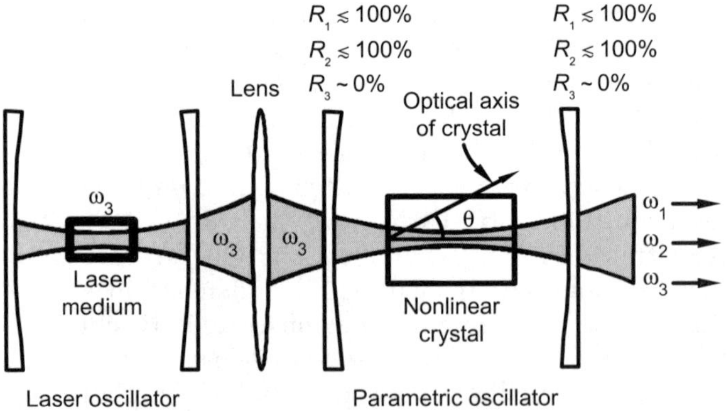

Figure 2.4 Schematic diagram of an optical parametric oscillator. The laser output at ω_3 acts as the pump for the nonlinear crystal, giving rise to oscillations at ω_2 and ω_1. The nonlinear medium is embedded in a Fabry-Perot cavity that provides exclusive feedback for ω_2 and ω_1 (reflectivities R are given for each mirror). The optical axis in the crystal is drawn as well. Rotating the optical axis with respect to the pump beam, one tunes the frequencies according to Eq. 2.2. (Figure 8.16 from *Optical Electronics in Modern Communications* by Amnon Yariv, copyright © 1997 by Oxford University Press, Inc. Used by permission of Oxford University Press, Inc.)

only in pulsed mode and only at cryogenic temperatures, but exhibit very narrow linewidths owing to the atomiclike nature of the optical transition. Finally, it should be mentioned that high-power terahertz radiation can be obtained from electron-beam sources such as the *free-electron laser* [125], relativistic accelerators [126], and synchrotrons [127, 128].

2.2.2 Detection of terahertz radiation

Detection of terahertz light represents a challenge because of the peculiar frequency region of this radiation. Direct coupling into circuitry is difficult because electronics typically fail at such high frequencies. On the other hand, the photon energy is very small, between 4 meV (1 THz) and 40 meV (10 THz), which makes the fabrication of photodiodes almost impossible. Additionally, most terahertz sources emit only moderate power levels compared to the relatively high thermal background levels.

Very sensitive broadband detectors are cryogenic bolometers, either based on semiconducting materials such as Si, Ge or InSb, or employing superconducting materials such as niobium. In the former case, noise-equivalent power levels reach pW/Hz$^{1/2}$ at liquid-He temperatures, and even fW/Hz$^{1/2}$ can be reached in ^{3}He cryostats [129]. Pyroelectric infrared detectors may also be used at terahertz frequencies, but typically have noise-equivalent powers in the microwatt range. For all these detectors, however, it must be noted that they possess a slow response; typical cutoff frequencies are below 1 kHz. These values can be increased at the expense of sensitivity. Recently, even a single photon detector for terahertz photons was demonstrated [130]. It relies on a single-electron transistor consisting of a quantum dot in a high magnetic field, and it offers unprecedented sensitivity. Moreover, the cutoff frequency of this detector should be in the gigahertz range.

Heterodyne sensors offer the advantage of providing a high spectral resolution and a high sensitivity but require a local oscillator emitting at the terahertz frequency of interest. The signal is then mixed with the local oscillator, e.g., in a Schottky diode, and the down-converted signal is detected by using conventional electronics. Higher sensitivity is achieved in superconductor mixers based on superconductor-insulator-superconductor tunnel junctions or even better on hot-electron bolometers [131].

Recently, the use of electron plasma waves in a field-effect transistor was also proposed and successfully demonstrated up to 600 GHz [132].

In certain cases such as the detection of pulsed terahertz radiation in time-domain spectroscopy systems, coherent detectors are necessary. The two most common methods are based on photoconductive sampling and free-space electrooptical sampling, both of which rely on ultrafast laser sources. Fundamentally, the electrooptical effect is a coupling between a low-frequency electric field (terahertz pulse) and a laser beam (optical pulse) in the sensor crystal. Simple tensor analysis indicates that using a <110>-oriented zinc-blend crystal provides the highest sensitivity. The terahertz electric field modulates the birefringence of the sensor; this in turn modulates the polarization ellipticity of the optical probe beam passing through the crystal. The ellipticity modulation of the optical beam can then be analyzed to provide information on both the amplitude and the phase of the applied electric field [21, 22]. Thin sensor crystals in combination with ultrashort laser pulses (duration < 15 fs) allow for a very large bandwidth and detection of up to tens of terahertz [42].

Photoconductive antennas are widely used for pulsed terahertz detection. An identical structure to the photoconductive antenna emitter may be used. Rather than applying a bias voltage to the electrodes of the antenna, a current amplifier and meter are used to measure a transient current. Ultrahigh bandwidth detection has been demonstrated by using photoconductive antenna detectors with detectable frequencies in excess of 60 THz [133].

2.3 Terahertz Quantum Cascade Laser

2.3.1 Intersubband spontaneous emission at terahertz frequencies

The ability to tailor the emission wavelength of a QC laser just by changing layer thicknesses and applied bias made this type of emitter a natural choice for the realization of a terahertz semiconductor laser. Such a laser in principle could achieve high-performance operation and face no fundamental limits in operating temperature. The extension of the QC principle to terahertz frequencies, i.e., to photon energies of 4 to 40 meV, is, however, not straightforward. In the design of the active region, the use of electron-phonon scattering to tailor lifetimes is normally blocked. Instead, electron-electron scattering begins to play an important role in determining the carrier dynamics.

The first far-infrared emission from intersubband transitions was obtained by Helm et al. in 1989 using a full superlattice structure [63]. In 1997, Xu et al. [134] were able to observe broadband far-infrared emission from an electrically pumped three-level system based on intersubband transitions. The first terahertz emitter based on the QC concept was demonstrated by Rochat et al. [80] in 1998, and it triggered the investigation of several similar structures [81, 83, 84]. Figure 2.5

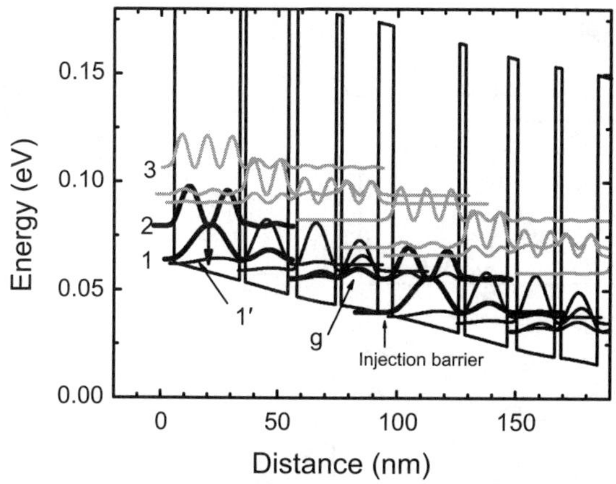

Figure 2.5 Conduction band structure of the first terahertz QC emitter. Also shown are the moduli squared of the wave functions in two periods. The optical transition takes place between states 2 and 1; electrons in state 1 are supposed to exit via state 1′ into the injector downstream. (Reprinted with permission from [80]. Copyright © 1998, American Institute of Physics.)

shows a portion of the conduction band structure of the prototypical design by Rochat et al. [80]. The optical transition of 14.1-meV energy ($\lambda \sim 88$ μm) is vertical in real space and takes place within a single quantum well between states 2 and 1. The vertical character of the optical transition offers the advantage of a high oscillator strength and a narrow linewidth due to the reduced sensitivity to interface roughness. Extraction of electrons from state 1 is intended to take place via a third state 1′, which is almost in resonance (1.4 meV) with state 1. The focus in this structure was on achieving a long lifetime of carriers in the upper laser level. To this end, a weak coupling between the injector and active region was chosen, and the miniband dispersion was kept small in order not to open a scattering path via the emission of optical phonons. Moreover, the small miniband width, well below the photon energy, prevents reabsorption of the emitted light. Electroluminescence spectra at different injection currents are shown in Fig. 2.6. At low temperatures, the emission displays a very narrow ($\Delta\nu = 0.7$ meV) line centered at 14.1 meV, in good agreement with the calculated value of 13.9 meV. The voltage-current (*V-I*) and the light-current (*L-I*) characteristics are shown in Fig. 2.7. The *V-I* characteristics at low currents are in line with those observed in mid-infrared QC structures; beyond a certain threshold, *negative differential resistance* (NDR) occurs, marking the end of resonant tunneling transport through the structure, and the optical spectra become strongly distorted (see inset of Fig. 2.6). As shown in Fig. 2.7, at low temperatures the device exhibits a square root dependence of the emitted light on injected current. Such a sublinear dependence can be

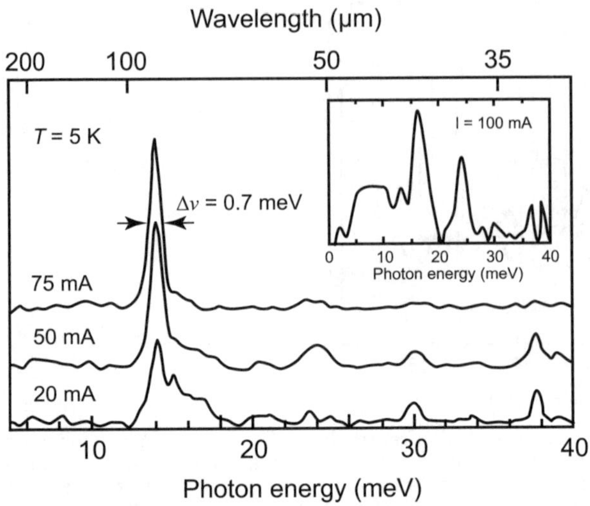

Figure 2.6 Electroluminescence spectra of the prototypical device of Fig. 2.5 for different drive currents taken at a temperature of 5 K. The inset shows a spectrum recorded after the onset of negative differential conductance. (Reprinted with permission from [80]. Copyright © 1998, American Institute of Physics.)

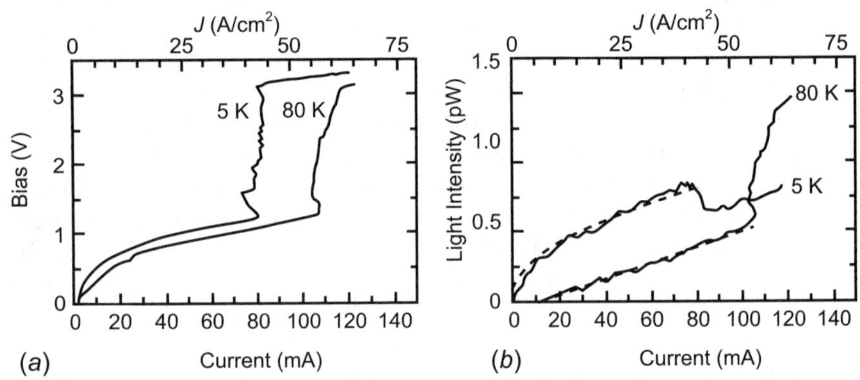

Figure 2.7 (*a*) Voltage-current characteristics of the device of Fig. 2.6 at heat sink temperatures of 5 and 80 K, respectively. (*b*) Light-current characteristics. At 5 K the characteristics are sublinear, which can be explained by a density-dependent nonradiative lifetime of the carriers in the upper level. (Reprinted with permission from [80]. Copyright © 1998, American Institute of Physics.)

understood from a density-dependent scattering rate of electrons in level 2. Raising the injection current leads to an increased density of electrons in level 2, thereby enhancing the electron-electron scattering rate. Such electron-electron scattering can then act as the activation mechanism for longitudinal optical (LO)–phonon emission, which is otherwise blocked by the lack of states at appropriate energy. In fact, LO-phonon emission in this structure is only possible for electrons with sufficient excess kinetic energy since no two subbands are separated by more than the LO-phonon energy. Besides this vertical design, also structures featuring a diagonal (interwell) transition were investigated [83, 134], aiming at a more favorable internal quantum efficiency owing to the reduced nonradiative recombination rates.

All the previously described design concepts, however, have limitations that compromise the achievement of population inversion. The small dispersion of the injector miniband as well as the small tunnel coupling between active region and injector states restricts efficient electric transport to a limited range of currents and voltages determined by the onset of negative differential resistance. Furthermore, the miniband width forbids the use of LO-phonon scattering as a means of fast depletion of the lower level. Finally, the small separation between the Fermi level and the states in the active region leads to devices sensitive to carrier heating, i.e., suffering from backfilling of the active region upstream.

For a deeper understanding of the operation of the above terahertz emitters, a semiclassical global simulation scheme was developed based on a Monte Carlo sampling of a coupled set of Boltzmann-like equations [135, 136]. For an appropriate modeling of the system, both *carrier-carrier* (c-c) and *carrier-LO phonon* (c-p) scattering have to be included in a fully three-dimensional framework. Imposing periodic boundary conditions the full voltage-current characteristics, population of the levels, and gain spectra can be computed without resorting to phenomenological parameters. A simulation of the structure by Rochat et al. performed at the design field of 2.7 kV/cm and 80 K lattice temperature showed that under these conditions, the steady-state populations of the upper and lower laser levels are 8.5×10^8 and 3.5×10^9 cm^{-2}, respectively [137]. This unfavorable population ratio is not a consequence of fast depopulation of the upper level, but of slow extraction of carriers from level 1. The bottleneck of this particular structure is in fact the weak coupling between the active region and the injector downstream. The calculated current density of 50 A/cm^2 was in good agreement with the measured value at a bias of 0.85 V, corresponding to an applied electric field of 2.7 kV/cm.

2.3.2 Active region of the first laser structure

As discussed previously, the design of terahertz QC emitters focused initially on avoiding detrimental nonradiative depletion of the upper laser level. A bit in contrast to this approach, the first structure that exhibited optical gain in the terahertz region follows the so-called chirped *superlattice* (SL) concept [90]; this scheme is well established

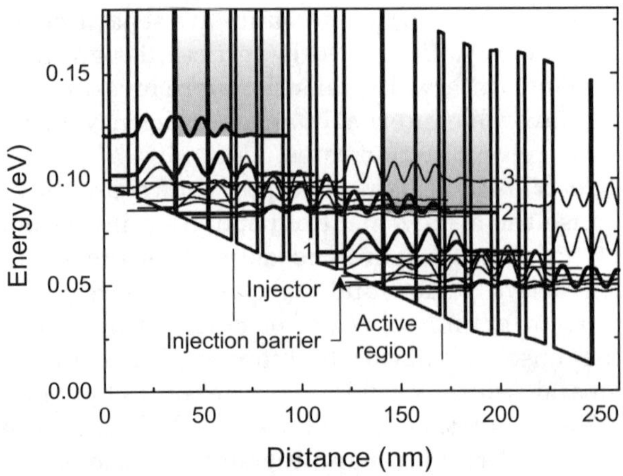

Figure 2.8 Conduction band energy diagram of the first terahertz "chirped" SL gain medium under an average electric field of 3.5 kV/cm. The layer thicknesses (in nanometers) are, from left to right, starting from the injection barrier **4.3**/18.8/ **0.8**/15.8/ **0.6**/11.7/ **2.5**/10.3/ **2.9**/10.2/ **3.0**/10.8/ **3.3**/9.9, where $Al_{0.15}Ga_{0.85}$ As layers are in boldface and the 10.2-nm-wide GaAs well is doped 4×10^{16} cm^{-3}. Also shown are the moduli squared of the wave functions; the optical transition occurs between states 2 and 1.

in the *mid-infrared* (mid-IR) as favoring population inversion even when precise tailoring of the lifetimes is too complex, thanks to its broad phase space of final states available for electron scattering processes. A portion of the conduction band diagram is shown in Fig. 2.8. The moduli squared of the relevant wave functions are displayed, with the laser transition taking place between the two levels drawn in boldface. The SL period and duty cycle are varied to realize a flat-band condition under applied bias, and the optical transition takes place between the first and second minibands rather than between individual electronic levels. The main characteristics of this structure are the large dipole matrix element of 7.8 nm at an emission energy of 18 meV (4.3 THz) and the wide injector miniband. The lower laser level 1 is strongly coupled to this miniband, which, in conjunction with the large miniband dispersion of 17 meV, facilitates the achievement of population inversion. The strong coupling leads to a rapid extraction of carriers from state 1 owing to the large scattering matrix elements between this level and the injector states [137]. Moreover, it reduces the nonradiative scattering rate τ_{21}^{-1} from the upper laser level 2 directly into 1, as electrons in the former can scatter to a whole dense miniband of states. Finally, the large dispersion of the miniband supports a wide range of currents and voltages in the operating characteristics, which is necessary for the achievement of high output powers and, additionally, suppresses thermal backfilling from the active region downstream. Also in this structure direct electron-LO phonon scattering from the lower laser level or within the miniband is not possible, as no levels are separated by more than the LO phonon energy. However, elastic scattering processes such as electron-electron scattering can provide sufficient excess kinetic energy to open that scattering path. This view is strongly supported by simulations [137]. In Table 2.1 we have summarized the steady-state populations of the individual subbands of one period, with and without c-c scattering included in the simulation. As expected, in the absence of c-c scattering, relaxation is considerably suppressed and carrier densities remain high in level 1 and the higher injector subbands. The inclusion of c-c scattering, however, opens a path for thermalization via the described processes, and a favorable redistribution of carriers is obtained. The population of state 1 decreases by 40% while that of state 2 is now 4 times larger. In fact, these calculations indicate that at 80 K the population inversion for the lasing transition is achieved. The computed lifetimes of states 2 and 1 are $\tau_2 = 0.8$ ps, $\tau_1 = 2.2$ ps, and $\tau_{21} = 8.3$ ps with τ_2 and τ_{21} dominated by c-p scattering processes. While the total lifetime τ_2 is shorter than τ_1, the necessary condition $\tau_1 < \tau_{21}$ is fulfilled. Moreover, note that the number of carriers in level 3 is negligible, indicating a low loss of carriers into higher lying states.

Figure 2.9 shows the current density and the population inversion between the upper and lower laser levels as a function of the applied bias as extracted from the Monte Carlo simulation [89]. A significant population inversion of $\delta N_S \approx 1.4 \times 10^9$ cm^{-2} is predicted for biases around ~3 kV/cm and a concomitant current range of ~500 A/cm^2. A large total current density on the order of 1 kA/cm^2 is also obtained before the onset of NDR that marks the end of resonant tunneling transport. It is worth noting that, contrary to mid-IR QC lasers, where

Subband index	Energy (meV)	Population (c-p) (10^9 cm^{-2})	Population (c-p and c-c) (10^9 cm^{-2})
3	51.7	0.02	0.06
2	36.1	0.94	3.99
1	17.9	4.72	2.83
F	14.1	7.13	4.72
E	11.0	8.08	6.08
D	7.7	7.34	5.45
C	4.5	5.14	5.98
B	3.2	1.99	4.62
g	0	6.19	7.87

TABLE 2.1 Population of the individual levels in the chirped SL, calculated at 80 K under the bias of Fig. 2.8 with only carrier-phonon (c-p) scattering and with both c-p and carrier-carrier (c-c) scattering. Letters refer to the injector states comprising the first miniband in Fig. 2.8; numbers refer to the states in the active region. Index *g* represents the injector ground state while capital letters label the other levels in ascending order.

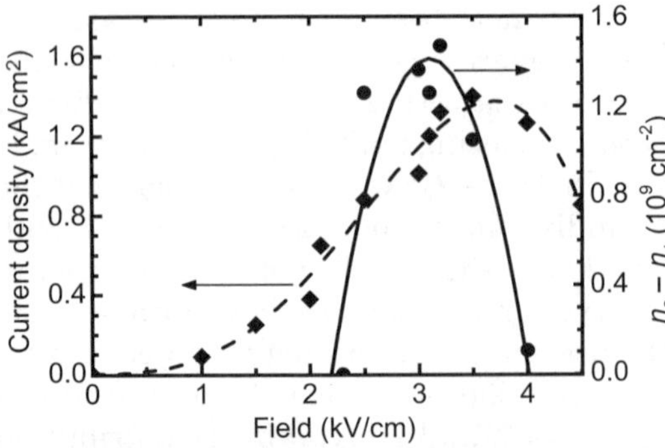

Figure 2.9 Calculated population inversion between states 2 and 1 of the active superlattice (circles), expressed in terms of electron sheet density (n_2 and n_1, respectively), and current density traversing the structure (diamonds). Both are computed as a function of bias field at 10 K. Solid lines are polynomial fits.

depletion of the lower laser state is controlled by optical phonon emission, in this case the transparency condition of the interminiband transition is reached at a current density significantly different from zero. By assuming a linewidth of 2 meV and taking the computed value for the dipole matrix element of 7.8 nm, the gain coefficient is estimated to be 39 cm^{-1}.

Electroluminescence measurements were performed in a sample grown by *molecular beam epitaxy* (MBE) with 40 periods of the active regions/injectors. To detect the strictly in-plane propagating intersubband emission in the direction orthogonal to the mesa, a metallic grating has to be used as the out-coupler for the emitted light. For optimum out-coupling efficiency, the grating period should be somewhat shorter than the wavelength of the light in the material [64, 138]. A microscope image of a typical mesa is shown in Fig. 2.10. Grating

Figure 2.10 Wet-etched mesa of size 500 × 400 μm^2 with a metallic Ge/Au-Au grating (period 21 μm) on top. The rather rough surface of the metal is a consequence of the thermal annealing process required for ohmic contacts. The curved shape of the lines is due to a distortion in the microscope camera.

periods were 15, 18, 21, and 24 μm. Samples were then soldered to copper blocks by using an In-Ag alloy and were wire-bonded.

Figure 2.11 shows two electroluminescence spectra of the first terahertz SL sample as recorded at 8 K heat sink temperature with a resolution of 4 cm^{-1} from a mesa with an 18-μm period grating [139]. The solid line corresponds to a spectrum taken with the correct bias polarity, applying 800-ns-long negative pulses of 1000-mA amplitude to the top contact. The dotted line is a spectrum obtained at the same level of dissipated power but under reverse bias. The two peaks at 243 and 292 cm^{-1} correspond to the GaAs LO phonon frequencies at the Γ point and the X point of the Brillouin zone. The main peak at 144 cm^{-1} (equal to 18 meV) can be unambiguously attributed to the intersubband transition. As expected, it is absent in the spectrum taken under reverse bias. In fact, owing to the asymmetry of the level structure, intersubband emission can only be recorded when the active-region field has the appropriate polarity. This is not the case for other spurious processes such as black-body emission or defect-related transitions. Moreover, this peak is present also in the spectra recorded from mesas with different grating periods ranging from 15 to 24 μm. Its intensity, however, decreases with longer grating periods whereas the 15-μm grating gives almost as good results as the 18-μm grating [139]. This is in good agreement with theoretical predictions [138] and the results of previous experiments [64]. The transition energy of 18 meV matches very well the calculated value at this applied field. At low temperature, the *full width at half maximum* (FWHM) is 2 meV, which is a little broader than that measured in samples employing a vertical

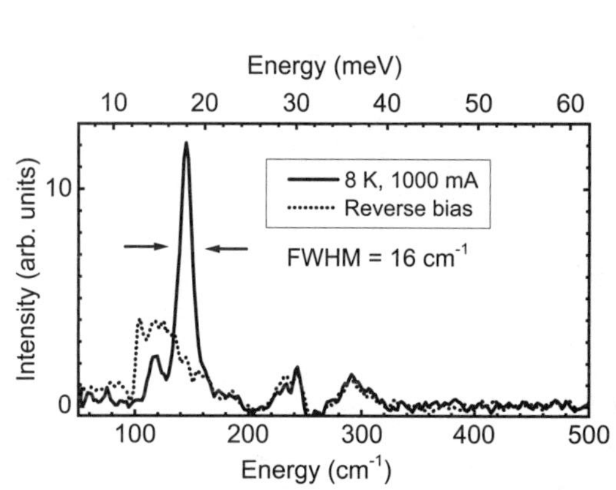

Figure 2.11 Electroluminescence spectra of the chirped SL structure recorded at 8 K. The solid line is taken in forward bias at a current of 1000 mA and with a pulse length of 800 ns (20% overall duty cycle). The dotted line is the reverse-bias spectrum at an equal dissipated power. The main peak at 144 cm^{-1} is attributed to the intersubband transition and is absent in the reverse-bias spectrum, as expected from the asymmetry of the structure. (From [139].)

transition within a single quantum well [80, 84]. However, this is a well-known property of SL active regions because the wave functions involved extend across several barriers and are more sensitive to interface roughness.

Figure 2.12 displays the light-current (*L-I*) and voltage-current (*V-I*) characteristics of the device. The data correspond to the power emitted within the intersubband peak and increase linearly with current up to 1000 mA (500 A/cm^2), where the maximum peak output power of 500 pW was measured [139]. The *V-I* characteristics are typical for transport through QC structures. At low voltages the resistance of the sample is high owing to level misalignment. At a voltage of about 1.1 V, corresponding to an electric field of 3 kV/cm (close to the design field of 3.5 kV/cm), levels align and injection into the upper SL miniband takes place, as indicated by the differential resistivity dropping to 0.7 Ω. Linear *V-I* characteristics are obtained from about 400 to 1100 mA. At a current of 1300 mA (~ 650 A/cm^2) NDR occurs, marking the suppression of resonant tunneling transport. The wide range of currents over which both the *V-I* and the *L-I* characteristics are linear confirms the notion that the large dispersion of the SL miniband ensures efficient electron transport through the structure. Further-more, the linearity of the *L-I* curve indicates that the radiative efficiency is not reduced by density-dependent mechanisms such as electron-electron or Auger scattering.

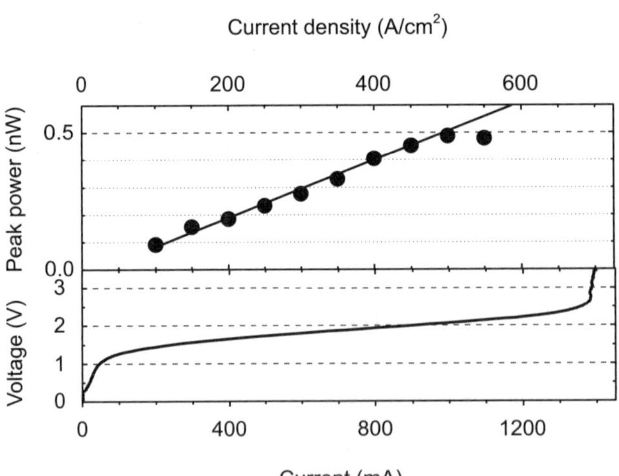

Figure 2.12 Light-current (top) and voltage-current (bottom) characteristics of the SL device of Fig. 2.11 as recorded in pulsed mode (800-ns-long pulses) at 8 K. The values of the peak output power were corrected for the absorption of the polyethylene window of the cryostat. The data represent the power of the intersubband emission peak. The solid line is a linear fit to the data from 200 to 1000 mA. (From [139].)

2.3.3 Waveguiding and lasing

As mentioned in Sec. 2.1, waveguiding of terahertz radiation is difficult owing to the long wavelength and the high optical losses caused by free-carrier absorption. In mid-infrared QC lasers, traditional dielectric waveguides are used, in which the optical mode is guided by sandwiching the active core between cladding layers with low refractive index. This solution, however, cannot be used at ever-longer wavelengths for at least two reasons. First, as a rule of thumb, the necessary cladding thickness is on the order of $\lambda/2$, resulting in an epilayer thickness incompatible with MBE growth. Second, the penetration of the optical mode into the necessarily doped claddings causes very high propagation losses owing to the strong absorption by free carriers. This problem was circumvented in long-wavelength QC lasers by the introduction of surface-plasmon-based waveguides [76, 140, 141]. Surface plasmons are solutions to Maxwell's equations existing at the interface between two materials with dielectric constants of opposite sign, such as a metal and a dielectric. The optical mode peaks at the interface and decays exponentially into both sides, with the decay constants controlled by the wavelength of the light and by the dielectric constant of the materials at each side of the interface.

Despite the successful implementation up to 24-μm wavelength, the single-surface-plasmon scheme is not applicable in the terahertz region. Although the penetration of light into the top metal layer is very low at such long wavelengths, only a very small overlap of the guided mode with any active stack of reasonable thickness is achieved. As an example, Fig. 2.13 shows the computation for a mode of $\lambda \sim$ 67-μm wavelength bound to the interface between a gold layer and an undoped GaAs substrate. A higher, close-to-unity confinement factor of the mode can be achieved in a double-plasmon waveguide, in which radiation is guided between two metallic layers [142]. Since this requires a complicated processing such as epilayer lift-off and wafer bonding, it was initially thought that the easiest approach to the fabrication of such waveguides is to grow the active core on highly doped ($n > 5 \times 10^{18}$ cm^{-3}), almost metallic, GaAs substrates. High confinement is achieved, but at the expense of high optical losses ($\alpha_W = 50 - 80$ cm^{-1}) resulting from penetration into the doped GaAs cladding layer [143]. This solution was then soon discarded; in contrast, as we will see in Sec. 2.4, double-metal waveguides have now become a mature technology, yielding terahertz QC lasers with excellent performance [94].

Since a major source of optical losses is caused by absorption in the claddings, solutions based on semi-insulating GaAs substrates represented an interesting alternative. To a good approximation, in fact, GaAs semi-insulating substrates can be assumed to be free of

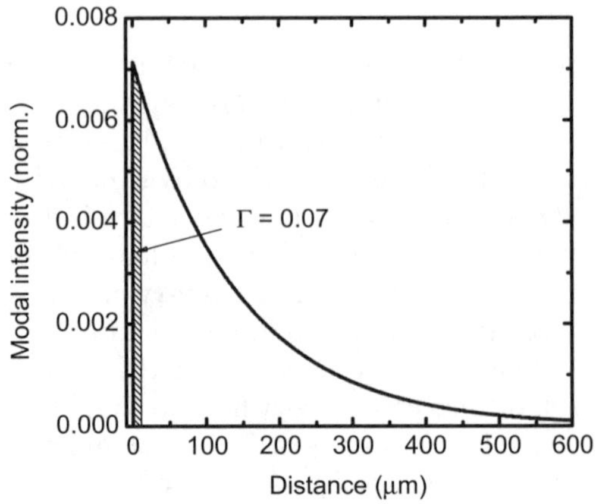

Figure 2.13 Computed profile of a surface-plasmon mode of wavelength $\lambda \sim 67$ μm bound to the interface between undoped GaAs and gold. The very long decay into the semiconductor renders the overlap with any active core of reasonable thickness very small; to highlight this, the 7% overlap with a 10-μm-thick stack has been shaded.

terahertz absorption. The waveguide developed for the first terahertz QC lasers relied on this approach and exploited the presence of a thin, highly doped layer (also used as bottom electrical contact), which was grown directly underneath the low-doped stack of active SLs [89]. Surface plasmons at the interfaces between the bottom contact layer and the active core on one side and the substrate on the other exist if the dielectric constant ϵ of the contact layer is negative. This can indeed be achieved by adjusting the doping concentration; in the case of GaAs and of the 67-μm wavelength for the SL active region just described, ϵ turns negative at a doping concentration of $\sim 2.3 \times 10^{17}$ cm^{-3}. The dielectric constant can then be further tailored to control the decay of the optical mode perpendicular to the interface and to find the appropriate compromise between confinement factor and waveguide losses. In essence, since the thickness of the inserted layer is on the order of the surface-plasmon skin depth, the two surface-plasmon modes existing at its interfaces merge into a single mode that is bound to the central layer. The penetration in the surrounding semiconductor is minimized, and a tight confinement with low optical losses is achieved. The actual mode profile of the complete waveguide is shown in Fig. 2.14, where the presence of a highly doped GaAs layer necessary for the top contact and of the final gold layer was included in the calculation. The waveguide employs a doping concentration of 2×10^{18} cm^{-3} in the central guiding layer, giving a confinement factor $\Gamma = 0.47$ and losses $\alpha_w = 13$ cm^{-1}, respectively.

Using *central-layer-guided* (CLG) waveguide concept and a chirped SL active region, the first QC laser at terahertz frequencies was

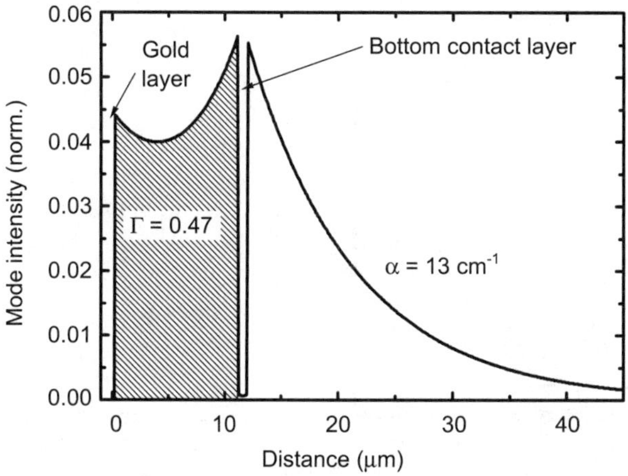

Figure 2.14 Calculated mode profile along the growth direction of the first terahertz laser structure [89]. The origin of the abscissa is at the top metal-semiconductor interface, and the shaded area indicates the waveguide active core.

demonstrated [89]. The sample was grown by MBE on a semi-insulating GaAs substrate, starting with the 800-nm-thick n-doped (2×10^{18} cm^{-3}) GaAs layer, followed by 104 repetitions of the superlattice (thicknesses are given in the caption of Fig. 2.8) and terminated by a 200-nm-thick n-doped (5×10^{18} cm^{-3}) GaAs layer to facilitate electrical contacting. Ridge-geometry mesas were defined by optical contact lithography and wet-chemical etching to a depth exposing the bottom contact layer. Evaporation of GeAu (65 nm) and Au (200 nm) and rapid thermal annealing provided ohmic contacts on top and at the sides of the ridges. Fabry-Perot cavities were formed by cleaving, and devices were soldered onto a copper block, using an In-Ag alloy and wire-bonded.

Figure 2.15 shows the emission of one laser for different drive currents at 8 K heat sink temperature. The characteristic narrowing of the emission spectrum and the nonlinear dependence of the intensity on drive current are clearly observed up to a current of about 880 mA, where the laser threshold is reached. Lasing takes place at ~4.4 THz, on the high-energy side of the luminescence line, probably owing to the reduced waveguide losses at shorter wavelengths. Single-mode emission is obtained, most likely a consequence of the relatively narrow gain spectrum and the wide Fabry-Perot mode spacing. The spontaneous emission at low drive currents (300 mA) displays a full width at half-maximum of 2 meV, similar to the value obtained from luminescence structures with only 40 periods (see Fig. 2.11). These latter spectra are best fitted with a lorentzian line shape which, together with the absence of any broadening when increasing the

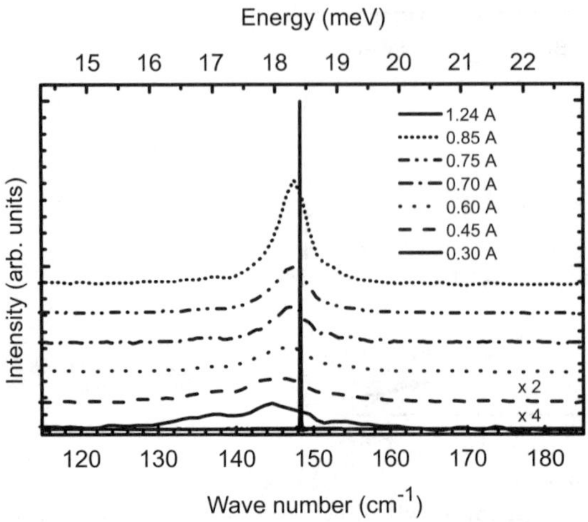

Figure 2.15 Emission spectra from a terahertz laser device for different drive currents, recorded at 8 K by applying 200-ns-long pulses at intervals of 2 μs [89]. The lowest curves have been multiplied by factors of 4 and 2, respectively, for clarity. The 1240-mA laser spectrum was scaled down by several orders of magnitude. The full width at half maximum decreases from 2 meV at 450-mA drive current to 0.5 meV at 850 mA, as the threshold current of 880 mA is approached. Subthreshold spectra were collected with a resolution of 2 cm^{-1} (60 GHz), while the 1240-mA spectrum was obtained with the maximum 0.125-cm^{-1} (3.75-GHz) resolution of the apparatus.

number of SL periods, indicates the high quality and uniformity of the growth throughout the whole 10-μm-thick active-material stack.

Figure 2.16 shows the *L-I* and *V-I* characteristics of the device. At a heat sink temperature of 8 K, the output peak power was estimated to be more than 2 mW, with a threshold current density of 290 A/cm^2 [89]. The latter was already a very small value for QC lasers, and it allowed operation of these first devices at duty cycles up to 10%. The threshold current increases with temperature up to the maximum operating temperature of 50 K. Current-voltage characteristics are similar to those observed in the luminescence structure. The electric field at threshold is 7.5 kV/cm. This is larger than the design value of 3.5 kV/cm, probably as a result of the nonnegligible spreading resistance in the bottom contact layer.

2.4 Progress in Performance

2.4.1 Optimized fabrication and continuous-wave operation

Soon after the first demonstration [89], another terahertz QC laser was developed with analogous characteristics but a different choice of

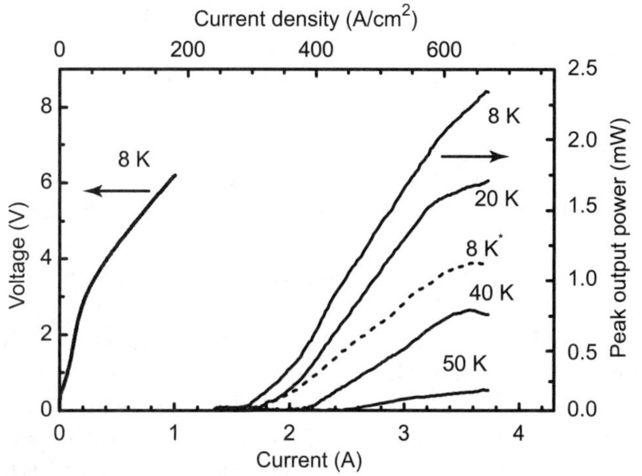

Figure 2.16 Light-current characteristics of a 180-μm-wide and 3.1-mm-long laser ridge [89]. Data were recorded in pulse mode, applying 100-ns-long pulses at a repetition rate of 333 Hz. The peak power values represent what is collected from a single facet onto the detector. At the lowest temperature the device emits more than 2 mW. By increasing the duty cycle to 0.5%, the peak power reduces to 1 mW due to the stronger heating, as is shown by the dashed curve labeled with an asterisk. At the maximum operating temperature of 50 K we observe still 120 μW of peak output power. The *V-I* curve at a temperature of 8 K is also displayed.

the waveguide parameters, optimized for the lowest losses rather than for the confinement factor [144]. Even lower thresholds were obtained (< 200 A/cm^2), but high reflection coatings had to be provided to the laser back facet owing to the smaller gain. These are, in general, easy to realize through deposition of thin insulation and Au metallization; however, they require care as they tend to peel off after thermal cycling. In any case, despite the excellent current densities, also these devices could be operated only in pulsed mode.

To reduce the detrimental joule heating of the laser and allow for continuous-wave operation, several main avenues were immediately pursued in improving the fabrication processes. First, a most relevant aspect is the distance of the lateral side contacts from the laser ridge [145]. As the bottom highly doped layer is relatively thin, it is also quite resistive, and it is therefore important to have the lateral contacts as close as possible to the active region, so as to minimize the applied voltage. On the other hand, the distance should still be sufficient to prevent a coupling of the optical mode to the contact metal, which would increase the waveguide losses. The best compromise for 4.4-THz lasers was found in a few tens of micrometers.

Second, the top metallization on the laser ridge was modified so that the annealed ohmic contact was restricted only to two narrow stripes (10 – 15 μm wide) at the edges of the ridge [146]. This reduces the absorption losses, which are believed to be higher in the annealed material; the interface quality is in fact quite relevant as the optical mode peaks precisely there. Another evaporation step of Cr/Au or Ti/Au was then used to cover the remaining portion of the laser and complete the waveguide. A schematic view of a processed laser is shown in Fig. 2.17. This type of device fabrication is presently the standard one commonly employed for terahertz QC lasers with CLG waveguides.

These changes in fabrication enabled continuous-wave operation of terahertz QC lasers, initially with relatively low emitted powers of a few hundred microwatts up to almost 40 K [146]. In parallel, cw operation was also obtained by mainly reducing the device size through narrowing of the ridges, which could then be contacted using extended metallic regions and polymide insulation to avoid shorting [145].

Soon afterward optimized devices were fabricated yielding similar cw and pulsed performance and output powers of various milliwatts [91]. The measured powers of a typical uncoated laser device and of a laser with a back-facet coating are reported in Fig. 2.18 as a function of cw drive current. For the uncoated device, almost 2.5 mW was obtained at low temperatures, with still more than 500 μW at 40 K. Lasing ceased at around a 45 K heat sink temperature. These values translate to 7.5 and 1.5 mW, respectively, of total emitted power per facet, using the mimimum 0.33 collection efficiency estimated for the setup. At 10 K, a low threshold current density of 210 A/cm^2 was recorded. In the linear regime from threshold to 1 A, the slope efficiency, again corrected with the factor of 0.33, is 17 mW/A. This can be compared with the expected value

$$\frac{\partial P}{\partial I} = \frac{1}{2} N_p \frac{\alpha_M}{\alpha_M + \alpha_W} \frac{h\nu}{q_0} \tau \tag{2.3}$$

where P is the optical power; I the injection current; α_M and α_W the outcoupling and waveguide loss, respectively; $h\nu$ the photon energy; $N_p = 104$ the number of periods; and q_0 the electronic charge. The factor of $^1/_2$ stems from the fact that only one facet is considered. The

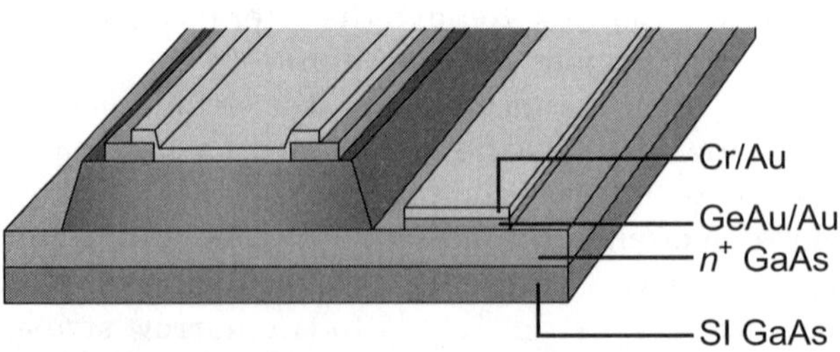

Figure 2.17 Schematic view of a processed and cleaved laser bar from the front facet.

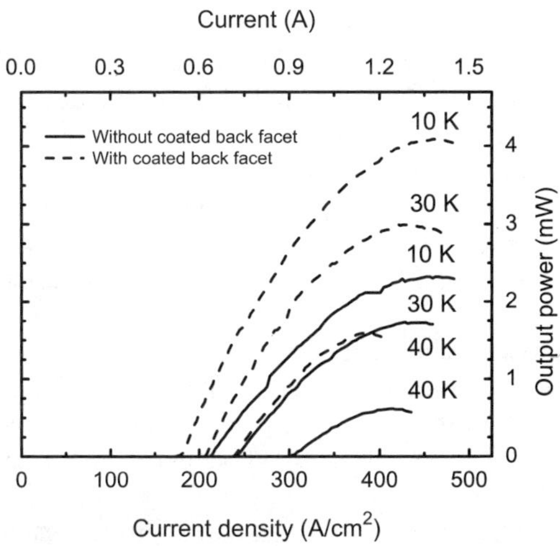

Figure 2.18 Light-current characteristics of a 1.96-mm-long and 150-μm-wide chirped SL laser, recorded in cw operation at different heat sink temperatures, are plotted with solid lines [91]. The power values represent what was collected from one facet onto a calibrated pyroelectric detector after correction for the transmittance of the polyethylene window (0.63). The collection efficiency was estimated to be 0.33. The device stopped operating in cw mode at a heat sink temperature of 45 K. Dashed lines represent data collected from a 2.23-mm-long device with a coated back facet. Output powers are in this case more than 4 mW at 10 K and still 1.6 mW at 40 K. The maximum operating temperature increased only slightly to about 48 K. Due to the slightly different device size, the scale of the current refers only to the dashed lines.

dimensionless constant τ takes into account the fact that in QC lasers the lower-state population differs significantly from zero and that the injection efficiency η_i of carriers into the upper laser state can deviate from unity. In a rate equation framework, τ can be derived as [147–149]

$$\tau = \left[1 - \frac{\tau_1}{\tau_2}\left(\frac{1}{\eta_i} - 1\right) - \frac{\tau_1}{\tau_{21}}\right] \frac{\tau_2}{\tau_1 + \tau_2\left(1 - \tau_1 / \tau_{21}\right)} \tag{2.4}$$

where τ_2 and τ_1 are the total lifetimes of the upper and lower laser levels, respectively, and τ_{21} denotes the scattering time of electrons from level 2 to level 1. The presence of a considerable population in the lower laser level at threshold [137], indicated by the ratio of lifetimes τ_1/τ_{21}, is the main reason for the reduction of the slope efficiency from its ideal value. Inserting lifetimes calculated by using Monte Carlo simulations [137, 150], a slope efficiency of 55 mW/A can be computed, assuming unity injection efficiency. The difference with the experimental value of 17 mW/A may be due to heating effects, uncertainties in the measurements, other scattering mechanisms not included in the simulations, and a nonunity injection efficiency.

Substantial improvements in threshold current density, output powers, and slope efficiency were obtained from a laser with its back facet covered with the high-reflection coating. Data recorded from this device are represented by the dashed lines in Fig. 2.18. A reduction of the threshold current density by about 10% (to 185 A/cm^2) is achieved, and simultaneously the estimated maximum output power increases to more than 4 mW [91]. In pulsed mode, the best devices showed threshold current densities as low as 130 A/cm^2 at 10 K and a highest operating temperature of 75 K [151].

Figure 2.19 shows the voltage and the differential resistance across the device as a function of the cw injected current. The *V-I* characteristics follow the well-known behavior of high resistance at low currents, and a turn-on toward lower differential resistance when minibands align. The measured threshold voltage differs from the design value of 3.8 V by 0.3 V, which is attributed to residual series resistance in the bottom contact layer and electrical contacts. The differential resistance is displayed as well, showing the characteristic drop at threshold when the onset of stimulated emission reduces the lifetime τ_2. Its clear visibility is in general a good indication of a high injection efficiency [147].

2.4.2 New designs and waveguides

The desire to improve the thermal behavior of terahertz QC lasers and the need of extending the operating frequency range toward even lower

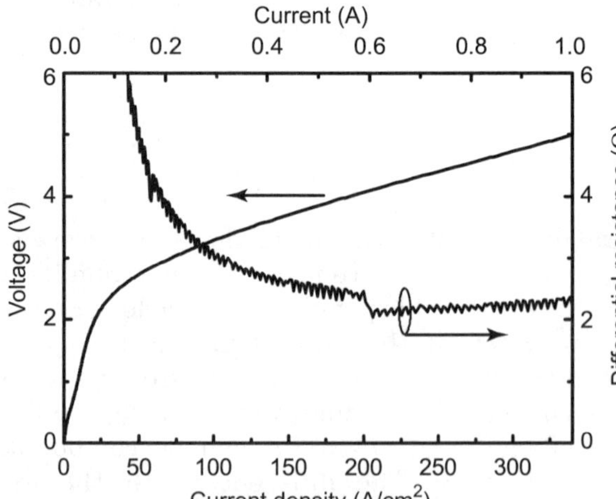

Figure 2.19 Voltage-current characteristics of the laser of Fig. 2.18, recorded in cw operation at a heat sink temperature of 7 K [91]. The derivative is also shown. The sharp drop in differential resistance at threshold is caused by the reduction of the upper-state lifetime by stimulated emission.

frequencies immediately fostered innovative ideas in the quantum design of the active region. Following a development path similar to that of the mid-infrared case, the improvement on the original chirped superlattice was the implementation of *bound-to-continuum (BTC) transitions* [152]. The band diagram of an exemplary structure is displayed in Fig. 2.20. In essence, the SL concept is reinterpreted by making use of a spatially diagonal transition to reduce the nonradiative loss of carriers from the upper laser level and improve the injection efficiency. The name stems from the fact that, contrary to the minibandlike lower state, the upper laser level is mostly localized in a "defect" well of the SL periodic potential. As such, one can speak of donorlike and acceptorlike states, according to their being closer in energy to the upper or lower miniband. For a given transition frequency, the former type provides larger oscillator strengths, owing to the wave function symmetry; on the other hand, the latter simplifies the design of the injection, since the upper miniband can be well separated in energy. The first implementation of the bound-to-continuum scheme in the terahertz range already yielded devices operating above liquid nitrogen temperature [92]; yet, as we will see, the concept enjoyed the greatest success in the realization of lasers with high output power and low emission frequencies.

The above approach tries to address the issue of the upper-state lifetime, but obviously reducing the lifetime of the lower lasing level is also very important to establish and maintain a significant population

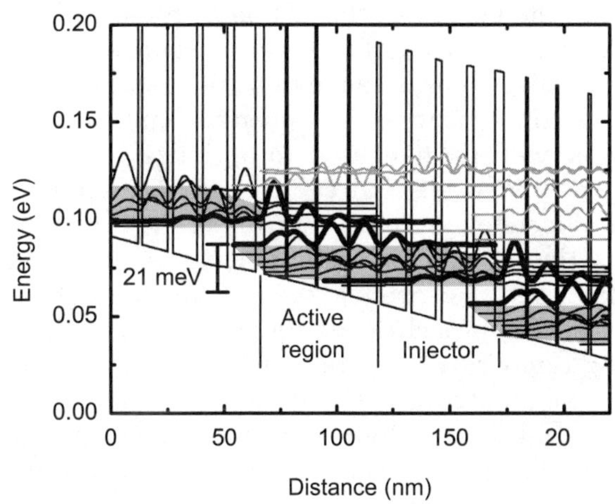

Figure 2.20 Self-consistent calculation of a portion of the conduction band structure of a 2.8-THz bound-to-continuum laser under an applied bias of 2.9 kV/cm [153]. The layer sequence, starting from the injection barrier to the left of the active region, is (in nanometers) **3.9**/9.6/**0.9**/12.4/**0.9**/13.4/**1.1**/11.7/**1.8**/11.2/**2.3**/10.7/**2.6**/10.9/**3.0**/9.3, where $Al_{0.15}$ $Ga_{0.85}$ As barriers are in boldface and the first 9 nm of the 10.9-nm-wide GaAs well is silicon doped at 5×10^{16} cm^{-3}.

inversion. In the BTC design as well as in the previously described chirped SL laser, extraction of carriers from the lower level mainly occurs through electron-electron-activated scattering into states of the injector miniband. Efficient depletion of the lower state in these terahertz QC lasers is hindered by the insufficient cooling of the electron distribution, and its *effective* lifetime, defined as in the inverse of the *net* scattering rate out, is strongly influenced by the electronic temperature due to the small dispersion of the injector miniband. In mid-infrared QC lasers, on the contrary, population inversion up to room temperature and above is maintained by exploiting resonant fast electron-phonon scattering as the extraction process. The idea, however, cannot be transferred in a straightforward manner to the terahertz range. In fact the small transition energy means that any process that is energetically resonant for electrons in the lower lasing subband is "almost" resonant also for those in the upper subband. Therefore, using emission of optical phonons as the depletion mechanism would seriously compromise the transition lifetime. To circumvent this difficulty, one possibility is to spatially separate the region where LO phonon emission takes place from the optical transition. This concept was already examined in Ref. 137, and the simulation predicted that a much larger population inversion than that attainable in superlattices was feasible. The first working structure was realized only a couple of years later using a simpler design targeting 3-THz emission frequency [154]. The band diagram is reported in Fig. 2.21. A single large well contains two subbands (labeled 2 and 3) roughly divided by the energy of the LO phonon (36 meV in GaAs); the terahertz transition, on the other hand, is realized using tunnel-split states in two coupled wells, so as to raise the energy of the lowest subband 4 enough to couple it with the upper subband 3 of the phonon stage. Just another well is then used to optimize injection. The first lasers based on this design already showed promising characteristics, with a low dependence on temperature of the threshold current [154]. A poor injection efficiency, however, compromised the overall performance. This issue was well addressed in subsequent designs—so much so that this approach holds the present records in terms of highest operating temperatures [94], as will be discussed in the following. Also in this case Monte Carlo simulations were employed to model the transport and gain characteristics, showing the importance of impurity scattering in accounting for large electronic temperatures [155, 156].

An alternative design based on optical phonon emission relies on interlacing photon and phonon emitting stages bridged by appropriate minibands [157]. Figure 2.22 shows a portion of the conduction band structure of an 80-μm wavelength device under an applied bias of

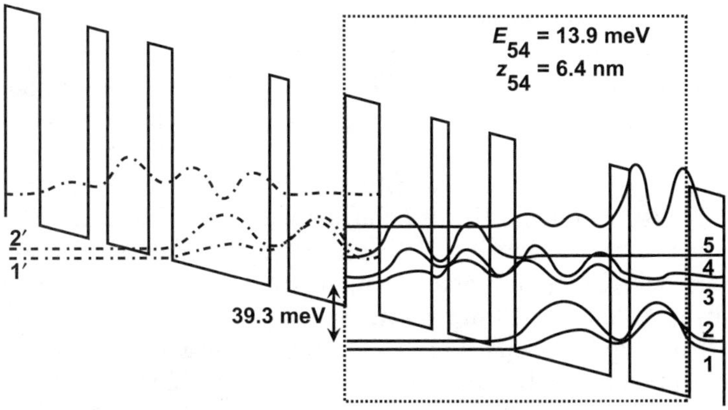

Figure 2.21 Conduction band profile of the structure based on phonon depopulation of Ref. 154 calculated by using a self-consistent Schrödinger and Poisson solver at 64 mV/module, which corresponds to a field of 12.2 kV/cm. The four-well module is outlined. Beginning with the left injection barrier, the layer thicknesses in angstroms are 54/78/24/64/38/148/24/94. The 148-Å well is doped at $n = 1.9 \times 10^{16}$ cm^{3}, yielding a sheet density of 2.8×10^{10} cm^{2}. (Reprinted with permission from [154]. Copyright © 2003, American Institute of Physics.)

5.4 kV/cm. While the minibands, labeled *A* and *B* in Fig. 2.22, ensure good control of the transport characteristics and allow for a wide range of currents and voltages, the LO phonon stage efficiently shortens the

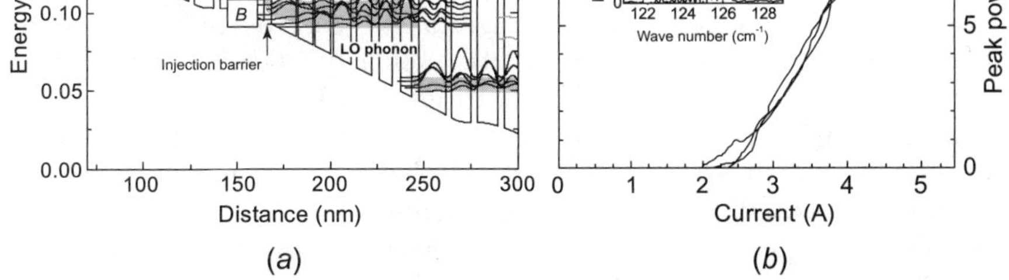

Figure 2.22 (*a*) Portion of the conduction band structure of the interlaced cascade design under an average electric field of 5.4 kV/cm. The optical transition is between the two states drawn in boldface and has a dipole moment of 7.2 nm. The layer thicknesses can be found in Ref. 157. The phonon emission is indicated by the dashed arrow. Minibands *A* and *B* are connecting the phonon to the photon and the photon to the phonon stage, respectively. (*b*) Light-current characteristics of a 4.72-mm-long and 150-μm-wide device, recorded in pulsed operation at a duty cycle of 3.5 μ%. The maximum output power of 10 mW is reached at 30 K, and still 4 mW is obtained at 80 K. The device stops operating at 95 K. Note the very weak temperature dependence of the slope efficiency.

effective lifetime τ_1 of the lower lasing state. Extraction of carriers from this level still occurs through *e-e* scattering into the states comprising miniband B; but the lower states of the miniband are strongly coupled to the phonon stage, and therefore rapid extraction from these states is provided by electron-phonon scattering. The expectation of a short lifetime τ_1 and negligible thermal backfilling is indeed confirmed in this structure, as is best seen from the *L-I* characteristics in Fig. 2.22. The slope efficiency, i.e., the derivative of the output power versus injection current, is virtually independent of temperature, which indicates a constant τ_1. In fact, lasing at temperatures higher than the currently achieved 95 K appears to be limited solely by the maximum current density that can be passed through the device. The residual threshold dependence on temperature seems, then, to be mainly determined by the nonradiative relaxation of upper-state electrons. Comparisons of this interlaced approach with the previous direct phonon extraction scheme, in samples with similar waveguide and growth conditions, showed analogous temperature dependence but possibly better absolute thresholds. Beyond these first preliminary results, however, no further implementations have yet been reported.

On the waveguide front, recent developments have seen the emergence of double-metal waveguides [93]. As can be seen in Fig. 2.23, they offer the clear advantage of providing unity confinement Γ. A price is paid on the absorption losses, of which, however, only a fraction of about one-third is due to the metallic claddings. In the end, the ratio Γ/α_w is very similar to that of CLG waveguides; however, the high confinement factor and hence modal gain is beneficial as the total losses to be compensated also include the mirror outcoupling α_{m}. Furthermore, the optical mode is strongly squeezed in a core definitely smaller than the wavelength; no cutoff frequency in fact exists for TM modes in this type of configuration. This translates into a large "impedance mismatch" with vacuum at the ridge facets, considerably increasing their reflectance. While this property is clearly useful in minimizing α_m even without high-reflection coatings, it also poses problems in extracting high output powers. Finally, as the mode is deeply linked to the two metallic interfaces, the ridge width can be considerably shrunk, yielding very small devices with low drive currents and excellent cw performance. All these characteristics are naturally even more helpful with longer emission wavelengths, and, in fact, QC lasers at 2.1 THz have been realized thanks to this approach [97]. The fabrication of the first double-metal lasers used an In-Au wafer bonding technique [93]. An $Al_{0.5}Ga_{0.5}$ As layer grown before the waveguide core was employed as etch stop in the substrate removal. An important aspect was the use of a thin, low-temperature-grown GaAs cap layer to avoid annealing of the ohmic contacts, which have to be deposited after the bonding

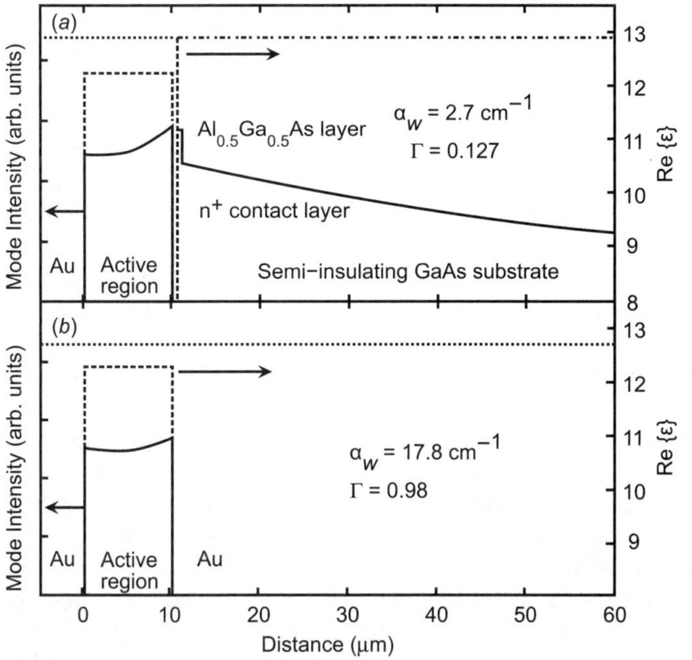

Figure 2.23 Mode intensity (solid lines) and real part of the dielectric constant (dashed lines) for (*a*) the SI surface-plasmon waveguide and (*b*) the double-sided metal waveguide of Ref. 93. The dotted lines represent the square of the modal effective index. (Reprinted with permission from [93]. Copyright © 2003, American Institute of Physics.)

procedure. Since then, the technique has been successively refined, in particular concerning the structural quality of the bonding interface which is relevant for the device thermal resistance.

At the same time, the standard CLG waveguide has undergone some interesting makeover. The highly doped layer buried under the core can be substituted by more layers with various doping densities [158]. This solution allows the possibility of controlling separately the boundary conditions of the surface plasmons on the two sides, thereby resulting in a better compromise between optical losses and confinement factor. In the case of a 2.5-THz laser, a confinement factor of about 48% is reached, with 10-cm^{-1} absorption losses. These values easily permit the use of such waveguides even at long wavelengths, providing very low threshold currents and high output powers [98].

2.4.3 High-temperature and long-wavelength operation

The first terahertz QC laser to operate above liquid nitrogen, at temperatures up to 95 K, was the BTC device demonstrated by Scalari et al. [92]. The luminescence intensity recorded from that structure indicated that the temperature dependence of the threshold current could be ascribed to the nonradiative relaxation of electrons in the upper level. Basically, through collisions, electrons can gain sufficient in-plane kinetic energy to be able to relax via emission of LO phonons to the lower subbands. This process is indeed very difficult to suppress by quantum design. The only solution is to make the optical transition more and more diagonal in space, to reduce the overlap of the states. This can be accomplished only at the expense of losing dipole matrix element for the optical transition, and up to now, very diagonal structures have not shown lasing in the terahertz range. Research has then focused on optimizing the injection by carefully tailoring the thickness of tunnel barriers and the doping concentration. It was in fact shown that the injection process also plays a fundamental role in degrading performance with increasing temperature [150]. This allowed reduction of parasitic conduction paths and improved the dynamic current range, but brought only marginal improvements in the maximum temperature, which was raised to about 100 K in pulsed and 70 K in cw operation [159]. On the other hand, as can be seen in Fig. 2.24, the output powers achieved are indeed quite spectacular, with slope efficiencies of 170 mW/A (once corrected for collection) and threshold currents as low as 100 A/cm^2. Very recently cw operation at liquid nitrogen temperature has been demonstrated at 2.9 THz using a buried waveguide for better heat dissipation [160].

Designs based on optical phonon emission for the depopulation of the lowest lasing subband have immediately shown good potential for high-temperature operation. Not only do they provide faster extraction of carriers from the active region, but also they possess a large energy separation of the optical transition from the ground state of each cascade period [154]. The latter is an important characteristic as it suppresses thermal backfilling from electrons in the injectors downstream. As previously explained for the interlaced design [157], the insensitivity of the output slope efficiency to temperature observed in these devices demonstrates that the lifetime of the lower lasing state is relatively unaffected by the increasing temperature. Furthermore, once fit with the usual empirical formula

$$J_{th} = J_1 + J_0 \exp \frac{T}{T_0} \tag{2.5}$$

the threshold current densities showed a characteristic temperature $T_0 > 30$ K, larger than superlattice or BTC structures, in both the interlaced and the coupled-well devices [94, 157]. Despite these useful characteristics, the maximum operating temperature was still hovering

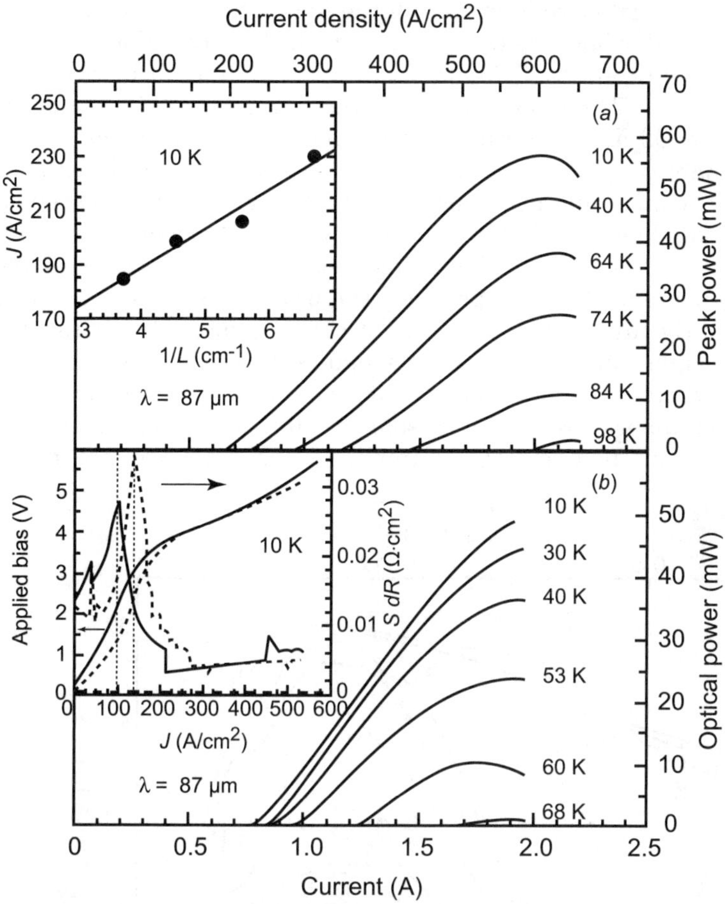

Figure 2.24 (*a*) Peak optical output power versus injected current in pulsed mode at a duty cycle of 2% at various temperatures, as measured from the 3.4-THz laser of Ref. 159. The sample was processed into a 1.6-mm-long and 210-μm-wide waveguide with back facet coating. The inset shows pulsed threshold current as a function of inverse laser length with back facet coating measured at 10 K. (*b*) CW optical power from a single facet of the same device as a function of drive current for various heat sink temperatures. Inset: Bias voltage as a function of injection current and differential resistance deduced from the *V-I* curve for two samples with different injection barriers. (Reprinted with permission from [159]. Copyright © 2004, American Institute of Physics.)

around 90 K, owing to the poor low-temperature thresholds [161]. Current injection was then improved by thickening the injection barrier and changing the spatial profile of the wave functions; double-metal waveguides were also implemented to exploit the larger confinement factor. As can be seen in Fig. 2.25, these advancements allowed the threshold current densities to be reduced to about 600 A/cm^2, resulting in maximum pulsed operating temperatures of about 140 K [94]. The introduction of further refinements in the injector design and a better control of the wafer-bonding technology of metallic waveguides enabled phonon-coupled terahertz QC lasers to reach a high performance level also in cw. Figure 2.26 displays cw *L-I* traces from Ref. 96. The lasers operated up to 93 K, with output powers in the milliwatt range.

The ultimate goal is represented by the possibility of reaching laser operation at room temperature, or at least on thermoelectrically cooled stages. It should be said that high-value applications, such as security scanners in airports or medical imaging systems in hospitals, could live with cryogenic temperatures, provided the practical system is reliable and sufficiently turn-key. In any case it can be expected that further advances in quantum design and waveguide technology will lead to optimized structures capable of lasing at higher temperatures. To really tackle room temperature, however, some breakthrough is needed. This could come from new materials (nitrides, for instance, are being

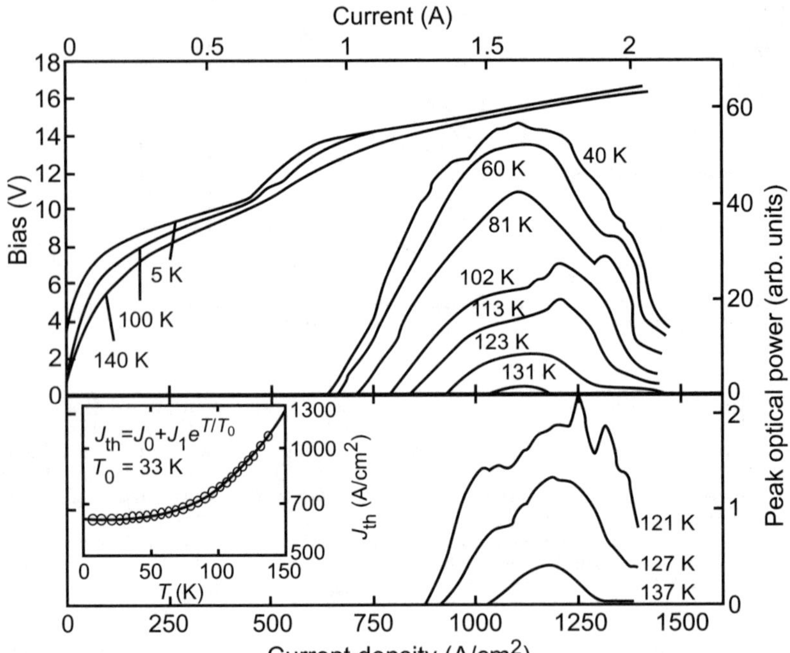

Figure 2.25 Emitted light and bias versus current at various temperatures for the metal-metal 3.8-THz devices of Ref. 94. They were measured using 200-ns pulses repeated at 1 kHz. The *L-I* characteristics in the upper panel are measured from a 60-μm-wide, 2.48-mm-long ridge. The *V-I* characteristics are measured using a smaller structure 100 μm wide, and 1.45 mm long. The lower panel displays *L-I* characteristics from a 150-μm-wide, 2.74-mm-long ridge, along with its threshold current density versus temperature in the inset. Note that the current axis is only applicable to the 60-μm-wide device. (Reprinted with permission from [94]. Copyright © 2003, American Institute of Physics.)

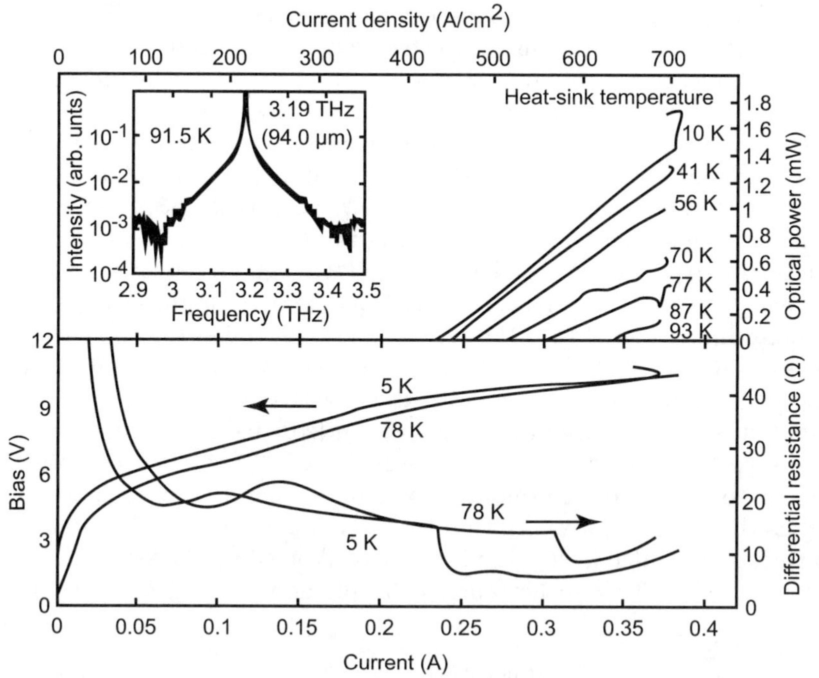

Figure 2.26 Continuous-wave characteristics measured from a 40-μm-wide, 1.35-mm-long ridge laser of Ref. 96. The upper panel shows the *L-I* characteristics for various heat sink temperatures. The inset shows the cw spectrum at a heat sink temperature of 91.5 K. The linewidth is limited by the spectrometer resolution of 0.125 cm^{-1}. The lower panel shows the cw *V-I* and *dV/dI–I* characteristics at heat sink temperatures of 5 and 78 K. (Reprinted with permission from [96]. Copyright © 2004, American Institute of Physics.)

proposed) with larger LO phonon energies, or from nanostructured devices (such as quantum dots) where the electronic motion is confined in all directions and the subband dispersion is broken into individual levels. In all instances these approaches aim at reducing the main source of nonradiative relaxation from the upper lasing state linked to LO phonon emission. They are, however, quite challenging from the point of view of present technical knowledge.

Several applications, particularly in astronomy and chemical recognition, involve detection at frequencies in the range of 1.5 to 2.5 THz. A chirped superlattice design similar to the one described in Sec. 2.3 led to lasing at 3.5 THz with comparable performance [162]. Moving toward longer wavelengths, however, the BTC concept appeared to be more favorable and formed the basis of lasers emitting at 3.5 THz [92] and then 2.8 THz [153]. Optimized designs were then developed for 2.9 THz [95] and 2.4 THz [159] with impressive output powers of tens of milliwatts and low threshold currents also in cw; see Fig. 2.27 for exemplary *L-I* curves. More recently, BTC lasers have begun to be realized at the desired emission frequencies required for specific sensing applications. In particular we show in Fig. 2.28 the spectra of two lasers emitting around 4.8 and 2.55 THz, important for the detection of atomic oxygen and of the OH radical, respectively [98]. The latter shows a record low threshold current density of about 75 A/cm^2 with a high cw output power of 6 mW per facet (33% collection efficiency) and a maximum cw operating temperature of 56 K.

All these devices were based on the CLG waveguide concept that was employed in the first 4.4-THz laser. It is clear, though, that as the wavelength increases (2 THz corresponds to 150 μm), it becomes progressively more difficult to maintain a large confinement factor without resorting to impractically thick active regions. As already explained, an improvement certainly comes from using multiple highly doped layers, as was done in the case of the 2.5-THz structure. In this direction, however, double-metal waveguides seem to have the advantage, as they can display Γ close to unity independently of wavelength and thickness. Indeed, they have recently allowed the lowest emission frequency of 2.1 THz to be reached by phonon-coupled devices [97].

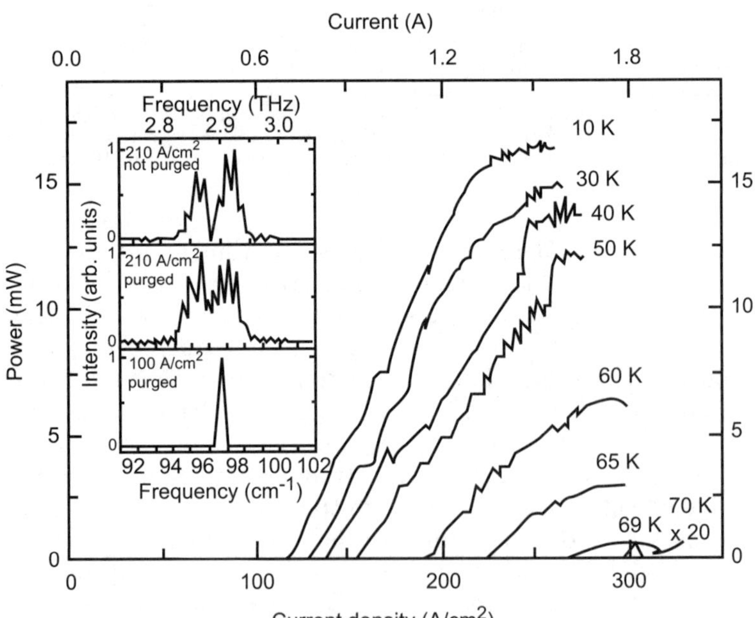

Figure 2.27 Emitted power as a function of current density for the 2.9-THz device of Ref. 95 operated cw. The ripples are due to atmospheric water absorption. Inset: Spectra at different current densities. To highlight the effect of water absorption, the top panel was measured with the optical path unpurged. (Reprinted with permission from [95]. Copyright © 2004, American Institute of Physics.)

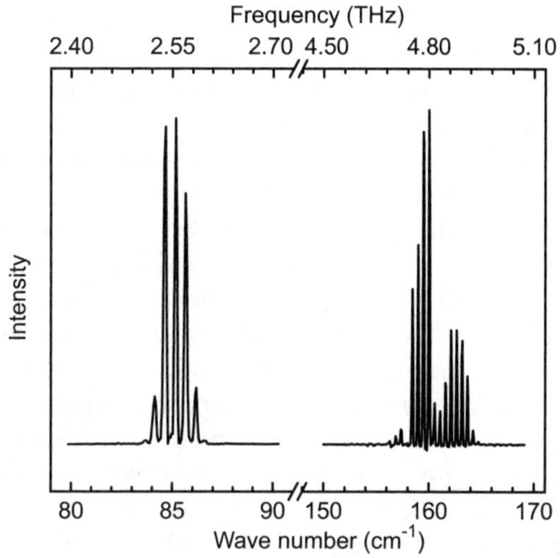

Figure 2.28 Fabry-Perot emission spectra of BTC quantum cascade lasers operating near 2.55 and 4.8 THz.

As discussed in Sec. 2.5, lasing at 1.9 THz has also been reported in high magnetic fields using single-well structures [163]. Beyond waveguiding, the major issue related to further extending the operation wavelength concerns the injection efficiency. When the transition energy is only a few millielectronvolts, tunnel injection has to be very selective, posing stringent conditions on the injector miniband width and coupling. The net result is usually a poorer conductance or small useful voltage range. This decreases the maximum current densities available to pump the active region. More and more diagonal transitions could possibly alleviate this problem, but how far down in frequency QC lasers will be able to go is still an open question.

2.5 New Research Directions and Outlook

2.5.1 Single-mode devices

The lasers described in the previous sections employed a multimode Fabry-Perot cavity; therefore, single-mode operation is observed only occasionally with very short resonators and over a limited range of injection currents [91, 162]. Stable single-mode emission at a precisely designed frequency is, however, highly desirable for most applications. For this purpose, the concept of *distributed feedback* (DFB) has been successfully implemented also in terahertz QC devices [164]. The peculiar surface-plasmon structure of the waveguide required the realization of rather unconventional DFB gratings for the diffraction of light propagating along the laser ridge. In the first devices a periodic corrugation was etched into the thin, highly doped semiconductor contact layer underneath the top metallization. This causes a modulation of the real part of the modal refractive index; lasers featuring such grating still displayed a regularly spaced Fabry-Perot spectrum, but with two modes strongly suppressed, as can be seen by the dashed curves in Fig. 2.29. This is a typical feature of index-coupled DFB lasers with only moderate coupling strength but strong reflectivity at the Fabry-Perot facets. To enhance the coupling constant and increase mode discrimination, an additional loss modulation was introduced by depositing and annealing GeAu/Au contacts [91] selectively on the crests of the etched grating. By comparing devices with an annealed contact across the entire ridge [89] and devices where only two narrow ridges have been contacted [91], this further processing step is estimated to introduce a difference in waveguide losses of ~8 cm^{-1}. Thus, complex-coupled DFB resonators, which are much more effective in achieving single-mode operation than bare index-coupled devices, are obtained with an estimated coupling constant of 4 cm^{-1}. Devices were accordingly fabricated with grating periods Λ of 9.2 and 9.4 μm on a sample nominally identical to that of Ref. 91. No other changes were introduced in the fabrication process. Figure 2.29 shows the spectra (solid lines) of two such devices (length = 3.2 mm and Λ = 9.2 μm in one device and length = 4.4 mm with Λ = 9.4 μm in the other). Lasing takes place within the previously observed stopband, as expected for a complex-coupled DFB laser. The shift of the emission wavelength corresponds precisely to the change of grating period, and an effective refractive index of 3.68 can be extracted, in good agreement with the calculated value of 3.73. The coupling is now sufficiently strong to achieve single-mode operation under all injection currents and operating temperatures. The tuning of the emission wavelength with temperature and current is below the resolution of the experimental apparatus. However, a tuning coefficient of 4 MHz/mA and an instantaneous linewidth of 30 kHz were recently measured for 4.66- and 3.69-THz Fabry-Perot lasers [165]. While we can expect a similar tuning coefficient for our device, the emission linewidth of a DFB laser is typically well below that of a Fabry-Perot structure.

The technique just described produces gratings with rather limited coupling coefficients, thereby making long laser cavities necessary for the achievement of single-mode operation. This affects the device performance, increasing the driving currents and hindering lasing in continuous-wave mode. The issue becomes more and more relevant with increasing emission wavelengths as the DFB grating period has to be proportionally enlarged. An elegant way of realizing a

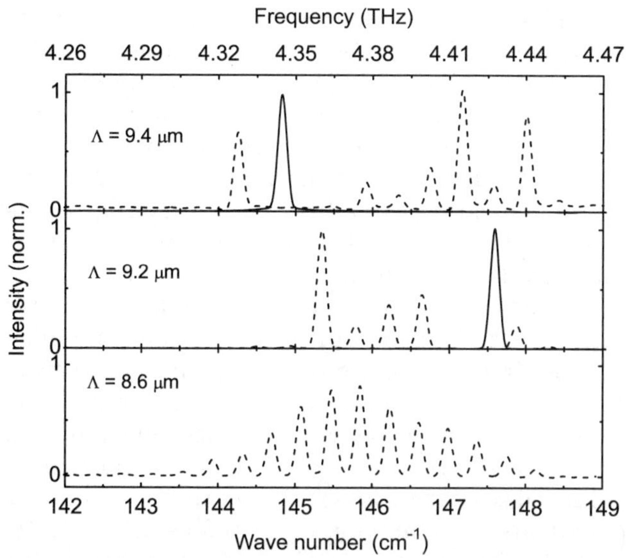

Figure 2.29 Spectra of five different DFB lasers with grating periods of 9.4, 9.2, and 8.6 μm, respectively, collected at an injection current close to the maximum output power [164]. In the upper two panels, the dashed lines correspond to spectra recorded from devices in which the distributed feedback was implemented only by removing 150 nm of the top contact layer. The appearance of a stop-band in the original Fabry-Perot spectrum indicates a predominant index-modulation. The solid lines are spectra recorded from devices in which an additional loss-modulation was introduced by an annealed-contact grating (see text). In this case, the stronger modulation leads to stable single-mode emission. In the bottom panel, the spectrum of a 150 nm etched device with no annealed-contact grating and Λ = 8.6 μm is shown.

one-dimensional photonic crystal for surface-plasmon modes has been recently demonstrated [166]. A slit opened in the metallic layer acts as a barrier for the wave propagation as no surface plasmon is supported there. Part of the light is reflected back, part is transmitted by tunneling across the slit, and part is scattered out of the interface mode. A periodic series of slits then acts as a photonic crystal structure, and if the slits are sufficiently thin (much narrower than the wavelength), this can be accomplished with minimal scattering losses. This idea was applied to the fabrication of DFB QC lasers emitting near 2.5 THz. Devices were processed from a sample featuring a BTC active region and the modified version of the CLG waveguide described in Sec.2.4. The procedure follows the one of Ref. 91 until the deposition of the last metallic layer. This is patterned by electron-beam lithography with a series of slits at the $\lambda/2$ DFB period (see inset of Fig. 2.31). Electron-beam lithography, while not necessary for resolution reasons, is very useful to easily vary the grating period and test different slit widths. For the latter, the best compromise between grating strength and diffraction losses was found at approximately 2-μm width (close to 10%

of the period length). Single-mode operation was obtained in pulsed and continuous-wave, with a side-mode suppression ratio of at least 20 dB, as can be seen in the spectrum of Fig. 2.30. The position of the emission peak shifts with the grating period, in reasonable agreement with the computed modal refractive index of 3.65. It is interesting to notice that, contrary to the one depicted in Fig. 2.31, the first devices based on this concept featured a top metallization extending to the very edge of the ridge, in order to leave a larger area for bonding. However, they frequently showed emission from more than one transverse mode, as exemplified in the right inset of Fig. 2.30; a narrower metallization has been adopted to solve this issue. The procedure can be further simplified by avoiding the liftoff step. In this case the resist used in the electron-beam lithography fills the slits, with the final gold layer covering everything. While the operating principle is analogous, this approach simplifies the bonding and in general yields better performance, probably owing to decreased scattering losses. The output power of one such DFB laser, approximately 2 mm long, is plotted as a function of drive current in Fig. 2.31. The emission is single mode in the whole current and temperature ranges of operation, with about 8 mW measured at 10 K and a low threshold current density of about

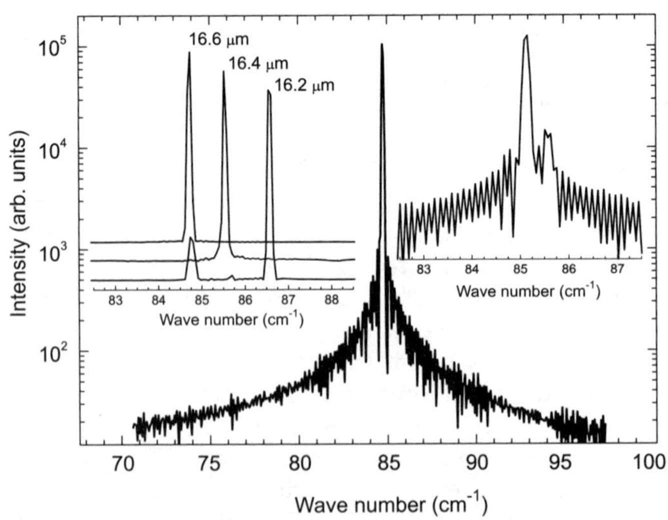

Figure 2.30 Single-mode emission spectrum in logarithmic scale of a 2.5-THz DFB laser with a 16.5-μm slit periodicity and narrow top metallization. It was collected in pulsed mode at 1% duty cycle and 0.92-A drive current, close to the maximum output power. In the left inset we plot the spectra of three lasers with different DFB periods and wide metallization. The upper two were collected close to threshold, while the lowest one is at a higher current in order to better show the DFB mode, here somewhat detuned from the maximum gain (hence the presence also of Fabry-Perot modes). The right inset shows a spectrum at higher current of the 16.4-μm laser of the left inset. The appearance of a second transverse mode is evident.

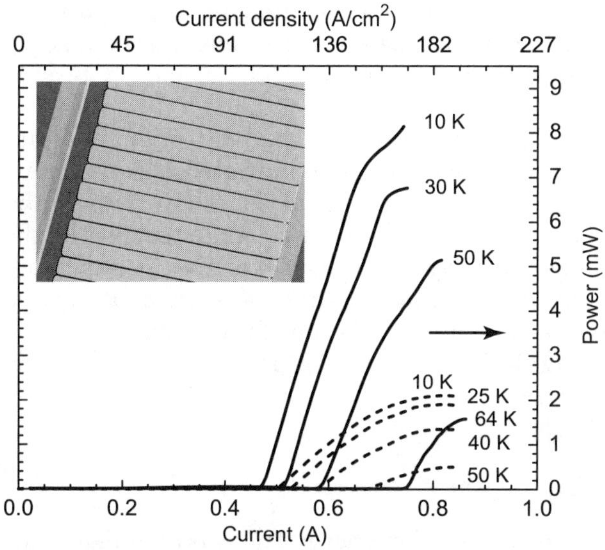

Figure 2.31 Light-current characteristics of a 2.5-THz DFB QC laser. Solid lines refer to pulsed operation with 5% duty cycle and were collected from one facet with *f*/1 parabolic optics on a calibrated pyroelelectric detector; dashed lines were obtained by mounting a Winston cone in front of the laser and using a pyroelectric power meter (33% estimated collection efficiency). The inset shows a scanning electron microscope picture of a surface-plasmon DFB grating.

100 A/cm^2. The performance decreases in continuous-wave mode with a slight increase in threshold and reduced maximum operating temperature (from 70 to 55 K). Note, however, that the difference in measured power might be ascribed in part to the different collection efficiencies of the two setups.

2.5.2 Terahertz lasers in high magnetic fields

When an intense magnetic field is applied perpendicular to the plane of a two-dimensional electron gas in a semiconductor heterostructure, the electronic energy spectrum is completely modified. The free in-plane motion of the electrons is now constricted into Landau orbits, breaking up the subband dispersion into discrete energy levels. The relevance of this effect to the operation of QC lasers stems from the possibility it offers to suppress LO-phonon emission, known to be the dominant non-radiative relaxation mechanism. Owing to energy conservation and thanks to the small phonon dispersion, the lifetime of a mid-infrared intersubband transition can be dramatically increased when the magnetic field intensity is such that no two states exist that are separated by the LO-phonon energy or, vice versa, resonantly lowered when the latter condition is verified. The light output intensity of mid-infrared QC lasers is then strongly modulated with varying magnetic field [167], reproducing the behavior expected from a zero-dimensional intersubband quantum box laser [168].

The situation at terahertz frequencies, i.e., transition energies below the optical phonon resonance, is more complicated, because many scattering mechanisms, such as carrier-carrier, impurities, and interface roughness, play a relevant role. Thus, the response to the application of a magnetic field may differ from structure to structure, depending on the design concept. It is true that, as in the mid-infrared case, a general characteristic is the modulation of the transition upper-state lifetime. In terahertz QC lasers, however, nonradiative relaxation is enhanced when

$$\Delta E = \frac{\hbar eB}{m^*} N \tag{2.6}$$

where ΔE is the unperturbed transition energy and N is an integer, i.e., when an excited Landau level of the lower subband comes into coincidence with the ground level of the upper subband. In this case resonant tunneling between Landau levels (mainly taking place through short-range disorder) opens up an additional relaxation channel, reducing the population inversion. On the contrary, when the Landau level ladders are off resonance, scattering rates are minimized. This behavior was actually first demonstrated by measuring the intensity of spontaneous emission in electroluminescence experiments [84, 169], but later it was confirmed in the operation of terahertz lasers [170, 171]. Data from a chirped superlattice structure identical to that of Sec. 2.3 are plotted in Fig. 2.32. As expected, oscillations deriving from the above mechanism are seen in both the threshold current and the output power [170]. A reduction of the threshold current density by about a factor of 2 is observed at 4.3 T with respect to the zero-field value. Together with the corresponding decrease of current seen at constant bias voltage, this result is a strong indication of reduced scattering rate out of the upper lasing state. Note, however, that the maximum output power is always smaller than without the magnetic field, and actually displays a decreasing trend with increasing field. This originates from the positive magnetoresistance of the injector miniband where relaxation is progressively quenched due to the in-plane quantization.

In general, a similar lifetime modulation should hold also for the lowest lasing subband, when in or out of resonance with the Landau ladders of the injector states. In this case, though, the result on the lasing properties is opposite, since a lifetime reduction corresponds to

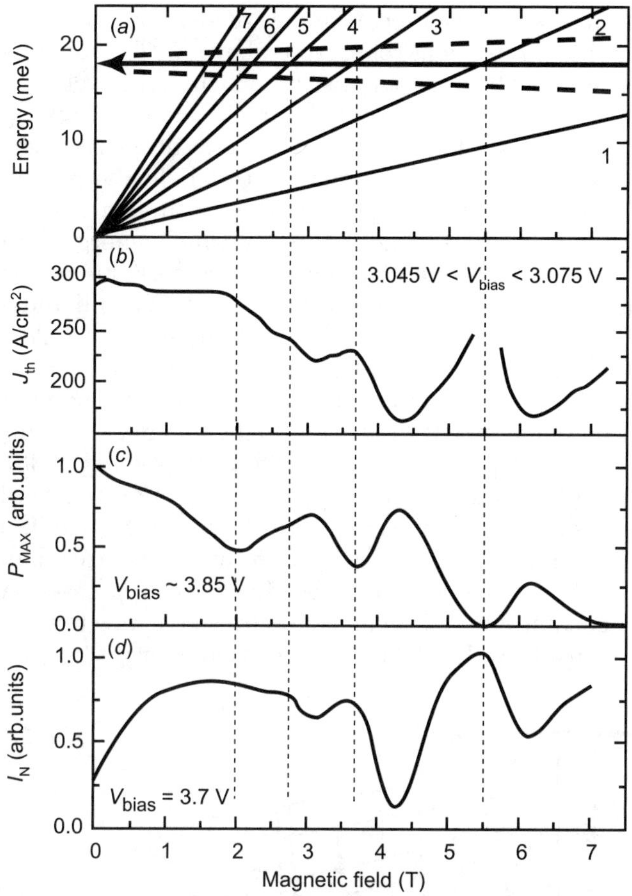

Figure 2.32 (*a*) Energy of the most relevant Landau levels for the laser of Ref. 170 plotted as a function of magnetic field with respect to the lowest subband ground state. Dashed lines show the typical Landau level broadening. (*b*) Threshold current density as a function of magnetic field. (*c*) Maximum output power as a function of magnetic field. (*d*) Normalized current at the constant bias voltage of 3.7 V. This curve was obtained from the experimental data by subtracting the decaying background current.

a more efficient extraction and therefore to improved population inversion. This effect appears to be less important in superlattices, where depletion of the lowest subband is already efficient [170]. In contrast, it was clearly observable in a structure lasing between the third and second subbands of a single-quantum-well active region [172].

This design concept proved invaluable in studying the high field regime, in which the cyclotron energy is larger than the photon energy. For this purpose a device was engineered for a low emission frequency of 1.9 THz, corresponding to a magnetic field of only 4.7 T [163]. The structure exhibited lasing only in a magnetic field of at least 2.5 T, and various peculiar features were observed at higher fields. First, a strong reduction of waveguide losses was observed above 6 T, which is attributed to the reduction of free carrier absorption when the cyclotron energy becomes comparable to level broadening. Second, a strong reduction of output power was found at 9.3 T, where no more inter-Landau-level resonances should be observed. At this field the photon energy is one-half the cyclotron energy, and the feature is a clear signature of carrier-carrier scattering events in which energy and momentum are conserved with one electron that scatters out of the upper lasing level to a lower Landau level of the second subband and a second electron that scatters up to a higher Landau level. This configuration in fact presents a maximum for the carrier-carrier scattering probability [173]. Most importantly a progressive reduction of threshold current density was seen at even higher magnetic fields, with a strong temperature dependence from 4.2 to 20 K (see Fig. 2.33). Under these conditions, as is known to happen in the quantum Hall effect, electrons localize in the potential fluctuations due to the residual sample disorder. Extremely long intersubband lifetimes are then observed, since all scattering mechanisms are efficiently suppressed. A threshold current density as low as 1 A/cm^2 was achieved, with a temperature activation compatible with a 1-meV localization energy [163].

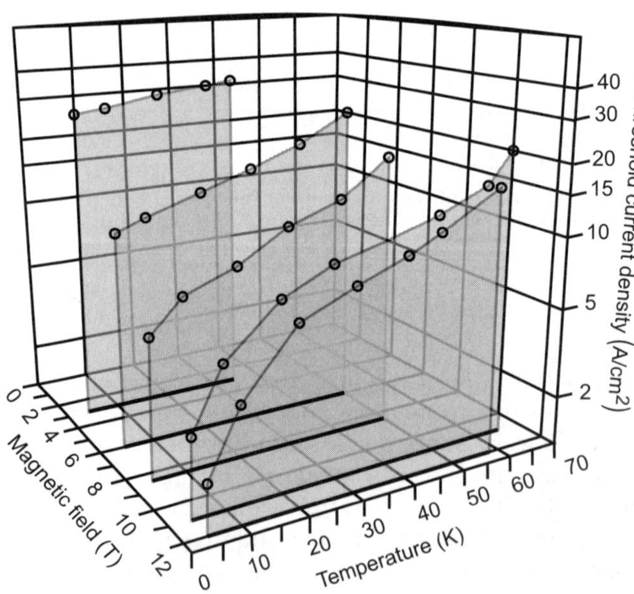

Figure 2.33 Dependence of threshold current density on temperature and magnetic field for a 3.5-mm, 320-μm-wide device of Ref. 163. The change in lasing regime is evident above 6 T and below 20 K. (Reprinted with permission from [163]. Copyright © 2004 by the American Physical Society.)

2.5.3 Open challenges and future research

Since their initial demonstration terahertz QC lasers have made rapid and substantial progress. They have reached a level of performance such that the first terahertz applications employing QC laser sources are appearing in imaging [174, 175] and spectroscopy [176]. They already display output power levels ideal for most conceivable uses, except maybe long-range remote sensing, and their narrow linewidth and frequency stability make them an ideal candidate for local oscillators in heterodyne systems. As already mentioned in Sec. 2.5.2, however, room temperature operation and frequencies below 2 THz remain as the main challenges left before widespread device exploitation can take place. While the latter objective is within reach of current technology, high-temperature operation would require some breakthrough development. The latest magnetic field results clearly encourage the investigation of zero-dimensional structures. A final aspect is the tunability of single-mode emission. Temperature (or current) tuning is presently limited to 3 to 4 GHz. This would be sufficient for high-sensitivity spectroscopy, but poses very stringent requirements on the fabrication of devices targeting a specific frequency. Solutions allowing broadband tunability, such as external cavities, are being explored [98]. On the other hand, terahertz QC lasers, with their peculiar waveguides and long emission wavelengths, also represent an ideal playground in which to experiment with new ideas for surface-plasmon-based photonics, such as electrically controllable optical circuits or disordered systems.

In the end, the commercial future of terahertz QC lasers is still hard to predict and will most likely be determined by the general success of terahertz-based technologies. Nowadays the possibility of performing molecular detection in biological environments (for instance, for DNA or protein sensing) or detection of hidden substances (explosives, drugs, biochemical agents) in security controls is attracting a huge interest in terahertz radiation, and the same can be said for imaging of biological tissues in medical applications (tooth cavities, superficial tumors, etc.). Nevertheless the commercial fortune will be decided mainly by market rules: Will terahertz photonics be able to provide unique high-value capabilities to justify a strong economic effort in its real-world implementation? Or will it be able to tackle the relevant applications at a lower cost than other competing technologies? The answers to these questions will be decided in the near future.

2.6 Acknowledgments

We would like to express our gratitude to all the people who have been working with us on the development of terahertz quantum cascade lasers; in particular, Harvey Beere and David Ritchie of the Cavendish Laboratory, University of Cambridge; Edmund Linfield and Giles Davies at the University of Leeds; Rita Iotti and Fausto Rossi at the Politecnico di Torino; Sukhdeep Dhillon and Carlo Sirtori at Thales Research and Technology; Stefano Barbieri, Jesse Alton, and John Fowler at Teraview; Lukas Mahler, Jihua Xu, Tonia Losco, Richard Green, Cosimo Mauro, and Fabio Beltram of the NEST CNR-INFM Center at the Scuola Normale Superiore. Further we would like to thank Jerome Faist and his group at the University of Neuchâtel and Qing Hu and his coworkers at the Massachusetts Institute of Technology for kindly allowing the use of their figures in this chapter and for contributing so much to the success of terahertz QC lasers.

Finally, we would like to acknowledge the financial support by the European Commission through the IST projects WANTED and TERANOVA, through the Marie Curie RTN POISE, and through the PASR project TERASEC; by the Fondazione Cassa di Risparmio di Pisa through the project Teralight; and by Physical Sciences Inc. of Andover.

References

1. L. Esaki, "New phenomenon in narrow germanium *p-n* junction," *Phys. Rev.* **109**, p. 603, 1958.
2. W. T. Read, "A proposed high-frequency negative resistance diode," *Bell Sys. Tech. J.* **37**, p. 401, 1958.
3. R. N. Hall, "Tunnel diodes," *IRE Trans. Electron Devices* **7**, p. 1, 1960.
4. J. B. Gunn, "Microwave oscillation of current in III-V semiconductors," *Solid State Commun.* **1**, p. 88, 1963.
5. J. B. Gunn, "Instabilities of current in III-V semiconductors," *IBM J. Dev. Res.* **8**, p. 141, 1964.
6. C. A. Lee, R. L. Batdorf, W. Wiegman, and G. Kaminsky, "The read diode and avalanche, transit-time, negative-resistance oscillators," *Appl. Phys. Lett.* **6**, p. 89, 1965.
7. R. L. Johnston, B. C. DeLoach, and B. G. Cohen, "A silicon diode oscillator," *Bell Sys. Tech. J.* **44**, p. 369, 1965.
8. R. N. Hall, G. E. Fenner, J. D. Kingsley, T. J. Soltys, and R. O. Carlson, "Coherent light emission from GaAs junctions," *Phys. Rev. Lett.* **9**, p. 366, 1962.
9. B. Senitzky and A. A. Oliner, "Submillimeter waves—a transition region," in *Proceedings of the Symposium on Submillimeter Waves*, J. Fox, ed., vol. 20, Press of the Polytechnic Institue of Brooklyn, New York, 1970.

10. K. H. Yang, P. L. Richards, and Y. R. Shen, "Generation of far-infrared radiation by picosecond light pulses in $LiNbO_3$," *Appl. Phys. Lett.* **19**, p. 320, 1971.

11. G. Mourou, C. V. Stancampiano, and D. Blumenthal, "Picosecond microwave pulse generation," *Appl. Phys. Lett.* **38**, p. 470, 1981.

12. D. H. Auston, "Subpicosecond electro-optic shock waves," *Appl. Phys. Lett.* **73**, p. 713, 1983.

13. D. H. Auston, K. P. Cheung, and P. R. Smith, "Picosecond photoconducting hertzian dipoles," *Appl. Phys. Lett.* **45**, p. 284, 1984.

14. C. Fattinger and D. Grischkowsky, "Point source terahertz optics," *Appl. Phys. Lett.* **53**, p. 1480, 1988.

15. M. Bass, P. A. Franken, J. F. Ward, and G. Weireich, "Optical rectification," *Phys. Rev. Lett.* **9**, p. 446, 1962.

16. G. A. A'skaryan, "Cherenkov radiation and transition radiation from electromagnetic waves," *Sov. Phys. JETP* **15**, p. 943, 1962.

17. B. B. Hu, X.-C. Zhang, D. H. Auston, and P. R. Smith, "Free-space radiation from electro-optic crystals," *Appl. Phys. Lett.* **56**, p. 506, 1990.

18. L. Xu, X.-C. Zhang, and D. H. Auston, "Terahertz beam generation by femtosecond optical pulses in electro-optic materials," *Appl. Phys. Lett.* **61**, p. 1784, 1992.

19. E. R. Brown, F. W. Smith, and K. A. McIntosh, "Coherent millimeter wave generation by heterodyne conversion in in low-temperature-grown GaAs photoconductors," *J. Appl. Phys.* **73**, p. 1480, 1993.

20. E. R. Brown, K. A. McIntosh, K. B. Nichols, and C. L. Dennis, "Photomixing up to 3.8 THz in low-temperature-grown GaAs," *Appl. Phys. Lett.* **66**, p. 285, 1995.

21. J. A. Valdmanis, G. A. Mourouand, and C. W. Gabel, "Subpicosecond electrical sampling," *IEEE J. Quantum Electronics* **19**, p. 664, 1983.

22. Q. Wu and X.-C. Zhang, "Free-space electro-optic sampling of terahertz beams," *Appl. Phys. Lett.* **67**, p. 3523, 1995.

23. A. Nahata, A. S. Weling, and T. F. Heinz, "A wideband coherent terahertz spectroscopy system using optical rectification and electro-optic sampling," *Appl. Phys. Lett.* **69**, p. 2321, 1996.

24. Q. Wu and X.-C. Zhang, "Ultrafast electro-optical field sensors," *Appl. Phys. Lett.* **68**, p. 1604, 1996.

25. A. Nahata, D. H. Auston, T. F. Heinz, and Q. Wu, "Coherent detection of freely propagating terahertz radiation by electro-optic sampling," *Appl. Phys. Lett.* **68**, p. 150, 1996.

26. D. H. Auston and K. P. Cheung, "Coherent time-domain far-infrared spectroscopy," *J. Opt. Soc. Am. B* **2**, p. 606, 1985.

27. K. P. Cheung and D. H. Auston, "Excitation of coherent phonon polaritons with femtosecond optical pulses," *Phys. Rev. Lett.* **55**, p. 2152, 1985.

28. K. P. Cheung and D. H. Auston, "A novel technique for measuring far-infrared absorption and dispersion," *Infrared Phys.* **26**, p. 23, 1986.

29. B. B. Hu and M. C. Nuss, "Imaging with terahertz waves," *Opt. Lett.* **20**, p. 807, 1995.

30. D. Mittleman, "Terahertz imaging," in *Sensing with Terahertz Radiation, Optical Sciences*, D. Mittleman, ed., vol. 95, p. 117, Springer, Berlin, 2002.

31. D. M. Mittleman, R. H. Jacobsen, and M. C. Nuss, "T-ray imaging," *IEEE J. of Selected Topics in Quantum Electronics* **2**, p. 679, 1996.

32. K. J. Siebert, H. Quast, R. Leonhardt, T. Löffler, M. Thomson, T. Bauer, and H. G. Roskos, "Continuous-wave all-optoelectronic terahertz imaging," *Appl. Phys. Lett.* **80**, p. 3003, 2002.

33. S. Wang and X.-C. Zhang, "Tomographic imaging with a terahertz binary lens," *Appl. Phys. Lett.* **82**, p. 1821, 2003.

34. D. M. Mittleman, S. Hunsche, L. Boivin, and M. C. Nuss, "T-ray tomography," *Opt. Lett.* **22**, p. 904, 1997.

35. J. L. Johnson, T. D. Dorney, and D. M. Mittleman, "Enhanced depth resolution in terahertz imaging using phase-shift interferometry," *Appl. Phys. Lett.* **78**, p. 835, 2001.

36. M. Koch, "Bio-medical applications of THz imaging," in *Sensing with Terahertz Radiation, Optical Sciences*, D. Mittleman, ed., vol. 95, p. 295, Springer, Berlin, 2002.

37. J. M. Chamberlain, N. N. Zinonev, A. J. Fitzgerald, E. Berry, G. C. Walker, J. W. Handley, M. A. Smith, and R. E. Miles, "What constitutes a useful and meaningful terahertz image?" in *2002 IEEE Tenth International Conference on Terahertz Electronics Proceedings*, J. M. Chamberlain, P. Harrison, R. E. Miles, A. G. Davies, E. H. Linfield, and S. Withington, eds., vol.10, p. 81, IEEE, Piscataway, 2002.

38. D. Mittleman, ed., *Sensing with Terahertz Radiation*, Springer, Berlin, 2002.

39. B. Ferguson and X.-C. Zhang, "Materials for terahertz science and technology," *Nature Materials* **1**, p. 26, 2002.

40. M. Nagel, F. Richter, P. Haring-Bolivar, H. Kurz, A. Bosserhoff, and R. Büttner, "Integrated THz biomolecular sensors for DNA," in *2002 IEEE Tenth International Conference on Terahertz Electronics Proceedings*, J. M. Chamberlain, P. Harrison, R. E. Miles, A. G. Davies, E. H. Linfield, and S. Withington, eds., vol. 10, p. 70, IEEE, Piscataway, 2002.

41. M. Nagel, P. Haring-Bolivar, M. Brucherseifer, H. Kurz, A. Bosserhoff, and R. Büttner, "Integrated THz technology for label-free genetic diagnostics," *Appl. Phys. Lett.* **80**, p. 154, 2002.

42. R. Huber, F. Tauser, A. Brodschelm, M. Bichler, G. Abstreiter, and A. Leitenstorfer, "How many-particle interactions develop after ultrafast excitation of an electron-hole plasma," *Nature* **414**, p. 286, 2001.

43. A. A. Andronov, I. V. Zverev, V. A. Kozlov, Y. N. Nozdrin, S. A. Pavlov, and V. N. Shastin, "Stimulated emission in the long-wavelength IR region from hot holes in Ge in crossed electric and magnetic fields," *JETP Lett.* **40**, p. 804, 1984.

44. A. A. Andronov, A. M. Belyantsev, V. I. Gavrilenko, E. P. Dodin, Z. F. Krasil'nik, V. V. Nikoronov, and S. A. Pavlov, "Induced hot-hole millimeter emission in germanium in fields E||H (cyclotron-resonance negative-effective-mass amplifier and generator)," *JETP Lett.* **40**, p. 989, 1984.

45. Y. B. Vasil'ev and Y. L. Ivanov, "Frequency of stimulated emission during transitions through light-hole Landau levels in germanium," *Sov. Tech. Phys. Lett.* **10**, p. 398, 1984.

46. A. A. Andronov, "Hot electrons in semiconductors and submillimeter waves," *Sov. Phys. Semicond.* **21**, p. 701, 1987.

47. Y. P. Gousev, I. V. Altukhov, K. A. Korolev, V. P. Sinis, M. S. Kagan, E. E. Haller, M. A. Odnoblyudov, I. N. Yassievich, and K.-A. Chao, "Widely tunable continuous-wave THz laser," *Appl. Phys. Lett.* **75**, p. 757, 1999.

48. P. Pfeffer, W. Zawadzki, K. Unterrainer, C. Kremser, C. Wurzer, E. Gornik, B. Murdin, and C. R. Pidgeon, "*p*-type Ge cyclotron-resonance laser: Theory and experiment," *Phys. Rev. B* **47**, p. 4522, 1993.

49. H. Eisele, "Conventional and novel approaches to RF power generation with two-terminal devices at terahertz frequencies," in *2002 IEEE Tenth International Conference on Terahertz Electronics Proceedings*, J. M. Chamberlain, P. Harrison, R. E. Miles, A. G. Davies, E. H. Linfield, and S. Withington, eds., vol. 10, p. 13, IEEE, Piscataway, 2002.

50. E. R. Brown, J. R. Söderström, C. D. Parker, L. J. Mahoney, K. M. Molvar, and T. C. McGill, "Oscillations up to 712 GHz in InAs/AlAs resonant-tunneling diodes," *Appl. Phys. Lett.* **58**, p. 2291, 1991.

51. A. Maestrini, J. Ward, J. Gill, H. Javadi, E. Schlecht, G. Chattopadhyay, F. Maiwald, N. R. Erickson, and I. Mehdi, "A 1.7 – 1.9 THz local oscillator source," *IEEE Microwave Wireless Comp. Lett.* **14**, pp. 253–255, 2004.

52. S. J. Allen, J. S. Scott, M. C. Wanke, K. Maranowski, A. C. Gossard, M. J. W. Rodwell, and D. H. Chow, "Terahertz dynamics in quantum structures: Towards a fundamental terahertz oscillator," in *Terahertz Sources and Systems, NATO Science Series*, R. Miles, P. Harrison, and D. Lippens, eds., vol. 27, p. 3, Kluwer Academic, Dordrecht, 2001.

53. R. P. Joshi, S. Viswanadha, P. Shah, and R. D. del Rosario, "Monte Carlo analysis of GaN-based Gunn oscillators for microwave power generation," *J. Appl. Phys.* **93**, p. 4863, 2003.
54. E. Starikov, P. Shiktorov, V. Gruzinskis, L. Reggiani, L. Varani, J. C. Vaissière, and J. H. Zhao, "Monte Carlo simulation of the generation of terahertz radiation in GaN," *J. Appl. Phys.* **89**, p. 1161, 2001.
55. L. N. Kurbatov, A. D. Britov, S. M. Karavaev, S. D. Sivachenko, S. N. Maksimovskii, I. I. Ovchinnikov, M. M. Rzaev, and P. M. Starik, "Far-IR heterojunction lasers tunable to 46.2 μm," *JETP Lett.* **37**, p. 499, 1983.
56. A. P. Shotov, O. I. Davarashvili, and A. V. Babushkin, "Laser heterostructures using $Pb_{1-x}Sn_xSe_{1-y}Te_y$ solid solutions," *Sov. Tech. Phys. Lett.* **5**, p. 629, 1979.
57. B. G. Girich, D. M. Gureev, I. I. Zasavitskii, T. I. Koropotkina, E. P. Kulygina, B. N. Matsonashvili, M. I. Nikolaev, O. V. Pelevin, G. V. Flusov, and A. P. Shotov, "Laser $Pb_{1-x}Sn_x$ Te heterostructures," *Sov. Physics Semic.* **12**, p. 70, 1978.
58. J. N. Walpole, A. R. Calawa, S. R. Chinn, S. H. Groves, and T. C. Harman, "CW, operation of distributed feedback $Pb_{1-x}Sn_x$ Te lasers," *Appl. Phys. Lett.* **30**, p. 524, 1977.
59. R. F. Kazarinov and R. A. Suris, "Possibility of the amplification of electromagnetic waves in a semiconductor with a superlattice," *Sov. Physics Semic.* **5**, p. 707, 1971.
60. R. F. Kazarinov and R. A. Suris, "Electric and electromagnetic properties of semiconductors with a superlattice," *Sov. Physics Semic.* **6**, p. 120, 1972.
61. K. L. Wang and P.-F. Yuh, "Theory and applications of band-aligned superlattices," *IEEE J. Quantum Electronics* **25**, p. 12, 1988.
62. F. Capasso, K. Mohammed, and A. Y. Cho, "Resonant tunneling through double barriers, perpendicular quantum transport phenomena in superlattices, and their device applications," *IEEE J. Quantum Electronics* **22**, pp. 1853–1869, 1986.
63. M. Helm, P. England, E. Colas, F. D. Rosa, and S. J. Allen, "Intersubband Emission from Semiconductor Superlattices Excited by Sequential Tunneling," *Phys. Rev. Lett.* **63**, p. 74, 1989.
64. M. Helm, E. Colas, P. England, F. DeRosa, and S. J. Allen, "Observation of grating-induced intersubband emission from GaAs/AlGaAs superlattices," *Appl. Phys. Lett.* **53**, p. 1714, 1988.
65. R. Ferreira and G. Bastard, "Evaluation of some scattering times for electrons in unbiased and biased single- and multiple-quantum-well structures," *Phys. Rev. B* **40**, p. 1074, 1989.
66. A. Seilmeier, H.-J. Hübner, G. Abstreiter, G. Weimann, and W. Schlapp, "Intersubband relaxation in GaAs-AlGaAs quantum well structures observed directly by an infrared bleaching technique," *Phys. Rev. Lett.* **59**, p. 1345, 1987.

67. J. Faist, F. Capasso, D. L. Sivco, C. Sirtori, A. L. Hutchinson, and A. Y. Cho, "Quantum cascade laser," *Science* **264**, pp. 553–556, 1994.

68. A. Evans, J. S. Yu, J. David, L. Doris, K. Mi, S. Slivken, and M. Razeghi, "High-temperature high-power continuous-wave operation of buried heterostructure quantum-cascade lasers," *Appl. Phys. Lett.* **84**, pp. 314–316, 2004.

69. J. Faist, C. Gmachl, F. Capasso, C. Sirtori, D. L. Sivco, J. N. Baillargeon, and A. Y. Cho, "Distributed feedback quantum cascade lasers," *Appl. Phys. Lett.* **70**, p. 2670, 1997.

70. C. Gmachl, F. Capasso, J. Faist, A. L. Hutchinson, A. Tredicucci, D. L. Sivco, J. N. Baillargeon, S. N. G. Chu, and A. Y. Cho, "Distributed feedback quantum cascade lasers," *Appl. Phys. Lett.* **72**, p. 1430, 1998.

71. R. Köhler, C. Gmachl, A. Tredicucci, F. Capasso, D. L. Sivco, S. N. G. Chu, and A. Y. Cho, "Single-mode tunable, pulsed, and continuous wave quantum-cascade distributed feedback lasers $\lambda \sim 4.6 - 4.7$ μm," *Appl. Phys. Lett.* **76**, p. 1092, 2000.

72. D. Hofstetter, M. Beck, T. Aellen, and J. Faist, "High-temperature operation of distributed feedback quantum-cascade lasers at 5.3 μm," *Appl. Phys. Lett.* **78**, p. 396, 2001.

73. A. Tredicucci, C. Gmachl, F. Capasso, A. L. Hutchinson, D. L. Sivco, and A. Y. Cho, "Single-mode surface-plasmon laser," *Appl. Phys. Lett.* **76**, p. 2164, 2000.

74. D. Hofstetter, J. Faist, M. Beck, A. Müller, and U. Oesterle, "Demonstration of high-performance 10.16 μm quantum cascade distributed feedback lasers fabricated without epitaxial regrowth," *Appl. Phys. Lett.* **75**, p. 665, 1999.

75. J. Faist, F. Capasso, D. L. Sivco, A. L. Hutchinson, S. N. G. Chu, and A. Y. Cho, "Short wavelength ($\lambda \sim 3.4$ μm) quantum cascade laser based on strained compensated InGaAs/AlInAs," *Appl. Phys. Lett.* **72**, p. 680, 1998.

76. R. Colombelli, F. Capasso, C. Gmachl, A. L. Hutchinson, D. L. Sivco, A. Tredicucci, M. C. Wanke, A. M. Sergent, and A. Y. Cho, "Far-infrared surface-plasmon quantum-cascade lasers at 21.5 μm and 24 μm wavelengths," *Appl. Phys. Lett.* **78**, pp. 2620–2622, 2001.

77. C. Sirtori, P. Kuck, S. Barbieri, P. C. J. Nagle, M. Beck, J. Faist, and U. Oesterle, "GaAs/Al_x Ga_{1-x} As quantum cascade lasers," *Appl. Phys. Lett.* **73**, p. 3486, 1998.

78. K. Ohtani and H. Ohno, "An InAs-based intersubband quantum cascade laser," *Jap. J. Appl. Phys. 2* **41**, p. L1279, 2002.

79. K. Ohtani and H. Ohno, "InAs/AlSb quantum cascade lasers operating at 10 μm," *Appl. Phys. Lett.* **82**, p. 1003, 2003.

80. M. Rochat, J. Faist, M. Beck, U. Oesterle, and M. Ilegems, "Far-infrared ($\lambda \sim 88$ μm) electroluminescence in a quantum cascade structure," *Appl. Phys. Lett.* **73**, pp. 3724–3726, 1998.

81. B. S. Williams, B. Xu, Q. Hu, and M. R. Melloch, "Narrow-linewidth terahertz intersubband emission from three-level systems," *Appl. Phys. Lett.* **75**, pp. 2927–2929, 1999.

82. J. Ulrich, R. Zobl, N. Finger, K. Unterrainer, G. Strasser, and E. Gornik, "Terahertz-electroluminescence in a quantum cascade structure," *Physica B* **272**, p. 216, 1999.

83. J. Ulrich, R. Zobl, W. Schrenk, G. Strasser, K. Unterrainer, and E. Gornik, "Terahertz quantum cascade structures: intra- versus interwell transition," *Appl. Phys. Lett.* **77**, pp. 1928–1930, 2000.

84. J. Ulrich, R. Zobl, K. Unterrainer, G. Strasser, and E. Gornik, "Magnetic-field-enhanced quantum-cascade emission," *Appl. Phys. Lett.* **76**, p. 19, 2000.

85. B. S. Williams and Q. Hu, "Optimized energy separation for phonon scattering in three-level terahertz intersubband lasers," *J. Appl. Phys.* **90**, p. 5504, 2001.

86. B. S. Williams, H. Callebaut, Q. Hu, and J. L. Reno, "Magnetotunneling spectroscopy of resonant anticrossing in terahertz intersubband emitters," *Appl. Phys. Lett.* **79**, p. 4444, 2001.

87. R. Colombelli, A. Straub, F. Capasso, C. Gmachl, M. I. Blakey, A. M. Sergent, S. N. G. Chu, K. W. West, and L. N. Pfeiffer, "Terahertz electroluminescence from superlattice quantum cascade structures," *J. Appl. Phys.* **91**, p. 3526, 2002.

88. S. A. Lynch, R. Bates, D. J. Paul, D. J. Norris, A. G. Cullis, Z. Ikonic, R. W. Kelsall, P. Harrison, D. D. Arnone, and C. R. Pidgeon, "Intersubband electroluminescence from Si/SiGe cascade emitters at terahertz frequencies," *Appl. Phys. Lett.* **81**, p. 1543, 2002.

89. R. Köhler, A. Tredicucci, F. Beltram, H. E. Beere, E. H. Linfield, A. G. Davies, D. A. Ritchie, R. C. Iotti, and F. Rossi, "Terahertz semiconductor-heterostructure laser," *Nature* **417**, pp. 156–159, 2002.

90. A. Tredicucci, F. Capasso, C. Gmachl, D. L. Sivco, A. L. Hutchinson, and A. Y. Cho, "High performance interminiband quantum cascade lasers with graded superlattices," *Appl. Phys. Lett.* **73**, pp. 2101–2103, 1998.

91. R. Köhler, A. Tredicucci, F. Beltram, H. E. Beere, E. H. Linfield, A. G. Davies, D. A. Ritchie, S. Dhillon, and C. Sirtori, "High-performance continuous-wave operation of superlattice terahertz quantum-cascade lasers," *Appl. Phys. Lett.* **82**, pp. 1518–1520, 2003.

92. G. Scalari, L. Ajili, J. Faist, H. E. Beere, E. H. Linfield, D. A. Ritchie, and A. G. Davies, "Far-infrared $\lambda \sim 85$ μm bound-to-continuum quantum-cascade lasers operating up to 90 K," *Appl. Phys. Lett.* **82**, pp. 3165–3167, 2003.

93. B. S. Williams, S. Kumar, H. Callebaut, Q. Hu, and J. L. Reno, "Terahertz quantum-cascade laser at $\lambda \sim 100$ μm using metal waveguide for mode confinement," *Appl. Phys. Lett.* **83**, pp. 2124–2126, 2003.

94. B. S. Williams, S. Kumar, H. Callebaut, Q. Hu, and J. L. Reno, "Terahertz quantum-cascade laser operating up to 137 K," *Appl. Phys. Lett.* **83**, pp. 5142–5144, 2003.

95. S. Barbieri, J. Alton, H. E. Beere, J. Fowler, E. H. Linfield, and D. A. Ritchie, "2.9 THz quantum cascade lasers operating up to 70 K in continuous wave," *Appl. Phys. Lett.* **85**, pp. 1674–1676, 2004.

96. S. Kumar, B. S. Williams, S. Kohen, Q. Hu, and J. L. Reno, "Continuous-wave operation of terahertz quantum cascade lasers above liquid-nitrogen," *Appl. Phys. Lett.* **84**, pp. 2494–2496, 2004.

97. B. S. Williams, S. Kumar, Q. Hu, and J. L. Reno, "Resonant phonon terahertz quantum cascade laser operating at 2.1 THz ($\lambda \simeq 141$ μm)," *Electron. Lett.* **40**, pp. 431– 433, 2004.

98. A. Tredicucci, L. Mahler, T. Losco, J. Xu, C. Mauro, R. Köhler, H. E. Beere, D. A. Ritchie, and E. H. Linfield, "Advances in THz quantum cascade lasers: fulfilling the application potential," in *Proc. SPIE, Novel In-plane Semiconductor Lasers IV*, C. Marmelstein and D. P. Bour, eds., vol. 5738, SPIE, in press.

99. H. Eisele, "Active two-terminal devices for terahertz power generation by multiplication," in *Terahertz Sources and Systems, NATO Science Series*, R. Miles, P. Harrison, and D. Lippens, eds., vol. 27, p. 69, Kluwer Academic, Dordrecht, 2001.

100. S. M. Duffy, S. Verghese, and K. A. McIntosh, "Photomixers for continuous-wave terahertz radiation," in *Sensing with Terahertz Radiation, Optical Sciences*, D. Mittleman, ed., vol. 95, p. 193, Springer, Berlin, 2002.

101. E. A. Michael, B. Vowinkel, R. Schieder, M. Mikulics, M. Marso, and P. Kordoš, "Large-area traveling-wave photonic mixers for increased continuous terahertz power," *Appl. Phys. Lett.* **86,** 111120, 2005.

102. J. Darmo, T. Müller, G. Strasser, K. Unterrainer, T. Le, G. Tempea, and A. Stingl, "Intra- and extra-cavity THz generation from optically and electrically confined photoconducting layers," in *2002 IEEE Tenth International Conference on Terahertz Electronics Proceedings*, J. M. Chamberlain, P. Harrison, R. E. Miles, A. G. Davies, E. H. Linfield, and S. Withington, eds., vol. 10, p. 67, IEEE, Piscataway, 2002.

103. E. Peytavit, S. Arscott, D. Lippens, G. Mouret, S. Matton, P. Masselin, R. Bocquet, J. F. Lampin, L. Desplanque, and F. Mollot, "Terahertz frequency difference from vertically integrated low-temperature-grown GaAs photodetector," *Appl. Phys. Lett.* **81**, p. 1174, 2002.

104. R. Kersting, K. Unterrainer, G. Strasser, H. F. Kauffmann, and E. Gornik, "FewCycle THz Emission from Cold Plasma Oscillations," *Phys. Rev. Lett.* **79,** p. 3038, 1997.

105. X.-C. Zhang, B. B. Hu, J. T. Darrow, and D. H. Auston, "Generation of femtosecond electromagnetic pulses from semiconductor surfaces," *Appl. Phys. Lett.* **56**, p. 1011, 1990.

106. X.-C. Zhang, Y. Jin, L. E. Kingsley, and M. Weiner, "Influence of electric and magnetic fields on THz radiation," *Appl. Phys. Lett.* **62,** p. 2477, 1993.

107. M. B. Johnston, D. M. Whittaker, A. Corchia, A. G. Davies, and E. H. Linfield, "Simulation of terahertz generation at semiconductor surfaces," *Phys. Rev. B* **65,** 165301, 2002.

108. A. Rice, Y. Jin, X. F. Ma, X.-C. Zhang, D. Bliss, J. Larkin, and M. Alexander, "Terahertz optical rectification from [110] zinc-blende crystals," *Appl. Phys. Lett.* **64**, p. 1324, 1994.

109. J. J. Carey, R. T. Bailey, D. Pugh, J. N. Sherwood, F. R. Cruickshank, and K. Wynne, "Terahertz pulse generation in an organic crystal by optical rectification and resonant excitation of molecular charge transfer," *Appl. Phys. Lett.* **81,** p. 4335, 2002.

110. Y. S. Lee, T. Meade, T. B. Norris, and A. Galvanauskas, "Tunable narrow-band terahertz generation from periodically poled lithium niobate," *Appl. Phys. Lett.* **78,** p. 3583, 2001.

111. E. Bründermann, D. R. Chamberlin, and E. E. Haller, "High duty cycle and continuous terahertz emission from germanium," *Appl. Phys. Lett.* **76,** p. 2991, 2000.

112. E. Bründermann, A. M. Linhart, H. P. Röser, O. D. Dubon, W. L. Hansen, and E. E. Haller, "Miniaturization of p-Ge lasers: progress toward continuous wave operation," *Appl. Phys. Lett.* **68,** p. 1359, 1996.

113. A. V. Muravjov, S. H. Withers, H. Weidner, R. C. Strijbos, S. G. Pavlov, V. N. Shastin, and R. E. Peale, "Single axial-mode selection in a far-infrared p-Ge laser," *Appl. Phys. Lett.* **76**, pp. 1996–1998, 2000.

114. A. V. Muravjov, S. H. Withers, R. C. Strijbos, S. G. Pavlov, V. N. Shastin, and R. E. Peale, "Actively mode-locked p-Ge laser in Faraday configuration," *Appl. Phys. Lett.* **75**, pp. 2882–2884, 1999.

115. M. A. Odnoblyudov, I. N. Yassievich, M. S. Kagan, Y. M. Galperin, and K. A. Chao, "Population inversion induced by resonant states in semiconductors," *Phys. Rev. Lett.* **83**, p. 644, 1999.

116. A. Yariv, *Optical Electronics in Modern Communications,* 5th ed., Oxford University Press, Oxford, 1997.

117. K. Kawase, J. Shikata, K. Imai, and H. Ito, "Transform-limited, narrow-linewidth, terahertz-wave parametric generator," *Appl. Phys. Lett.* **78**, p. 2819, 2001.

118. K. Imai, K. Kawase, H. Minamide, and H. Ito, "Achromatically injection-seeded terahertz-wave parametric generator," in *2002 IEEE Tenth International Conference on Terahertz Electronics Proceedings*, J. M. Chamberlain, P. Harrison, R. E. Miles, A. G. Davies, E. H. Linfield, and S. Withington, eds., vol. 10, p. 56, IEEE, Piscataway, 2002.

119. K. Imai, K. Kawase, H. Minamide, and H. Ito, "Achromatically injection-seeded terahertz-wave parametric generator," *Opt. Lett.* **27**, p. 2173, 2002.

120. T. Tanabe, K. Suto, J. Nishizawa, K. Saito, and T. Kimura, "Tunable terahertz wave generation in the 3- to 7-THz region from GaP," *Appl. Phys. Lett.* **83**, p. 237, 2003.

121. T. Tanabe, K. Suto, J. Nishizawa, T. Kimura, and K. Saito, "Frequency-tunable high-power terahertz wave generation from GaP," *J. Appl. Phys.* **93**, p. 4610, 2003.

122. S. G. Pavlov, R. K. Zhukavin, E. E. Orlova, V. N. Shastin, A. Kirsanov, H.-W. Hübers, K. Auen, and H. Riemann, "Stimulated emission from donor transitions in silicon," *Phys. Rev. Lett.* **84**, p. 5220, 2000.

123. S. G. Pavlov, H.-W. Hübers, M. H. Rümmeli, R. K. Zhukavin, E. E. Orlova, V. N. Shastin, and H. Riemann, "Far-infrared stimulated emission from optically excited bismuth donors in silicon," *Appl. Phys. Lett.* **80**, p. 4717, 2002.

124. V. N. Shastin, R. K. Zhukavin, E. E. Orlova, S. G. Pavlov, H.-W. Hübers, M. H. Rümmeli, J. N. Hovenier, T. O. Klaassen, H. Riemann, I. V. Bradley, and A. F. G. van der Meer, "Stimulated terahertz emission from group-V donors in silicon under intracenter photoexcitation," *Appl. Phys. Lett.* **80**, p. 3512, 2002.

125. P. G. O'Shea and H. P. Freund, "Free-electron lasers: status and applications," *Science* **292**, p. 1853, 2001.

126. G. L. Carr, M. C. Martin, W. R. McKinney, K. Jordan, G. R. Neil, and G. P. Williams, "High-power terahertz radiation from relativistic electrons," *Nature* **420**, p. 153, 2002.

127. M. Abo-Bakr, J. Feikes, K. Holldack, G. Wstefeld, and H.-W. Hübers, "Steady-state far-infrared coherent synchrotron radiation detected at BESSY II," *Phys. Rev. Lett.* **88**, 254801, 2002.

128. M. Abo-Bakr, J. Feikes, K. Holldack, P. Kuske, W. B. Peatman, U. Schade, G. Wstefeld, and H.-W. Hübers, "Brilliant, coherent far-infrared (THz) synchrotron radiation," *Phys. Rev. Lett.* **90**, 094801, 2003.

129. P. Helminger, J. K. Messer, and F. C. de Lucia, "Continuously tunable coherent spectroscopy for the 0.1 to 1.0 THz region," *Appl. Phys. Lett.* **42**, p. 309, 1983.

130. S. Komiyama, O. Astafiev, V. Antonov, T. Kutsuwa, and H. Hirai, "A single-photon detector in the far-infrared range," *Nature* **403**, p. 405, 2000.

131. J. J. A. Baselmans, M. Hajenius, J. R. Gao, T. M. Klapwijk, P. A. J. de Korte, B. Voronov, and G. Gol'tsman, "Doubling of sensitivity and bandwidth in phonon cooled hot electron bolometer mixers," *Appl. Phys. Lett.* **84**, pp. 1958–1960, 2004.

132. W. Knap, Y. Deng, S. Rumyantsev, and M. S. Shur, "Resonant detection of subterahertz and terahertz radiation by plasma waves in submicron field-effect transistors," *Appl. Phys. Lett.* **81**, p. 4637, 2002.

133. S. Kono, M. Tani, P. Gu, and K. Sakai, "Detection of up to 20 THz with a low-temperature-grown GaAs photoconductive antenna gated with 15 fs light pulses," *Appl. Phys. Lett.* **77**, p. 4104, 2001.

134. B. Xu, Q. Hu, and M. R. Melloch, "Electrically pumped tunable terahertz emitter based on intersubband transitions," *Appl. Phys. Lett.* **71**, p. 440, 1997.

135. R. C. Iotti and F. Rossi, "Carrier thermalization versus phonon-assisted relaxation in quantum-cascade lasers: a Monte Carlo approach," *Appl. Phys. Lett.* **78**, p. 2902, 2001.

136. R. C. Iotti and F. Rossi, "Nature of Charge Transport in Quantum-Cascade Lasers," *Phys. Rev. Lett.* **87**, 146603, 2001.

137. R. Köhler, R. C. Iotti, A. Tredicucci, and F. Rossi, "Design and simulation of terahertz quantum cascade lasers," *Appl. Phys. Lett.* **79**, pp. 3920–3922, 2001.

138. B. Xu and Q. Hu, "Grating coupling for intersubband emission," *Appl. Phys. Lett.* **70**, p. 2511, 1997.

139. R. Köhler, A. Tredicucci, F. Beltram, H. E. Beere, E. H. Linfield, A. G. Davies, and D. A. Ritchie, "High-intensity interminiband terahertz emission from chirped superlattices," *Appl. Phys. Lett.* **80**, pp. 1867–1869, 2002.

140. C. Sirtori, C. Gmachl, F. Capasso, J. Faist, D. L. Sivco, A. L. Hutchinson, and A. Y. Cho, "Long-wavelength ($\lambda \sim 8 - 11.5$ μm) semiconductor lasers with waveguides based on surface plasmons," *Opt. Lett.* **23**, pp. 1366–1368, 1998.

141. A. Tredicucci, C. Gmachl, M. C. Wanke, F. Capasso, A. L. Hutchinson, D. L. Sivco, S. N. G. Chu, and A. Y. Cho, "Surface plasmon quantum cascade lasers at $\lambda \sim 19$ μm," *Appl. Phys. Lett.* **77**, p. 2286, 2000.

142. K. Unterrainer, R. Colombelli, C. Gmachl, F. Capasso, H. Y. Hwang, A. M. Sergent, D. L. Sivco, and A. Y. Cho, "Quantum cascade lasers with double metal-semiconductor waveguide resonators," *Appl. Phys. Lett.* **80**, pp. 3060–3062, 2002.

143. M. Rochat, M. Beck, J. Faist, and U. Oesterle, "Measurement of far-infrared waveguide loss using a multisection single-pass technique," *Appl. Phys. Lett.* **78**, p. 1967, 2001.

144. M. Rochat, L. Ajili, H. Willenberg, J. Faist, H. E. Beere, A. G. Davies, E. H. Linfield, and D. A. Ritchie, "Low-threshold terahertz quantum cascade lasers," *Appl. Phys. Lett.* **81**, p. 1381, 2002.

145. S. Barbieri, J. Alton, S. S. Dhillon, H. E. Beere, M. Evans, E. H. Linfield, A. G. Davies, D. A. Ritchie, R. Köhler, A. Tredicucci, and F. Beltram, "Continuous-wave operation of terahertz quantum-cascade lasers," *IEEE J. Quantum Electronics* **39**, p. 586, 2003.

146. L. Ajili, G. Scalari, D. Hofstetter, M. Beck, J. Faist, H. Beere, G. Davies, E. Linfield, and D. Ritchie, "Continuous-wave operation of far-infrared quantum cascade lasers," *Electron. Lett.* **38**, pp. 1675–1676, 2002.

147. C. Sirtori, F. Capasso, J. Faist, A. L. Hutchinson, D. L. Sivco, and A. Y. Cho, "Resonant tunneling in quantum cascade lasers," *IEEE J. Quantum Electronics* **34**, p. 1722, 1998.

148. V. B. Gorfinkel, S. Luryi, and B. Gelmont, "Theory of gain spectra for quantum cascade lasers and temperature dependence of their characteristics at low and moderate carrier concentration," *IEEE J. Quantum Electronics* **32**, p. 1995, 1996.

149. J. Faist, F. Capasso, C. Sirtori, D. L. Sivco, A. L. Hutchinson, and A. Y. Cho, "Vertical transition quantum cascade laser with Bragg-confined excited state," *Appl. Phys. Lett.* **66**, p. 538, 1995.

150. D. Indjin, P. Harrison, R. W. Kelsall, and Z. Ikonic, "Mechanisms of temperature performance degradation in terahertz quantum-cascade lasers," *Appl. Phys. Lett.* **82**, pp. 1347–1349, 2003.

151. A. Tredicucci, R. Köhler, F. Beltram, H. E. Beere, E. H. Linfield, A. G. Davies, and D. A. Ritchie, "Terahertz quantum cascade lasers," *Physica E* **21**, pp. 846–851, 2004.

152. J. Faist, M. Beck, T. Aellen, and E. Gini, "Quantum-cascade lasers based on a bound-to-continuum transition," *Appl. Phys. Lett.* **78**, pp. 147–149, 2001.

153. R. Köhler, A. Tredicucci, F. Beltram, H. E. Beere, E. H. Linfield, D. A. Ritchie, and A. G. Davies, "Quantum cascade lasers emitting at ($\lambda > 100$ μm)," *Electron. Lett.* **39**, pp. 1254–1255, 2003.

154. B. S. Williams, H. Callebaut, S. Kumar, Q. Hu, and J. L. Reno, "3.4-THz quantum cascade laser based on LO-phonon scattering for depopulation," *Appl. Phys. Lett.* **82**, pp. 1015–1017, 2003.

155. H. Callebaut, S. Kumar, B. S. Williams, Q. Hu, and J. L. Reno, "Analysis of transport properties of tetrahertz quantum cascade lasers," *Appl. Phys. Lett.* **83**, p. 207, 2003.

156. H. Callebaut, S. Kumar, B. S. Williams, Q. Hu, and J. L. Reno, "Importance of electron-impurity scattering for electron transport in terahertz quantum-cascade lasers," *Appl. Phys. Lett.* **84**, p. 645, 2004.

157. R. Köhler, A. Tredicucci, C. Mauro, F. Beltram, H. E. Beere, E. H. Linfield, D. A. Ritchie, and A. G. Davies, "Terahertz quantum-cascade lasers based on an interlaced photon-phonon cascade," *Appl. Phys. Lett.* **84**, pp. 1266–1268, 2004.

158. D. Indjin, Z. Ikonić, P. Harrison, and R. W. Kelsall, "Surface plasmon waveguides with gradually doped or NiAl intermetallic compound buried contact for terahertz quantum cascade lasers," *J. Appl. Phys.* **94**, pp. 3249–3252, 2003.

159. L. Ajili, G. Scalari, J. Faist, H. E. Beere, E. H. Linfield, D. A. Ritchie, and A. G. Davies, "High power quantum cascade lasers operating at $\lambda \simeq 87$ and 130 μm," *Appl. Phys. Lett.* **85**, pp. 3986–3988, 2004.

160. J. Alton, S. S. Dhillon, C. Sirtori, A. de Rossi, M. Calligaro, S. Barbieri, H. E. Beere, E. H. Linfield, and D. A. Ritchie, "Buried waveguides in

terahertz quantum cascade lasers based on two-dimensional surface plasmon modes," *Appl. Phys. Lett.* **86**, 071109, 2005.

161. B. S. Williams, S. Kumar, H. Callebaut, Q. Hu, and J. L. Reno, "3.4 THz terahertz quantum-cascade laser operating above liquid nitrogen temperature," *Electron. Lett.* **39**, p. 915, 2003.

162. R. Köhler, A. Tredicucci, F. Beltram, H. E. Beere, E. H. Linfield, A. G. Davies, and D. A. Ritchie, "Low-threshold quantum-cascade lasers at 3.5 THz ($\lambda \sim 85$ μm)," *Opt. Lett.* **28**, pp. 810–812, 2003.

163. G. Scalari, S. Blaser, J. Faist, H. E. Beere, E. H. Linfield, D. A. Ritchie, and A. G. Davies, "Terahertz emission from quantum cascade lasers in the quantum hall regime: evidence for many body resonances and localization effects," *Phys. Rev. Lett.* **93**, 237403, 2004.

164. L. Mahler, R. Köhler, A. Tredicucci, F. Beltram, H. E. Beere, E. H. Linfield, D. A. Ritchie, and A. G. Davies, "Single-mode operation of terahertz quantum cascade lasers with distributed feedback resonators," *Appl. Phys. Lett.* **84**, pp. 5446–5448, 2004.

165. A. Barkan, F. K. Tittel, D. M. Mittleman, R. Dengler, P. Siegel, G. Scalari, L. Ajili, J. Faist, H. E. Beere, E. H. Linfield, A. G. Davies, and D. A. Ritchie, "Linewidth and tuning characteristics of THz quantum cascade lasers," *Opt. Lett.* **29**, p. 575, 2004.

166. J. C. Weeber, Y. Lacroute, A. Dereux, E. Devaux, T. Ebbesen, C. Girard, M. U. González, and A. L. Baudrion, "Near-field characterization of Bragg mirrors engraved in surface plasmon waveguides," *Phys. Rev. B* **70**, 235406, 2004.

167. D. Smirnov, C. Becker, O. Drachenko, V. V. Rylkov, H. Page, J. Leotin, and C. Sirtori, "Control of electron-optical-phonon scattering rates in quantum box cascade lasers," *Phys. Rev. B* **66**, 121305, 2002.

168. C. Becker, C. Sirtori, O. Drachenko, V. V. Rylkov, D. Smirnov, and J. Leotin, "GaAs quantum box cascade lasers," *Appl. Phys. Lett.* **81**, pp. 2941–2943, 2002.

169. S. Blaser, M. Rochat, M. Beck, D. Hofstetter, and J. Faist, "Terahertz intersubband emission in strong magnetic fields," *Appl. Phys. Lett.* **81**, pp. 67–69, 2002.

170. J. Alton, S. Barbieri, J. Fowler, H. E. Beere, J. Muscat, E. H. Linfield, D. A. Ritchie, A. G. Davies, R. Köhler, and A. Tredicucci, "Magnetic field in-plane quantization of population inversion in a THz superlattice quantum cascade laser," *Phys. Rev. B* **68**, 081303, 2003.

171. V. Tamosiunas, R. Zobl, J. Ulrich, K. Unterrainer, R. Colombelli, C. Gmachl, K. West, L. Pfeiffer, and F. Capasso, "Terahertz quantum cascade lasers in a magnetic field," *Appl. Phys. Lett.* **83**, p. 3873, 2003.

172. G. Scalari, S. Blaser, L. Ajili, J. Faist, H. E. Beere, E. H. Linfield, D. A. Ritchie, and A. G. Davies, "Population inversion by resonant magnetic confinement in terahertz quantum-cascade lasers," *Appl. Phys. Lett.* **83**, pp. 3453–3455, 2003.

173. K. Kempa, P. Bakshi, A. Scorupsky, C. G. Du, G. Feng, R. Zobl, G. Strasser, C. Rauch, C. Pacher, K. Unterrainer, and E. Gornik, "Intersubband transport in quantum wells in strong magnetic fields mediated by single- and two-electron scattering," *Phys. Rev. Lett.* **88**, 226803, 2002.

174. J. Darmo, V. Tamosiunas, G. Fasching, J. Kröll, K. Unterrainer, M. Beck, M. Giovannini, J. Faist, C. Kremser, and P. Debbage, "Imaging with a terahertz quantum cascade laser," *Opt. Express* **12**, pp. 1879–1884, 2004.

175. D. R. Chamberlin, P. R. Robrish, W. R. Trutna, G. Scalari, M. Giovannini, L. Ajili, and J. Faist, "Imaging at 3.4 THz with a quantum-cascade laser," *Appl. Opt.* **44**, pp. 121–125, 2005.

176. D. C. Larrabee, G. A. Khodaparast, F. K. Tittel, J. Kono, G. Scalari, L. Ajili, J. Faist, H. E. Beere, A. G. Davies, E. H. Linfield, D. A. Ritchie, Y. Nakajima, M. Nakai, S. Sasa, M. Inoue, S. Chung, and M. B. Santos, "Application of terahertz quantum-cascade lasers to semiconductor cyclotron resonance," *Opt. Lett.* **29**, pp. 122–124, 2004.

Chapter

3

High-Speed Operation and Ultrafast Pulse Generation with Quantum Cascade Lasers

Roberto Paiella

Department of Electrical and Computer Engineering and Photonics Center
Boston University
Boston, Massachusetts

Rainer Martini

Department of Physics and Engineering Physics
Stevens Institute of Technology
Hoboken, New Jersey

Alexander Soibel

Jet Propulsion Laboratory
California Institute of Technology
Pasadena, California

H. C. Liu

Institute for Microstructural Sciences
National Research Council
Ottawa, Ontario, Canada

Federico Capasso

Division of Engineering and Applied Sciences
Harvard University
Cambridge, Massachusetts

3.1 Introduction

The investigation of the high-speed dynamics of semiconductor lasers has long been one of the main thrusts of research in the development of these devices, due to favorable properties and important applications [1–7]. With their electric injection pumping mechanism, semiconductor lasers can be used to directly convert high-speed signals from the electrical to the optical domain—a function of central importance in the field of optical communications. Their small size and relatively low power consumption allow for the efficient delivery to the active region of electric signals at speeds of several tens of gigahertz, when properly designed device structures and packaging are used. Furthermore, their short cavity lengths compared to other types of lasers allow for passive mode-locked operation at uniquely high repetition rates [8–10], which can be extended into the terahertz region [11].

A key parameter in determining the dynamic properties of a laser is the relaxation lifetime τ of the laser population inversion. In conventional semiconductor lasers based on interband recombination, the population inversion (i.e., the density of conduction band electrons and valence band holes) primarily decays through spontaneous radiative recombination and band-to-band carrier-carrier scattering (Auger recombination). The resulting carrier lifetime is on the order of a few nanoseconds [7]. While this is faster than in most other types of lasers, it effectively limits the *direct-current* (dc) modulation bandwidth to ~10 to 20 GHz [12]. In addition, the practical use of directly modulated interband lasers as high-speed optical transmitters is severely limited by their large wavelength chirp [13, 14].

In *quantum cascade* (QC) lasers, the population inversion results from a nonequilibrium electronic distribution in the conduction subbands of properly designed coupled quantum wells or superlattices [15, 16]. Nonradiative electronic decay from a higher-energy subband to a lower-energy one is primarily mediated by the emission of *longitudinal optical* (LO) phonons, as illustrated schematically in Fig. 3.1. This relaxation mechanism can be extremely fast; the corresponding carrier lifetime τ ranges from a few picoseconds to a few hundred femtoseconds [17], depending on the mutual wave function overlap and energy separation between the initial and final electronic states. Thus, in QC lasers the dynamics of the population inversion are significantly faster compared to all other laser systems, by 3 orders of magnitude or more. Incidentally, this also means that intersubband active materials have intrinsically low radiative efficiencies; high-performance lasing is nonetheless obtained with these materials, through the design of active stages with particularly short-lived lower laser levels and through the use of the cascading scheme.

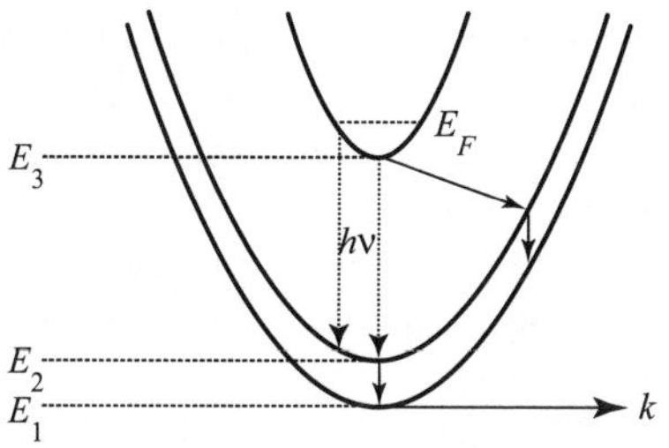

Figure 3.1 In-plane energy dispersion of the bottom three subbands of a typical QC laser active stage. The optical transitions are denoted by the dotted arrows. The solid arrows illustrate nonradiative decay processes mediated by the emission of LO phonons. The time scale of these processes ranges from a few picoseconds to a few hundred femtoseconds. The energy separation of the bottom two subbands is by design approximately equal to the LO phonon energy, which makes the intersubband relaxation rate between these two subbands particularly fast.

From the point of view of high-speed laser dynamics, the ultrafast carrier lifetime of QC lasers is an attractive feature with important implications. As described below, it leads to an ultrawide modulation response with no relaxation oscillation resonance—unlike all other laser systems—and with an intrinsic cutoff frequency that can exceed 100 GHz [18–21]. Combined with the negligibly small chirp parameter of intersubband transitions at their line center [15, 22–24], these properties make QC lasers ideally suited for use as high-speed optical data sources. Some work to explore this application has already been reported, in particular in the area of optical wireless communications [25–27] where the use of mid-infrared (mid-ir) light has potential advantages [28]. In addition, these properties provide a major motivation to the development of QC lasers operating at near-ir wavelengths, for use in ultra-broadband fiber-optic communications.

Another unique feature of intersubband devices—with important implications to their dynamic properties—is the giant ultrafast optical nonlinearities associated with real electronic transitions between subbands. These are among the strongest optical nonlinearities ever observed, due to their resonant nature and to the large oscillator strengths of intersubband transitions [29–33]. At the same time, unlike most other resonant nonlinearities, they exhibit ultrafast response times, given by the picosecond intersubband relaxation lifetimes. This combination of ultrafast response and giant nonlinear optical coefficients is unique and has significant device applications. For example, it plays an important role in the operation of mode-locked QC lasers [34, 35], where a large self-phase modulation due to a refractive index nonlinearity of the active material itself is observed. Furthermore, this combination is extremely attractive for all-optical switching and information processing applications.

This chapter describes the dynamic properties of QC lasers and their applications. In particular, Sec. 3.2 is devoted to the response of these devices to high-speed current modulation and to the use of directly modulated QC lasers for free-space optical communications. Section 3.3 presents various techniques that have been demonstrated to generate picosecond pulses with mid-ir QC lasers, namely, gain switching, active mode locking, and passive mode locking. Finally, in Sec. 3.4 a brief outlook for future directions of research in this area is presented.

3.2 High-Speed Modulation Properties of Quantum Cascade Lasers

3.2.1 Absence of relaxation oscillations and ultrawide bandwidth

The dynamic properties of a laser manifest themselves most directly in its modulation response, i.e., the laser modulation efficiency as a function of modulation frequency. This is obtained by adding a small-signal time-harmonic component of frequency Ω to the pumping level above threshold and measuring the corresponding harmonic component of the output power as a function of Ω. The modulation response of a typical high-speed interband diode laser is plotted in Fig. 3.2 (dashed line); its main feature is a pronounced resonance peak associated with relaxation oscillations. These are coupled damped oscillations of the population inversion and the laser field toward their steady-state values, which follow any dynamic change in the laser pumping level [36]. The coupling and damping mechanisms are provided by stimulated emission and gain saturation. As shown in Fig. 3.2, these transient oscillations produce a resonance enhancement of the laser modulation response at their characteristic frequency f_R, typically a few gigahertz in interband lasers. Furthermore, the modulation response becomes increasingly weaker as the drive frequency is increased above this resonance, so that f_R provides a measure of the laser intrinsic modulation bandwidth.

In QC lasers the situation is radically different, as illustrated by the continuous line in Fig. 3.2. Due to the ultrafast carrier relaxation dynamics in these devices, transient oscillations of the population inversion and hence of the photon density are overdamped, and no resonance appears in the frequency response. The modulation dynamics of QC lasers are then, to a large extent, simply limited by their picosecond carrier lifetime, so that intrinsic bandwidths in excess of 100 GHz can be obtained. We emphasize that this is the only laser system demonstrated so far that does not exhibit relaxation oscillations.

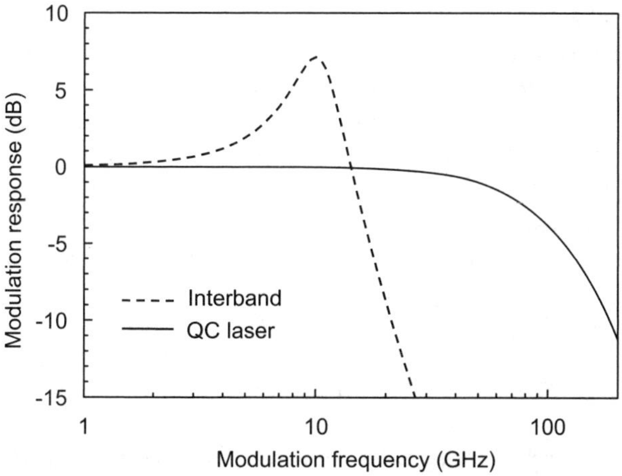

Figure 3.2 Schematic modulation response of a typical interband diode laser (dashed line) and QC laser (solid line). The latter is characterized by the absence of the relaxation oscillation resonance and by a cutoff frequency that can exceed 100 GHz.

This property of QC lasers has been extensively investigated in the past few years, both theoretically [18–21] and experimentally [20]. Early work [18] even predicted the possibility of terahertz cutoff frequencies, although this conclusion has been subsequently disputed [21].

To put this discussion in a quantitative framework, here we use a simple argument based on the "textbook" expression for the laser modulation response [36]. This is characterized by two poles at the following complex frequencies

$$\omega_{\pm} = \frac{i}{2}\left(\frac{1}{\tau} + \frac{cg_N P}{n}\right) \pm \sqrt{\frac{1}{\tau_p}\frac{cg_N P}{n} - \frac{1}{4}\left(\frac{1}{\tau} + \frac{cg_N P}{n}\right)^2} \tag{3.1}$$

where τ is the lifetime of the upper laser state, τ_p is the photon lifetime, c is the speed of light in vacuum, g_N is the differential gain, n is the modal refractive index, and P is the intracavity photon density [the quantity $n/(cg_N P)$ is often referred to as the stimulated emission lifetime τ_{st}]. This expression is very general provided that the lower laser state has negligible population, an assumption that is quite valid in QC lasers. In interband devices, the second term in the square root is negligible, so that the two poles of Eq. 3.1 have a nonzero real part. Correspondingly, a resonance is introduced in the modulation response at the frequency $f_R = |\mathrm{Re}(\omega_{\pm})|/(2\pi)$ (the relaxation oscillation frequency).

On the other hand, in QC lasers, due to their ultrashort carrier lifetime τ, the second term in the square root of Eq. 3.1 typically dominates, so that the two poles are pure imaginary and no such resonance exists. The intrinsic cutoff frequency is then given by $f_{3\text{ dB}} = \text{Im}(\omega_-)/(2\pi)$ and is maximized by decreasing the lifetimes τ, τ_p, and τ_{st}. The latter is inversely proportional to the output power and therefore can be minimized by operating the laser well above threshold. An important limit is obtained in the case $\tau_p \approx \tau > \tau_{st}$, where $f_{3\text{ dB}}$ is simply equal to $1/(2\pi\tau)$. Using a typical calculated value $\tau = 1.5$ ps gives $f_{3\text{ dB}} \approx 106$ GHz. Even larger bandwidths can be achieved by decreasing the photon lifetime, e.g., shortening the cavity length, and by decreasing the carrier lifetime, e.g., increasing the wave function overlap between the upper and lower laser states as in superlattice active regions.

The predicted absence of relaxation oscillations in QC lasers was experimentally verified in Ref. 20 through direct measurements of the modulation response of several devices. To minimize parasitic effects, a cryogenic high-speed laser package based on a semirigid coaxial cable and a 50-Ω microstrip line was developed. Furthermore, the devices were processed with relatively thick insulating layers underneath the top metal contacts, to reduce the chip bypass capacitance. A chalcogenide glass ($Ge_{0.25}Se_{0.75}$), deposited over the QC laser ridge by pulsed laser ablation, was used for this purpose [37]. A cross section of the resulting laser waveguide is shown in the inset of Fig. 3.3. Figure 3.4 shows exemplary modulation response traces of two QC lasers for different values of the dc drive current, measured using a high-speed *quantum well infrared photodetector* (QWIP) [38]. Both lasers emit near 8 μm and are based on the "vertical" transition design in a three-coupled-quantum-well GaInAs/AlInAs active region [39]. These devices can be operated in *continuous-wave* (cw) mode up to about 120 K; therefore, all the data presented are for cryogenic temperatures. The dc light-current-voltage characteristics of one of the lasers are plotted in Fig. 3.3.

As illustrated by the data of Fig. 3.4, no relaxation oscillation resonance peak appears in the modulation response of these devices. Some features are observed in these traces that resemble weak resonance peaks (e.g., near 6 and 8 GHz), but they remain at exactly the same frequencies as the laser optical power is varied by over one order of magnitude, whereas f_R strongly varies with optical power according to Eq. 3.1. Thus these features are attributed to parasitic effects. It is also important to mention that no sign of relaxation oscillations is found even for values of dc bias approaching the cw threshold for which $P \rightarrow 0$. In this limit, no matter how fast the laser intrinsic response, the relaxation oscillation frequency is expected from Eq. 3.1 to eventually enter the relatively low-frequency range tested. Therefore these results provide conclusive evidence of the absence of relaxation oscillations in the QC lasers tested.

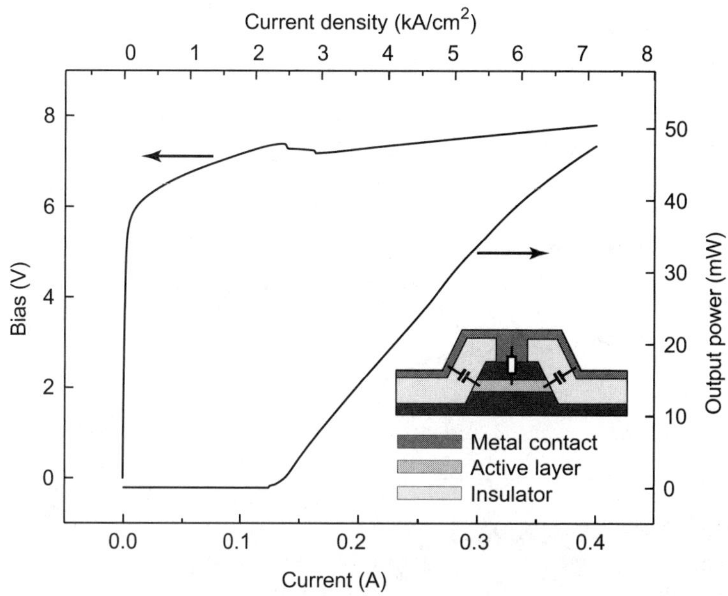

Figure 3.3 CW light-current-voltage characteristics of one of the QC lasers used to demonstrate the absence of relaxation oscillations in these devices, measured at a temperature of 20 K. Inset: schematic cross section of the laser waveguide. The shaded areas represent, in order of increasingly dark shading, the chalcogenide dielectric blocking layer, the active material, the metal contact, and the cladding regions. The capacitance and resistance symbols indicate the main chip parasitics limiting the laser high-frequency response.

Regarding the laser modulation bandwidth, all traces in Fig. 3.4 exhibit a low-frequency shoulder up to about 2 GHz and a cutoff frequency near 7 GHz. Both of these features are attributed to parasitic effects; in particular, from the estimated capacitance of the dielectric blocking layers $C \approx 10$ pF and the measured differential resistance above threshold $R \approx 2\ \Omega$, it is concluded that the cutoff frequency is to a large extent RC-limited. The experimental observation of the predicted ultrawide modulation bandwidth of QC lasers will require more extensive microwave engineering of the device package and chip layout, or alternative parasitic-free measurement techniques such as all-optical modulation by active-layer photomixing [40].

3.2.2 Negligible linewidth enhancement factor

A characteristic feature of interband semiconductor lasers is a relatively large variation of the refractive index of the lasing mode(s) with injection current [4, 7, 36]. This property is commonly described by a parameter variously referred to as the linewidth enhancement

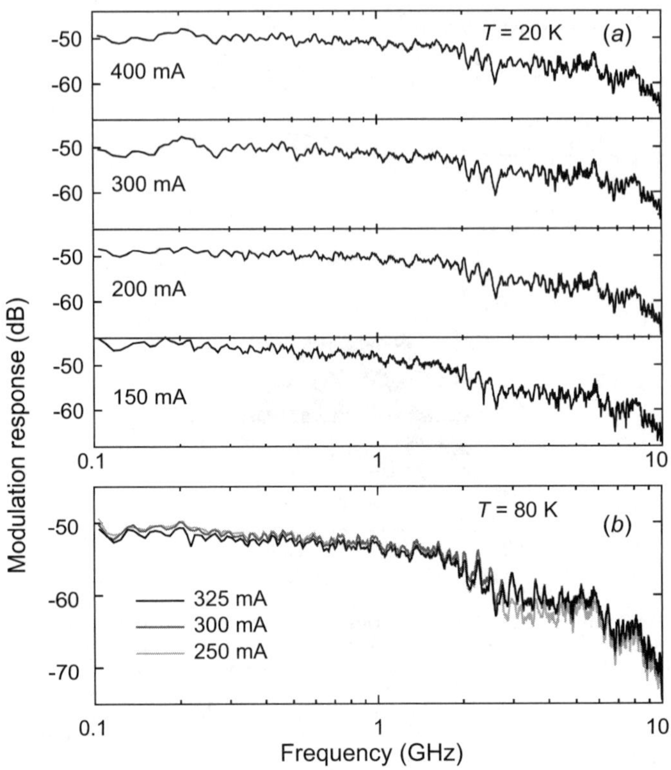

Figure 3.4 (*a*) Exemplary high-frequency modulation response traces of the QC laser of Fig. 3.3 at 20 K, for different values of the drive current ranging from very near threshold (150 mA) up to over one order of magnitude higher photon density (400 mA). These traces were normalized with the experimental frequency response of the receiver and reflect only the modulation response of the QC laser. (*b*) High-frequency modulation response of a different laser, for different values of the laser bias at 80 K.

factor, the chirp parameter, or Henry's α parameter after the seminal paper by C. H. Henry describing its impact on the laser linewidth [41]; it is defined as

$$\alpha = -\frac{4\pi}{\lambda_0}\frac{d\delta n/dN}{dg/dN}\bigg|_{\lambda=\lambda_0} \tag{3.2}$$

where λ_0 is the free-space emission wavelength, δn is the carrier contribution to the refractive index, g is the gain coefficient, and N is the carrier density in the active region.

The relatively large value of α in interband lasers, most typically in the range of 4 to 6, has several important—and mostly undesirable—effects [4, 7, 36]. Most notably, it leads to a broadening of the laser linewidth by the factor of $1 + \alpha^2$ [41], since it causes a strong coupling of carrier density fluctuations to variations in the index and hence in the emission frequency. This linewidth enhancement is undesirable in applications requiring a high degree of temporal coherence, such as high-resolution sensing. Similarly, a large value of α causes a significant wavelength chirping upon direct modulation [13, 14]. Combined with the large dispersion of standard optical fibers, this latter effect severely limits the use of directly modulated diode lasers as data sources in fiber-optic communications.

Physically, the nonzero α parameter of interband lasers can be traced back to the asymmetric gain spectra associated with optical transitions between two bands of opposite curvature (the conduction and valence bands) [36]. This argument is illustrated in Fig. 3.5. The carrier-induced refractive index spectrum $\delta n(\lambda)$ is related to the gain spectrum $g(\lambda)$ through the Kramers-Kronig relations; by the nature of these relations, if $g(\lambda)$ is symmetric about the wavelength λ_0 of the lasing mode, then $\delta n(\lambda_0)$ is zero independent of the carrier density; and vice versa, if $g(\lambda)$ is asymmetric about λ_0, then $\delta n(\lambda_0)$ is nonzero and varies with N. In QC lasers the gain spectrum is symmetric to a high degree of approximation due to the very similar in-plane dispersion of the lasing subbands. Thus, unlike the case of interband semiconductor lasers, $\delta n(\lambda_0) \equiv 0$ and the chirp parameter is nearly zero in these devices. Incidentally, it should also be mentioned that at wavelengths sufficiently detuned from λ_0, the index dependence on carrier density $d\delta n/dN$ is actually very large in intersubband transitions, leading to giant index nonlinearities associated with these transitions [30].

The negligible chirp parameter of intersubband lasers was predicted in the original report of QC laser action [15], and since then it has been verified in a few experiments [22–24]. In Ref. 22, the gain spectrum of a GaInAs/AlInAs 4.6-μm device was first measured as a function of bias

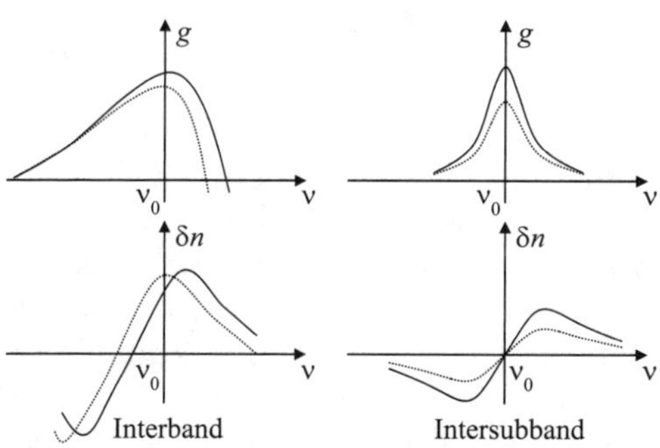

Figure 3.5 Schematic gain coefficient g and carrier-induced refractive index δn versus optical frequency ν for the case of interband (left) and intersubband transitions (right). The dotted and solid lines correspond to two different values of the injected carrier density N. In the case of intersubband transitions, at the frequency ν_0 of maximum gain δn is zero independent of N.

current with the Hakki-Paoli technique [42]; the Kramers-Kronig relations were then used to calculate the corresponding refractive index spectra. From these data, a very small value of 0.1 was obtained for the α parameter. It should be mentioned that the numerical accuracy of this technique is limited due to the limited frequency range of the gain spectrum used in the Kramers-Kronig integrals. In any case, this measurement provided the first experimental indication of the very small chirp parameter associated with intersubband laser transitions.

A more accurate procedure has been used recently [23, 24] in which the refractive index change with bias current was determined from the measured wavelength shifts of the Fabry-Perot modes. In these measurements, a slightly larger—and negative—value of $\alpha \approx -0.5$ was obtained for a GaInAs/AlInAs 8.2-μm device, which was attributed mainly to carrier heating effects. Finally, an indirect but very significant confirmation of the small α parameter of QC lasers is provided by measurements of their emission linewidth: in particular, a value of 12 kHz was inferred for an 8.5-μm QC-DFB (distributed feedback) laser in a frequency stabilization feedback loop [43], which is extremely small for a semiconductor device.

3.2.3 Applications of high-speed quantum cascade lasers

As discussed in the previous sections, QC lasers can be directly modulated at speeds of several tens of gigahertz, with no spectral distortion caused by the relaxation oscillation resonance and with negligible chirp. Thus they are ideally suited to be used as data sources in broadband optical communications. The electron cascading scheme is also favorable in this respect, as it allows for slope efficiencies that are larger than the conventional limit of one emitted photon per injected electron. This is advantageous for applications in microwave-photonics communication links, whose *radio-frequency* (RF) gain and signal-to-noise ratio are proportional to the square of the transmitter slope efficiency [44].

These considerations strongly motivate the development of QC lasers emitting at near-ir wavelengths—in the low-loss transmission window of optical fibers—for use as ultra-high-speed data sources in fiber-optic communications. In the more traditional QC-laser heterostructure materials, i.e., GaInAs/AlInAs lattice matched to InP and GaAs/AlGaAs, the conduction band offset ΔE_c is too small (≤ 0.5 eV) to accommodate near-ir intersubband transitions. Therefore, novel quantum-well systems with larger ΔE_c are currently being explored toward this goal, including antimony-based compounds [45] and group-III nitrides [46–50].

In recent years, high-speed data transmission using mid-ir QC lasers has also been investigated [51], in light of its potential application in the field of free-space optical communications [25–27]. Optical wireless is a promising approach for the development of rapidly deployable, cost-effective, and secure communication links, e.g., for special events, in the battlefield, or in disaster recovery sites. Due to its favorable atmospheric transmission properties, mid-ir light at 3 to 5 or 8 to 13 μm is, in principle, better suited to this application compared to near-ir radiation. In particular, both light scattering and scintillation effects, which provide the dominant sources of propagation losses in optical wireless systems [52], exhibit a strong dependence on the laser frequency that favors longer wavelengths, especially under conditions of low visibility [28].

The potential of QC lasers as high-speed digital data sources was first investigated in the *bit error rate* (BER) measurements reported in Ref. 51. Mid-ir devices similar to the ones discussed in Sec. 3.2.1 were directly modulated with a non-return-to-zero pseudorandom bit stream at 2.5 Gb/s, and their output was measured with a high-speed QWIP. Error-free data transmission (BER $< 10^{-12}$) was obtained for a wide range of operating temperatures and bias currents. The corresponding eye diagrams are clear and open, as illustrated in Fig. 3.6*a* and *b* for an 8.1-μm device at 20 and 85 K, respectively, although some fluctuations in the average "1" and "0" levels are observed. These are likely due to heating of the laser during long "on" times, and therefore they can be

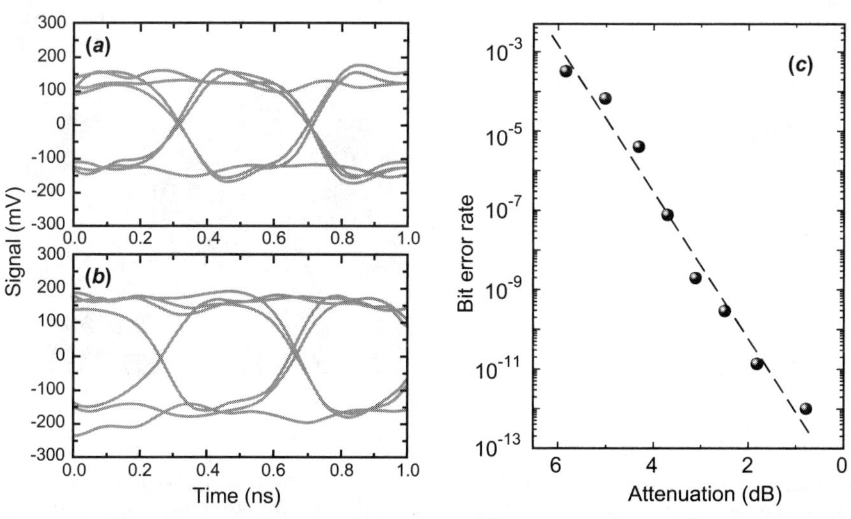

Figure 3.6 Measured eye diagram of an 8.1-μm QC laser at a heat sink temperature of (*a*) 20 K and (*b*) 85 K. The bit rate is 2.5 Gb/s. (*c*) Corresponding bit error rate versus received power at 85 K. An attenuation of 0 dB (i.e., full transmission) corresponds to a received power of about 1 mW in this measurement.

expected to vanish if a return-to-zero modulation format is used, which is especially suitable for higher data rates. The minimum received power (where BER = 10^{-9} by convention) of this all-intersubband transmission system is about 0.5 mW, or 50% of the total detected power, as shown in Fig. 3.6*c*. This is quite large compared to typical fiber-optic systems, mainly due to the lower quantum efficiency and higher noise equivalent power of the QWIP detector used and to the above-mentioned variations in the detected digital levels.

A similar high-speed QC laser was also used to demonstrate free-space optical transmission of complex digital data (multimedia satellite channels) over a distance of about 200 m [27]. The device was modulated with a relatively high-frequency signal (750 MHz to 1.45 GHz) received from a satellite dish, containing about 800 television channels and 100 radio channels. The modulated beam was transmitted across an outdoor 100-m path to a retroreflector, mounted on a neighboring building in the Murray Hill, N.J., campus of Bell Laboratories, Lucent Technologies. The reflected beam was then collected with an *f*/8 telescope and detected with a high-responsivity mercury cadmium telluride detector. A total transmission loss of about 10 dB was measured, which was attributed to beam divergence and losses in the optical elements. A typical RF spectrum of the transmitted data is plotted in Fig. 3.7*a* (continuous line), together with the original signal from the satellite dish (dashed line). The transmission loss is found to increase with increasing frequency due to the limited bandwidth of the detector (~1 GHz); as a result, the number of actually decoded channels was reduced to around 650. A screenshot of one such channel (an

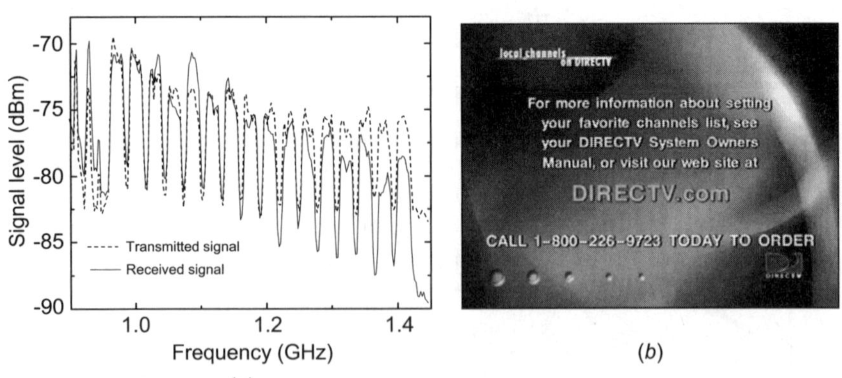

Figure 3.7 (*a*) RF spectrum of a digital signal consisting of a few hundred television and radio channels as received from a satellite dish with a low-noise block down-converter (dashed line) and after transmission over a 200-m free-space optical link (solid line). The link consists of an 8.1-μm QC laser directly modulated with the signal from the down converter, bulk optics, and a HgCdTe detector. (*b*) Screenshot of one television channel after transmission over the same link.

advertising page from the digital satellite provider) obtained with a standard satellite set-top box is shown in Fig. 3.7*b*. Excellent video and audio quality was observed with no degradation after several hours of nonstop operation.

Finally, to investigate the benefits of using mid-ir light for optical wireless, a second beam was included in the optical path, originating from a 0.85-μm diode laser and detected with a standard Si detector. Similar performance was obtained from the two links under typical weather conditions. However, in the presence of dense fog, the mid-ir link still displayed adequate transmission even as the near-ir signal was below its detection limit (Fig. 3.8), as expected from the reduced atmospheric propagation losses at longer wavelengths. A more systematic comparison of the performance of near-ir and QC-based mid-ir optical wireless communication links has been carried out recently, using well-controlled atmospheric conditions in the Pacific Northwest National Laboratory's Aerosol Wind Tunnel Research Facility [53]. Again the QC-laser link consistently showed a higher level of performance, especially in the presence of large densities of atmospheric scatterers.

3.3 Ultrafast Pulse Generation

QC lasers are also excellent candidates for the generation of ultrashort pulses of mid-ir light due to their well-established strengths, such as

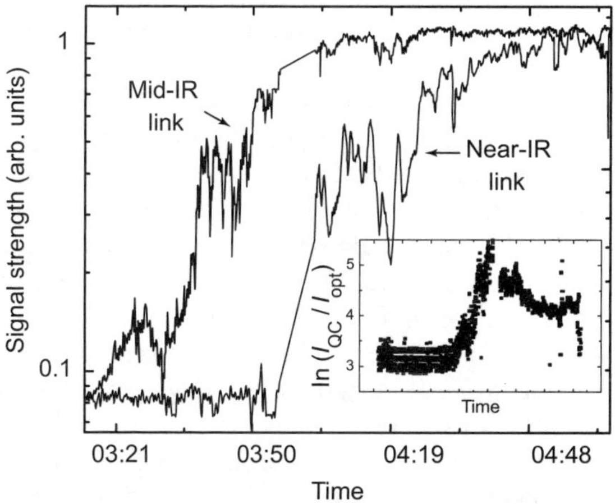

Figure 3.8 Comparison of the received intensities of the mid-ir (λ = 8.1μm) free-space optical link of Fig. 3.7 and a near-ir (λ = 0.85μm) link employing the same optics, as a function of time during a foggy night. Fog came up at around 2 a.m. and progressively dissipated during the measurement. The inset is a logarithmic plot of the ratio of the two curves in the main graph and represents the time evolution of the difference in optical losses at the two wavelengths as the structure of the fog changed.

compactness, low power consumption, and wavelength tunability by design, as well as their favorable dynamic properties. The development of such mid-ir ultrafast laser sources is motivated by several applications, including time-resolved spectroscopy and coherent control of the many chemical and biological species exhibiting resonant absorption in the mid-ir. In the following, we will review various techniques that have been employed to generate picosecond light pulses with GaInAs/AlInAs QC lasers—again based on the "vertical" transition design in three-coupled-quantum-well active regions. These experiments were the first demonstrations of semiconductor-based ultrafast optical sources emitting in the mid-ir.

3.3.1 Gain switching

Gain switching is a technique for the generation of picosecond light pulses from injection lasers, which relies on the large buildup and subsequent rapid depletion of the population inversion following a short-lived electrical excitation [5, 6]. Its main advantage lies in its inherent simplicity, since, for instance, it does not require specially fabricated devices or external cavity configurations. In Refs. 54 and 55, several QC lasers mounted in a relatively high-speed cryogenic setup were gain-switched with an integrated step recovery diode (HP 33002A comb generator), as shown in Fig. 3.9*a*. This device, when driven with a sinusoidal voltage waveform of the appropriate frequency and power, generates a train of electrical pulses having *full width at half maximum* (FWHM) as short as 90 ps, at a repetition rate of about 100 MHz. The laser emission was detected with a high-speed QWIP and measured with a sampling oscilloscope.

A typical pulse obtained from a gain-switched 7.7-μm QC laser is shown in Fig. 3.9*b*. The measured pulse FWHM in this figure is 89 ps, corresponding to an estimated actual FWHM $\Delta\tau$ of about 45 ps if the response time of the detection electronics (~ 61 ps) is accounted for. This is shorter than the width of the electrical pulses used to drive the laser by roughly a factor of 2 or more, since the electrical pulses are likely broadened by the laser parasitics. The reason is the finite time required for stimulated emission to build up following the initial creation of a large nonequilibrium population inversion, which effectively shortens the rise time of the optical pulses compared to the electrical ones. This is similar to what happens in gain-switched interband diode lasers. In the latter devices, each pulse essentially corresponds to the first spike of relaxation oscillations, and as a result the optical fall time can also be significantly shortened [5, 6]. In QC lasers, relaxation oscillations are overdamped, as described in Sec. 3.2.1, so that the optical pulse shape is expected to follow more closely the electrical one, except for the

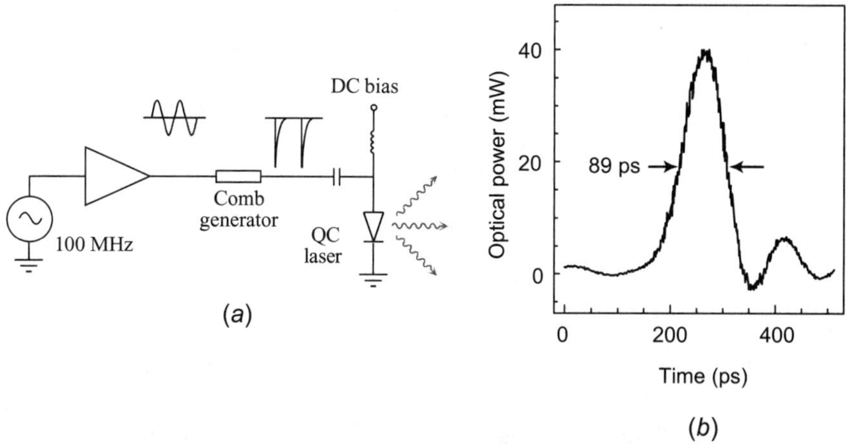

Figure 3.9 (*a*) Circuit diagram of the experimental setup used to demonstrate gain switching of QC lasers. (*b*) Oscilloscope trace of a pulse generated by a gain-switched 7.7-μm QC laser at 10 K, measured with a high-speed QWIP. The oscillations observed after the main peak are attributed to microwave reflections between the comb generator and the laser.

initial delay in photon emission. Thus, somewhat counterintuitively, the faster carrier dynamics of intersubband devices can actually lead to longer pulses in this case.

3.3.2 Active mode locking

Active mode locking of semiconductor lasers is achieved by modulating the laser current at the cavity round-trip frequency f_{rt} (i.e., the frequency separation between neighboring longitudinal modes) so that a large number of modes are driven above threshold by the modulation sidebands of their neighbors [5, 6, 36]. This forces all the lasing modes to oscillate with their phases locked to one another; as a result, their amplitudes add coherently to produce a train of short optical pulses with repetition rate f_{rt}. Typical interband lasers are sufficiently short that the corresponding round-trip frequencies are prohibitively large (> 100 GHz), so that external cavity configurations or multisection devices are commonly used to reduce the modulation frequency. On the other hand, optimum performance of QC lasers is achieved with relatively long devices (typically 1.5 to 3 mm) [16], so that a simple monolithic approach becomes suitable.

In the first demonstration of actively mode-locked QC lasers [35], 3.75-mm-long devices were used, with corresponding round-trip frequencies close to 12 GHz. These devices were processed and packaged for high-speed operation as described in Sec. 3.2.1; in the experiment, they were biased with a direct current slightly above threshold and modulated with the output of a low-phase-noise synthesized RF signal generator. The initial evidence of mode locking is presented in Fig. 3.10*a*, where several optical spectra of the laser output are shown, corresponding to different values of the modulation frequency f_m. When the detuning between f_{rt} (11.653 GHz in this case) and f_m is sufficiently large (~70 MHz), as in the top and bottom traces, the spectrum consists of a single lasing mode with its modulation sidebands. As the detuning is decreased, more and more modes appear in the spectrum as they are brought above threshold by their neighbors' sidebands, indicating effective mode locking.

The dependence of the spectral response on the RF power in actively mode-locked QC lasers was studied in Ref. 56. Figure 3.11 shows optical spectra of a QC laser operating at a constant dc bias $I = 1.6$ A and modulated at a frequency near f_{rt}, for different values of the RF power P_{RF}. When $P_{RF} \le 14$ dBm, the active modulation results only in the excitation of two weak neighboring modes or RF sidebands. A small increase in the RF modulation above 14 dBm triggers a considerable change in the emitted spectrum and results in the transition from cw to pulsed emission. In the pulsed emission regime, the spectral width $\Delta\nu$ (and pulse duration) depends weakly on the RF power, as shown in the inset of Fig. 3.11. A fit to the data at high RF powers gives the following spectral width dependence on the RF power: $\Delta\nu \sim I_{RF}^{0.18}$, where $I_{RF} \approx \sqrt{P_{RF}/50\ \Omega}$ is the RF current before the 50-Ω-impedance bias tee. This result is in good agreement with theoretical predictions for actively mode-locked lasers, giving $\Delta\nu \sim I_{RF}^{0.25}$ [36].

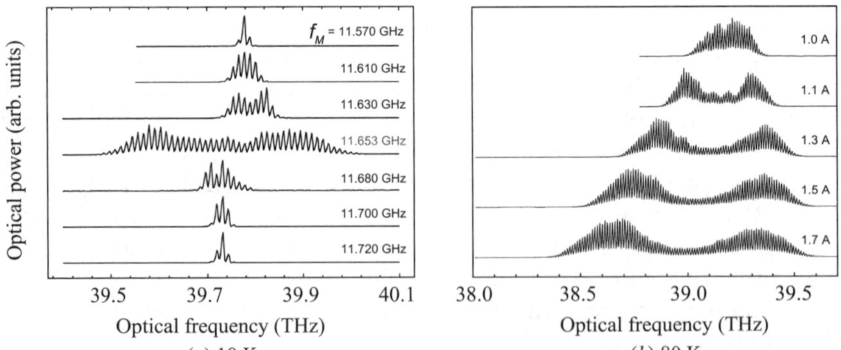

Figure 3.10 (*a*) Optical spectra of a 7.6-μm QC laser under large-signal RF modulation causing active mode locking, for different values of the modulation frequency f_M close to the laser round-trip frequency (11.653 GHz). The heat sink temperature is 10 K, and the dc bias is 900 mA (approximately 11% above the laser threshold). (*b*) Optical spectra of the same actively mode-locked device at 80 K, for different values of the dc bias. The RF drive frequency was varied at each bias to maximize the spectral width; the optimum frequencies were found to lie in the small range of 11.611 to 11.614 GHz.

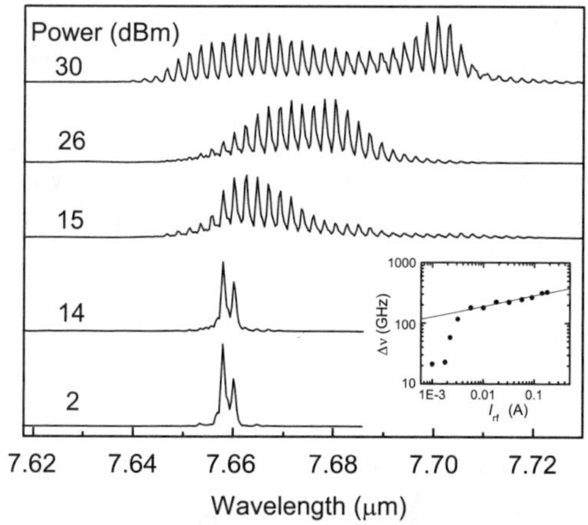

Figure 3.11 Optical spectra of a modulated QC laser at constant dc bias I = 1.6 A and temperature T = 20 K for different values of the RF modulation power. The RF frequency is f = 11.46 GHz, close to the cavity round-trip frequency. Inset: spectral width of the QC laser emission plotted versus RF modulation current. The dc laser bias is I = 1.3 A, the temperature T = 20 K, and the modulation frequency f = 11.486 GHz. The straight line is a quadratic least-squares fit to the data at high RF powers (data points above P_{RF} = 2 dBm or I_{RF} = 5.6 mA are included in this fit).

The resulting time-varying optical waveforms are too fast to be properly resolved via direct detection. An accurate characterization of their temporal features, such as pulse width and modulation depth, requires second-order autocorrelation measurements [6]. Unfortunately, at the time of the work described in this chapter, no suitable second-order autocorrelation setup was available for operation in the mid-ir at the relatively low peak power levels of these devices, on the order of a few hundred milliwatts. Incidentally, work is ongoing in the labs of one of the coauthors (Capasso) to develop such a setup based on two-photon absorption in a properly designed QWIP [57]. In any case, in Ref. 35 evidence of QC laser pulsed emission and an estimate of the pulse duration were obtained from a detailed analysis of the optical spectra at resonance (i.e., for $f_m = f_{rt}$). As shown in Fig. 3.10*b*, a large dip develops at the center of these spectra as the bias current is increased. This feature cannot be explained in terms of unlocked multimode cw emission, given the primarily homogeneously broadened narrow gain curves of the QC lasers used, as evidenced by subthreshold electroluminescence measurements. On the other hand, the spectral shapes of Fig. 3.10*b* are a classic signature of short laser pulses undergoing strong *self-phase modulation* (SPM) [58]. Namely, in the presence of a quadratic (Kerr) nonlinearity in the refractive index, the optical field acquires a time-varying phase proportional to the pulse intensity profile. The observation of SPM in this context is not surprising, given the very large nonlinear index n_2 of intersubband transitions [30], estimated at $\sim 10^{-9}$ cm^2/W in the present devices at optical frequencies slightly detuned (~100 GHz) from the line center.

These considerations form the basis for a technique to estimate the duration of ultrafast laser pulses first outlined in Ref. 59. As is well known, e.g., from nonlinear optical studies in fiber optics [58], in the presence of SPM the spectrum of a train of optical pulses develops an oscillatory envelope and broadens to

$$\Delta\omega_{\text{rms}} = \frac{\sqrt{2\log 2}}{\Delta\tau}\sqrt{1 + \frac{4}{3\sqrt{3}}\phi_{\max}^2} \tag{3.3}$$

where $\Delta\omega_{\text{rms}}$ is the root-mean-square spectral width, $\Delta\tau$ is the pulse FWHM, and a gaussian pulse shape is assumed. Finally, $\phi_{\max}$ is the maximum nonlinear phase shift which in the present context is given by

$$\phi_{\max} = \frac{4\pi L}{\lambda}\Gamma n_2 I_{\max} \tag{3.4}$$

where L is the laser cavity length, Γ is the confinement factor, and $I_{\max}$ is the pulse peak intensity. Numerical studies [58] indicate that a first pronounced dip appears in the optical spectrum when $\phi_{\max}$ reaches the value 1.5π; and additional dips are predicted to occur at higher values of $\phi_{\max}$. Referring to the spectra in Fig. 3.10*b*, a dip is observed starting approximately at the bias of 1.1 A, for which $\Delta\omega_{\text{rms}} = 148$ GHz; by using this value in Eq. 3.3 with $\phi_{\max} = 1.5\pi$, an estimate for the pulse duration of about 5 ps is obtained. Incidentally, values of $\phi_{\max}$ of order π in these mode-locked QC lasers are consistent with the pulse peak intensities—estimated from the measured average powers—and with the calculated nonlinear index n_2 of the lasing transition. Numerical simulations of the amplification of picosecond pulses in QC active media also predict spectral shapes similar to the data of Fig. 3.10*b* [60].

In general, the minimum pulse duration of a mode-locked laser is limited by the width of the gain spectrum Δf, which is relatively narrow in QC lasers. Under ideal active mode-locking conditions, the pulse FWHM $\Delta\tau$ is given by [36]

$$\Delta\tau = \frac{\sqrt{2\log 2}}{\pi}\frac{1}{M^{1/4}\sqrt{f_m \Delta f}} \tag{3.5}$$

where M is the gain modulation depth, estimated to be near 5% for the experimental conditions of Ref. 35. Using $\Delta f \approx 3$ THz for a typical 8-μm QC laser and $f_m = 12$ GHz, Eq. 3.5 gives $\Delta\tau = 4$ ps, which is close to the estimate above. To generate shorter—and hence more energetic—pulses, intersubband active media can be designed with larger gain bandwidths. In particular, ultra-broadband QC lasers have been demonstrated based on a cascade of three-well active stages of different emission wavelengths, capable of simultaneous multimode emission from 6 to 8 μm ($\Delta f \approx 12$ THz) [61]. An investigation of active mode locking in similar devices optimized for cw operation [62] has also been reported [56]. Very broad (> 2 THz) emission spectra indicative of self-phase modulation were observed; however, these consisted of two separate groups of modes, which suggested the possibility of two independent trains of pulses simultaneously excited in the laser cavity.

3.3.3 Passive mode locking

In the case of passive or self-mode locking, a laser emits a train of ultrafast pulses with repetition rate equal to the round-trip frequency even without any external modulation. This requires a sufficiently fast saturable loss mechanism (i.e., a source of optical loss that decreases with increasing intensity) within the laser cavity. In the presence of such a mechanism, it becomes favorable for the laser to emit ultrashort pulses because of their higher instantaneous intensity, and hence lower losses, relative to cw emission [5, 6, 36]. Typical passively mode-locked interband diode lasers employ saturable absorbers, e.g., introduced by proton bombardment or ion implantation, or consisting of properly designed multiple-quantum-well structures [5, 6]. In Ref. 34, passive mode locking was observed in several QC lasers based on relatively standard active-material designs and without any intentionally introduced saturable absorbers.

The experimental evidence of self-mode-locked operation in these devices is summarized in Figs. 3.12 and 3.13. Several QC lasers emitting near 5 or 8 μm were used, whose main distinctive feature is a particularly long cavity length (≥ 3.5 μm). When biased with a direct current over a wide range, these devices emit an extremely broad multimode spectrum (up to one-half of the gain spectrum) characterized by a smooth multi-peaked envelope; examples are shown in Fig. 3.12. As discussed in Sec. 3.3.2, this spectral shape is indicative of pulsed emission in the presence of strong SPM, and cannot otherwise be explained if a cw laser output is assumed. To confirm that the devices were correspondingly self-pulsating at their cavity round-trip frequency f_{rt}, their output was detected with a fast QWIP and measured with an electrical spectrum analyzer. As shown in Fig. 3.13*a*, a sharp feature centered at f_{rt} (~ 13 GHz in this case) is observed, whose peak power corresponds to a modulation amplitude on the laser beam on the order of the measured average optical power. This feature results from

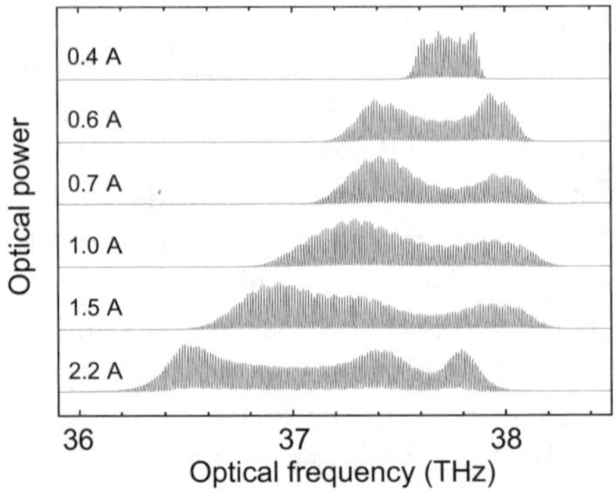

Figure 3.12 Optical spectra of a 3.5-mm-long 8-μm QC laser under conditions of self-mode locking for different values of the dc bias current. The laser heat sink temperature is 80 K.

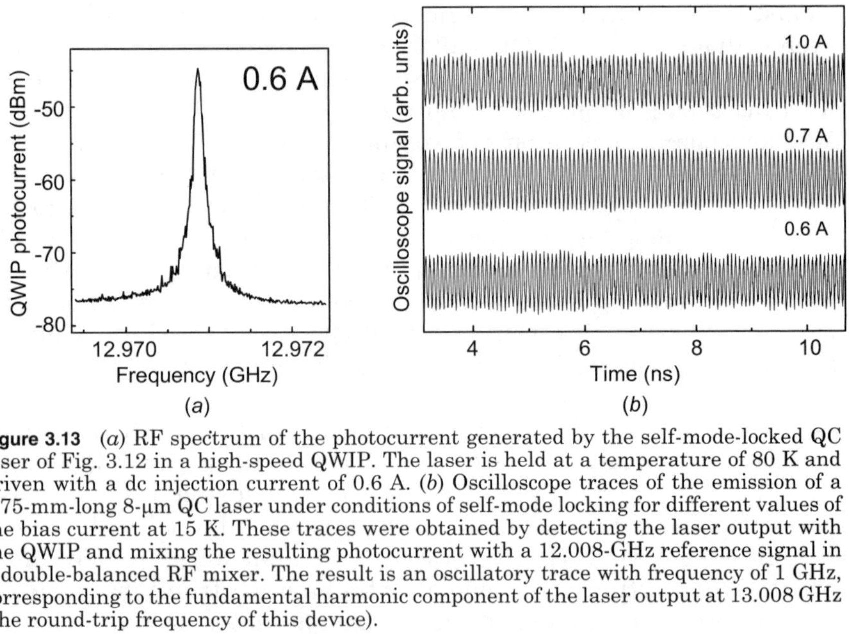

Figure 3.13 (*a*) RF spectrum of the photocurrent generated by the self-mode-locked QC laser of Fig. 3.12 in a high-speed QWIP. The laser is held at a temperature of 80 K and driven with a dc injection current of 0.6 A. (*b*) Oscilloscope traces of the emission of a 3.75-mm-long 8-μm QC laser under conditions of self-mode locking for different values of the bias current at 15 K. These traces were obtained by detecting the laser output with the QWIP and mixing the resulting photocurrent with a 12.008-GHz reference signal in a double-balanced RF mixer. The result is an oscillatory trace with frequency of 1 GHz, corresponding to the fundamental harmonic component of the laser output at 13.008 GHz (the round-trip frequency of this device).

the mutual beating of adjacent modes in the optical spectrum. Its large magnitude, strong stability, and narrow width (typically less than 100 kHz) indicate the presence of at least partial phase locking among the lasing modes. Incidentally, we note that to infer complete phase locking, a stable and narrow microwave beat note at higher harmonics of the round-trip frequency and a nonlinear autocorrelation trace with a peak value of 8:1 with respect to the background are required. In a subsequent report [63], the QWIP photocurrent from a self-mode-locked

QC laser was also analyzed directly in an oscilloscope using an RF down-conversion technique with a 12-GHz reference signal. Exemplary results are shown in Fig. 3.13*b*: periodic oscillations with relatively constant amplitude are observed, which is indicative of stable QC laser pulsations.

The exact mechanism responsible for the observed self-mode locking is still the subject of experimental and theoretical investigations. In Ref. 34 it was attributed to Kerr lensing, which is well known to cause self-mode locking in Ti:sapphire and other solid-state lasers [64]. This phenomenon arises from a refractive index nonlinearity inside the laser cavity with positive nonlinear index n_2, which leads to a narrowing of the laser beam diameter with increasing optical power. If the laser optical losses increase with increasing beam diameter (e.g., due to coupling of the optical mode with the top metal contacts in QC lasers), the net result is a saturable loss mechanism. Kerr lensing in self-mode-locked QC lasers is consistent with the observation of strong SPM in these devices (which is indicative of a large intracavity n_2) and with their long cavity lengths (for which the above-mentioned saturable loss contribution is large relative to the non saturable mirror losses). Furthermore, this interpretation was substantiated by far-field beam profile measurements, in which evidence of mode narrowing during pulsed emission was observed [34]. On the other hand, a complete theoretical description of self-mode-locked QC lasers is complicated by their very fast gain relaxation dynamics, and will likely require the inclusion of more complex and less conventional phenomena than the simple Kerr-lensing mechanism just described. What is particularly unusual is the fact that the gain recovery time in these devices is shorter than their cavity round-trip times; in existing theories of passive mode locking, the opposite is true [65]. Capasso has recently established collaboration with Franz Kaertner's group at M.I.T. to further investigate the physics of these devices.

Similarly, the temporal features of these self-mode-locked optical waveforms, i.e., their pulse duration and modulation depth, have not yet been fully characterized. As previously mentioned, the accurate measurement of these quantities requires second-order autocorrelation techniques that are not presently available in the mid-ir for the power levels of these devices. In the initial report of Ref. 34, the indirect technique described in Sec. 3.3.2 (based on the shape of the optical spectra) was used to estimate the pulse width; an approximate value of 3 ps was obtained for the device of Fig. 3.12. Subsequently, a different approach has been employed based on QC lasers specially designed for intracavity *second harmonic generation* (SHG) [63].

A portion of the conduction band diagram of these devices is given in Fig. 3.14*a* together with the wave functions squared of the relevant

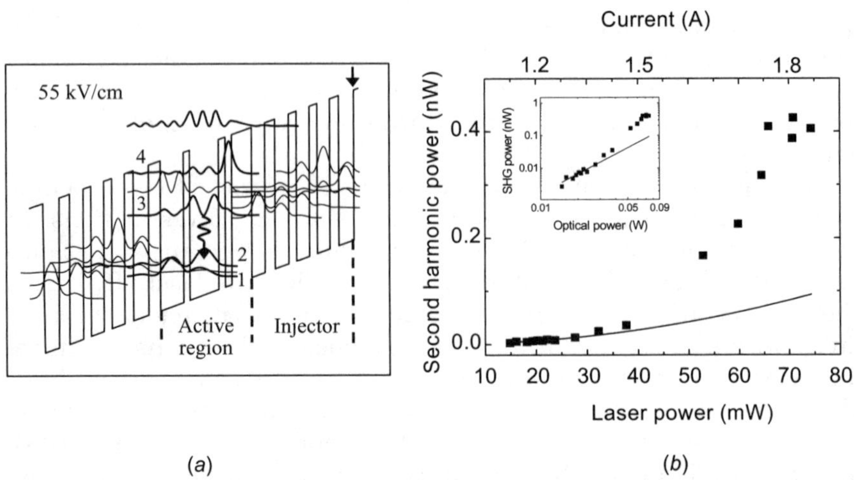

Figure 3.14 (*a*) Conduction band diagram of one active stage sandwiched between two injectors of a QC laser designed for intracavity SHG. The wave functions squared of the relevant bound states, referenced to their respective energy levels, are also shown. The laser transition is between levels 3 and 2, as indicated by the wavy arrow. A resonant nonlinearity for SHG results from the cascaded intersubband transitions 4-3-2 and 5-4-3. (*b*) Time-average power of the SHG signal versus time-average power of the fundamental signal, for a 4.5-mm-long QC laser based on the design of (*a*). The solid line is a quadratic least-squares fit to the data at low laser powers (up to 40 mW), where the laser emission is single-mode and cw. At larger power levels, the device exhibits self-mode locking. The same data on a logarithmic scale are shown in the inset.

bound states [66, 33]. The laser transitions occur between the states labeled 3 and 2 in the active stages; at the same time, two triplets of equally spaced levels (3-4-5 and 2-3-4) allow for resonant SHG. As a result, the emission of these devices contains a fundamental signal near 7.2 μm and its second harmonic near 3.6 μm. QC lasers based on this design—and with sufficiently long cavities—were also found to self-mode-lock, and an estimate of their pulse width was obtained from the dependence of the time-average SHG power on the power of the fundamental signal. This is plotted in Fig. 3.14*b* for a 4.4-mm-long device. At fundamental powers below ~40 mW, the emission is single-mode and cw, and the SHG power is well described by the square-law fit shown by the solid line. At larger fundamental power levels ≥ 65 mW, the laser reaches a regime of pulsed emission, and a substantial increase in SHG power relative to the solid line is observed, by a factor k of about 5.6. This increase is due to the larger instantaneous fundamental power—and hence larger SHG efficiency—during pulsed emission and can be written as

$$k = \sqrt{\frac{2\ln 2}{\pi}}\frac{T}{\Delta\tau} \tag{3.6}$$

where T is the cavity round-trip time, $\Delta\tau$ is the pulse FWHM, and a gaussian pulse shape is assumed [63]. From the measured value of $k \approx 5.6$ and $T \approx 104$ ps, a pulse duration of about 12 ps was inferred for this device. This result is in fair agreement with the value (9 ps) estimated for the same device by using Eq. 3.3.

3.4 Conclusions and Outlook

In conclusion, we have described the high-speed modulation properties of QC lasers and the use of these devices to generate ultrashort pulses of mid-ir radiation. Intersubband laser transitions are unique in their very fast relaxation lifetimes of order 1 ps—much faster than even the typical cavity round-trip times. In addition they are characterized by a negligible linewidth enhancement factor and simultaneously by a very large nonlinear index at optical frequencies sufficiently detuned (~ 100 GHz) from their line center. These features lead to interesting and favorable dynamic properties, including the absence of relaxation oscillations, ultrawide modulation bandwidths, and the lack of wavelength chirping upon direct modulation, all of which have been reviewed in this chapter. Furthermore we have described the experimental demonstration of gain-switched, actively mode-locked, and self-mode-locked QC lasers, which represent the first semiconductor-based ultrafast lasers in the mid-ir.

An important implication of these unique dynamic features is the prediction that directly modulated QC lasers can, in principle, provide superior high-speed performance compared to all existing optical data sources, both direct and externally modulated (the latter having the disadvantage of added optical losses, higher RF power consumption, and larger footprint). In addition, the combination of large optical nonlinearities and ultrafast response is very attractive for all-optical information processing, a key enabling ingredient of future ultra-broadband communication networks which is not adequately covered by existing device technologies. As a result, significant interest currently exists in the development of intersubband devices at near-ir wavelengths, for operation in fiber-based optical communications. Toward this goal, novel quantum well systems with sufficiently large conduction band offsets are being explored [45–50].

A second important direction of future work in this area is the further investigation of mode-locked emission from QC lasers. So far the temporal characteristics of their output waveforms—most notably their pulse widths—have only been estimated by using indirect methods. The development of high-performance ultrafast QC lasers would certainly benefit from the availability of a sensitive second-order autocorrelation measurement system at mid-ir wavelengths. The use of QWIPs based on two-photon absorption is particularly promising in this respect [57]. In addition, a complete theoretical description of the observed self-mode-locked operation has yet to be reported. Its development will be extremely valuable to clarify the physics of these devices and perhaps to provide guidelines for the design of QC lasers particularly well suited for ultrafast pulsed emission.

3.5 Acknowledgments

Several past and present collaborators have made significant contributions to the research described in this chapter, and their work is gratefully acknowledged. We would especially like to thank the following researchers: J. N. Baillargeon, C. G. Bethea, A. Y. Cho, S. N. G. Chu, L. Diehl, C. Gmachl, A. L. Hutchinson, H. Y. Hwang, F. X. Kartner, M. L. Peabody, A. M. Sergent, D. L. Sivco, and E. A. Whittaker.

References

1. K. Y. Lau and A. Yariv, *IEEE J. Quantum Electron.* **21**, 121 (1985).
2. R. S. Tucker, *J. Lightwave Technol.* **3**, 1180 (1985).
3. J. E. Bowers, *Solid State Electron.* **30**, 1 (1987).
4. M. Osinski and J. Buus, *IEEE J. Quantum Electron.* **23**, 9 (1987).
5. K. Y. Lau, *J. Lightwave Technol.* **7**, 400 (1989).
6. P. Vasil'ev, *Ultrafast Diode Lasers: Fundamentals and Applications* (Artech House, Norwood, MA, 1995).
7. L. A. Coldren and S. W. Corzine, *Diode Lasers and Photonic Integrated Circuits* (Wiley, New York, 1995), chap. 5.
8. P. Vasil'ev and A. B. Sergeev, *Electron. Lett.* **25**, 1049 (1989).
9. S. Sanders, L. Eng, J. Paslaski, and A. Yariv, *Appl. Phys. Lett.* **56**, 310 (1990).
10. Y. K. Chen, M. C. Wu, T. Tanbun-Ek, R. A. Logan, and M. A. Chin, *Appl. Phys. Lett.* **58**, 1253 (1991).
11. S. Arahira, Y. Matsui, and Y. Ogawa, *IEEE J. Quantum Electron.* **32**, 1211 (1996).
12. J. E. Bowers, B. R. Hemenway, A. H. Gnauck, and D. P. Wilt, *IEEE J. Quantum Electron.* **22**, 833 (1986).
13. C. Harder, K. J. Vahala, and A. Yariv, *Appl. Phys. Lett.* **42**, 328 (1983).
14. T. L. Koch and J. E. Bowers, *Electron. Lett.* **20**, 1038 (1984).
15. J. Faist, F. Capasso, D. L. Sivco, C. Sirtori, A. L. Hutchinson, and A. Y. Cho, *Science* **264**, 553 (1994).
16. C. Sirtori and R. Teissier, Chap. 1 of this book.

17. T. Elsaesser, Chap. 4 of this book.
18. N. Mustafa, L. Pesquera, C. Y. L. Cheung, and K. A. Shore, *IEEE Photon. Technol. Lett.* **11**, 527 (1999).
19. J. Faist, F. Capasso, C. Sirtori, D. L. Sivco, and A. Y. Cho, *Quantum Cascade Lasers in Intersubband Transition in Quantum Wells: Physics and Device Application II*, (Academic Press, London, 2000), chap. 8.
20. R. Paiella, R. Martini, F. Capasso, C. Gmachl, H. Y. Hwang, D. L. Sivco, J. N. Baillargeon, A. Y. Cho, E. A. Whittaker, and H. C. Liu, *Appl. Phys. Lett.* **79**, 2526 (2001).
21. F. Rana and R. J. Ram, *Phys. Rev.* **B65**, 125313 (2002).
22. J. Faist, F. Capasso, C. Sirtori, D. L. Sivco, A. L. Hutchinson, and A.Y. Cho, *Appl. Phys. Lett.* **67**, 3057 (1995).
23. M. Lerttamrab, S. L. Chuang, C. Gmachl, D. L. Sivco, F. Capasso, and A. Y. Cho, *J. Appl. Phys.* **94**, 5426 (2003).
24. J. Kim, M. Lerttamrab, S. L. Chuang, C. Gmachl, D. L. Sivco, F. Capasso, and A. Y. Cho, *IEEE J. Quantum Electron.* **40**, 1663 (2004).
25. R. Martini, C. Gmachl, J. Falciglia, F. G. Curti, C. G. Bethea, F. Capasso, E. A. Whittaker, R. Paiella, A. Tredicucci, A. L. Hutchinson, D. L. Sivco, and A. Y. Cho, *Electron. Lett.* **37**, 191 (2001).
26. S. Blaser, D. Hofstetter, M. Beck, and J. Faist, *Electron. Lett.* **37**, 778 (2001).
27. R. Martini, C. Bethea, F. Capasso, C. Gmachl, R. Paiella, E. A. Whittaker, H. Y. Hwang, D. L. Sivco, J. N. Baillargeon, and A. Y. Cho, *Electron. Lett.* **38**, 181 (2002).
28. V. E. Zuev, in *Laser Monitoring of the Atmosphere*, E. D. Hinklet ed. (Springer, Heidelberg, 1976).
29. M. M. Fejer, S. J. B. Yoo, R. L. Byer, A. Harwit, and J. S. Harris, Jr., *Phys. Rev. Lett.* **62**, 1041 (1989).
30. M. Segev, I. Grave, and A. Yariv, *Appl. Phys. Lett.* **61**, 2403 (1992).
31. F. Capasso, C. Sirtori, and A. Y. Cho, *IEEE J. Quantum Electron.* **30**, 1313 (1994).
32. E. Rosencher, A. Fiore, B. Vinter, V. Berger, Ph. Bois, and J. Nagle, *Science* **271**, 168 (1996).
33. C. Gmachl, O. Malis, and A. A. Belyanin, Chap. 5 of this book; A. A. Belyanin, M. Troccoli, and F. Capasso, Chap. 6 of this book.
34. R. Paiella, F. Capasso, C. Gmachl, D. L. Sivco, J. N. Baillargeon, A. L. Hutchinson, A. Y. Cho, and H. C. Liu, *Science* **290**, 1739 (2000).
35. R. Paiella, F. Capasso, C. Gmachl, H. Y. Hwang, D. L. Sivco, A. L. Hutchinson, A. Y. Cho, and H. C. Liu, *Appl. Phys. Lett.* **77**, 169 (2000).
36. A. Yariv, *Quantum Electronics*, 3d ed. (Wiley, New York, 1989).
37. C. Gmachl, H. Y. Hwang, R. Paiella, D. L. Sivco, J. N. Baillargeon, F. Capasso, and A. Y. Cho, *IEEE Photon. Technol. Lett.* **13**, 182 (2001).

38. H. C. Liu, J. Li, M. Buchanan, and Z. R. Wasilewski, *IEEE J. Quantum Electron.* **32**, 1024 (1996).

39. C. Gmachl, A. Tredicucci, F. Capasso, A. L. Hutchinson, D. L. Sivco, J. N. Baillargeon, and A. Y. Cho, *Appl. Phys. Lett.* **72**, 3130 (1998).

40. M. A. Newkirk and K. J. Vahala, *Appl. Phys. Lett.* **52**, 770 (1988).

41. C. H. Henry, *IEEE J. Quantum Electron.* **18**, 259 (1982).

42. B. W. Hakki and T. L. Paoli, *J. Appl. Phys.* **46**, 1299 (1975).

43. R. M. Williams, J. F. Kelly, J. S. Hartman, S. W. Sharpe, M. S. Taubman, J. L. Hall, F. Capasso, C. Gmachl, D. L. Sivco, J. N. Baillargeon, and A. Y. Cho, *Optics Lett.* **24**, 1844 (1999).

44. C. H. Cox, H. V. Roussell, R. J. Ram, and R. J. Helkey, *Technical Digest for International Topical Meeting on Microwave Photonics 1998,* 157 (1998).

45. A. Neogi, Chap. 10 of this book.

46. C. Gmachl, H. M. Ng, S.-N. G. Chu, and A. Y. Cho, *Appl. Phys. Lett.* **77**, 3722 (2000).

47. K. Kishino, A. Kikuchi, H. Kanazawa, and T. Tachibana, *Appl. Phys. Lett.* **81**, 1234 (2002).

48. D. Hofstetter, S.-S. Schad, H. Wu, W. J. Schaff, and L. F. Eastman, *Appl. Phys. Lett.* **83**, 572 (2003).

49. N. Iizuka, K. Kaneko, and N. Suzuki, *Electron. Lett.* **40**, 962 (2004).

50. I. Friel, K. Driscoll, E. Kulenica, M. Dutta, R. Paiella, and T. D. Moustakas, *J. Crystal Growth* **278**, 387 (2005).

51. R. Martini, R. Paiella, C. Gmachl, F. Capasso, E. A. Whittaker, H. C. Liu, H. Y. Hwang, D. L. Sivco, J. N. Baillargeon, and A. Y. Cho, *Electron. Lett.* **37**, 1290 (2001).

52. P. F. Szajowski, G. Nykolak, J. J. Auburn, H. M. Presby, G. E. Tourgee, E. J. Korevaar, J. Schuster, and I. I. Kim, *Proc. SPIE, Optical Wireless Communications*, E. J. Korevaar, ed., vol. 3552, 29 (1998).

53. R. Martini, C. Glazowski, E. A. Whittaker, W. W. Harper, Y. F. Su, J. F. Shultz, C. Gmachl, F. Capasso, D. L. Sivco, and A. Y. Cho, *Proc. SPIE, Quantum Sensing and Nanophotonic Devices*, M. Razeghi and G. J. Brown, eds., vol. 5359, 196 (2004).

54. R. Paiella, F. Capasso, C. Gmachl, C. G. Bethea, D. L. Sivco, J. N. Baillargeon, A. L. Hutchinson, and A. Y. Cho, *Appl. Phys. Lett.* **75**, 2536 (1999).

55. R. Paiella, F. Capasso, C. Gmachl, C. G. Bethea, D. L. Sivco, J. N. Baillargeon, A. L. Hutchinson, A. Y. Cho, and H. C. Liu, *IEEE Photon. Technol. Lett.* **12**, 780 (2000).

56. A. Soibel, F. Capasso, C. Gmachl, M. L. Peabody, A. M. Sergent, R. Paiella, H. Y. Hwang, D. L. Sivco, A. Y. Cho, H. C. Liu, C. Jirauschek, and F. X. Kartner, *IEEE J. Quantum Electron.* **40**, 844 (2004).

57. H. C. Liu and H. Schneider, Chap. 7 of this book.

58. G. P. Agrawal, *Nonlinear Fiber Optics* (Academic Press, San Diego, 1995), chap. 4.

59. C. H. Lin and T. K. Gustafson, *IEEE J. Quantum Electron.* **8**, 429 (1972).

60. J. M. Tang, P. S. Spencer, and K. A. Shore, *Appl. Phys. Lett.* **77**, 2449 (2000).

61. C. Gmachl, D. L. Sivco, R. Colombelli, F. Capasso, and A. Y. Cho, *Nature* **415**, 883 (2002).

62. A. Soibel, F. Capasso, C. Gmachl, D. L. Sivco, M. L. Peabody, A. M. Sergent, and A. Y. Cho, *Appl. Phys. Lett.* **83,** 24 (2003).

63. A. Soibel, F. Capasso, C. Gmachl, M. L. Peabody, A. M. Sergent, R. Paiella, D. L. Sivco, A. Y. Cho, and H. C. Liu, *IEEE J. Quantum Electron.* **40**, 197 (2004).

64. D. E. Spence, P. N. Kean, and W. Sibbett, *Optics. Lett.* **16**, 42 (1991).

65. H. Haus, *IEEE J. Quantum Electron.* **12**, 169 (1976).

66. N. Owschimikow, C. Gmachl, A. A. Belyanin, V. Kocharovsky, D. L. Sivco, R. Colombelli, F. Capasso, and A. Y. Cho, *Phys. Rev. Lett.* **90**, 043902 (2003).

Chapter

4

Ultrafast Dynamics of Intersubband Excitations in Quantum Wells and Quantum Cascade Structures

Thomas Elsaesser

Max-Born-Institut für Nichtlineare Optik und Kurzzeitspektroskopie
Max-Born-Str. 2 A
D-12489 Berlin, Germany

4.1 Introduction

Intersubband (IS) excitations in low-dimensional semiconductors have been a topic of research for more than 30 years [1–4]. Starting from early work on heterostructures [5], linear IS absorption spectra have been studied for a large variety of quasi-two-dimensional semiconductors made from III-V and II-VI material systems. Such experimental work has been complemented by detailed theoretical calculations, addressing electronic subband structure, transition energies, dipole strength, and selection rules as well as Coulomb mediated many-body effects among carriers, the latter mostly for electrons.

The dynamics of intersubband excitations and related carrier relaxation can be inferred only indirectly from experiments under steady-state conditions. Such dynamics occur in the ultrashort time domain and are relevant for understanding both the basic physical properties of IS excitations and the high-speed, high-frequency response of optoelectronic devices making use of IS transitions [6]. Femtosecond nonlinear spectroscopy in the mid-to far-infrared spectral range represents a powerful method for studying such phenomena [7].

It allows for a real-time observation of the fundamental nonequilibrium dynamics, in this way separating different types of processes. There are three different regimes of ultrafast IS response:

1. Resonant excitation of an IS transition by a coherent electric field transient generates a quantum-coherent superposition of electron wave functions in the optically coupled subbands, resulting in a coherent macroscopic IS polarization. As long as such polarization exists, it has a well-defined phase relationship with the external electric field, allowing for its nonlinear control by ultrashort pulses. The IS polarization is subject to phase-destroying scattering processes of carriers, leading to a dephasing on a (sub)picosecond time scale. In the linear IS absorption spectra, this dephasing gives rise to a homogeneous broadening of the spectral envelope.
2. Coherent optical excitation can also induce coherent electron motion in IS devices under bias. Electron wave packets consisting of a nonstationary superposition of subband wave functions are relevant for coherent tunneling in quantum cascade strucures and other devices based on vertical transport.
3. After the decay of quantum coherence by dephasing, the ultrafast IS dynamics are entirely determined by transient carrier distributions in the different subbands and their relaxation. Coulomb interactions within the carrier plasma as well as different types of carrier-phonon interactions lead to carrier thermalization, i.e., the formation of a quasi-equilibrium distribution, and subsequent cooling by phonon emission.

Femtosecond studies of IS excitations have provided detailed insight into the phenomena outlined above and, in particular, have allowed for dissecting the different nonequilibrium processes in the time-resolved response. In addition, first experiments on coherent nonlinear control of IS excitations and coherent carrier transport in quantum cascade structures have been reported. Such work is reviewed in this chapter with the main emphasis on ultrafast IS dynamics of single-component carrier plasmas, i.e., electrons *or* holes, in quantum wells and quantum cascade structures. The chapter is organized as follows. In Sec. 4.2, the basic properties of IS excitations in semiconductor quantum wells are introduced, followed by a description of experimental techniques applied in ultrafast spectroscopy (Sec. 4.3). Coherent optical IS excitations and phase relaxation are discussed in Sec. 4.4. Section 4.5 gives an overview of carrier relaxation following IS excitation, including inter- and intrasubband carrier relaxation and cooling. In Sec. 4.6, recent optical studies of electron transport in quantum cascade structures are presented, providing evidence for the quantum-coherent character of electron transport into the optically active region of such device structures. Conclusions are given in Sec. 4.7.

4.2 Optical Intersubband Excitations in Semiconductor Quantum Wells

Optical intersubband transitions of electrons have been investigated in a variety of quasi-two-dimensional semiconductors, i.e., heterostructures, *quantum wells* (QWs), and superlattices, mainly by measuring the linear IS absorption bands and, to a lesser extent, IS emission spectra. As most ultrafast time-resolved studies of IS excitations have concentrated on QWs, the following discussion is restricted to such systems [8–12]. For a more extended discussion of IS transitions, see Ref. 4.

Quantum confinement of electrons and holes results in the formation of conduction and valence subbands which are characterized by a quantization of the carrier **k** vector and energy in the direction perpendicular to the QW plane and quasi-free carrier motion in the QW plane with a continuous energy dispersion as a function of the in-plane **k** vector $\mathbf{k}_{\parallel} = (k_x, k_y)$ (Fig. 4.1). Doping the QW or modulation-doping the barriers introduces a quasi-stationary electron or hole plasma populating the lowest subband. IS absorption of light induces a transition between different subbands, i.e., changes the quantized $\mathbf{k}_z$ value, whereas the in-plane wave vector remains unaffected (apart from the very small photon momentum). Consequently, the IS dipole points into the direction perpendicular to the QW layer. In a single-particle picture, the dipole moment

$$M_{\mathrm{IS}} \propto e\int_{\mathrm{crystal}} \zeta(z)\eta\cdot \mathbf{z}\zeta'(z)\,dz\int_{\mathrm{cell}} u(\mathbf{r})u^*(\mathbf{r})\,d^3r \tag{4.1}$$

is taken between the envelope functions $\zeta(z)$ and $\zeta'(z)$ of the optically coupled subbands (in the first integral, e is elementary charge and η is polarization vector of light) weighted by the (second) integral over the cell-periodic Bloch functions [13]. The second integral yields unity when normalized Bloch functions are used. For an infinitely deep QW of thickness L, the dipole moment of the $n = 1 \rightarrow 2$ transition is given by $M_{\mathrm{IS}} = (16/9\pi^2)eL$. The large spatial extension of the envelope functions on a nanometer length scale L results in large IS dipole moments and absorption cross sections of up to 10^{-13} cm^2. The polarization selection rules for symmetric QWs allow IS transitions between subbands of different parity, that is, $n = 1 \rightarrow 2$, $n = 2 \rightarrow 3$, $n = 1 \rightarrow 4$, but not $n = 1 \rightarrow 3$.

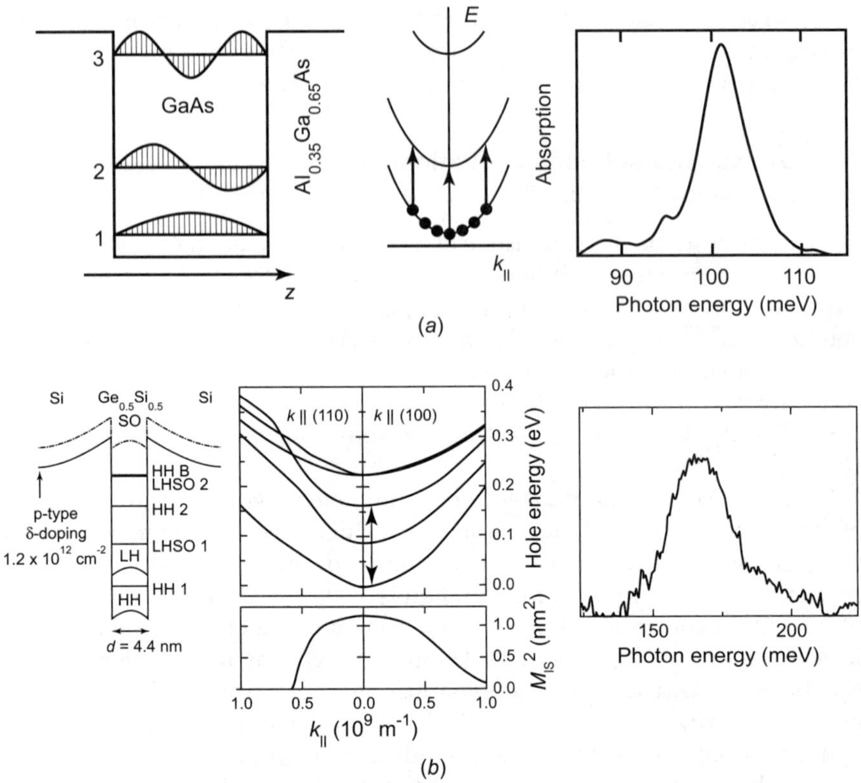

Figure 4.1 (*a*) Schematic of the conduction subband structure of a quantum well with quantized electron energies for motion in the direction perpendicular to the QW plane and a quasi-free continuum of states for in-plane motion. The panel on the right-hand side shows the IS absorption line of an *n*-type modulation-doped GaAs/$Al_{0.3}Ga_{0.7}$ As multiple-QW structure containing 51 QWs of 10-nm width (electron concentration 5×10^{10} cm^{-2}, sample temperature 10 K). (*b*) Schematic of the valence subband structure of a 4.4-nm-wide $Si_{0.5}Ge_{0.5}$/Si quantum well. The in-plane dispersion of the strongly mixed heavy-hole (HH), light-hole (LH), and split-off (SO) bands and the IS dipole moment M_{IS}^2 was calculated with the $\mathbf{k}\cdot\mathbf{p}$ method. The HH1 → HH2 absorption line measured with a sample containing 10 QWs is shown on the right-hand side (hole concentration 1.2×10^{12} cm^{-2}, sample temperature 15 K).

In general, the line shape of IS transitions is determined by the in-plane subband dispersion, many-body effects in the carrier plasma, and scattering processes resulting in a broadening of the transition. In a single-particle picture, IS transitions originating from different $k_{\parallel}$ states are uncoupled, and the overall IS absorption spectrum represents a superposition of transitions from all populated initial states. For a nearly parallel dispersion of conduction subbands (Fig. 4.1*a*), the transition energy is independent of $k_{\parallel}$, resulting in a delta-like joint density of states and a narrow IS absorption line centered at $E_{IS} = E_j - E_i$, where $E_{j,i}$ are the band minima of the optically coupled subbands. For nonparallel in-plane dispersions of the optically coupled bands, the transition energies cover a certain range resulting in an inhomogeneous broadening of the IS absorption. In a similar way, sample imperfections, e.g., fluctuations of the QW thickness, alloy fluctuations, and carrier localization in disordered QWs lead to a distribution of transition energies and inhomogeneous broadening. For valence subbands with their strong mixing of heavy and light hole states, the distribution of transition energies is highly complex and the IS dipole moment becomes k-dependent (Fig. 4.1*b*).

Many-body effects in the carrier plasma modify this single-particle picture substantially [1, 14–20]. Coulomb interactions result in a coupling of IS transitions from different $k_{\|}$ states, affecting the energy position, shape, and width of the absorption line. A coherent excitation of coupled IS transitions leads to collective charge oscillations in the z direction. The corresponding charge separation during such oscillations generates additional restoring forces, i.e., makes the plasma more rigid, and thus blue-shifts the absorption line [1]. This depolarization shift is accompanied by a substantial spectral narrowing of the IS absorption [17]. In addition to the depolarization shift, exchange interactions occur between an electron excited to a higher subband and the quasi-hole left behind in the original carrier distribution. This interaction partly compensates the depolarization blue shift. There are theoretical predictions of excitonic IS features due to the attractive Coulomb interaction between the excited electron and the quasi-hole which have, so far, not been identified in experiments [21, 22]. Many-body effects are particularly pronounced at high carrier concentrations in wide QWs. For narrow GaAs/AlGaAs QWs of sub-10-nm width and low carrier concentrations $N \leq 5 \times 10^{10}\text{cm}^{-2}$, they are less relevant, and one observes a behavior closer to that of the single-particle picture.

Scattering processes that destroy the quantum mechanical phase between the optically coupled subbands lead to a dephasing of the macrosopic coherent IS polarization and a corresponding broadening of the IS absorption line [7, 23]. Carrier-carrier and carrier-phonon scattering and scattering from the QW disorder potential are processes relevant for dephasing. In the absence of memory effects, i.e., in the Markov limit, dephasing results in a homogeneous broadening $\Delta\omega = 2/T_2$ characterized by the dephasing time T_2, the decay time of the macroscopic polarization. Dephasing is discussed in greater detail in Sec 4.4.

So far, IS transitions in individual QWs have been considered. In multiple-QW systems containing a high-density carrier plasma, radiative coupling effects of IS excitations in different QWs may occur [24, 25]. Under such conditions, the local electric field acting on a particular QW consists of a superposition of the external light field and the internal optical field generated by excitation of the other QWs. Such coupling leads to strong changes of both the linear and nonlinear IS response.

The IS absorption spectrum in Fig. 4.1*a* was recorded with an *n*-type modulation-doped GaAs/$Al_{0.3}Ga_{0.7}As$ multiple-QW sample of high structural quality and low electron concentration (51 QWs of $L = 10$ nm width, electron density $N = 5 \times 10^{10} cm^{-2}$) [26]. The sample was processed into a prism to achieve interaction of the IS dipole and the *z* component of the incident electric field. The absorption line displays a small width of approximately 4 meV, pointing to a negligible inhomogencous broadening. In Fig. 4.1*b*, the IS absorption band of holes in $Si_{0.5}Ge_{0.5}$/Si QWs is shown ($N = 1.2 \times 10^{12} cm^{-2}$). Calculations suggest a *k*-dependent dipole moment and a substantial inhomogencous broadening of this line [27].

Steady-state linear spectroscopy does not allow for a separation of the different broadening mechanisms and the underlying couplings between different elementary excitations, resulting in complex carrier dynamics. The basic nonequilibrium dynamics of carriers occur in the ultrafast time domain below 1 ps, and thus nonlinear infrared spectroscopy with femtosecond time resolution has developed into an important tool to unravel such phenomena. This has provided detailed insight into both coherent IS polarizations and carrier dynamics following IS excitation, as will be discussed in the following.

4.3 Experimental Techniques

Femtosecond studies of IS dynamics are based on nonlinear spectroscopy using ultrashort pulses in the mid- to far-infrared spectral range. In this section, the state of the art in infrared pulse generation is briefly reviewed, followed by an introduction to the spectroscopic methods for monitoring ultrafast coherent and incoherent carrier dynamics.

4.3.1 Femtosecond mid-infrared pulses

Femtosecond mid-infrared generation relies mainly on nonlinear frequency conversion of laser pulses at shorter wavelength in the near-infrared. Free-electron lasers represent another type of infrared source that is presently undergoing a rapid development [28] but has found limited application in IS spectroscopy so far [29].

The second-order nonlinearity $\chi^{(2)}$ of transparent bulk crystals is exploited to generate femtosecond mid-infrared pulses by parametric amplification, difference-frequency mixing, and/or optical rectification [30, 31]. Pulses at center frequencies ω_p, ω_s, and ω_i with $\omega_p = \omega_s + \omega_i$ and

$\omega_i < \omega_s < \omega_p$ (p: pump, s: signal, i: idler) interact with one another in the nonlinear medium [30] under nonresonant conditions; i.e., absorption in the nonlinear material is negligible at all three frequencies. In parametric generation and amplification, pulses at ω_s and ω_i are generated and amplified by transferring energy from an intense pump pulse at ω_p. In difference-frequency mixing, pulses at $\omega_i = \omega_p - \omega_s$ are derived from two input pulses at ω_p and ω_s. Optical rectification represents a special case of difference-frequency mixing where $\omega_p \simeq \omega_s$ and $\omega_i \simeq \Delta\omega$, the spectral bandwidth of the input pulses.

The phase relationship between the electric fields of the three interacting pulses is set by the phase-matching condition $\mathbf{k}_p = \mathbf{k}_s + \mathbf{k}_i$, where $|\mathbf{k}_{p,s,i}| = n_{p,s,i}(\omega_{p,s,i}/c)$ are the respective wave vectors. This phase-matching condition can be fulfilled by adjusting the refractive indices $n_{p,s,i}$ in birefringent nonlinear media (c: vacuum velocity of light) [30]. Changing the phase-matching angle through rotation or variation of the temperature of the nonlinear crystal allows for frequency tuning of the generated pulses. Even for phase-matched parametric frequency conversion, the mismatch of the group velocities of the three interacting pulses limits the effective interaction length in the nonlinear medium and determines the minimum pulse duration achieved [32].

For idler wavelengths $\lambda_i = 2\pi c / \omega_i$ of up to 5 μm, bulk and periodically poled $LiNbO_3$, $LiIO_3$, $KNbO_3$, beta-barium borate (BBO), and $KTiOPO_4$ (KTP) have been used as nonlinear materials. $AgGaS_2$ represents a standard material for λ_i up to 12 μm [33], and $AgGaSe_2$ and GaSe allow parametric mixing at even longer wavelengths up to $\lambda_i \simeq 20$μm. Data on nonlinear materials and their application for frequency conversion have been collected in Ref. 31. In most sources of femtosecond infrared pulses, the peak intensities of the input pulses are between 1 GW/cm^2 and 1 TW/cm^2, resulting in an energy conversion efficiency into the mid-infrared between several 10^{-5} and several percent.

There are (1) parametric generation schemes for femtosecond mid-infrared pulses at megahertz repetition rates and comparably low peak intensities and energies per pulse, and (2) sources for intense pulses working at much lower kilohertz repetition rates. For generating pulses of 30- to 300-fs duration at high repetition rate, optical rectification of the broadband output of mode-locked Ti:sapphire oscillators [34–38], optical parametric oscillators synchronously pumped by femtosecond pulse trains [39–41] and a combination of both techniques [42–44] have been applied. Such sources have found widespread applications in nonlinear spectroscopy of solids. Intense near-infrared pulses generated in regenerative Ti:sapphire and, in some cases, Cr:forsterite amplifiers at kilohertz repetition rates have been used for pumping a variety of parametric conversion schemes. For a recent overview, see Refs. 45 and 46. In the wavelength range from

2.5 to approximately 10 μm, such sources provide pulses of 50- to 150-fs duration and energies up to 50 μJ [47–49]. Very recently, tunable infrared pulses with an electric field amplitude of up to megavolts per centimeter have been generated between 10 and 20 μm by difference-frequency mixing of spectral components of an amplified 25-fs pulse in a thin GaSe crystal [50]. This technique allows for the generation of well-defined few-cycle pulses of very high peak intensity.

A complete characterization of femtosecond pulses requires a measurement of their time-dependent electric field in amplitude and absolute carrier phase. In contrast to the visible and near-infrared, such measurements are possible for mid-infrared electric field transients by electrooptic sampling, a technique that was originally developed for the far-infrared around 300 μm [51]. The infrared pulse induces a change of the refractive index in a birefringent electrooptic crystal, e.g., ZnTe, which is proportional to its momentary electric field and monitored through the polarization rotation of an ultrashort probe pulse (Fig. 4.2*a*). By changing the delay between infrared and probe pulse, the time-dependent electric field of the infrared pulse is determined. For a quantitative measurement, the probe pulse has to be short compared to the period of the infrared field, and the group velocity dispersion in the electrooptic crystal has to be limited. Using 10-fs pulses at 800 nm and ZnTe crystals typically of 10-μm thickness, infrared pulses down to a wavelength of approximately 7 μm have been characterized [51]. This technique has been extended to characterize intense mid-infrared pulses generated at a 1-kHz repetition rate [50]. In such recent experiments, 13-fs pulses from the 75-MHz repetition rate oscillator of an amplified Ti:sapphire laser system sample the electric field of the mid-infrared pulses at a 1-kHz repetition rate. An electronic gating technique is applied to measure the polarization rotation of the probe pulse overlapping in time with the mid-infrared pulse. The time-dependent electric field of an intense pulse generated in a 30-μm-thick GaSe crystal is plotted in Fig. 4.2*b*, together with its intensity spectrum centered at 15 μm (Fig. 4.2*c*). The electric field transient gives a width of the temporal intensity envelope [full width at half maximum (FWHM), pulse duration] of approximately 50 fs.

In addition to electrooptic sampling, other standard techniques for characterizing ultrashort pulses have been applied in the mid-infrared. Measurements of interferometric autocorrelations and *frequency-resolved optical gating* (FROG) have allowed for an analysis of the optical phase of the pulses relative to a reference pulse, information sufficient to determine the frequency chirp of the pulses [52–54].

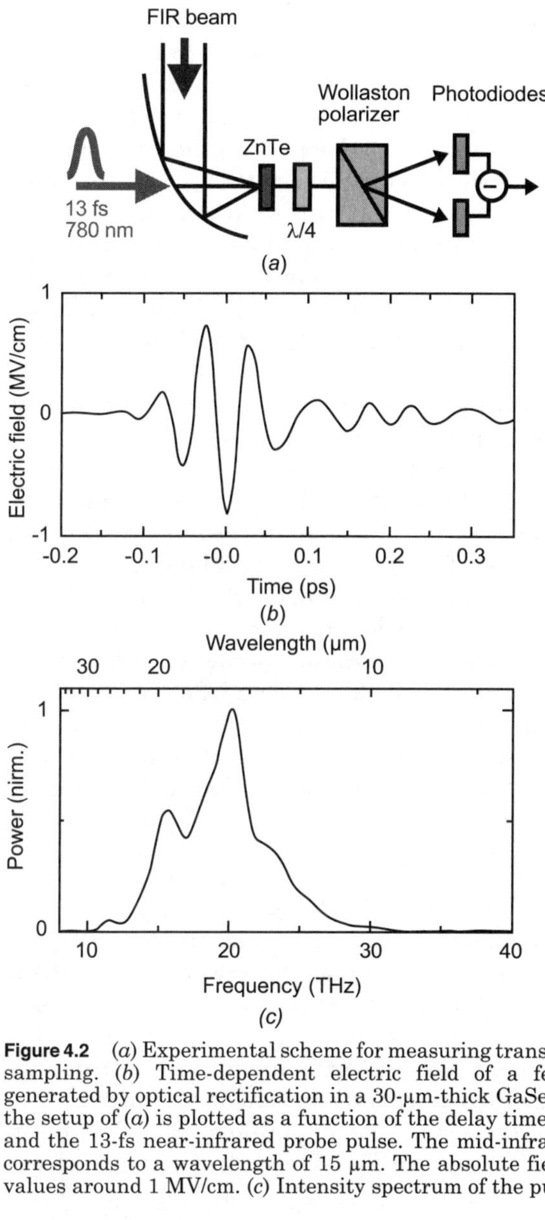

Figure 4.2 (*a*) Experimental scheme for measuring transient electric fields by electrooptic sampling. (*b*) Time-dependent electric field of a femtosecond mid-infrared pulse generated by optical rectification in a 30-μm-thick GaSe crystal. The signal measured in the setup of (*a*) is plotted as a function of the delay time between the mid-infrared pulse and the 13-fs near-infrared probe pulse. The mid-infrared center frequency of 20 THz corresponds to a wavelength of 15 μm. The absolute field amplitude reaches very high values around 1 MV/cm. (*c*) Intensity spectrum of the pulse in (*b*).

4.3.2 Spectroscopic techniques

In the following, techniques for monitoring coherent IS excitations and carrier relaxation are outlined briefly. A coherent IS polarization induced by femtosecond excitation gives rise to a coherent emission with a carrier frequency determined by the IS transition energy (Fig. 4.3*a*).

The temporal envelope of the emitted electric field is determined by both the driving pulse envelope and the dynamics of the IS polarization. Coherent IS emission has been studied with different phase-resolving detection schemes, e.g., based on electrooptic sampling [55, 56] or (linear) interferometry with a reference pulse [57]. In the first scheme, the emission is sampled by near-infrared pulses which are synchronized with the infrared pulse generating the transient IS polarization. To record interferograms, a synchronized reference pulse in the mid-infrared is brought to interference with the coherent emission from the sample, and the resulting signal is measured by a time-integrating detector as a function of delay between the excitation and the reference pulse.

Nonlinear spectroscopic techniques are required for a separation of the IS broadening mechanisms and the underlying microscopic interactions in the time domain. For this purpose, four-wave-mixing methods exploiting the third-order nonlinearity of the semiconductor are a powerful tool which has been developed to high sophistication [7]. The most elementary degenerate four-wave-mixing scheme is depicted schematically in Fig. 4.3*b*. Femtosecond pulses of identical photon energy propagating into the directions $\mathbf{k}_1$ and $\mathbf{k}_2$ generate a transient grating in the sample from which a small fraction of the respective pulse is (self-)diffracted into the directions $2\mathbf{k}_2 - \mathbf{k}_1$ and $2\mathbf{k}_1 - \mathbf{k}_2$. These coherent third-order signals depend on the time delay Δt_{12}, the so-called coherence time, between the first and the second pulses and thus reflect the time evolution of the coherent polarization in the sample. There are two main methods to analyze the diffracted signal:

1. In the simplest case, the diffracted intensity is recorded with a time-integrating detector as a function of the delay time Δt_{12} (TI-FWM). The signals diffracted into the directions $2\mathbf{k}_2 - \mathbf{k}_1$ and $2\mathbf{k}_1 - \mathbf{k}_2$ are symmetric in time with respect to zero delay between the incident pulses, the temporal separation of their maxima, the so-called peak shift, reflecting the degree of spectral inhomogeneity in the ensemble of optically driven transitions. Most experiments in the literature are based on this technique, which gives information on the buildup and decay, i.e., dephasing, of the macroscopic nonlinear polarization in the sample. For an optically thin sample, small diffraction efficiency, and perfect phasematching, the slowly varying amplitude approximation gives the following expression for the homodyne detected intensity I_{HOM} of the diffracted signal [58]:

$$I_{\mathrm{HOM}} = \frac{\pi\,\omega^2}{2n\,(\omega)\,c}\,l^2 \int_{-\infty}^{\infty} dt\,|P^{(3)}(t, \Delta t_{12})|^2 \tag{4.2}$$

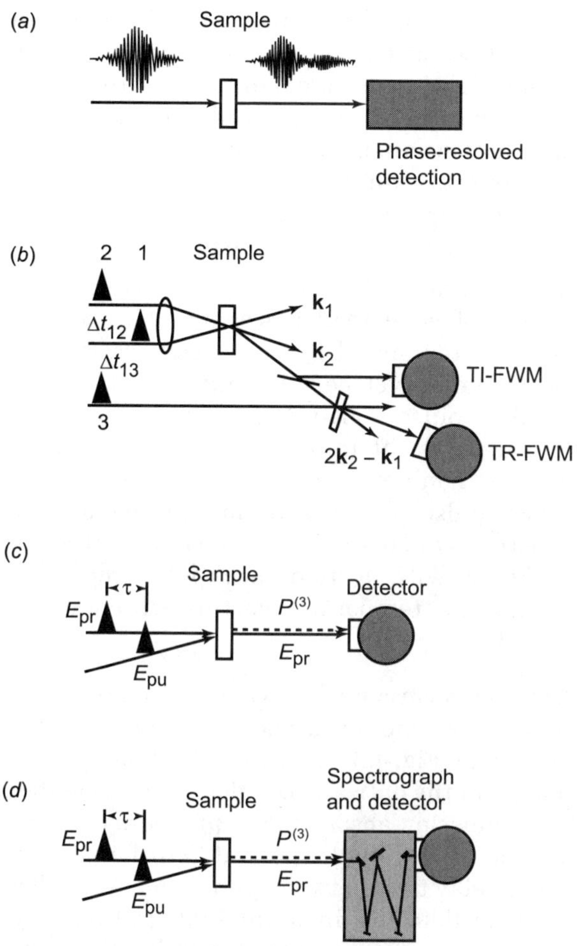

Figure 4.3 Experimental techniques of nonlinear spectroscopy. (*a*) Scheme of a propagation measurement. A coherent infrared pulse interacts with the sample, and the transmitted light is detected in a phase-resolved way, e.g., by electrooptic sampling, (*b*) Degenerate four-wave-mixing scheme. Pulses 1 and 2 of mutual delay Δt_{12} generate a transient grating in the sample from which a four-wave-mixing signal is (self-)diffracted into the direction $2\mathbf{k}_2 - \mathbf{k}_1$. This signal is either detected by a time-integrating detector (TI-FWM) or time-resolved by convolution with a third reference pulse in a nonlinear crystal (TR-FWM, time delay Δt_{13}). (*c*) Pump-probe configuration with spectrally integrated detection of the probe pulse. (*d*) Spectrally resolved pump-probe experiment.

for the transition frequency ω, the refractive index $n(\omega)$, the sample thickness l, and the third-order polarization $P^{(3)}(t, \Delta t_{12})$. In the simplest theoretical approximation, the nonlinear polarization has been analyzed with the optical Bloch equations for an ensemble of independent two-level systems [7]. This approach gives a single exponential decay $\exp\left(-\Delta t_{12}/t_{\text{decay}}\right)$ of I_{HOM}. The decay time t_{decay}

has values of $T_2/2$ and $T_2/4$ for a homogeneously and an inhomogeneously broadened set of two-level systems, respectively (T_2 is the dephasing time). For semiconductors, this frequently applied model does not represent an adequate description, as it neglects the Coulomb interactions among the carriers and thus does not account for many-body effects strongly influencing the nonlinear response of a semiconductor. Here, more sophisticated models, e.g., based on the semiconductor Bloch equations, have to be used [7, 59].

2. Information on the inherent time evolution of the diffracted signal can be obtained in time-resolved detection schemes [7, 26]. For instance, gating of the diffracted signal by sum-frequency mixing with a third pulse gives the time envelope of the diffracted intensity convoluted with the third pulse (TR-FWM, Fig. 4.3*b*). Such measurements have provided insight into many-body and/or local field effects influencing the nonlinear response, as discussed in detail elsewhere [7]. Three-pulse photon echo measurements and interferometric techniques of signal detection in which the diffracted electric field is heterodyned with a reference pulse represent other methods to analyze the temporal structure of nonlinear polarizations [60].

The incoherent redistribution of carriers following IS excitation has mainly been studied by pump-probe methods. Such methods are based on a two-pulse interaction scheme (Fig. 4.3*c*) in which a first pump pulse generates a population change in the optically coupled subband states. The resulting change of transmission/absorption is probed by a weak second pulse at the same or a different spectral position and is measured as a function of the delay time between pump and probe. The transmitted probe intensity is detected in a time-integrating way, either by detecting the full probe spectrum or by dispersing the probe light to measure transient spectra.

In the perturbative limit, pump-probe studies of nonlinear absorption are third-order in the electric field. A third-order polarization is generated in the sample by two interactions with the pump and one interaction with the probe field. This polarization is detected through the field that it radiates into the propagation direction of the probe, and the two electric fields interfere on the time-integrating detector. The measured signal reflects transient polarizations and population changes; i.e., both real and imaginary parts of the nonlinear susceptibility contribute. For a pulse envelope varying slowly in time compared to the period of the optical field, the spectrally integrated absorption change $\Delta\alpha(\omega_{pr})$ measured by a probe pulse of frequency ω_{pr} is given by [58]

$$\Delta\alpha(\omega_{pr}) = -\log\frac{T(\omega_{pr})}{T_0(\omega_{pr})}$$
$$= \frac{4\pi\omega_{pr}}{cn(\omega_{pr})}\mathrm{Im}\left[\int_{-\infty}^{\infty} dt\, E^*_{pr}(t)P^{(3)}(t,\tau) \Big/ \int_{-\infty}^{\infty} dt\, |E_{pr}(t)|^2\right] \tag{4.3}$$

Here, $T(\omega_{pr})$ and $T_0(\omega_{pr})$ represent the transmission of the sample with and without excitation, c is the velocity of light, $n(\omega_{pr})$ is the refractive index of the sample, and $E_{pr}(t)$ and $P^{(3)}(t, \tau)$ are the time-dependent electric field of the probe pulse and the third-order polarization, the latter containing the nonlinear response of the sample and depending on the pump-probe delay τ [58]. For a spectrally resolved detection of the probe pulse (Fig. 4.3*d*), the measured signal is given by

$$\Delta\alpha(\omega) = \frac{4\pi\omega_{pr}}{cn(\omega)}\frac{|E_{pr}(\omega)|^2\ \mathrm{Im}\left[P^{(3)}(\omega)/E^*_{pr}(\omega)\right]}{\int_0^{\infty} d\omega\ |E_{pr}(\omega)|^2} \tag{4.4}$$

Here $E_{pr}(\omega) = \int_{-\infty}^{\infty} dt \exp(i\omega t)\, E_{pr}(t)$ is the Fourier transform of the time-dependent probe field, and $P^{(3)}(\omega)$ is the nonlinear polarization component at the frequency ω.

Pump-probe schemes for investigating IS excitations are displayed in Fig. 4.4. In Fig. 4.4*a*, the pump pulse promotes carriers from the $n = 1$ to the $n = 2$ subband by resonant absorption on the IS transition. The resulting nonlinear change of IS absorption is probed by a second pulse, either at the same spectral position as the pump or tunable in the range of the IS absorption line [61]. The nonlinear absorption changes observed in such experiments give information on the transient populations of optically coupled states in the lower and the upper subbands. In the simplest case of parallel in-plane dispersions and negligible many-body effects, the absorption change is proportional to the difference of the total populations in the two subbands, and the IS relaxation time can be extracted. In general, however, intrasubband redistribution and many-body coupling of carriers contribute to the absorption change, requiring an in-depth analysis of the measured signal.

Changes of the carrier distribution by IS excitation also lead to changes of the near-infrared interband absorption (Fig. 4.4*b*). In particular, one expects an absorption decrease by state-filling on interband transitions whose final **k** states are populated after the IS excitation. Similarly, initially blocked interband transitions whose final states are depopulated by IS excitation exhibit an increase of absorption (Fig. 4.4*c*). This fact and the selection rules for interband

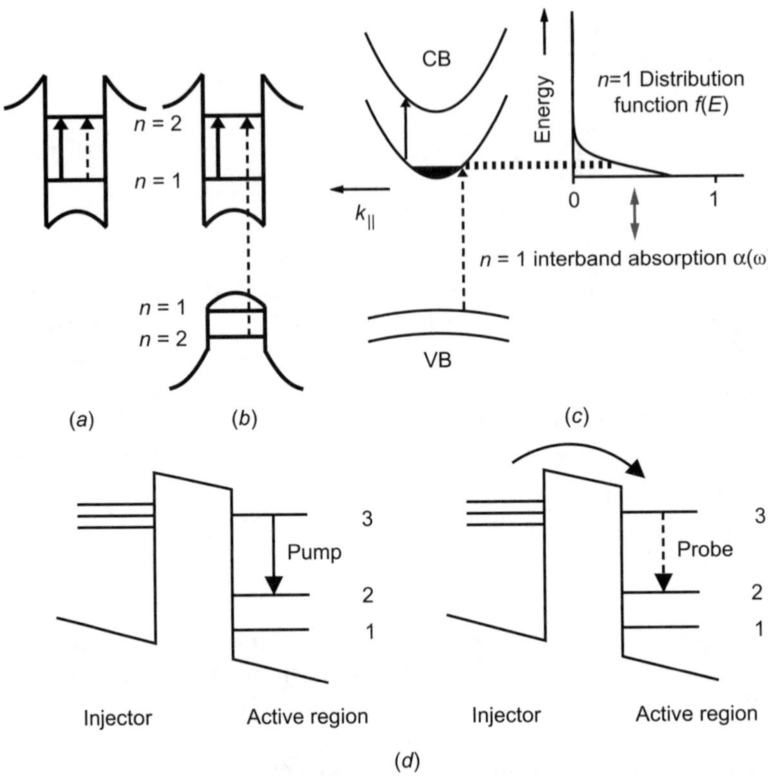

Figure 4.4 Pump-probe schemes for measuring carrier dynamics after IS excitation. (*a*) The probe pulse (dashed line) measures nonlinear changes of IS absorption induced by the pump pulse (solid line). (*b*) Probing of the transient ($n = 2$) interband absorption after resonant IS excitation. (*c*) Schematic of the in-plane dispersion for scheme (*b*) with a probe pulse on the $n = 1$ interband transition and the $n = 1$ electron distribution. (*d*) Scheme for studying carrier transport in quantum cascade devices. The pump pulse induces a depletion of the IS population inversion on the $3 \rightarrow 2$ transition. The motion of electrons through the injector barrier is monitored via probing transmission changes (gain recovery) on the same transition.

transitions enable one to selectively monitor the $n = 1$ and $n = 2$ carrier distributions with probe pulses in the near-infrared [62]. In addition to absorption changes caused by population effects, the measured signal might be influenced by many-body effects. In particular, transient absorption features related to the Coulomb enhancement of the interband absorption and the Fermi edge singularity of the plasma could play a role. A recent experimental study [63] demonstrates, however, that such signals are limited to a narrow spectral range of a few millielectronvolts and show a small amplitude for moderate carrier densities at low temperatures. For 100-fs probe pulses, the spectral bandwidth of the probe is much larger than this narrow energy interval, and thus population effects dominate the signals.

A pump-probe technique has also been applied to study electron transport in quantum cascade structures [64]. In such experiments, the structure was electrically biased with a quasi-stationary carrier distribution in the injectors and the active regions. Under conditions of population inversion between subbands 3 and 2 in each active region, a femtosecond pump pulse resonant to the IS transition $3 \rightarrow 2$ depletes the population of subband 3 (Fig. 4.4*d*), leading to a decrease of the quasi-stationary optical gain. A probe pulse on the same transition measures the recovery of the gain as a function of pump-probe delay, in this way monitoring electron transport from the injector into subband 3.

All pump-probe schemes of Fig. 4.4 involve a single carrier species, i.e., electrons or holes. Moreover, the total carrier concentration is constant in the schemes of Fig. 4.4*a* to *c*, and IS excitation results exclusively in a redistribution of carriers. This makes the analysis of the data much simpler than in the case of interband excitation, where both electrons and holes contribute to the optical signal.

Selective observation of electron IS dynamics is also possible with interband Raman excitation, as demonstrated in Refs. 65 and 66. In such experiments, population of the $n = 2$ conduction subband is created by a Raman process and probed via the anti-Stokes Raman signal generated with a probe pulse at a different photon energy. The strength of the anti-Stokes signal shifted by the intersubband transition energy relative to the center of the probe pulse is measured as a function of time delay to the pump and reflects the population dynamics of the upper subband.

4.4 Coherent Intersubband Dynamics

Ultrafast coherent spectroscopy has concentrated on measuring the dynamics of macroscopic IS polarizations to unravel the time scale and mechanisms of dephasing [23, 26]. More recently, the coherent infrared emission originating from IS polarizations in QWs has been analyzed in phase- and amplitude-resolved experiments in both the linear [55, 57] and nonlinear [55, 56] regimes. The latter work has led to the demonstration of IS Rabi flopping and, in a more general sense, coherent control of IS polarizations.

4.4.1 Intersubband dephasing

Four-wave-mixing techniques have been applied to study IS dephasing. Initially, experiments were performed on the $n = 1 \rightarrow 2$ IS transition of electrons in n-type modulation-doped GaInAs/AlInAs multiple QWs of 6-nm width [23]. In a degenerate four-wave-mixing experiment, two infrared pulses of 130-fs duration excited the IS transition resonantly, and the (self-)diffracted intensity was measured as a function of the coherence time (TI-FWM; see Fig. 4.3*b*). This signal rises within the time resolution of the experiment and displays a subsequent decay with a time constant depending on the electron concentration in the sample. For $N = 1.5 \times 10^{11}\ \mathrm{cm}^{-2}$, a decay time of 80 fs was measured, whereas a sub-50-fs decay occurs for $N = 1.5 \times 10^{12}\ \mathrm{cm}^{-2}$. A theoretical model based on the time-dependent Hartree-Fock equations accounting for many-body effects was used to analyze the experimental results. A dephasing time of $T_2 \simeq 300$ fs and $T_2 \leq 200$ fs was derived for an electron concentration of $N = 1.5 \times 10^{11}\ \mathrm{cm}^{-2}$ and $N = 1.5 \times 10^{12}\ \mathrm{cm}^{-2}$, respectively. Such dephasing times are substantially shorter than the lifetime of electrons in the $n = 2$ subband of approximately 1 ps. The homogeneous line widths corresponding to the measured dephasing times are smaller than the overall width of the IS absorption lines in the different samples, pointing to a pronounced inhomogeneous broadening of the IS transitions. Such inhomogeneity originates mainly from structural imperfections of the QWs.

In a second series of experiments, an *n*-type modulation-doped GaAs/$Al_{0.3}Ga_{0.7}As$ multiple-QW sample (51 QWs) of high structural quality and $L = 10$ nm width was studied at a low electron density of $N = 5 \times 10^{10}\ \mathrm{cm}^{-2}$ [26]. The IS absorption spectrum of this sample is shown in Fig. 4.1*a*. Degenerate four-wave-mixing experiments were performed with 130-fs pulses resonant to this absorption line. In addition to time-integrated detection of the signal diffracted into the direction $2\mathbf{k}_2 - \mathbf{k}_1$, its time structure was analyzed by mixing it with a 100-fs reference pulse in a nonlinear crystal and measuring the resulting sum-frequency signal as a function of the delay between the reference pulse and the signal (TR-FWM).

In Fig. 4.5*b*, the four-wave-mixing signal recorded with time-integrated detection is plotted as a function of the time delay Δt_{12} between the two pulses generating the transient grating. Data are presented for excitation ≤ 10 percent of the $n = 1$ electrons to the $n = 2$ subband (solid circles) and for excitation of 30 percent of the electrons (open circles). The signal rises within the time resolution of the experiment, shows a maximum at ≈ 100 fs, and decays afterward on a time scale of 1 ps, reflecting the decay of the macroscopic IS polarization. One derives a decay time $\tau = 160 \pm 15$ fs, independent of the excitation density.

The lifetime T_1 of electrons in the $n = 2$ subband was determined independently by pump-probe measurements in which the nonlinear change of IS absorption induced by the pump pulses was monitored by weak delayed probe pulses. In Fig. 4.5*a*, the spectrally integrated absorption change ΔA is plotted versus the delay time between pump

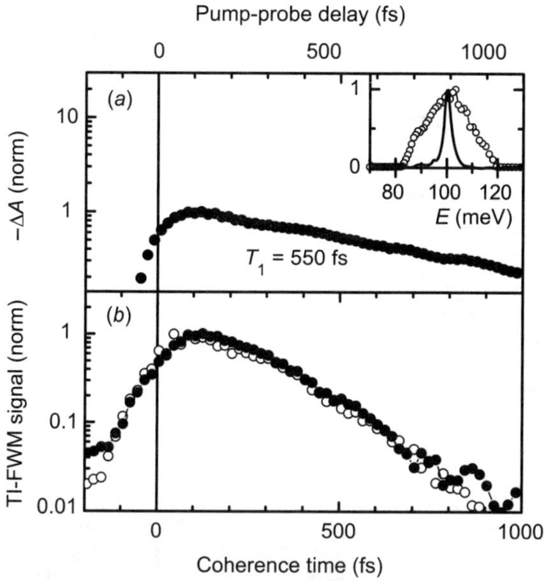

Figure 4.5 (a) Bleaching of the IS absorption in an n-type modulation-doped GaAs/AlGaAs multiple-QW sample measured with resonant excitation (sample temperature of 15 K). Inset: pulse spectrum (open circles) and IS absorption line (solid line). The spectrally integrated absorption change $-\Delta A = \ln(T/T_0)$ is plotted versus the delay time between pump and probe (T and T_0 are sample transmission with and without excitation respectively). The absorption change displays a single exponential decay with a time constant of $T_1 = 550$ fs. (b) Time-integrated four-wave-mixing signal diffracted into the direction $2\mathbf{k}_2 - \mathbf{k}_1$ for excitation of $\leq 10\%$ (solid circles) and 30% (open circles) of the $n = 1$ electrons as a function of the time delay Δt_{12} between the incident pulses (logarithmic ordinate scale). The signal decays with a time constant of 160 fs.

and probe. The strong IS bleaching decays with a time constant $T_1 = 550$ fs, i.e., markedly slower than the polarization decay in Fig. 4.5b, by relaxation of the $n = 2$ electrons back to the $n = 1$ subband. Thus, population relaxation makes a minor contribution to the dephasing of the IS polarization.

Four-wave-mixing experiments with time-integrated detection give no direct information on the predominant type of spectral broadening of the IS transition, i.e., homogeneous versus inhomogeneous broadening. In the homogeneous case, one expects a free induction decay whereas a photon echolike signal occurs for an inhomogeneously broadened transition [7]. To distinguish between these two cases, we performed an experiment with time-resolved detection of the four-wave-mixing signal [26].

In Fig. 4.6a, the time-resolved four-wave-mixing signals are shown for different fixed values of the delay Δt_{12} between the pulses generating the transient grating. With increasing Δt_{12}, the maximum of the signal shifts to later times. A plot of the position of the respective maximum

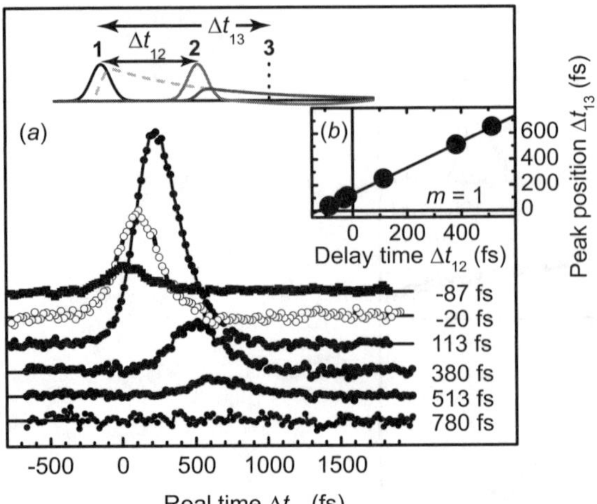

Figure 4.6 (*a*) Time-resolved four-wave-mixing signals for different fixed time delays Δt_{12} of the two pulses generating the transient grating in the sample. The intensity of the up-converted signal is plotted versus the real-time delay Δt_{13} between pulse 1 and the gating pulse 3. (*b*) Peak positions of the transient signals for different Δt_{12} (symbols) compared to a linear dependence $\Delta t_{13} = \Delta t_{12}$ + 130 fs with a slope of 1 (solid line). Inset: schematic of the pulse sequence, the polarization induced by pulse 1 (dashed line), and the FWM signal (thick solid line).

as a function of Δt_{12} displays a linear dependence $\Delta t_{13}^{\max} = \Delta t_{12}$ + 130 fs with a slope of 1 (Fig. 4.6*b*). This demonstrates that the four-wave-mixing signal reaches a maximum at a fixed time (of 130 fs) after the second pulse for all values of t_{12}. Such behavior is indicative of a free-induction decay of the IS polarization, i.e., homogeneous broadening of the IS transition, for which the signal is emitted immediately after the second pulse [7]. In contrast, emission of a photon echo from an inhomogeneously broadened transition requires a rephasing of the polarization components at different frequencies within the line width and leads to a signal emission at Δt_{12} after the second pulse. In the plot of Fig. 4.6*b*, this would lead to a slope of 2, which can be clearly ruled out on the basis of the experiment.

The time-integrated four-wave-mixing signal originating from a homogeneously broadened system decays with a time constant of $T_2/2$ [7]. The data in Fig. 4.5*b* thus give an IS dephasing time $T_2 = 2\tau = 320$ fs corresponding to a homogeneous line width of 4 meV. This number is in excellent agreement with the measured line width of 3.7 meV (see Fig. 4.1*a*) and confirms the predominant homogeneous broadening of the IS transition in this high-quality sample.

Intrasubband electron-electron scattering represents the main mechanism by which the quantum mechanical phase between the $n = 1$ and $n = 2$ wave functions is destroyed. For the small electron concentration in the sample studied, the scattering rate and thus T_2 are independent of the excitation density as the phase space available for scattering remains essentially unchanged. For a similar sample with an electron concentration of $N = 6 \times 10^{11}$ cm^{-2}, a pronounced increase of the dephasing rate with excitation density has been found, which is related to the unblocking of phase space upon excitation of carriers to the $n = 2$ subband. Such behavior is discussed in greater detail in Ref. 26.

4.4.2 Coherent dynamics of intersubband polarizations

Coherent IS polarizations give rise to coherent light emission on the IS transition. Measuring such electric field transients with an amplitude- and phase-resolved detection scheme, e.g., by electrooptic sampling, provides direct access to the polarization dynamics and the underlying couplings in the carrier plasma.

Coherent IS emission was first studied in the linear regime where the generated IS polarization depends linearly on the field of an ultrashort driving pulse. In Ref. 57, femtosecond excitation from the valence to the $n = 1$ and $n = 2$ conduction subbands of undoped asymmetric QWs has been applied to generate a coherent IS polarization. The IS emission was detected interferometrically, as discussed in Sec. 4.3.2. The emission displays a carrier frequency determined by the IS transition energy and an envelope decaying on a time scale of several hundreds of femtoseconds. The origin of this decay has not been clarified in such linear measurements, and both dephasing and inhomogeneous broadening may contribute.

In the following, studies with resonant IS excitation of electrons by femtosecond mid-infrared pulses are presented, covering both the linear and the nonlinear regimes of material response. Coherent IS polarizations were generated with the femtosecond infrared source described in Sec. 4.3.1 [50], and the transmitted light was detected by electrooptic sampling, allowing for a measurement of absolute optical phases rather than just relative phases with respect to a reference field. The transmitted light consists of the transmitted part of the excitation pulse plus the coherent emission from the IS polarization. To isolate the IS emission, the transmitted pulse was subtracted from the total signal. In Figs. 4.7 and 4.8, results for a GaAs/AlGaAs multiple-QW sample with an electron concentration of $N = 5 \times 10^{11}$ cm^{-2} (IS absorption spectrum in Fig. 4.1a) are shown [56]. For amplitudes of the driving field ≤ 5 kV/cm (Fig. 4.7), the sample response is in the linear regime, and the emission reflects the gradual buildup of the macroscopic IS polarization followed by the free induction decay with the dephasing

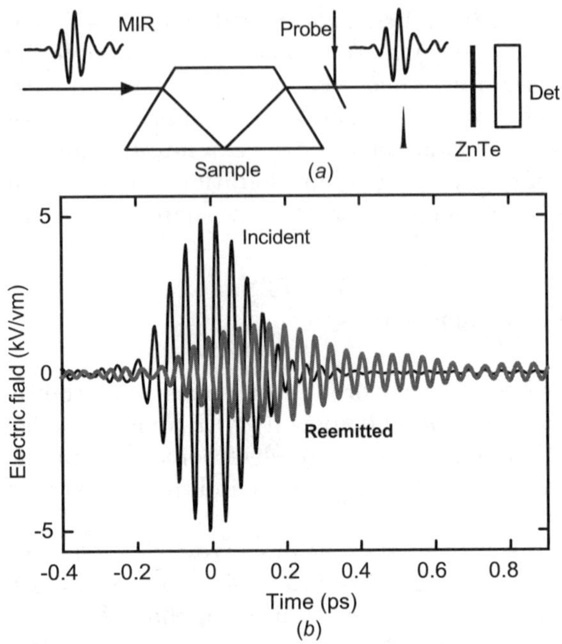

Figure 4.7 (*a*) Experimental geometry for studying coherent IS emission. A mid-infrared pulse excites the sample of prism shape, and the transmitted light is analyzed by electrooptic sampling in a ZnTe crystal. (*b*) Incident (thin line) and reemitted (thick line) electric field as a function of time. The sample response is in the linear regime. The reemitted field displays a free induction decay with a 180 degree phase shift with respect to the driving field.

time of 320 fs. In the resonant excitation scheme applied here, the IS polarization is 90 degrees out of phase with the driving field. The field reemitted from the layered sample is proportional to $\partial P/\partial t$ [67, 68], adding another phase shift of 90 degrees. As a result, the emitted field is 180 degrees out of phase with the driving field, corresponding to an overall absorptive response of the sample.

The characteristics of the emitted field change drastically on increasing the driving field. In Fig. 4.8*a* and *b*, the envelope of the emitted field displays a strong modulation and zeros in amplitude during the driving pulse. The number of emission periods during the pulse increases from two in Fig. 4.8*a* to three in Fig. 4.8*b* for an increase of the maximum amplitude of the driving field by a factor of 2. The maximum emission amplitude shows a sublinear increase, i.e., pronounced saturation, with the amplitude of the driving field. In addition, there is a change of the phase of the emitted field with respect to the driving field in successive emission periods. In Fig. 4.8*a*, emission in the first period is out of phase with the driving fields, similar to the linear case, whereas emitted and driving fields are in phase in the

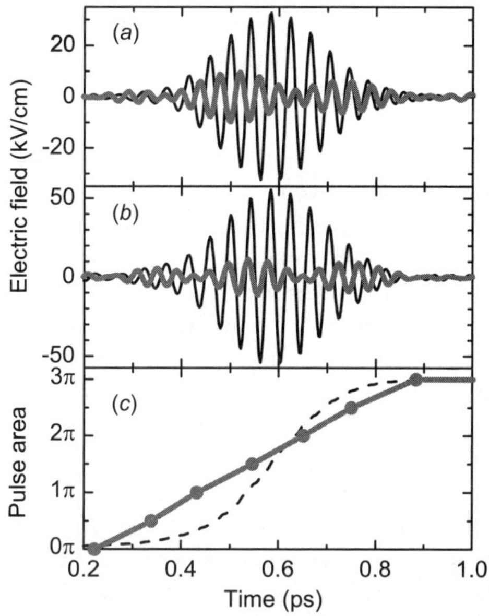

Figure 4.8 Coherent IS emission in the nonlinear regime. (*a*, *b*) Driving field and reemitted field for different amplitudes of the excitation pulse. The reemitted field displays a time evolution characteristic of Rabi flopping with several emission periods and phase jumps of 180 degrees between such periods, (*c*) Pulse area derived from the time evolution of the excitation pulse (dashed line) and from the reemitted field in (*b*) (symbols, solid line).

second period. In Fig. 4.8*b*, one observes a sequence of out-of-phase, in-phase, and out-of-phase emission.

The results summarized in Fig. 4.8 are a clear manifestation of IS Rabi oscillations induced by the strong driving field resonant to the IS transition. Both the amplitude of the macroscopic polarization and the population inversion between the optically coupled subbands undergo oscillations at the Rabi frequency $\Omega = M_{IS}E/\hbar$ where M_{IS} is the IS dipole moment (Eq. 4.1) and E is the driving field. The optical Bloch equations for noninteracting two-level systems represent the most elementary description of this phenomenon [7, 69]. In this density matrix approach, the macroscopic polarization $P(t)$, which is connected to the off-diagonal element $\rho_{12}(t)$ of the density matrix, is described by a differential equation containing a nonlinear term proportional to the population inversion between the two optically coupled subbands. This term leads to a saturation of the polarization amplitude and determines its phase. In Fig. 4.8*a*, there is a first period in which a complete population inversion builds up $\left[\partial\rho_{22}(t)/\partial t > 0\right]$, followed by a second period in which this inversion decreases by stimulated emission $\left[\partial\rho_{22}(t)/\partial t < 0\right]$. Correspondingly, the emitted field $E_{em}(t) \propto \partial P(t)/\partial t$ and the driving field are out of phase in the first period and in phase in the second period. The amplitude of the emitted field is zero for maximum inversion, i.e., at $t = 0.65$ ps in Fig. 4.8*a*. Increasing the driving field even further (Fig. 4.8*b*) results in an increase of the Rabi frequency and an additional emission period.

The basic features of the results of Fig. 4.8 are reproduced by the optical Bloch equations. Closer inspection, however, reveals discrepancies between this elementary picture and the experiment. In Fig. 4.8*c*, the pulse area $\Theta = \pi/2 \int |(\Omega t)|\, dt$ is plotted as a function of time. The dashed line gives Θ as derived from the time evolution of the driving field. On the other hand, the pulse area can be derived from the measured progress of the Rabi oscillations. For a pulse area of $\pi/2$, the amplitude of the coherent emission reaches a maximum, followed by a zero at π where the population inversion is maximum. At $3\pi/2$, the (stimulated) emission is maximum and then disappears at 2π where all electrons are back in the original subband (zero inversion). Based on such predictions of the optical Bloch equations, the symbols in Fig. 4.8*c* were derived from the emission transient in Fig. 4.8*b*. There is a substantial discrepancy between the pulse areas from this analysis and from the time evolution of the driving field. In other words, the measured Rabi frequency is essentially constant throughout the driving pulse, although the amplitude of the driving field changes strongly. This behavior clearly points to a failure of the simple Bloch picture.

There are essentially two different mechanisms that can modify the Bloch picture. First, local field effects may occur; i.e., the local field acting on a particular QW may consist of the external driving field plus the field generated by the polarizations in the other QWs of the sample. For the low electron concentration of the present sample, such effects are, however, of minor importance. Second, Coulomb mediated many-body effects enhance the IS dipole moment by introducing a coupling between IS transitions originating from different $k_{\|}$ states. For small driving fields, this coupling prevails, and the collective response leads to a stronger transition dipole and higher Rabi frequencies during the leading and trailing wings of the driving pulse. At, high driving fields around the pulse maximum, the many-body interaction is weaker than the coupling to the external field, and thus the behavior is closer to that of the single-particle picture with a smaller transition dipole and lower Rabi frequency. The interplay of many-body couplings and the interaction with the driving field make the instantaneous Rabi frequency less sensitive to the external field, in qualitative agreement with the behavior found in Fig. 4.8*b*.

Radiative coupling effects are present in samples with higher electron concentrations. In the linear regime, radiative coupling in multiple-QW structures leads to a strong modification of the IS absorption strength of the sample. More specifically, the total absorption cross section is no longer proportional to the electron concentration. For a particular QW, the field reemitted by the other QWs compensates the external field, resulting in a smaller local field and, as a consequence, a strongly reduced IS absorption. In the nonlinear regime, the time structure of the field transmitted through the sample and the overall nonlinear transmission are strongly affected by radiative coupling, as is evident from a very recent study [70].

In conclusion, the results discussed here demonstrate how the nonlinear IS response can be controlled by external optical fields, allowing for a change of IS polarizations and populations over a wide range. Such schemes are highly attractive for optical switching and display all the features required for implementing quantum bits. Femtosecond excitation allows for tailoring the phase-coherent response and, in particular, for subpicosecond switching times.

4.5 Intersubband Relaxation Processes

IS relaxation processes of carriers have been studied under a variety of conditions including both interband excitation of electron-hole plasmas [71, 72] and direct IS excitation of a single-component electron or hole plasma in doped QW systems and heterostructures. In the latter case, the total carrier concentration is constant; i.e., IS excitation and the subsequent relaxation are equivalent to redistributing a single type of carrier on a femtosecond- to picosecond time scale. This fact facilitates the analysis of the relaxation scenario substantially and enables one to extract much more specific information on microscopic processes than in the case of photogenerated electron-hole plasmas.

Resonant IS excitation generates an initial nonequilibrium distribution of electrons or holes populating states in the optically coupled subbands. As time evolves, different relaxation processes transform such nonequilibrium populations into a quasi-equilibrium distribution, eventually reestablishing the initial situation. The relevant relaxation processes are

1. IS scattering of carriers from higher subbands back to the initially populated subband
2. Carrier thermalization, i.e., the formation of quasi-equilibrium carrier distributions
3. Carrier cooling, i.e., the transfer of excess energy from the carrier plasma to the lattice by phonon emission

The time scales of such processes depend on the energy spacing of the subbands, the carrier density, the excitation density, and microscopic coupling parameters. IS scattering, thermalization, and cooling are not strictly sequential but occur in overlapping time windows determined by the underlying carrier-carrier and carrier-phonon scatterings.

In Fig. 4.9, the relaxation of electrons after resonant IS excitation at an energy $E_{IS} > \hbar\omega_{LO}$ ($\hbar\omega_{LO}$ is the optical phonon energy) is illustrated schematically. The femtosecond excitation pulse promotes electrons from the $n = 1$ to the $n = 2$ subband (Fig. 4.9*a*). This creates an excess population of $n = 2$ electrons *and* a nonequilibrium distribution of $n = 1$ electrons. Due to the nearly parallel in-plane dispersion of the two subbands, the excitation depletes $n = 1$ states in a broad range of energies from the bottom of the $n = 1$ subband up to the quasi-Fermi level of the initial distribution. The $n = 2$ carriers undergo IS scattering by interaction with optical phonons into high-lying states of the $n = 1$ subband (Fig. 4.9*b*). Such energetic electrons scatter with the cold sea of unexcited $n = 1$ electrons and with phonons, in this way relaxing toward the bottom of the subband. This relaxation and simultaneous scattering of the initially unexcited electrons result in a thermalization of the overall electron distribution toward a hot quasi-equilibrium distribution (Fig. 4.9*c*) which subsequently cools down to lattice temperature.

In general, the observation of such a relaxation scenario requires a measurement of transient electron distributions in the femtosecond time domain. IS scattering, i.e., the depopulation of the $n = 2$ subband,

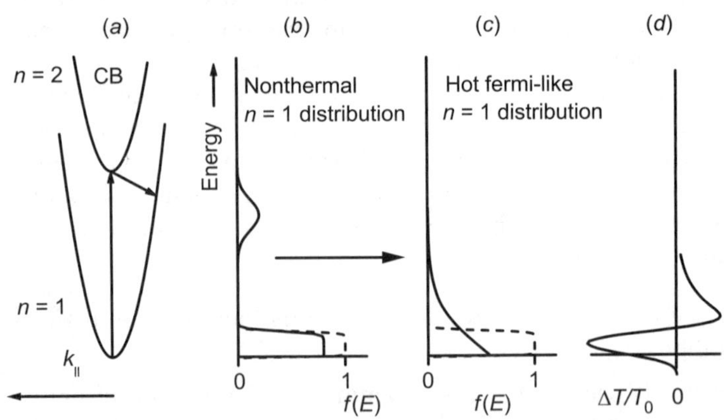

Figure 4.9 Schematic of relaxation processes occurring after IS excitation of electrons. (*a*) In-plane dispersion of the two optically coupled conduction subbands (CB) indicating the optical transition (vertical arrow) and the backscattering by phonons (diagonal arrow). (*b*) Schematic electron distribution $f(E)$ in the $n = 1$ subband after IS scattering. The distribution consists of the backscattered electrons at high energy and the cold sea of unexcited electrons at the bottom of the subband. Dashed line: initial Fermi distribution of electrons. (*c*) Hot electron distribution after thermalization of the $n = 1$ electrons displaying a pronounced high-energy tail. (*d*) Schematic change of transmission on the $n = 1$ interband transition due to the changes of the electron distribution in (*c*).

has been studied in pump-probe experiments monitoring the nonlinear changes of IS absorption (see Fig. 4.4*a*). In this scheme, the probe pulse averages over a broad range of $k_{\|}$ states and IS scattering and intrasubband thermalization in the $n = 1$ subband are difficult to separate. For nearly parallel subband dispersions, however, all backscattered electrons contribute to IS absorption independent of their in-plane wave vector, making the measured absorption change quite insensitive to intrasubband redistribution. In this limit, the absorption change is proportional to the difference of the total $n = 1$ and $n = 2$ electron populations. Based on this assumption, such measurements have been interpreted in terms of a dominant contribution from IS relaxation with IS scattering times between 100 fs and several picoseconds for QWs made from GaAs/AlGaAs (see Fig. 4.5*a*), GaInAs/InP, GaN/AlGaN, and other III-V and II-VI material systems.

Much more detailed insight into carrier relaxation is possible with the pump-probe scheme of Fig. 4.4*b* and *c* in which changes of the electron distribution are monitored via changes of the interband absorption [62]. Transient population and depopulation of $k_{\|}$ states by electrons result in a decrease and an increase of interband absorption into those states, respectively (Fig. 4.9*d*). In Figs. 4.10 and 4.11, results of such measurements are shown for n-type modulation-doped GaInAs/AlInAs multiple QWs ($L = 8$ nm, 50 QWs) with an electron concentration of $N = 5 \times 10^{11} \mathrm{cm}^{-2}$. In Fig. 4.10, the time-dependent nonlinear transmission change is shown for probe photon energies at the onset of the $n = 2$ interband absorption between 1.130 and 1.150 eV. One finds a transient increase of transmission, i.e., a bleaching, which rises within the time resolution of the experiment and decays with a time constant $\tau_{IS} = 1$ ps. The bleaching is caused by the transient electron population of the $n = 2$ conduction subband, leading to a blocking of the corresponding interband transitions. The decay of the signal is due to the depopulation of the $n = 2$ subband by IS scattering back to the $n = 1$ subband.

The transient behavior of the $n = 1$ electrons is evident from absorption spectra recorded in the range of the $n = 1$ absorption edge around 900 meV. Here, one probes states at the bottom of the $n = 1$ subband which are initially populated by the electron plasma. In Fig. 4.11, spectra are displayed for various delay times after the IS excitation (solid circles). At 0.2 ps (Fig. 4.11*a*), a decrease of transmission in the range of the initially populated electron states is found which is due to the depletion by the excitation pulse. At later times (Fig. 4.11*b* to *e*), the amplitude of this signal rises, and in addition a delayed decrease of absorption occurs at higher photon energies. For delay times between 0.4 and 1.1 ps, the maximum amplitude of this bleaching signal is substantially smaller than that of the absorption

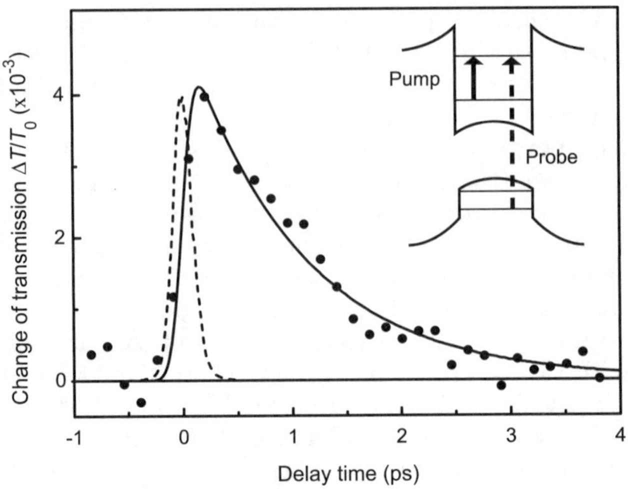

Figure 4.10 IS relaxation of electrons in n-type modulation-doped GaInAs/AlInAs quantum wells (sample temperature of 8 K). The change of transmission $\Delta T/T_0 = (T - T_0)/T_0$ on the $n = 2$ interband transition is plotted as a function of pump-probe delay (T and T_0 are sample transmission with and without excitation, respectively, on the IS transition at 200 meV). The solid line represents the result of an EMC simulation of electron dynamics, giving an IS relaxation time of 1 ps. Dashed line: cross-correlation of pump and probe pulses.

increase occurring below the initial Fermi level, whereas similar amplitudes are approached after 2 ps.

These changes of interband absorption occur on a time scale which is substantially slower than the dephasing of IS polarizations discussed in Sec. 4.4.1. Furthermore, many-body effects have a minor influence on the transient spectra because of the moderate electron concentration. The spectra are determined by the difference of the transient and the initial equilibrium electron distribution, and thus their time evolution reflects IS scattering and electron redistribution. Under such conditions, *Ensemble Monte Carlo* (EMC) techniques to solve the Boltzmann equation for the electron distribution functions are an appropriate tool to analyze the relaxation scenario. The experiments were simulated by an EMC approach described in Refs. 73 and 74. IS and intrasubband electron-electron and electron-phonon scattering, the Pauli exclusion principle, and nonequilibrium phonons were taken into account. IS excitation was simulated by promoting about 15% of the electrons from the $n = 1$ to the $n = 2$ subband with a 130-fs pulse that was resonant to the IS transition between the $n = 1$ and $n = 2$ conduction subbands of parallel in-plane dispersion. For comparison with the measured absorption changes, the difference between the transient and the initial electron distribution in each subband was multiplied by the steplike absorption coefficient of the respective valence to conduction

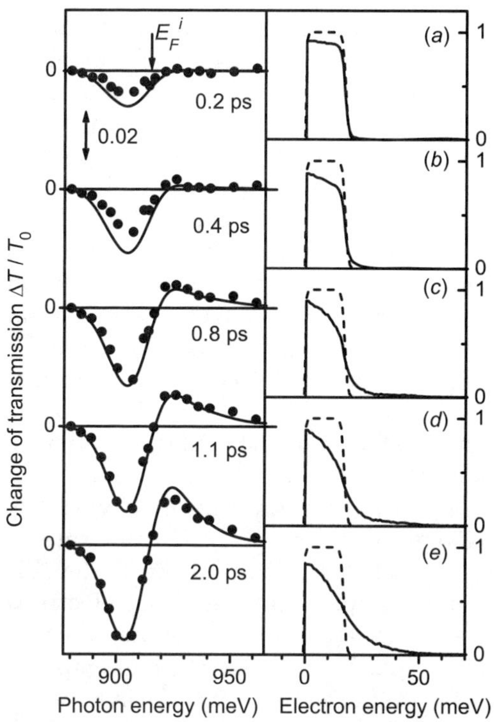

Figure 4.11 Transient spectra on the $n = 1$ interband transition measured after IS excitation of electrons in GaInAs/AlInAs quantum wells (E_F^i is initial Fermi level of the electrons). Left column: The change of transmission is plotted versus the center photon energy of the probe pulses of 13-meV bandwidth. Data are shown for delay times of (*a*) 0.2 ps, (*b*) 0.4 ps, (*c*) 0.8 ps, (*d*) 1.1 ps, and (*e*) 2.0 ps. Solid lines: calculated spectra from EMC simulations. Right column: calculated electron distributions showing the formation of a quasi-equilibrium distribution within 2 ps (solid lines). Dashed lines: initial electron distribution.

band transition and convoluted with the temporal and spectral envelope of the 100-fs probe pulses (spectral bandwidth 13 meV).

IS relaxation of the $n = 2$ electrons occurs with a characteristic time constant of 1 ps, as is evident from the decay of the signal in Fig. 4.10. This decay is very well reproduced by the EMC simulation (solid line), indicating that emission of confined and interface optical phonons via the polar interaction represents the main scattering mechanism. IS scattering requires emission of optical phonons with large **q** vectors. Due to the q dependence of the coupling matrix element, IS scattering rates are substantially smaller than the intraband scattering rates connected with optical phonon emission at much smaller **q** vectors.

IS scattering transfers nonequilibrium electrons into high-lying $n = 1$ states. No bleaching due to accumulation of backscattered carriers in those high-lying states was detected, demonstrating that spreading of the backscattered electrons over a broad energy range occurs much faster than the supply of nonequilibrium carriers with the IS scattering time of 1 ps. The redistribution of nonequilibrium electrons of high energy is part of the thermalization process within the $n = 1$ subband. Thermalization requires a transfer of such electrons toward the bottom of the subband. This relaxation involves both emission of optical phonons, transferring excess energy to the lattice, and Coulomb scattering between high and low-energy carriers, the latter process heating the cold electron plasma. The transient interband absorption spectra of Fig. 4.11 reveal the nonthermal character of the $n = 1$ electron distribution for a surprisingly long time interval of 2 ps. Just after the IS excitation (delay time 0.2 ps, Fig. 4.11*a*), one observes a transmission decrease from the onset of the $n = 1$ interband absorption up to the Fermi edge E_F^i of the initially cold distribution. The EMC simulation for 0.2 ps (solid lines) gives a constant depletion of the initial Fermi distribution (dashed line) by 15% of the carriers consistent with the excitation conditions. The EMC simulation shows that the rate of electron-electron scattering into states below the initial Fermi level is reduced by (1) the small fraction of unoccupied states to which electrons can be transferred, i.e., Pauli blocking, and (2) screening of the Coulomb interaction. As a result, the carrier depletion between the bandgap and the initial Fermi level persists for hundreds of femtoseconds, as is directly evident from the data in Fig. 4.11*b*. The inelastic Coulomb scattering between cold and backscattered electrons results in a transfer of cold electrons to states above the initial Fermi level E_F^i, thus forming an enhanced high-energy tail of the distribution. In Fig. 4.11, this is evident from the enhancement of the induced absorption below E_F^i and the bleaching, i.e., population, above E_F^i. Between 0.2 and 1 ps, the very broad high-energy tail results in an amplitude of the bleaching much smaller than that of the induced absorption. After a time delay of about 2 ps, this distinctly nonthermal distribution has evolved into a hot Fermi distribution with similar amplitudes for the enhanced absorption and bleaching. The main energy transfer to the cold plasma occurs at times between 500 fs and 2 ps, due to the delayed supply of energetic carriers from the $n = 2$ subband. The quantitative analysis shows that about 50% of the excess energy of the backscattered electrons goes into the cold electron plasma whereas the other 50% is transferred to the lattice via LO phonon emission.

The IS scattering and thermalization processes evident from the results in Figs. 4.10 and 4.11 are followed by electron cooling via phonon emission. Monitoring the transient interband spectra up to delay times of 50 ps gives detailed information on the cooling kinetics and allows one to derive the instantaneous carrier temperature, as discussed in detail in Ref. 75.

The III-V material systems discussed so far display a limited polarity of the crystal lattice and thus a moderate coupling of electrons and optical phonons via the polar-optical interaction. In contrast, group III nitrides are characterized by a very high polarity and strong electron-phonon coupling, properties which should lead to polaronic signatures in the line shape of IS absorption [76] and to different nonequilibrium dynamics of electrons. Very recently, evidence for such phenomena has been found in a femtosecond pump-probe study of IS dynamics of electrons in a $GaN/Al_{0.8}Ga_{0.2}N$ heterostructure [77]. The linear $n = 1 \rightarrow 2$ IS absorption of this sample displays a structureless envelope extending over more than 200 meV (Fig. 4.12*a*).

Transient IS absorption spectra were measured in two-color pump-probe experiments where a 100-fs pump pulse centered at 355 meV excites the sample and the resulting change of transmission is measured by independently tunable probe pulses of 50-fs duration. In Fig. 4.12*b* and *c*, transient spectra are shown for time delays of 0 and 100 fs. The spectrum for zero delay exhibits a transmission increase over a spectral range much wider than that of the pump spectrum,

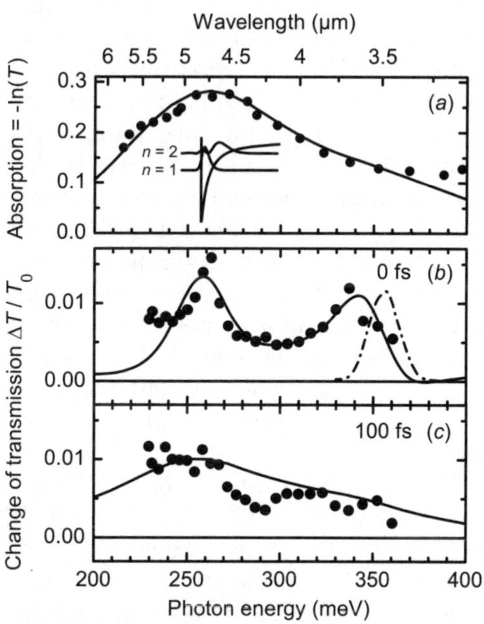

Figure 4.12 (*a*) Measured (symbols) and calculated (solid lines) linear $n = 1 \rightarrow 2$ IS absorption of a $GaN/Al_{0.8}Ga_{0.2}N$ heterostructure (sample temperature of 10 K). Inset: schematic potential energy diagram of the heterostructure and probability densities of electrons in the $n = 1$ and $n = 2$ subbands. (*b*, *c*) Measured (symbols) and calculated (solid lines) transient IS spectra for two different time delays after excitation by pump pulses centered at 355 meV. Clear signatures of optical phonon sidebands of IS absorption are observed. Dash-dotted line: spectrum of the pump pulse.

pointing to a substantial homogeneous broadening of the band. There are two pronounced transmission maxima at 260 and 340 meV, both red-shifted with respect to the maximum of the pump spectrum. After 100 fs, this substructure has essentially disappeared, pointing to a very rapid spectral diffusion within the absorption band. Time-resolved measurements at fixed spectral positions have been presented in Ref. 77. At the maximum of the linear IS absorption, one observes a decay of the transmission increase with a time constant of 380 fs. Around 340 meV, the transmission kinetics display a first 80-fs decay followed by a slower 380-fs decay. The amplitude ratio of the 380-fs components is close to the ratio of linear IS absorption at the two spectral positions.

The separation of the two transmission maxima in Fig. 4.12*b* is close to the LO phonon energy of GaN of 92 meV. This finding points to a vibronic substructure of the spectrum which originates from the strong coupling of electrons and LO phonons in GaN. This behavior can be described within the independent boson model by considering the energy states and the dynamics of the IS polaron, a quasi-particle of mixed electron-LO-phonon character [58, 78]. An analysis based on such a model is presented in Ref. 77 and accounts for the linear and nonlinear spectra shown in Fig. 4.12 (solid lines). The pump pulse overlaps with the $|\, n = 1, \nu = 0\rangle$ to $|\, n = 2, \nu = 1\rangle$ transition ($\Delta \nu = 1$, ν is vibrational quantum number) around 340 meV. The slight red shift of the transmission maximum with respect to the maximum of the pump spectrum is due to the initial motion of the vibronic wave packet in the $n = 2$ state. In addition, the pump induces a transmission increase on the $\Delta\nu = 0$ transition through depletion of the $|\, n = 1, \nu = 0\rangle$ state and optical gain from the $|\, n = 2, \nu = 1\rangle$ state. Such pronounced spectral features are washed out by spectral diffusion, i.e., a modulation of the transition energies due to coupling of the polaron to other fluctuating excitations in the system. This process gives rise to the fast 80-fs decay of the transmission change observed at 340 meV. All transmission changes decay by population relaxation from the $n = 2$ states with a time constant of 380 fs.

So far, energy separations of the optically coupled subbands larger than the optical phonon energies have been considered. Work on IS scattering for subband energy spacings smaller than the optical phonon energy has led to controversial results, and IS scattering times between 500 fs and 750 ps have been claimed [65, 79–82] . For small subband spacings, both carrier density and temperature have a strong influence on IS relaxation. For degenerate carrier distributions, i.e., a quasi-Fermi level within the subband, Pauli blocking substantially reduces the free phase space available for scattering, in this way affecting both carrier-carrier and carrier-phonon scattering rates. For high carrier temperatures, optical phonon emission by the most energetic carriers can result in an acceleration of IS scattering, an effect suggested by EMC simulations of the early experiments of Ref. 65. Thus, the measurement of IS scattering times for $E_{IS} < \hbar\omega_{LO,TO}$ requires a careful characterization of the initial and transient carrier distributions [82].

In contrast to electron relaxation, IS scattering of holes has remained nearly unexplored [27, 83]. The in-plane dispersion of valence subbands is quite complex with pronounced nonparabolicities and mixing between *heavy-hole* (HH), *light-hole* (LH), and *split-off* (SO) bands (cf. Fig. 4.1*b*). For SiGe/Si QWs, the in-plane dispersion of the HH1 and HH2 subbands is different with an IS energy spacing depending on the in-plane wave vector. This results in a distribution of IS transition energies and a broadening of the IS absorption line, making the transient IS absorption sensitive to carrier redistribution in the HH1 subband.

IS relaxation of holes has been studied in femtosecond pump-probe experiments with a *p*-type modulation-doped SiGe/Si QW sample consisting of 10 $Si_{0.5}Ge_{0.5}$ QWs of 4.4-nm width separated by 18-nm-thick Si barriers [27]. The holes introduced by doping populate the HH1 subband with a density of $N = 1.2 \times 10^{12}\text{cm}^{-2}$. The femtosecond pump pulse is resonant with the HH1 $\rightarrow$ HH2 transition and excites approximately 30% of the holes to the HH2 band. Transient spectra measured with a probe pulse on the same transition are shown in Fig. 4.13. At early time delays, a strong increase of transmission occurs around the center of the IS absorption line. As time evolves, this bleaching decreases and an additional absorption increase is observed at small photon energies, i.e., on the low-energy tail of the stationary IS absorption line. The kinetics of bleaching and enhanced absorption at early times are different: The bleaching shows a rise within the time resolution and a pronounced decay within the first 2 ps (Fig. 4.14*a*, dashed line), whereas a delayed rise and a much slower decay are found for the enhanced absorption (Fig. 4.14*a*, solid line). Both signal components decay completely on a time scale of 50 ps (not shown).

The transient spectra in Fig. 4.13 reflect both IS scattering of excited holes from the HH2 back to the HH1 subband and intra-HH1 subband redistribution of carriers. The strong bleaching at early delay times is mainly due to a depletion of the optically coupled HH1 states and stimulated emission from the HH2 subband. The stimulated emission component decays through the depopulation of the HH2 subband by IS scattering of holes into high-lying states of the HH1 subband. The depletion of HH1 states around $k_{\|} = 0$ and the interaction of the backscattered holes at high energy with the unexcited holes in the HH1 subband initiate an intraband thermalization process by which a hot quasi-equilibrium distribution is formed. To establish the high-energy

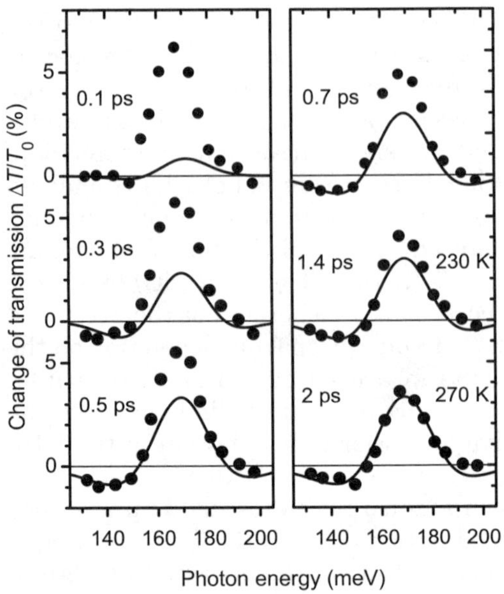

Figure 4.13 Transient IS spectra of holes in *p*-type modulation-doped SiGe/Si quantum wells (sample temperature of 10 K). The change of transmission induced by a pump pulse centered at 170 meV (bandwidth of 20 meV) is plotted versus photon energy within the spectrum of the probe pulse (symbols). Data are shown for different pump-probe delays. Solid lines: calculated difference spectra for thermalized hole distributions of a given hole temperature.

tail of this distribution, holes are scattered from states around $k_{\|} = 0$ into states at larger **k** vectors. This process leads to an additional delayed bleaching at the center of the steady-state IS line and a delayed enhanced absorption on the red wing as the HH1-HH2 energy separation decreases with increasing $k_{\|}$ vector. Eventually, the hot quasi-equilibrium distribution cools down to the lattice temperature by phonon emission.

A quantitative analysis of the data on the basis of this picture allows one to extract the pure IS scattering dynamics [27]. The solid lines in Fig. 4.13 represent spectra calculated for a hot quasi-equilibrium distribution in the HH1 subband. At a 2-ps delay, this modeling is in good agreement with the data, giving a transient hole temperature of 270 K. At earlier delays, the measured bleaching signal is stronger than predicted by this model, reflecting the contribution due to IS scattering. To determine the latter component, the calculated thermal spectra with an amplitude fitting the enhanced absorption at lower photon energies were subtracted from the overall signal. The time evolution of the resulting IS scattering signal is shown in Fig. 4.14*b* and gives an IS scattering time of 250 fs.

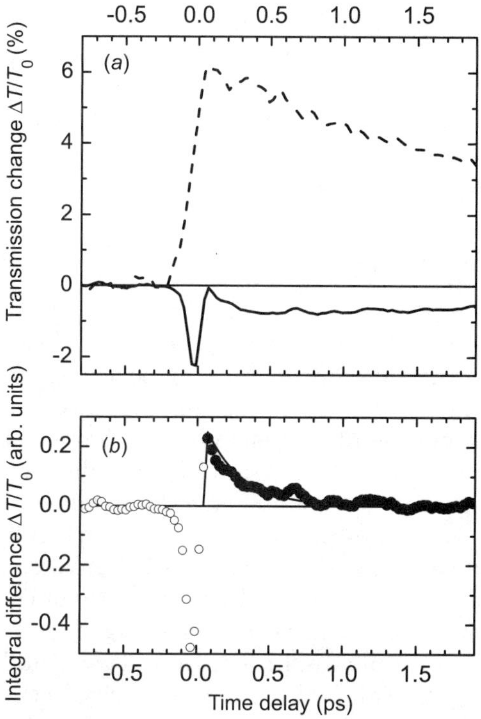

Figure 4.14 (*a*) Change of IS transmission at 168 meV (dashed line) and 140 meV (solid line) as a function of pump-probe delay. (*b*) Fast bleaching signal derived by the subtraction procedure explained in the text (symbols). At delays longer than 70 fs, the kinetics reflect IS scattering of holes with a time constant of 250 fs (solid line).

IS scattering of holes in SiGe/Si QWs is dominated by emission of optical phonons via the deformation potential interaction. There are two relaxation channels: a direct backscattering from the HH2 to the HH1 subband and a sequential process in which a hole scatters into the LHSO1 subband and from there into the HH1 subband. Calculations give an overall scattering time of 225 fs with a dominant contribution from the cascaded process [27]. Such a number is in very good agreement with the experimental results.

In conclusion, IS excitation of carriers induces a complex relaxation scenario by which the original equilibrium distribution of electrons or holes is reestablished. For IS transition energies higher than the optical phonon energy, IS scattering times range from 100 fs up to about 1 ps, depending on the strength of the carrier-phonon interactions and the details of the in-plane subband dispersion. The dominant relaxation mechanism consists in the emission of optical phonons via the polar-optical interaction or the optical deformation potential. For subband energy spacings below the optical phonon energy, carrier-carrier scattering and, in hot carrier distributions, phonon emission result in somewhat longer IS scattering times. IS scattering is followed by carrier thermalization in the lower subband on a femtosecond- to picosecond time scale, leading to a hot quasi-equilibrium distribution. Eventually, carrier cooling by phonon emission transfers the excess energy to the lattice on a time scale of 1 to 50 ps.

4.6 Quantum Transport in Quantum Cascade Structures

Since the first demonstration in 1994 [84], impressive progress has been made in extending the spectral range and optimizing the performance parameters of quantum cascade lasers. This fact is documented in other chapters of this book. From the viewpoint of basic physics, quantum cascade structures are interesting model systems to study unipolar carrier transport in different regimes, in particular for a wide range of carrier concentrations. In the quantum-coherent regime, electronic wave packets representing a nonstationary superposition of envelope wave functions can move through such structure by a sequence of coherent tunneling processes [85–94]. Coherent transport is highly relevant for the operation of quantum cascade light-emitting devices and is connected with ultrafast propagation and relaxation dynamics. In the following, recent experiments on coherent electron transport in GaAs/AlGaAs quantum cascade structures are discussed [64, 95].

The electronic structure of a quantum cascade laser under forward bias is shown in Fig. 4.15*a* [84, 96–98]. The bias is chosen in such a way that at least one of the injector subbands (state *g*) is in resonance with the upper laser subband 3 of the active region downstream. Electrons are transferred from the injector region, which represents a reservoir of electrons, through a barrier into the active region, which consists of three coupled quantum wells. Under steady-state conditions, a quasi-stationary population exists in subband 3 and gives rise to IS emission on the 3 → 2 optical IS transition with an energy larger than the energy of the optical phonons. Electrons in subband 3 have an intrinsic lifetime of approximately 1 ps. In contrast, the lifetime of subband 2 is much shorter (≃ 100 fs) due to efficient electron relaxation to subband 1 via resonant LO phonon emission and electron transfer from both subbands 2 and 1 into the next injector region. As a result, a population inversion builds up between subbands 3 and 2, resulting in optical gain and laser action on the 3 → 2 IS transition.

The nature of the electronic transport from the injector into subband 3, a process crucial for the performance of the device, has led to some controversy in the literature. On the one hand, this transport has been described on the basis of resonant tunneling pictures, invoking

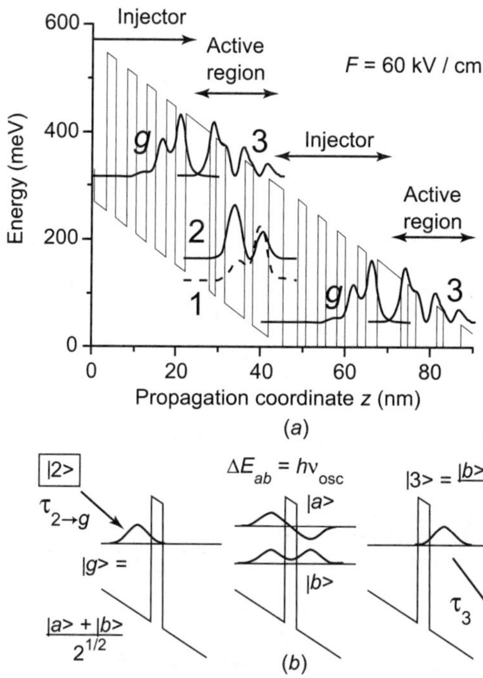

Figure 4.15 (*a*) Conduction band diagram of a GaAs/$Al_{0.3}Ga_{0.7}$ As quantum cascade structure with an active region consisting of three coupled quantum wells under forward bias. The probability densities of electrons in the injector subband *g* and 3 subbands of the active region are also shown. The laser transitions occur between subbands 3 and 2. Such probability densities are based on envelope wave functions calculated without the tunnel coupling between subbands *g* and 3. (*b*) Scheme of coherent electron transport from the injector ground subband *g* into the upper laser subband 3 in the picture of resonant tunneling. The levels $|b\rangle$ and $|a\rangle$ are bonding and antibonding eigenstates of the electronic hamiltonian including the tunnel coupling between $|g\rangle$ and $|3\rangle$. The latter represent coherent superpositions of $|b\rangle$ and $|a\rangle$.

a quantum-coherent superposition of states in subbands *g* and 3 [97–99]. Yet recent Monte Carlo simulations of incoherent electron dynamics, which neglect coherent transport phenomena, give rather high electron-electron scattering rates, resulting in quasi-thermal carrier distributions in each subband under steady-state operation conditions [89, 100, 101]. The high scattering rates suggest a minor role of quantum coherence for electron transport.

In Sec. 4.3 (Fig. 4.4*d*), a pump-probe scheme for studying such electron transport via the depletion and recovery of IS gain has been described. In these experiments, the quasi-stationary gain in GaAs/AlGaAs quantum cascade structures with 10 periods of injector and active region [96, 102] was depleted by a 150-fs pump pulse resonant to the $3 \rightarrow 2$ lasing transition at a wavelength of 10 μm (photon energy 124 meV). The subsequent time evolution of the optical gain was monitored by weak probe pulses which were spectrally dispersed after interaction with the sample. Such measurements were performed for several samples in a wide range of injection currents, sample temperatures, and pump intensities [64, 95].

In Figs. 4.16 through 4.18, results for a sample with 33% Al in the barriers and an injection barrier of 6.2-nm width are presented [96]. The time-resolved transmission changes in Fig. 4.16*a* to *d* represent the difference between a pump-probe curve with an injection current density $j = 7$ kA/cm^2 applied and a pump-probe curve without current as a function of the time delay between pump and probe pulses. The quasi-stationary transmission increase before zero delay is due to the current-induced gain present in absence of the femtosecond pulses. Upon femtosecond excitation, the gain is depleted both on the short-wavelength wing (Fig. 4.16*a*) and at the maximum (Fig. 4.16*b*) of the gain spectrum (Fig. 4.16*f*). The gain recovers completely within several hundreds of femtoseconds. During this period, the transmission changes consist of a decay transient superimposed by pronounced

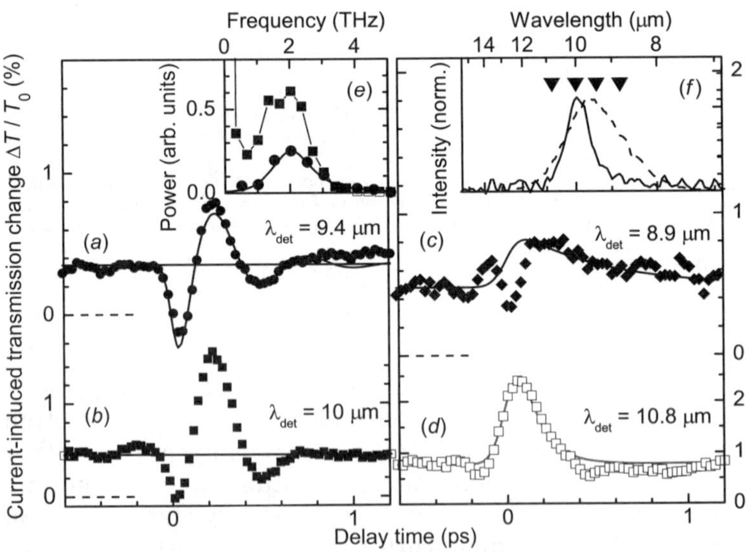

Figure 4.16 (*a–d*) Time-resolved transmission changes of a GaAs/AlGaAs quantum cascade structure under forward bias (injection current density $j = 7$ kA/cm^2, nominal sample temperature of 10 K). The change of transmission $\Delta T/T_0 = [T(j, t) - T(j = 0, t)]/T(j = 0, t)$ is plotted as a function of pump-probe delay t for different wavelength positions in the probe spectrum (symbols); $T(j, t)$ is the transmission through the sample for a current density j. The energy of pump and probe pulse resonant to the $3 \rightarrow 2$ transition was 7500 and 75 pJ, respectively. Solid lines: exponentially decaying cosine function (*a*) or decaying exponential functions (*c, d*) convoluted with the cross-correlation between pump and probe. (*e*) Fourier transforms of the transients (*a, b*). (*f*) Electroluminescence spectrum of the sample for $j = 7$ kA/cm^2 (solid line) and spectrum of the pump and probe pulses (dashed line). The arrows indicate the wavelength positions at which the transients (*a–d*) were recorded.

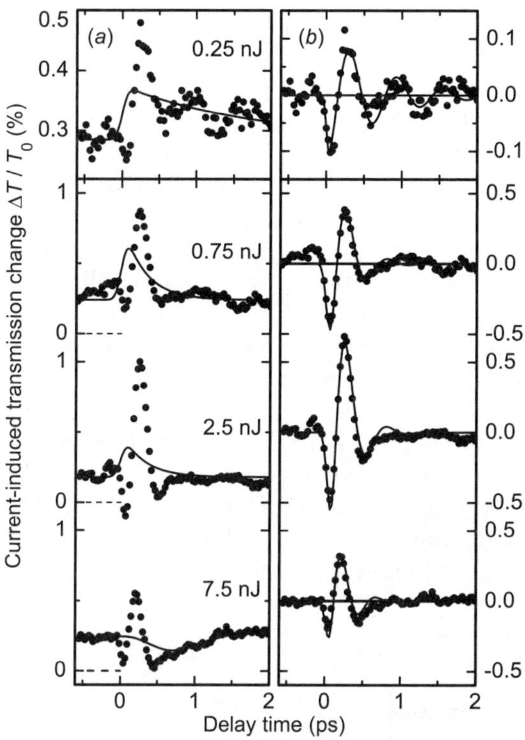

Figure 4.17 (*a*) Transmission changes of a quantum cascade structure measured for different energies of the pump pulse at a probe wavelength of $\lambda = 10$ μm (symbols, current density of 7 kA/cm^2, sample temperature of 10 K). Solid lines: slowly varying signal component. (*b*) Oscillatory component of the signal, i.e., total signal minus slowly varying component (symbols), and exponentially decaying cosine fit curves convoluted with the cross-correlation of pump and probe pulses (solid lines).

oscillations which are particularly strong at the gain maximum. The Fourier transforms of the oscillatory components (Fig. 4.16*e*) show a maximum at 2 THz corresponding to an oscillation period of 500 fs.

Transients taken outside the emission spectrum (Fig. 4.16*e* and *d*) display a different time evolution: There is a transmission increase at early time delays followed by a decay depending on the particular probe wavelength. The modulation in Fig. 4.16*c*, which starts at negative delay times, is caused by nonlinear optical effects such as the perturbed free induction decay and coherent pump-probe coupling. Oscillatory features are absent at positive time delays. Such transients reflect carrier dynamics in the injector and are not discussed further [95]. Similarly, oscillations are absent for reverse bias.

In Fig. 4.17, time-resolved transmission changes measured at $\lambda = 10$ μm, the gain maximum, are presented for different energies of the pump pulse, i.e., different levels of gain saturation (current density

j = 7 kA/cm^2, lattice temperature is 10 K). For weak saturation (Fig. 4.17*a*), one finds several oscillations of small amplitude and a long dephasing time of approximately 700 fs. With increasing pump energy, the oscillation amplitude grows linearly up to a pulse energy of 2.5 nJ and then starts to saturate. Concomitantly, dephasing becomes faster whereas the oscillation frequency of 2 THz remains unchanged.

Data taken for different injection currents and a pump energy of 7.5 nJ are summarized in Fig. 4.18. Transmission oscillations occur in a wide range of injection currents, starting from j = 70 A/cm^2 well below the regime of quasi-stationary gain up to j = 8 kA/cm^2. Between j = 0.3 and 8 kA/cm^2, the oscillation amplitude remains essentially unchanged whereas dephasing becomes faster with increasing current.

Gain saturation by the intense mid-infrared pump pulses leads to a strong reduction of the electron population of subband 3, which is subsequently compensated for by the transfer of electrons from the injector region through the injection barrier into subband 3. The electrons dumped by the excitation pulse into subband 2 leave the active region very quickly by LO phonon scattering to subband 1 and

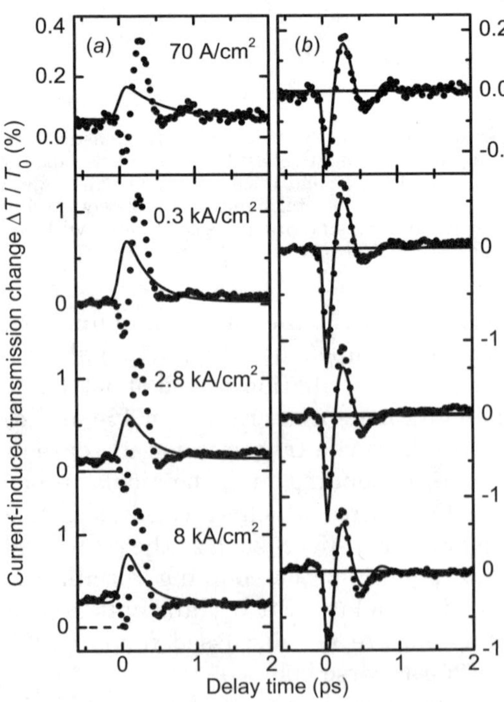

Figure 4.18 Transmission changes of a quantum cascade structure at a probe wavelength of λ = 10 μm for different injection current densities. (*a*) Measured transmission change (symbols) and slowly varying component of the signal with frequencies ≤ 1 THz (solid lines). (*b*) Oscillatory signal component as in Fig. 4.17*b*.

transfer into the next injector. Thus, they do not contribute to the pump-probe signal for delay times longer than 100 fs. As a result, the gain saturation and recovery dynamics are dominated by the population dynamics in the upper laser subband 3.

The results demonstrate the oscillatory character of electron motion after gain saturation. This behavior gives evidence of coherent resonant tunneling of electrons through the injection barrier, as depicted schematically in Fig. 4.15*b*. Under bias, the subbands $|g\rangle$ and $|3\rangle$ are in resonance, and the tunnel coupling forms bonding $|b\rangle$ and antibonding $|a\rangle$ energy eigenstates. These are energetically split by an energy ΔE_{ab} which is determined by the width and height of the injection barrier and by the voltage applied. Coherent superpositions of states $|a\rangle$ and $|b\rangle$ represent the basis for wave packet propagation from the injector ground state $|g\rangle$ into the upper laser state $|3\rangle$. In the experiment, the femtosecond pump pulse depletes the subband $|3\rangle = (1/2^{1/2})(|b\rangle - |a\rangle)$, i.e., both eigenstates $|a\rangle$ and $|b\rangle$, and initiates a coherent wave packet motion by which electrons tunnel from the injector through the injection barrier into the active region. The tunneling carriers arrive after one-half the oscillation period $T_{osc} = h/\Delta E_{ab}$ in state $|3\rangle$ of the active region. As the electron lifetime in state $|3\rangle$ $(\tau_3 \simeq 1\text{ ps})$ is longer than the oscillation period T_{osc}, most of the electrons will move back into the injector, resulting in a gain depletion at delay times $t_D \approx T_{osc}$. This picture of coherent transport is strongly supported by comparing the results for samples with different injection barriers. Such studies reveal a change of the measured oscillation frequencies proportional to ΔE_{ab} [64]. In principle, this picture implies a doublet structure of the electroluminescence spectrum. However, the strong broadening of the electroluminescence spectra (Fig. 4.16*f*) smears out such a line structure.

The oscillatory electron motion is damped on a time scale of several hundreds of femtoseconds; i.e., the dephasing time is comparable to the period of the oscillations. Two mechanisms contribute to such damping: (1) the inhomogeneous broadening of the tunnel couplings due to injection barrier width and height fluctuations and (2) the homogeneous dephasing of the quantum coherence between the states $|a\rangle$ and $|b\rangle$. Mechanism 1 leads to a destructive interference of oscillations of different frequencies. Homogeneous dephasing is mainly caused by electron-electron scattering in the injector which destroys the quantum mechanical phase between the states making up the wave packet. For electron densities of several 10^{11} cm^{-2} in the injector, Monte Carlo simulations suggest sub-100-fs scattering times, substantially shorter than the observed dephasing times of the electron wave packets. Thus, only a fraction of the scattering events result in dephasing. Moreover, electron-electron scattering allows one to generate transport coherences [103–105]: Among the injector states between which electrons can be scattered are the coherent superpositions $|g\rangle$ of the electronic eigenstates $|a\rangle$ and $|b\rangle$. If a scattering event populates $|g\rangle$ on a time scale short compared to the oscillation period T_{OSC}, a coherent superposition of such eigenstates is generated which is in phase with the electron oscillation between the injector and the active region. In this way, the oscillation is enhanced, a mechanism which is also behind the pronounced gain overshoot after depletion by the pump pulse (cf. Figs. 4.16 to 4.18). A more detailed discussion of such a mechanism is given in Ref. 95.

The amplitude of the oscillatory electron transport depends linearly on the pump energy up to the value of 2.5 nJ, where the optical gain is fully saturated. With increasing pump level, the dephasing of the oscillations becomes faster, reflecting enhanced scattering rates in the injector. This enhanced scattering has been attributed to a heating of the electron distribution in the injector region by the carriers transferred from the preceding active region [95].

It is interesting to note that the oscillation frequency of the electron wave packet remains essentially unchanged upon increasing the injection current (Fig. 4.18). First, the bias on the device changes only weakly with injection current, as is evident from its current-voltage characteristics. For a change of injection current of over two orders of magnitude (Fig. 4.18), the electric field on the structure varies between 30 and 65 kV/cm. Calculations of the electronic band structure in this bias range predict three anticrossings between injector subbands and the upper laser subband with nearly identical tunnel splittings of 7 meV, corresponding to an oscillation frequency of 1.7 THz. This is close to the measured oscillation frequency of 2 ± 0.5 THz.

The results presented here demonstrate that quantum coherences between coupled subbands play a central role in the electron transport through quantum cascade structures; i.e., a description in terms of incoherent drift diffusion processes is incomplete. After femtosecond gain depletion, quantum coherence manifests itself in pronounced gain oscillations due to coherent wave packet motion between the injector and the upper laser state. Such results explain the empirical finding that the design of the injection barrier is crucial to the performance of a quantum cascade laser. A full theoretical understanding of this complex scenario is still missing and requires a model combining quantum transport and decoherence processes by electron-electron scattering.

4.7 Conclusions

The dynamics of IS excitations occur in the ultrafast time domain, covering a range from several tens of femtoseconds up to picoseconds.

Optical IS excitation induces coherent IS polarizations and/or transport coherences. Such quantum coherences decay by a variety of dephasing processes which, in most cases, are governed by carrier-carrier and, to a lesser extent, carrier-phonon scattering. Optical IS polarizations can be generated and manipulated in a wide parameter range by nonlinear interaction with ultrashort infrared pulses. Such schemes may lead to novel methods for processing information, e.g., by exploiting the properties of coupled IS transitions. Transport coherences play a central role in quantum cascade devices, in particular for the injection of carriers into the optically active regions. Although the first studies of quantum transport in these devices have been reported, much more experimental and theoretical work is required to understand the transport behavior in detail, e.g., the role of different scattering processes for decoherence. Incoherent carrier relaxation following IS excitation has been studied extensively for electrons, and the processes of IS scattering, carrier thermalization, and cooling are understood in quite some detail. Open questions in this area involve carrier dynamics in closely spaced subbands, e.g., in terahertz quantum cascade lasers, and IS scattering of holes in a wider range of systems. The combination of sophisticated growth methods for quantum wells and quantum cascade devices of high structural quality and ultrafast nonlinear optics to unravel their properties will remain a powerful approach for clarifying the microscopic physics of intersubband excitations.

I would like to acknowledge the important contributions of my present and former coworkers Michael Woerner, Klaus Reimann, Robert A. Kaindl, Felix Eickemeyer, and Stephan Lutgen. It is my pleasure to thank Paolo Lugli, Munich, Tilmann Kuhn, Münster, and Andreas Knorr and their coworkers for a theoretical analysis of experimental results and many discussions. I am indebted to Klaus Ploog, and Richard Hey, Paul Drude Institute, Berlin; Harald Künzel, Heinrich Hertz Institute, Berlin; Karl Brunner and Gerhard Abstreiter, Walter Schottky Institute, Munich; Daniel Hofstetter, Neuchatel; Lester Eastman and his coworkers, Cornell; Carlo Sirtori at Thales, Orsay; as well as Karl Unterrainer and Gregor Strasser, TU Vienna; for the high-quality samples studied in the experiments. I also thank the Deutsche Forschungsgemeinschaft for financial support.

References

1. T. Ando, A. B. Fowler, and F. Stern, *Rev. Mod. Phys.* **54**, 437 (1982).
2. E. Rosencher et al. (eds.), *Intersubband Transitions in Quantum Wells*, Plenum Press, New York, 1992.

3. S. S. Li, and Y. K. Su (eds.), *Intersubband Transitions in Quantum Wells: Physics and Devices,* Kluwer Academic Publishers, Boston, 1998.
4. M. Helm, in *Intersubband Transitions in Quantum Wells: Physics and Device Applications I,* F. Capasso and H. C. Liu (eds.), *Semiconductors and Semimetals*, vol. 62, 2000, p. 1.
5. A. Kamgar, P. Kneschaurek, G. Dorda, and J. F. Koch, *Phys. Rev. Lett.* **32**, 1251 (1974).
6. M. Woerner and T. Elsaesser, in *Ultrafast Phenomena in Semiconductors,* K. T. Tsen (ed.), Springer, Berlin, 2001, p. 93.
7. J. Shah, *Ultrafast Spectroscopy of Semiconductors and Semiconductor Nanostructures,* 2d ed., Springer, Berlin, 1999.
8. L. C. West and S. J. Eglash, *Appl. Phys. Lett.* **46**, 1156 (1985).
9. M. Zachau, P. Hegelsen, F. Koch, D. Grützmacher, R. Mayer, and P. Balk, *Semicond. Sci. Technol.* **3**, 1029 (1988).
10. H. Lobentanzer, W. König, W. Stolz, K. Ploog, T. Elsaesser, and R. J. Bäuerle, *Appl. Phys. Lett.* **53**, 571 (1988).
11. T. Fromherz, E. Koppensteiner, M. Helm, G. Bauer, J. F. Nützel, and G. Abstreiter, *Phys. Rev.* **B50**, 15073 (1994).
12. S. J. Allen, G. Brozak, E. Colas, F. DeRosa, P. England, J. Harbison, M. Helm, L. Florez, and M. Leadbeator, *Semicond. Sci. Technol.* **7**, B1 (1992).
13. C. Weisbuch and B. Vinter, *Quantum Semiconductor Structures: Fundamentals and Applications*, Academic Press, Boston, 1991.
14. A. Pinczuk, S. Schmitt-Rink, G. Danan, J. P. Valladares, L. N. Pfeiffer, and K. W. West, *Phys. Rev. Lett.* **63**, 1633 (1989).
15. S. L. Chuang, M. S. C. Luo, S. Schmitt-Rink, and A. Pinczuk, *Phys. Rev.* **B46**, 1897 (1992).
16. M. S. C. Luo, S. L. Chuang, S. Schmitt-Rink, and A. Pinczuk, *Phys. Rev.* **B48**, 11086 (1993).
17. M. Zaluzny, *Phys. Rev.* **B43**, 4511 (1991).
18. M. Zaluzny, *Phys. Rev.* **B49**, 2923 (1994).
19. P. von Allmen, *Phys. Rev.* **B46**, 13351 (1992).
20. R. J. Warburton, K. Weilhammer, J. P. Kotthaus, M. Thomas, and H. Kroemer, *Phys. Rev. Lett.* **80**, 2185 (1998).
21. B. Vinter, *Phys. Rev. Lett.* **35**, 598 (1975).
22. D. E. Nikonov, A. Imamoglu, L. V. Butov, and H. Schmidt, *Phys. Rev. Lett.* **79**, 4633 (1997).
23. R. A. Kaindl, S. Lutgen, M. Woerner, T. Elsaesser, B. Nottelmann, V. M. Axt, T. Kuhn, A. Hase, and H. Künzel, *Phys. Rev. Lett.* **80**, 3575 (1998).
24. S. Nojima, *Phys. Rev.* **B41**, 10214 (1990).
25. A. Liu, *Phys. Rev.* **B55**, 7796 (1997).

26. R. A. Kaindl, K. Reimann, M. Woerner, T. Elsaesser, R. Hey, and K. H. Ploog, *Phys. Rev.* **B63**, 161308(R) (2001).

27. R. A. Kaindl, M. Wurm, K. Reimann, M. Woerner, T. Elsaesser, C. Miesner, K. Brunner, and G. Abstreiter, *Phys. Rev. Lett.* **86**, 1122 (2001).

28. G. M. H. Knippels, R. F. X. A. M. Mols, A. F. G. van der Meer, D. Oepts, and P. W. van Amersfoort, *Phys. Rev. Lett.* **75**, 1755 (1995).

29. B. N. Murdin, W. Heiss, C. J. G. M. Langerak, S. C. Lee, I. Galbraith, G. Strasser, E. Gornik, M. Helm, and C. R. Pidgeon, *Phys. Rev.* **B55**, 5171 (1997).

30. Y. R. Shen, *The Principles of Nonlinear Optics,* chap. 5, Wiley, New York, 1984.

31. V. G. Dmitriev, G. G. Gurzadyan, and D. N. Nikogosyan, *Handbook of Nonlinear Optical Crystals*, Springer-Verlag, Berlin, 1991.

32. A. M. Weiner, *IEEE J. Quantum Electron.* **19**, 1276 (1983).

33. T. Elsaesser, A. Seilmeier, W. Kaiser, P. Koidl, and G. Brandt, *Appl. Phys. Lett.* **44**, 383 (1984).

34. M. Joffre, A. Bonvalet, A. Migus, and J. L. Martin, *Opt. Lett.* **21**, 964 (1996).

35. A. Bonvalet, M. Joffre, J. L. Martin, and A. Migus, *Appl. Phys. Lett.* **67**, 2907 (1995).

36. R. A. Kaindl, D. C. Smith, M. Joschko, M. P. Hasselbeck, M. Woerner, and T. Elsaesser, *Opt. Lett.* **23**, 861 (1998).

37. R. A. Kaindl, F. Eickemeyer, M. Woerner, and T. Elsaesser, *Appl. Phys. Lett.* **75**, 1060 (1999).

38. R. Huber, A. Brodschelm, F. Tauser, and A. Leitenstorfer, *Appl. Phys. Lett.* **76**, 3191 (2000)

39. E. S. Wachman, D. C. Edelstein, and C. L. Tang, *Opt. Lett.* **15**, 136 (1990).

40. Q. Fu, G. Mak, and H. M. van Driel, *Opt. Lett.* **17**, 1006 (1992).

41. K. C. Burr, C. L. Tang, M. A. Arbore, and M. M. Fejer, *Opt. Lett.* **22**, 1458 (1997).

42. A. Lohner, P. Kruck, and W. W. Rühle, *Appl. Phys.* **B59**, 211 (1994).

43. J. M. Fraser, D. Wang, A. Hache, G. R. Allan, and H. M. vanDriel, *Appl. Opt.* **36**, 5044 (1997).

44. S. Ehret, and H. Schneider, *Appl. Phys.* **B66**, 27 (1998).

45. For an overview, see W. R. Bosenberg, and R. C. Eckardt (eds.), Special issue on parametric devices, *J. Opt. Soc. Am.* **B12**, 2084 (1995).

46. V. Petrov, F. Rotermund, and F. Noack, *J. Opt. A: Pure Appl. Opt.* **3**, R1 (2001).

47. F. Seifert, V. Petrov, and M. Woerner, *Opt. Lett.* **19**, 2009 (1994).

48. R. A. Kaindl, M. Wurm, K. Reimann, P. Hamm, A. M. Weiner, and M. Woerner, *J. Opt. Soc. Am.* **B17**, 2086 (2000).

49. P. Hamm, R. A. Kaindl, and J. Stenger, *Opt. Lett.* **25**, 1798 (2000).
50. K. Reimann, R. P. Smith, A. M. Weiner, T. Elsaesser, and M. Woerner, *Opt. Lett.* **28**, 471 (2003).
51. Q. Wu, and X. C. Zhang, *Appl. Phys. Lett.* **71**, 1285 (1997).
52. J. A. Gruetzmacher and N. P. Scherer, *Rev. Sci. Instrum.* **73**, 2227 (2002).
53. B. A. Richman, K. W. DeLong, and R. Trebino, *Nucl. Instrum. Meth.* **A358**, 268 (1995).
54. D. T. Reid, P. Loza-Alvarez, C. T. A. Brown, T. Beddard, and W. Sibbett, *Opt. Lett.* **25**, 1478 (2000).
55. F. Eickemeyer, M. Woerner, A. M. Weiner, T. Elsaesser, R. Hey, and K. H. Ploog, *Appl. Phys. Lett.* **79**, 165 (2001).
56. C. W. Luo, K. Reimann, M. Woerner, T. Elsaesser, R. Hey, and K. H. Ploog, *Phys. Rev. Lett.* **92**, 047402 (2004).
57. A. Bonvalet, J. Nagle, V. Berger, A. Migus, J. L. Martin, and M. Joffre, *Phys. Rev. Lett.* **76**, 4392 (1996).
58. S. Mukamel, *Principles of Nonlinear Optical Spectroscopy,* Oxford, 1995.
59. H. Haug, and S. W. Koch, *Quantum Theory of the Optical and Electronic Properties of Semiconductors*, World Scientific, Singapore 1993.
60. D. M. Jonas, *Annu. Rev. Phys. Chem.* **54**, 425 (2003).
61. A. Seilmeier, H. J. Hübner, G. Abstreiter, G. Weimann, and W. Schlapp, *Phys. Rev. Lett.* **59**, 1345 (1987).
62. S. Lutgen, R. A. Kaindl, M. Woerner, T. Elsaesser, A. Hase, H. Künzel, M. Gulia, D. Meglio, and P. Lugli, *Phys. Rev. Lett.* **77**, 3657 (1996).
63. H. Wang, J. Shah, T. C. Damen, S. W. Pierson, T. L. Reinecke, L. N. Pfeiffer, and K. West, *Phys. Rev.* **B52**, R17013 (1995).
64. F. Eickemeyer, K. Reimann, M. Woerner, T. Elsaesser, S. Barbieri, C. Sirtori, G. Strasser, T. Müller, K. Bratschitsch, and K. Unterrainer, *Phys. Rev. Lett.* **89**, 047402 (2002).
65. D. Y. Oberli, D. R. Wake, M. V. Klein, J. Klem, T. Henderson, and H. Morkoc, *Phys. Rev. Lett.* **59**, 696 (1987)
66. M. C. Tatham, J. F. Ryan, and C. T. Foxon, *Phys. Rev. Lett.* **63**, 1637 (1989).
67. T. Stroucken, A. Knorr, P. Thomas, and S. W. Koch, *Phys. Rev.* **B53**, 2026 (1996).
68. J. E. Sipe, *J. Opt. Soc. Am.* **B4**, 481 (1987).
69. F. Bloch, *Phys. Rev.* **70**, 460 (1946).
70. T. Shih, C. W. Luo, K. Reimann, M. Woerner, T. Elsaesser, I. Waldmüller, A. Knorr, R. Hey, and K. H. Ploog, *Phys. Rev.* **B72,** 195338 (2005).

71. J. A. Levenson, G. Dolique, J. L. Oudar, and I. Abram, *Phys. Rev.* **B41**, 3688 (1990).

72. S. Hunsche, K. Leo, H. Kurz, and K. Köhler, *Phys. Rev.* **B50**, 5791 (1994).

73. S. M. Goodnick and P. Lugli, *Phys. Rev.* **B37**, 2587 (1988).

74. A. Tomita, J. Shah, J. E. Cunningham, S. M. Goodnick, P. Lugli, and S. L. Chuang, *Phys. Rev.* **B48**, 5708 (1993).

75. S. Lutgen, R. A. Kaindl, M. Woerner, A. Hase, and H. Künzel, *Solid State Commun.* **106,** 425 (1998).

76. S. Butscher, J. Förstner, I. Waldmüller, and A. Knorr, *Phys. Status Solidi* **B241,** R49 (2004).

77. Z. Wang, K. Reimann, M. Woerner, T. Elsaesser, D. Hofstetter, J. Hwang, W. J. Schaff, and L. F. Eastman, *Phys. Rev. Lett.* **94**, 037403 (2005).

78. S. Schmitt-Rink, D. A. B. Miller, and D. S. Chemla, *Phys. Rev.* **B35**, 8113 (1987).

79. B. N. Murdin, G. M. H. Knippels, A. F. G. van der Meer, C. R. Pidgeon, C. J. G. M. Langerak, M. Helm, W. Heiss, K. Unterrainer, E. Gornik, K. K. Geerinck, N. J. Hovenier, and W. T. Wenckebach, *Semicond. Sci. Technol.* **9**, 1554 (1994).

80. J. N. Heyman, K. Unterrainer, K. Craig, B. Galdrikian, M. S. Sherwin, K. Campman, P. F. Hopkins, and A. C. Gossard, *Phys. Rev. Lett.* **74**, 2683 (1995).

81. K. Craig, B. Galdrikian, J. N. Heyman, A. G. Markelz, M. S. Sherwin, K. L. Campman, P. F. Hopkins, and A. C. Gossard, *Phys. Rev. Lett.* **76**, 2382 (1996).

82. M. Hartig, S. Haacke, P. E. Selbmann, B. Deveaud, R. A. Taylor, and L. Rota, *Phys. Rev. Lett.* **80**, 1940 (1998).

83. Z. Xu, P. M. Fauchet, C. W. Rella, B. A. Richman, H. A. Schwettmann, and G. W. Wicks, *Phys. Rev.* **B51**, 10631 (1995).

84. J. Faist, F. Capasso, D. L. Sivco, C. Sirtori, A. L. Hutchinson, and A. Y. Cho, *Science* **264,** 553 (1994).

85. F. Capasso, K. Mohammed, and A. Y. Cho, *IEEE J. Quantum Electron.* **22,** 1853 (1986).

86. J. Rammer, *Rev. Mod. Phys.* **63**, 781 (1991).

87. H. Haug and A. P. Jauho, *Quantum Kinetics in Transport and Optics of Semiconductors*, Springer, Berlin, 1996.

88. C. Rauch, G. Strasser, K. Unterrainer, W. Boxleitner, E. Gornik, and A. Wacker, *Phys. Rev. Lett.* **81**, 3495 (1998).

89. R. C. Iotti, and F. Rossi, *Phys. Rev. Lett.* **87**, 146603 (2001).

90. P. Kleinert and V. V. Bryksin, *Physica* **B304**, 60 (2001).

91. S. C. Lee and A. Wacker, *Phys. Rev.* **B66**, 245314 (2002).

92. K. Leo, J. Shah, E. O. Göbel, T. C. Damen, S. Schmitt-Rink, W. Schäfer, and K. Köhler, *Phys. Rev. Lett.* **66**, 201 (1991).

93. J. Feldmann, K. Leo, J. Shah, D. A. B. Miller, J. E. Cunningham, T. Meier, G. von Plessen, A. Schulze, P. Thomas, and S. Schmitt-Rink, *Phys. Rev.* **B46**, 7252 (1992).

94. C. Waschke, H. G. Roskos, R. Schwedler, K. Leo, H. Kurz, and K. Köhler, *Phys. Rev. Lett.* **70**, 3319 (1993).

95. M. Woerner, K. Reimann, and T. Elsaesser, *J. Phys.: Condens. Matter* **16**, R25 (2004).

96. C. Sirtori, P. Kruck, S. Barbieri, P. Collot, J. Nagle, M. Beck, J. Faist, and U. Oesterle, *Appl. Phys. Lett.* **73**, 3486 (1998).

97. R. F. Kazarinov and R. A. Suris, *Fiz. Tekh. Poluprovodn.* **5**, 797 [*Sov. Phys. Semicond.* **5**, 707 (1971)].

98. R. F. Kazarinov and R. A. Suris, *Fiz. Tekh. Poluprovodn.* **6**, 148 [*Sov. Phys. Semicond.* **6**, 120 (1972)].

99. C. Sirtori, F. Capasso, J. Faist, A. L. Hutchinson, D. L. Sivco, and A. Y. Cho, *IEEE J. Quantum Electron.* **34**, 1722 (1998).

100. R. C. Iotti and F. Rossi, *Appl. Phys. Lett.* **76**, 2265 (2000).

101. R. C. Iotti and F. Rossi, *Appl. Phys. Lett.* **78**, 2902 (2001).

102. H. Page, C. Becker, A. Robertson, G. Glastre, V. Ortiz, and C. Sirtori, *Appl. Phys. Lett.* **78**, 3529 (2001).

103. H. Vaupel, P. Thomas, O. Kühn, V. May, K. Maschke, A. P. Heberle, W. W. Rühle, and K. Köhler, *Phys. Rev.* **B53**, 16531 (1996).

104. S. A. Gurvitz, I. Bar-Joseph, and B. Deveaud, *Phys. Rev.* **B43**, 14703 (1991).

105. F. Wolter, H. G. Roskos, P. Haring-Bolivar, G. Bartels, H. Kurz, K. Köhler, H. T. Grahn, and R. Hey, *Phys. Status Solidi* **B204**, 83 (1997).

Chapter

5

Optical Nonlinearities in Intersubband Transitions and Quantum Cascade Lasers

Claire Gmachl

Department of Electrical Engineering and PRISM
Princeton University
Princeton, New Jersey

Oana Malis

Bell Laboratories, Lucent Technologies
Murray Hill, New Jersey

Alexey Belyanin

Department of Physics
Texas A&M University
College Station, Texas

5.1 Introduction to Nonlinear Intersubband Optical Transitions

5.1.1 Introduction

Nonlinear optics is an extensive field of research at the intersection of optics and material science with appeal to both pure scientific pursuit as well as application-focused engineering. Nonlinear optics encompasses all effects describing a change of the optical properties of a material under incident light. Such changes are usually very minute in naturally occurring materials, and the field of nonlinear optics was hence tightly coupled to the invention of the laser and continuing advances in laser science [1]. Lasers are the only (practical and convenient) light sources capable of generating sufficiently large optical fields to noticeably change the material's optical properties. Over the last several decades, nonlinear optics has matured from being a descriptive science, measuring, calculating, and understanding the magnitude of the optical nonlinearities in naturally occurring and purified materials (such as vapors or solid-state crystals), to becoming an engineering science, capable of designing and fabricating new (quasi-) materials displaying giant optical nonlinearities.

A clear theoretical, quantum mechanical understanding of nonlinear optics, written into easy-to-use computer codes, as well as a compelling intuitive picture has helped to quickly advance the field. Within the intuitive picture, an electron in a confining potential, such as around an excited atom or molecule in a vapor or near a dopant ion in a solid-state crystal, is accelerated in this potential under an incident electromagnetic field. The accelerated electron oscillates and radiates according to its motion in the potential. This motion is the origin of the observed optical properties of the material, such as the refractive index. If the potential is asymmetric (anharmonic) and the driving field is large enough to force the electron into the noticeably nonharmonic portion of the potential, then the reradiated light will contain noticeable fractions of higher harmonics, with the actual shape of the potential determining the relative strength of the ensuing harmonics, such as in *second harmonic generation* (SHG) or *third harmonic generation* (THG). In analogy, one can infer that an electron in this anharmonic potential driven by two different incident electric fields will radiate the *sum-frequency* and/or *difference-frequency generation* (SFG and/or DFG) of the two incident fields. For an engineer of large optical nonlinearities, the intuitive message is then to design materials with as large an anharmonicity as possible of the electronic potential, and the reward is efficient nonlinear light generation.

The above intuitive picture is instructive as a first means of gauging an optical nonlinearity; this classical model does not, however, account properly for electronic potentials with discrete energy levels, for which a quantum mechanical treatment is needed. The latter is especially essential as the optical nonlinearities are usually greatly enhanced near resonances with these discrete energy levels and at frequencies resonant with optical transitions between such quantized energy levels. Nevertheless, also the quantum mechanical models have fully been worked out, including several excellent textbook discussions [2–4], and result in workable solutions, based on the density matrix approach, for the nonlinear susceptibilities of quantum mechanical systems.

Concurrently with nonlinear optics, the fields of semiconductor materials, devices, and especially lasers have equally matured. Semiconductor devices for either microelectronics or optoelectronics are highly advanced, and excellent modeling tools as well as fabrication methods exist, the combination of both allowing for swift device optimization and fast innovation cycles. While most optoelectronic devices use interband optical transitions across the material bandgap for electron–light interaction, recently another class of optoelectronic devices has attracted much (partly renewed) interest [5]. These are devices based on *intersubband* (ISB) optical transitions, mostly of electrons within the conduction band of a semiconductor heterostructure, such as single and coupled *quantum wells* (QWs), and *superlattices* (SLs). Much of this volume is dedicated to the physics and engineering of such devices, including *quantum cascade* (QC) lasers [6, 7], *quantum well infrared photodetectors* (QWIPs) [8], and modulators. Intersubband transitions are best understood and documented for the two-dimensional QW systems; however, ISB transitions have also been discussed in zero-dimensional quantum dots [9] and one-dimensional quantum wire systems [10]. Intersubband-based semiconductor structures are prime examples of engineered quasi-materials for which both design and fabrication methods (i.e., epitaxial layer-by-layer growth with atomic layer precision) are extremely well developed, and nearly any conceivable energy-level combination, and hence effective electron potential, can be designed and fabricated. This is a most promising situation for the design of large optical nonlinearities based on semiconductor heterostructures.

5.1.2 Intersubband transitions and optical nonlinearities

Gurnick and DeTemple were first to argue in their seminal paper in 1983 [11] that the then rapidly progressing technologies of semiconductor heterostructure growth (especially molecular beam epitaxy and metal-organic chemical vapor deposition) could be harnessed for the design and fabrication of large optical nonlinearities based on asymmetric QWs. In their work they predicted between 10 and 100 times larger second-order optical nonlinearities compared to the lattice value for a specially shaped, "graded," asymmetric potential $Al_xGa_{1-x}As$ heterostructure. This enhancement would come on top of an already relatively large optical nonlinearity for GaAs bulk material resulting from its zinc-blende crystal structure. Following the work by Gurnick and DeTemple, a continuous, still ongoing stream of work has been dedicated to the exploitation of resonantly enhanced optical nonlinearities in specially shaped heterostructure potentials, including ISB optical structures. Giant, i.e., by several orders of magnitude enhanced, optical nonlinearities have been engineered and demonstrated [12, 13], the large increase in the nonlinearity being the result of the resonant enhancement of the nonlinearity at the discrete energy levels, and also the result of the large optical dipole matrix elements that are achievable for ISB transitions (typically several nanometers, compared to angstroms for interband transitions), and which enter into the optical nonlinearity with higher order.

A single, undisturbed QW still constitutes a highly symmetric potential without much nonlinearity; thus, several strategies have evolved since the proposal by Gurnick and DeTemple to break the symmetry of the single well, and achieve large potential asymmetry, manifested in the symmetry breaking of the electronic wave functions, a loosening of the selection rules for transitions between states, and—related to that—large optical dipole matrix elements between all confined energy levels. Figure 5.1 shows a collection of such potentials and strategies for symmetry breaking. Figure 5.1*a* is the proposed asymmetric potential by Gurnick and DeTemple. This proposal was quickly followed by asymmetry induced by an applied electric field [14] (Fig. 5.1*b*, which allows for field tunability of the nonlinearity, both by changing the dipole matrix elements and by adjusting the resonance conditions), by step QWs [15] (Fig. 5.1*c*) and coupled QWs [16] (Fig. 5.1*d*). The primary material systems employed for such bandgap engineering are GaAs/$Al_xGa_{1-x}As$, $In_xGa_{x-1}As/Al_yIn_{y-1}As$ grown strained,

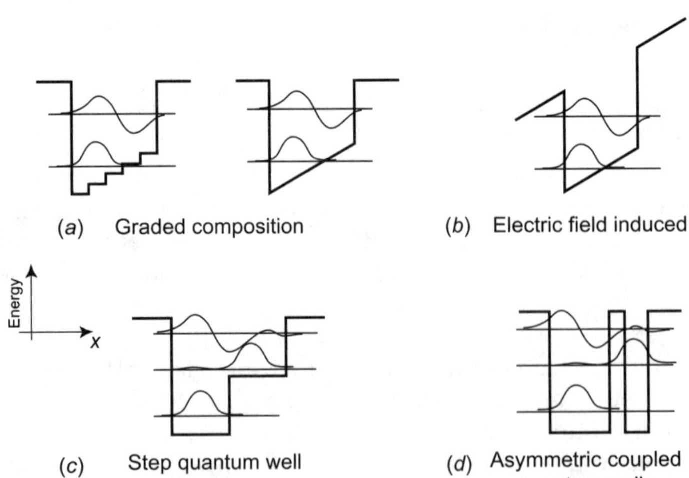

Figure 5.1 Schematics of approaches to QW potentials with an asymmetry leading to observable optical nonlinearities. (*a*) Graded-composition QWs as first proposed by Gurnick and De Temple [11]; (*b*) an asymmetry induced by an applied electric field [14]; (*c*) step QW [15]; and (*d*) asymmetric coupled QWs [16]. (The energy levels and wave functions are schematic only.)

strain-compensated, or lattice-matched on InP, and more recently also GaN/$Al_xGa_{1-x}N$ [17], Sb- [18], and II-VI [19] related materials; and the nonlinear optical effects most commonly investigated are SHG, THG, DFG, SFG, Raman scattering, and the optical Kerr effect.

Following the original proposal by Gurnick and DeTemple, several more theoretical discussions of nonlinear effects based on ISB transitions were published, with the one by J. Khurgin [20] being an excellent and rather complete example. The first experimental evidence followed in 1989. Fejer et al. demonstrated the first evidence for large second-order susceptibility and SHG in a GaAs/AlGaAs QW under an electric field and pumped by an CO_2 laser, in the mid-infrared (mid-ir) wavelength range of 9.6 to 10.8 μm [21]. Rosencher et al. reported the first experimental demonstration of SHG and optical rectification in a step-QW in 1989 [22] and 1990 [23], also in the mid-ir; and Sirtori et al. demonstrated the first SHG in an asymmetric coupled double-QW in 1991 [24]. Third-order optical nonlinearities were first reported in experiments by Segev et al. [25]. Nonlinear light generation in the far-ir and terahertz regime was reported first by Bewley et al. [26] for a semiparabolic graded QW, and by Sirtori et al. for coupled QWs [27]. Heyman et al. reported the first evidence of dynamical screening and the depolarization shift in ISB transitions [28]. Most of these examples use resonant ISB transitions in the conduction band of direct-gap GaAs/AlGaAs and InGaAs/AlInAs QWs. In the early 1990s, the interest shifted partially also to the valence band, with theoretical discussions by Tsang and Chuang [29] and Li and Khurgin [30], as well as to inclusions of additional conduction band valleys such as the L valley in Sb-based QWs by Xie et al. [31]. Good examples of publications summarizing this progress can be found in Ref. 5 or 13.

Aside from the intellectual challenge of designing and fabricating synthetic, engineered, novel materials with new giant optical nonlinearities, much of this work is also motivated by applications, and by supporting functions in a wider range of optoelectronic devices. A general appeal of this work follows from the wavelength coverage of ISB transitions in III-V semiconductor materials, which coincides with the mid-ir and far-ir wavelength range, for which few practical alternative materials exist.

The first application, at the very heart of nonlinear optics, is wavelength conversion, i.e., the use of high-performance lasers operating in one wavelength range to generate light in another wavelength range that is not easily accessible otherwise. Excellent examples of wavelength conversion have been reported in Ref. 32 where the light of near-ir semiconductor laser diodes is used to generate tunable mid-ir light; or in Ref. 33 where near-ir light is converted to terahertz light. One extremely beneficial aspect of wavelength conversion is that significant features of the pump lasers, such as a wide tuning range, are translated to the new wavelength range. As a result, nonlinear light sources often have superior tuning characteristics compared to linear light sources; this fact is amply demonstrated by the commercial availability of *optical parametric amplifiers* and *oscillators* (OPAs and OPOs). For this application, wavelength conversion, ISB-based nonlinearities are of interest simply for their giant nonlinearities, which should lead to very efficient wavelength conversion and hence compact optical systems.

A second group of applications relates to refractive index modulators [34], saturable absorbers [35], optical limiters [36], and Kerr switches [37], essentially a wide range of light modulators in which the transmission of one light signal through the material is controlled by another pump signal or light beam (self-modulation is also possible). In these applications, ISB transitions are of particular interest due to their ultrafast (about picosecond), nonlinear relaxation rates [38]. Recent work on ISB optical transitions for the communications wavelength range (~1.55 μm) is a prime example of this concentration [39].

A third application is nonlinear detectors, such as nonlinear QWIPs [40]. The interest stems from the operating wavelength in the mid-ir and the ability of phase-sensitive light detection, such as up-conversion autocorrelation.

Another recent interest stems from the interplay between quantum optics and nonlinear optics as ISB resonances have become narrow enough (as a result of improved material quality) to display the strong coupling effects previously reserved for vapors or ions in solid-state crystals [41]. This is a recent field of interest, which includes ideas as varied as photon pair generation, electromagnetic-induced transparency [42], lasing without inversion [43], and similar effects.

Aside from the considerable successes that have been documented for ISB-based optical nonlinearities (such as the ones given above), several drawbacks are also inherent to them, which need specific attention.

Leading among those difficulties is the fact that the nonlinearities are only strongly enhanced for light traveling in-plane with the QWs, as significant optical dipole matrix elements only exist for an electric field polarized normal to the QW plane, i.e., the plane of electron confinement. As a result, for efficient nonlinear light generation or manipulation, light needs to be efficiently coupled into the in-plane configuration, which is inherently more difficult than normal incidence coupling (but not impossible).

A second issue is the resonant absorption accompanying most ISB-based optical nonlinearities. The strength of the optical nonlinearity is about proportional to the density of electrons in the asymmetric QW potential, assuming only one energy level, usually the lowest one, is occupied. As this lowest energy level is a real state with real electron population, it induces strong resonant absorption of either the pump light or the nonlinear light, or even both, depending on the nature (real or virtual) and resonance condition of each subsequent energy level. Real and near-resonant energy levels show resonant enhancement of both the nonlinearity and the absorption, with the latter outweighing the former under most conditions [12, 13]. Hence, only near-resonant optical nonlinearities are used, or new methods of populating the nonlinearities with charge carriers have to be found and implemented. We show in the subsequent sections how a quantum cascade approach can be used for this implementation.

The third challenge is related to phase matching. As semiconductor materials have noticeable nonzero material dispersion across the wide wavelength spans of nonlinear optics, the copropagation of linear and nonlinear light generally leads to destructive interference of the nonlinearly generated light along the propagation direction of the primary pump beam. Additionally, the fact that the optical nonlinearity is strong mostly for the in-plane light favors copropagating pump and nonlinear light within a waveguide structure; this adds complexity to the situation in the form of waveguide dispersion. The commonly used III-V materials are furthermore not (sufficiently) birefringent such that birefringence phase matching is not a viable option. Nevertheless, quasi-phase matching has been demonstrated [44], and modal phase matching has also been implemented [45] and is described in greater detail in a later section.

Overall, the initial shortcomings of in-plane propagation, resonant absorption, and the need for phase matching have been repeatedly addressed, and with good success. The most recent solution comes from the integration of the ISB optical nonlinearities with ISB QC lasers, the main topic of the second part of this chapter.

5.1.3 Theoretical model

Before focusing on the implementation of optical nonlinearities into QC lasers, we will briefly present the theoretical framework used here to calculate the strength of the optical nonlinearities and the expected nonlinear optical output power. This theoretical model employs density matrix formalism following the calculation of the energy levels and wave functions from solving Schrödinger's equation in the envelope function approximation. The electromagnetic fields are treated classically with Maxwell's equations. Excellent textbook discussions of these formalisms can be found in Refs. 2, 3, and 4. We here reproduce the derivation that we presented in Ref. 48, which in itself follows the textbook of Ref. 2.

Figure 5.2 shows a schematic of the conduction band diagram of a coupled QW structure under an applied electric field, which was previously designed for efficient SHG [46]. This nonlinearity is populated with electrons following a cascade approach, such that under the appropriate bias energy level 3, the center level of the nonlinear level triplet is populated. This is one of the approaches that strongly reduce the absorption of the nonlinear light; it does, however, carry the penalty of pump-light absorption. Level 2 is considered practically empty of electrons, which quickly scatter through resonant longitudinal optical phonon emission into level 1 [6]. The structure is designed with large optical dipole matrix elements between all levels 2, 3, and 4; this structure "D2912" is discussed in greater detail in Ref. 48 and Sec. 5.4.2.

Equations for the density matrix elements ρ_{mn} for this multilevel system can be written in the following general form [2]:

$$\frac{d\rho_{mn}}{dt} + \left(\gamma_{mn} + i\omega_0^{mn}\right)\rho_{mn} = \frac{ie}{\hbar}\,\mathbf{E}\sum_q\left(z_{mq}\rho_{qn} - z_{qn}\rho_{mq}\right) \tag{5.1}$$

$$\frac{d\rho_{mm}}{dt} + \hat{R}_m = \frac{ie}{\hbar}\,\mathbf{E}\sum_q\left(z_{mq}\rho_{qm} - z_{qm}\rho_{mq}\right) \tag{5.2}$$

Here γ_{mn} is the relaxation rate of the off-diagonal element ρ_{mn} of the density matrix, ω_0^{mn} is the transition frequency, z_{mn} is the optical dipole moment of the transition $m - n$, and $\hat{R}_m$ denotes all relaxation and pumping terms that determine the population $n_m = \rho_{mm}$ of the mth state in the absence of a radiation field; e is the electron charge, and $\hbar$ is the reduced Planck constant.

The electric component $\mathbf{E}(\mathbf{r}, t)$ of the radiation field is considered to be a sum over quasi-monochromatic components with slowly varying

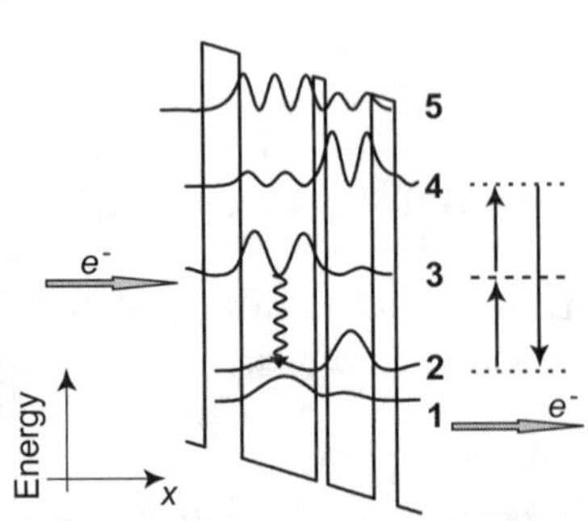

Figure 5.2 Conduction band diagram and moduli squared of the essential wave functions of a two-well active region design suitable for laser emission (wavy line) between energy levels 3 and 2, and nonlinear light generation from the level triplet 2–3–4 (thin arrows). The thick gray arrows indicate the injection and exit of the electron flow. (Source: Ref. 46.)

amplitudes $\mathbf{F}_{mn}$ and frequencies ω_{mn} nearly resonant to the transition frequencies ω_0^{mn}:

$$\mathbf{E}(\mathbf{r}, t) = \sum \mathbf{F}_{mn}(\mathbf{r}, t) e^{-i\omega_{mn} t} + c.c. \tag{5.3}$$

In the SHG process (see Fig. 5.2) $\omega_{32} = \omega_{43} = \omega$, and $\omega_{42} = 2\omega$, where ω is the fundamental frequency of the pump laser. Similarly, $F_{32} = F_{43} = E_\omega$ and $F_{42} = E_{2\omega}$.

Off-diagonal elements of the density matrix can be written as

$$\rho_{mn}(t) = \sigma_{mn}(t) e^{-i\omega_{mn} t} \tag{5.4}$$

where the amplitude σ_{mn} is a slowly varying function of time. For the diagonal elements, $\rho_{mm} = \sigma_{mm}$.

Using the expressions in Eqs. 5.1 and 5.2, we arrive at the truncated density matrix equations

$$\frac{d\sigma_{mn}}{dt} + \Gamma_{mn}\sigma_{mn} = \frac{ie}{\hbar}\sum_q \left(z_{mq}\mathbf{F}_{mq}\sigma_{qn} - z_{qn}\mathbf{F}_{qn}\sigma_{mq}\right) \tag{5.5}$$

$$\frac{d\sigma_{mm}}{dt} + \hat{R}_m = \frac{ie}{\hbar}\sum_q \left(z_{mq}\mathbf{F}_{mq}\sigma_{qm} - z_{qm}\mathbf{F}_{qm}\sigma_{mq}\right) \tag{5.6}$$

Here $\Gamma_{mn} = \gamma_{mn} + i\left(\omega_0^{mn} - \omega_{mn}\right)$, $\sigma_{mn} = \sigma^*_{mn}$, $z_{mn} = z^*_{mn}$, and $\mathbf{F}_{mn} = \mathbf{F}^*_{mn}$. The nonlinear polarization at the second harmonic can be calculated from Eqs. 5.5 and 5.6 as

$$P(\mathbf{r}, t) = eN_e z_{24}\rho_{42} \tag{5.7}$$

where N_e is the volume density of electrons in the active mixing region. This polarization serves as a source for the electromagnetic field at the second harmonic. [Here we have neglected the potential presence of nonresonant lattice nonlinearities and other near-resonant nonlinearities, such as the cascade 3–4–5 in Fig. 5.2. If the latter were included (as is done later in Sec. 5.4.2 and in Ref. 46), then $P(\mathbf{r}, t) = eN_e(z_{24}\rho_{42} + z_{35}\rho_{53})$.]

In solving for the steady-state solution, all time-dependent terms in the above equations are set to zero, and the calculation of σ_{42} is reduced to algebra. In principle, the expression for the nonlinear polarization defined by Eq. 5.7 is valid for arbitrarily strong field amplitudes and includes all resonant nonlinear effects such as saturation and power broadening. However, for weak enough fields, we can expand the solution in powers of the field amplitudes and arrive at the following expression for the amplitude of nonlinear polarization in the $\chi^{(2)}$

approximation $P(2\omega) = \chi^{(2)}_{2\omega} E_x^2(\omega)$, where $\chi^{(2)}$ is the second-order nonlinear susceptibility:

$$\chi^{(2)}_{2\omega} = \frac{P(2\omega)}{E_x^2(\omega)} = \frac{e^3 N_e}{\hbar^2}\left[\frac{z_{23}\, z_{34}\, z_{24}}{\Gamma_{42}}\left(\frac{n_3 - n_4}{\Gamma_{43}} + \frac{n_3 - n_2}{\Gamma_{32}}\right)\right] \tag{5.8}$$

Here $E_x(\omega)$ is the x component of the electric field at fundamental frequency ω; and x is the coordinate in the growth direction [001], i.e., normal to the QW plane. Increasing the field intensity beyond the saturation value leads to the decrease in the population differences $\propto [1/|E_x(\omega)|^2]$ and broadening of the line widths $\Gamma_{42} \propto |E_x(\omega)|^2$, so that the value of $\chi^{(2)}$ decreases. The saturation intensity is on the order of 1 MW/cm^2, assuming exact resonance, and increases with detuning. In the same weak-field approximation, the nonlinear polarization amplitude at the third harmonic is given by

$$\begin{aligned} P(\omega) &= eN_e z_{25}\sigma_{52} \\ &\cong \frac{ie^4 z_{23}z_{34}z_{45}z_{25}N_e E_x^3(\omega)}{\hbar^3\Gamma_{52}} \\ &\quad \left[\frac{1}{\Gamma_{42}}\left(\frac{n_3 - n_4}{\Gamma_{43}} + \frac{n_3 - n_2}{\Gamma_{32}}\right) - \frac{1}{\Gamma_{53}}\left(\frac{n_4 - n_5}{\Gamma_{54}} - \frac{n_3 - n_4}{\Gamma_{43}}\right)\right] \end{aligned} \tag{5.9a}$$

$$\begin{aligned} \chi^{(3)}_{3\omega} &= \frac{P(3\omega)}{E_x^3(\omega)} \\ &\cong \frac{ie^4 z_{23}z_{34}z_{45}z_{25}N_e}{\hbar^3\Gamma_{52}} \\ &\quad \left[\frac{1}{\Gamma_{42}}\left(\frac{n_3 - n_4}{\Gamma_{43}} + \frac{n_3 - n_2}{\Gamma_{32}}\right) - \frac{1}{\Gamma_{53}}\left(\frac{n_4 - n_5}{\Gamma_{54}} - \frac{n_3 - n_4}{\Gamma_{43}}\right)\right] \end{aligned} \tag{5.9b}$$

5.2 Application of Intersubband Optical Nonlinearities to Quantum Cascade Lasers

5.2.1 Monolithic integration of nonlinearity and pump laser

As shown in Sec. 5.1, nonlinear optics with ISB transitions is a broad and promising field of science and engineering. Recently, these transitions have been implemented into QC lasers for the first time [47]. This integration of QC pump lasers and ISB-based optical nonlinearities has allowed the mitigation of several of the earlier drawbacks of combinations of separate ISB-based nonlinear crystals and external pump lasers, such as a CO_2 laser.

The realization that ISB optical nonlinearities are straightforward to implement into QC lasers stems from the fact that both are based on ISB transitions, both obey the same polarization selection rules, and hence both would couple efficiently to each other via the light field.

In monolithically integrating the optical nonlinearity and the QC lasers into one laser waveguide and cavity, several clear advantages arise. First, the intracavity optical field inside a QC laser can be very high (up to few megavolts per centimeter) [48], essential for an efficient presentation of the nonlinear effects; and the overlap of the nonlinear material and the pump light field can be made large through deliberate semiconductor waveguide design [49], which is a highly mature field, for which excellent solutions and computational modeling tools exist.

The integration of the optical nonlinearity with the QC laser can essentially be accomplished in one of two different ways, both of which are outlined in Fig. 5.3. First, the nonlinear region can be made a separate stack [50] of coupled or otherwise asymmetric QWs inserted into the QC laser cavity wherever the laser field is large (Fig. 5.3*a*). In this case, the nonlinear structure can be separately optimized from the QC laser structure, which is advantageous for both. Yet, the nonlinear region should still be made capable of conducting the current density necessary for QC laser operation. In principle, a multiterminal device can be envisioned (see Fig. 5.3*b*) [50], which bypasses the nonlinear region; nevertheless, despite having been demonstrated, such multiterminal devices are cumbersome to fabricate, especially for large current-carrying devices such as the QC laser. Conversely, two-terminal devices with heterogeneous cores have repeatedly been shown to work well, even when the substacks required different local electric fields for operation [50]. A particular type of such heterogeneous integration is interleaved "interdigitated" active and nonlinear regions with matching injector regions [51].

The requirement for the nonlinear region to carry current has both an advantage and a disadvantage. On one hand, under bias it is fairly straightforward to concentrate electrons in an excited energy level by implementing a cascade scheme similar to that of QC lasers. This is done though the appropriate tailoring of the interlevel scattering times (longitudinal optical phonon scattering for mid-ir ISB transitions). As a result, the nonlinear optical transition can be made transparent for the nonlinear light, which is clearly advantageous. On the other hand, a concomitant requirement is that the tunneling times for electrons through the heterostructure be sufficiently short to allow for a low differential resistance and hence good current and voltage stability [52]. This design criterion implies that the wave functions overlap and mix

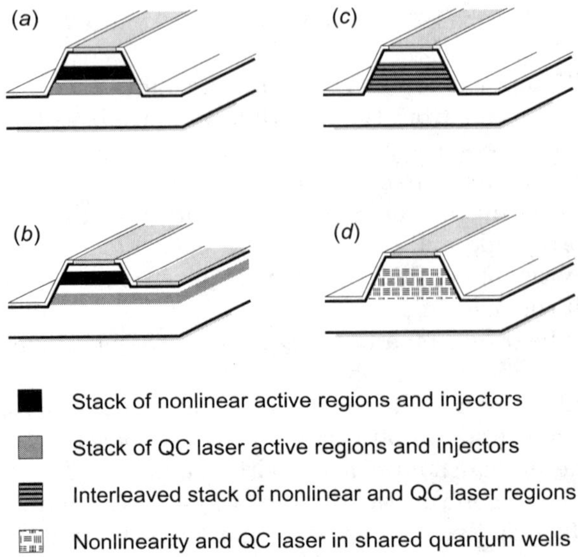

Figure 5.3 Schematics of various approaches to integrate an optical nonlinearity with a QC laser. (*a*) Separate serial stacks (black and gray bands) for the nonlinear and laser materials, and current is flowing through both stacks; (*b*) separate stacks and separate contacts for the laser and nonlinear regions; (*c*) interleaved laser active regions and nonlinear active regions; (*d*) the nonlinearity and laser active regions are implemented within the same active region (and injector) QWs.

across the structure, which generally leads to somewhat smaller optical dipole matrix elements and concomitant nonlinearities than could be achieved by using isolated asymmetric, coupled QWs.

The second method of monolithically integrating the nonlinearity into the QC laser entails integrating it directly into the QC laser structure, taking advantage of the QWs in the latter. Here again, one may use QWs in the injector regions of the QC laser to form the optical nonlinearity—an approach that has, to our knowledge, not yet been tried—or directly employ the laser active region to build the nonlinear region with significant level and population sharing between the two. This approach is highly advantageous because it practically guarantees near resonance for one "leg" of the nonlinear optical transition. Conversely, because lasers and nonlinearity share significant design resources, some compromises in design may have to be made for both. Furthermore, at high optical power, high optical nonlinearity, and high linear-to-nonlinear conversion efficiency, the pump laser transition may be bleached with direct detrimental consequences for laser operation. Nevertheless, this type of monolithic integration has been used predominantly in the literature so far and with good results [46, 47]; the nonlinear power levels are still far from where bleaching would be expected to occur.

5.2.2 Phase (mis)match and nonlinear optical power

The monolithic integration of the optical nonlinearity together with the pump laser into one waveguide and cavity raises a new challenge related to the phase mismatch between fundamental and nonlinear light. Given the (usually) linear geometry of the laser cavity, both waves necessarily are copropagating in guided modes along the same waveguide. First, the waveguide (or, potentially, twin waveguide [53]) must be capable of supporting the modes of all light fields, often at very different wavelengths, involved in the nonlinear process. Yet even then, the material and modal dispersion lead to different propagation constants, and effective modal refractive indices, for the different modes.

One can estimate the power of the nonlinear signal, as we demonstrate here for SHG, that can be obtained from such an integrated waveguide geometry for various conditions of mode (mis)match. As a starting point, the guided modes at the various wavelengths of the "cold" waveguide are calculated. Only the *transverse magnetic* (TM) modes are excited efficiently since the polarization associated with electronic ISB transitions contains only the x component, i.e., the field component normal to the QW plane. In conventional QC laser waveguides, a small number of lateral transverse modes are usually supported; i.e., the laser cavities are rarely transverse single-mode waveguides. Given the dominance of free-carrier absorption in the waveguide loss, which increases about quadratically with wavelength, the multiwavelength waveguides need to be optimized for the longest wavelength to be supported in them. As a result, for the shorter wavelengths the waveguide is "oversized," and a large number of lateral and vertical transverse modes exist, each with its own propagation constant.

For the TM modes it is convenient to use the magnetic field of the modes as a description; the magnetic field amplitude $H_y(\mathbf{r})$ of a given TM mode with frequency ω and longitudinal wave number $k = \omega\mu/c$, where μ is the mode effective refractive index and c is the speed of light in vacuum, can be represented as a product of functions $A(z)$ exp ikz, where the complex function $A(z)$ varies slowly with coordinate z along the waveguide direction, and the transverse distribution $H_\omega(x, y)$, which satisfies the transverse Helmholtz equation for the "cold" waveguide:

$$(\omega, x, y)\left(\frac{\partial}{\partial x}\frac{\partial}{\varepsilon\partial x} + \frac{\partial}{\partial y}\frac{\partial}{\varepsilon\partial y}\right)H + \left(\frac{\omega^2\,\varepsilon\,(\omega, x, y)}{c^2} - k^2\right)H = 0 \qquad (5.10)$$

The x component of the electric field is related to H_y as $\partial Hy/\partial z = i\omega\varepsilon$ $(\omega, x, y)E_x/c$. Substituting the above expression for H_y into the wave equation for the second harmonic polarization, using the orthogonality of the functions H for modes of different order, and solving the resulting equation for the amplitude $A_{2\omega}$, we arrive at the expression for the nonlinear signal power

$$W_2 \sim \frac{128\pi^5 \, \Sigma \, W_1^2 \left[R_1^2 + e^{-\alpha_2 L} - 2R_1 e^{-aL_2/2} \cos(\Delta k L) \right] (1 - R_2)}{\mu_1^2 \mu_2 c \lambda_2^2 \left(\Delta k^2 + \tilde{\alpha}^2 \right) (1 - R_1)^2} \tag{5.11}$$

where $\tilde{\alpha} = \alpha_2 / 2(1/L) \ln(1/R_1)$; α_2 stands for the waveguide intensity losses at wavelength $\lambda_2 = \lambda_1/2$ of the SHG signal; L is the cavity length; $R_{1,2}$ and $\mu_{1,2}$ are the intensity reflection factors of one cavity facet and effective refractive indices of modes at wavelengths $\lambda_{1,2}$, respectively; $\Delta k = k_{2\omega} - 2k_\omega = 2\omega(\mu_2 - \mu_1)/c$ is the phase mismatch; and W_1 is the power in the fundamental mode. We took into account only the interaction of copropagating modes and assumed that the intracavity intensity of the pump has an unsaturated exponential profile along the cavity. The nonlinear overlap factor Σ, which defines an effective interaction cross section $\hat{A} = |\chi^{(2)}|^2/\Sigma$ of two TM modes, is given by

$$\Sigma = \mu_1^4 \mu_2^2 \left(\left| \int \chi^{(2)}(x, y) \frac{1}{\varepsilon_\omega^2} H_\omega^2 H_{2\omega} \, dx \, dy \right|^2 \int \frac{1}{\varepsilon_{2\omega}} H_{2\omega}^2 \, dx \, dy \right) \Bigg/ \left(\int H_{2\omega}^2 \, dx \, dy \int \frac{1}{\varepsilon_\omega} H_\omega^2 dx \, dy \right)^2 \tag{5.12}$$

where $\varepsilon_\omega(x, y)$ and $\varepsilon_{2\omega}(x, y)$ are the dielectric permittivities of the waveguide at frequencies ω and 2ω, respectively. The values of $\hat{A}$ are between 300 and 1000 μm^2 for typical QC laser waveguides, depending on the thickness of the laser core, i.e., the number of linear/nonlinear active regions and injector pairs. Using Eqs. 5.8, 5.11, and 5.12, one can now estimate the expected second-order optical susceptibility $\chi^{(2)}$ and the nonlinear conversion efficiency $\eta = W_2/W_1^2$.

While the particular geometry and coupling of pump laser and optical nonlinearity are evidently highly sensitive to the phase mismatch Δk, the large number of available transverse modes allows for true modal phase matching for a particular choice of engineered waveguide and pair of modes. We discuss phase matching in Sec. 5.7.

The intracavity wave mixing described here for nonlinear QC lasers has been previously proposed and recently demonstrated for interband semiconductor lasers [54] as an efficient mechanism to combine pump and mixing nonlinearity.

Overall, the implementation of nonlinear active regions into QC lasers is highly advantageous as an alternative way to produce very short-wavelength (λ < 4 μm) light in QC lasers [55]. Such short wavelengths are of importance, e.g., for trace gas sensing of lightweight molecules [56]. Furthermore, many of the concepts discussed here are similarly valid for a much wider range of nonlinear optical effects including SFG, DFG, and higher harmonics generation, which in their entirety will significantly broaden the wavelength range and versatility of QC lasers.

5.3 Sum-Frequency Generation in Quantum Cascade Lasers Using Resonant Intersubband Nonlinearities

The first implementation of an optical nonlinearity into a QC laser was reported in Owschimikow et al. [47]. The original goal of this work was SFG based on two basic, rather practical considerations. First, up-conversion into the shortwave (3 to 5 μm) mid-ir would allow using the highly sensitive InSb detector, hence improving the chances of detecting the expectedly small signals; second, SFG, given its multiple wavelengths, would prove interesting and adaptable enough to evaluate the interplay of nonlinearity and QC lasers. In fact, given the nanowatt signal levels of these early structures, the choice of up-conversion was clearly fortuitous, and in addition to the sum-frequency, this early structure surprised by displaying SHG, which was later understood from the width of the ISB resonances and overall resonance conditions.

The structure of Ref. 47 was chosen to be a two-stack, two-wavelength ($\lambda \sim 7.1$ and 9.5 μm) QC laser [50] including a superlattice (SL) section sandwiched between the two laser stacks. The laser waveguide core contained a 500-nm-thick, low-doped InGaAs buffer layer followed by a stack of 16 periods of alternating active and injector regions [7] designed to emit at 7.1-μm wavelength. Each injector region was doped to a sheet density of 2.0×10^{11} cm^{-2}. Separated from this stack by 25 nm of low-doped InGaAs, an 18-period SL of alternating 4.1-nm-thick InGaAs wells and 2.3-nm-thick AlInAs barriers was grown into which nine equally spaced QWs, each 2.4 nm thick, were inserted. The entire SL totals 27 QWs and is doped homogeneously to 3×10^{16} cm^{-3}. After another 25-nm-thick InGaAs spacer layer, a second QC laser stack containing 19 periods of active regions and injectors designed for emission at 9.5 μm followed, doped to a sheet density of 1.9×10^{11} cm^{-2} in each injector. A 200-nm-thick InGaAs layer completes the waveguide core. A conventional, ternary top waveguide cladding, and the InP substrate as bottom cladding are used [49]. Figure 5.4 shows the mode-intensity profiles at the two laser wavelengths as well as at the SFG wavelength, and it also shows the arrangements of the lasers and SL in the waveguide core. The effective modal indices of this structure at the various wavelengths (calculated using a

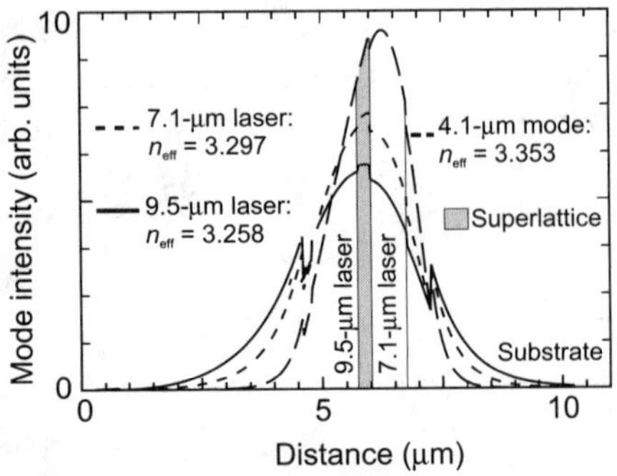

Figure 5.4 Calculated mode-intensity profiles at the two laser wavelengths as well as at the SFG wavelength of the first implementation of a nonlinear QC laser, from Ref. 47. The effective modal indices at the various wavelengths (calculated using a one-dimensional model for infinitely wide waveguides) are given.

one-dimensional model for infinitely wide waveguides) are clearly different; i.e. the modes are noticeably phase *mis*-matched.

This structure provides two distinct sources of resonant second-order ISB nonlinearities, one in the SL and one in the layer structure of the 7.1-μm-wavelength laser. The designs of both structures are shown in Figure 5, and the optical transitions are indicated by the arrows.

In the SL the first transition from the energy levels of the lowest miniband (1 in Fig. 5.5*a*) to the localized states of the inserted QWs (2) is in near resonance with the radiation of the 9.5 μm laser. The second transition, from these localized states (2) to the bottom of the second miniband (3), is near resonant with the 7.1-μm radiation. We calculate the transition frequencies ω_{ij} ($i, j = 1, 2, 3$) of $\hbar\cdot\omega_{21} = 134$ meV, $\hbar\cdot\omega_{32} = 171$ meV, $\hbar\cdot\omega_{31} = 305$ meV, and the corresponding optical dipole moments $z_{21} = 0.4$ nm, $z_{32} = 1.3$ nm, and $z_{31} = 0.02$ nm.

The second optical nonlinearity can be found in the 7.1-μm QC laser (Fig. 5.5*b*). While QC laser action is taking place between energy

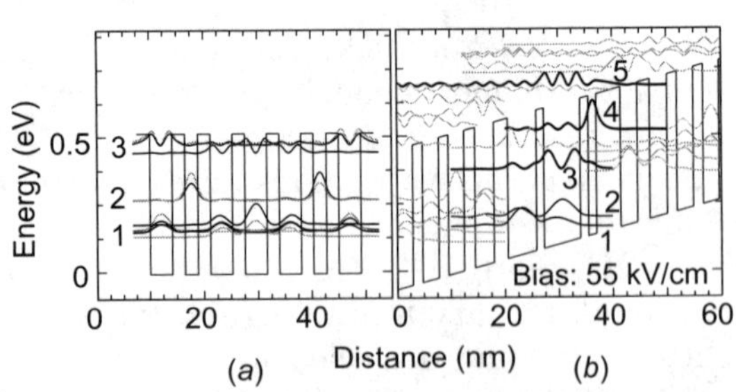

Figure 5.5 Schematics of the conduction band diagrams of the mixing (*a*) *SL* and (*b*) QC laser active region incorporating the optical nonlinearity. In (*a*) the resonant nonlinearities result from transitions 1–2–3. In (*b*) the QC laser process is supported by levels 3, 2 and 1; the optical nonlinearities stem from resonant transitions 3–4–5. (Source: Ref. 47.)

levels 3 and 2 [6, 7], near-resonant optical transitions of 9.5 and 7.1 μm can be found from level 3 to level 4 and from the latter on to level 5, respectively. Also here we calculate the relevant parameters as $\hbar\cdot\omega_{43}$ = 143 meV, $\hbar\cdot\omega_{54}$ = 164 meV, $\hbar\cdot\omega_{53}$ = 307 meV, z_{43} = 0.7 nm, z_{54} = 1.8 nm, and z_{53} = 0.4 nm. The line broadening at the pump frequencies is about 10 meV, and at a mixing frequency it is 20 to 30 meV. The presence of the nonlinearity in the QC laser region was somewhat fortuitous, and it enhanced the overall optical nonlinearity. In addition to the wafer above, a second wafer was grown, in which the SL was entirely homogeneous (alternating 4.1-nm-thick InGaAs wells and 2.3-nm-thick AlInAs barriers) and did not contain energy levels 2; in this wafer only the nonlinearity within the 7.1-μm QC laser was active.

The second-order nonlinear susceptibilities were estimated by using the density matrix approach for the above nonlinearities, with SFG mixing at 4.1 μm resulting from the 7.1-μm QC laser region providing the largest optical nonlinearity. Including all ISB transitions, a second-order optical susceptibility of $|\chi^{(2)}| \approx 3.4 \times 10^{-10}$ $(N_3/10^{16}\ \mathrm{cm}^{-3})$ m/V was estimated, where the population N_3 of state 3 changes between ~2 to ~5 × 10^{16} cm^{-3} as the current increases. This value of $\chi^{(2)}$ is approximately one order of magnitude greater than the lattice nonlinearity of the waveguide core.

Deep-etched ridge waveguide lasers were processed in conventional fashion [6, 7] with ridge widths ranging from 10 to 20 μm and cleaved cavity lengths from ~1 to 3 mm. The lasers were mounted in a temperature-controlled He flow cryostat and were operated in pulsed mode with pulse widths ranging from 50 to 600 ns and at a 1.8- to 85-kHz repetition rate. The laser output power was measured by using a calibrated, room temperature HgCdTe (MCT) photovoltaic detector. The laser spectra were measured by using the rapid-scan mode of a Nicolet 860 *Fourier transform infrared (FTIR) spectrometer* and a liquid-nitrogen-cooled HgCdTe photocurrent detector. The short-wavelength radiation was measured by using a calibrated, liquid-nitrogen-cooled InSb photovoltaic detector fitted with a sapphire flat to suppress the long-wavelength radiation. The short-wavelength spectra were obtained with the same detector and by using the FTIR spectrometer in step scan mode. We furthermore used a set of interference filters in the 4- to 6-μm-wavelength range and a wire-grid polarizer to discriminate between the various components of the emitted radiation. We performed similar measurements with other two-wavelength lasers [50] to rule out spurious effects in the detection optics and electronics.

Figure 5.6*a* shows high-resolution spectra of a laser obtained at various levels of pumping current and at 10 K heat sink temperature.

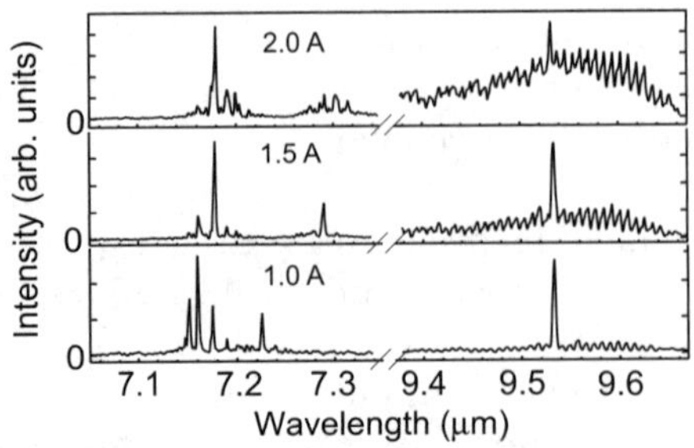

(a)

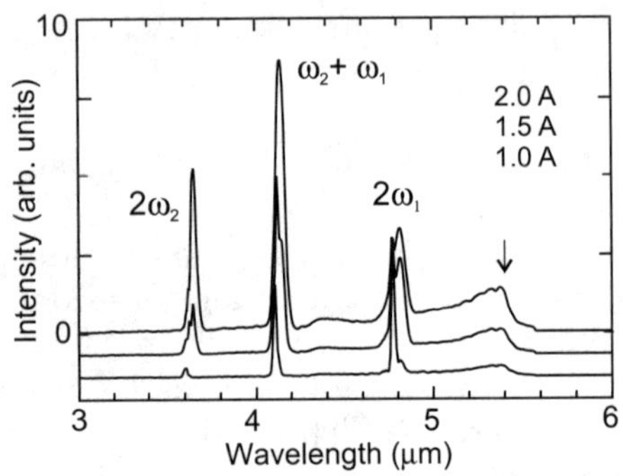

(b)

Figure 5.6 (*a*) High-resolution spectra measured of a 10-μm-wide, 1.6-mm-long laser under pulsed operation at various levels of peak current, as indicated. (*b*) Low-resolution step-scan spectra of the same device measured under the same operating conditions in the short-wavelength region. The spectra are offset with respect to one another, and the peaks are labeled in accordance with their origin, SFG and two SHG signals. The small arrow indicates the spectral cutoff of one filter used in the experiment. (Source: Ref. 47.)

Figure 5.6*b* shows the corresponding short-wavelength spectra. We observe SFG and SHG at 4.1 ($\omega_1 + \omega_2$) and 3.6 ($2\omega_2$) and 4.75 μm ($2\omega_1$), respectively. With increasing pumping level the relative strengths of the various components change, but we consistently observe all three components. The broad, slowly rising background, which we interpret as spontaneous emission from electrons excited into high-lying energy levels and the continuum above the barriers, can also be found in conventional QC lasers under high pumping conditions.

In Fig. 5.7 the light output power and voltage versus current (*L-I-V*) characteristics of the laser of Fig. 5.2 are shown, as well as the power of the nonlinearly generated radiation. The measurements have been taken at 10 K heat sink temperature. Laser threshold is reached at 550 mA approximately simultaneously for both wavelengths, and peak power is reached at 2.6 A with 90 and 60 mW for 9.5- and 7.1-μm wavelength, respectively. Close to peak power, a sudden, strong

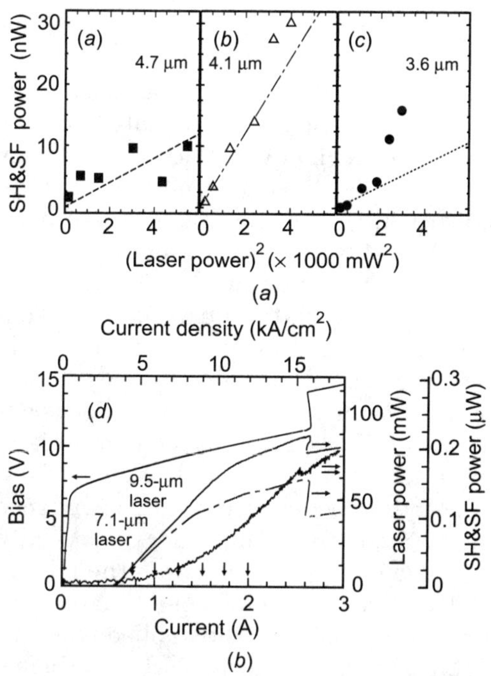

Figure 5.7 (*a* – *c*) Optical power measured for SFG signals [open triangles in (*b*)] and SHG signals [filled circles and squares for (*c*) 7.1 and (*a*) 9.5 μm, respectively] versus the power squared of the respective fundamental pump light. The symbols represent data, and the straight lines are from calculations. (*d*) Light output and voltage versus current characteristic and combined short wavelength power of the laser of Fig. 5.6. The little arrows pointing to the current axis indicate the positions where the spectra of Fig. 5.2 and the data points of parts (*a*) to (*c*) have been taken. (Source: Ref. 47).

increase in operating voltage occurs, indicative of a negative differential resistance region. We interpret this behavior as domain formation in the SL [52], when the low doping level of the latter is not anymore capable of screening the applied external electric field. Such occurrence can be mitigated by higher doping or a modified design of the SL. Nevertheless, as can be seen in Fig. 5.7*d*, the combined short-wavelength emission reaches a power level of 180 nW. Figure 5.7*a* through *c* show a detailed measurement of the optical power in each mixing component versus the power product of its generating laser sources. These data have been deduced from step-scan spectra and the laser data of Fig. 5.7*d*. The nonresonant background has been subtracted; yet the limited resolution of the step-scan spectra likely leads to a slight overestimation of the power at low pumping levels.

The results of our theoretical calculations are also shown in Fig. 5.7*a* to *c*. The power of each mixing signal has been obtained from solving the wave equation with the nonlinear polarization as a source, as outlined in Sec. 5.2.2, taking into account the phase mismatch of the waveguide modes and their losses due to diffraction and free-carrier absorption. Resonant losses due to resonant intersubband absorption have been neglected, first, because of the nonresonant nature of the nonlinearities, and, second, because of the dominant factor of the non-phase-matched structure. Similar formulas have been obtained for the second-harmonic signals. They are all shown in Fig. 5.7*a* to *c* as theoretical curves, evaluated for $N_3 = 4 \times 10^{16}$ cm^{-3}, and are in good agreement with the data. However, at high pumping levels, the data display noticeable deviations from the calculated linear curves. This can be understood from several effects. First, as the current flow is increased through the structure, also the relative electron population of the upper laser level (3 in Fig. 5.5*b*) increases; e.g., an increase of N_3 to 5×10^{16} cm^{-3}, consistent with QC laser operation, can already explain the deviation of the experimental curves (Fig. 5.7*b* and *c*) from straight lines at high powers. Another effect concerns the dependence of the dipole moments on the applied bias. This effect, especially important for dipole moment z_{54}, also leads to an excess of the mixing signal power as compared to the linear dependence. Finally, the frequencies of the ISB transitions also depend on the applied bias. We calculate an increase of ω_{43} with bias, while ω_{54} decreases, both by a few millielectronvolts over the range of applied electric fields deduced from the experiment (approximately 50 to 65 kV/cm). The change in ω_{43} is more important since the line width γ_{43} is the smallest of all three transitions. The susceptibility $\chi^{(2)}$ at 3.6 μm is tuned to resonance with increasing ω_{43}, while the one at 4.75 μm becomes detuned from resonance. This can explain the different behavior of the second-harmonic signals in Fig. 5.7*a* and *c*.

From this early work on SFG, important lessons were learned for the continued development of the field. First, the proof of principle had been made, and a general trend of robustness (especially with respect to the width of the resonance) was discerned; i.e., even though the structure was designed for resonant SFG, both SHG signals also were measured, Second, however, it was evident from the smallness of the signal that the nonlinearities had to be improved (which was very feasible—given that the nonlinear electron potential in the SL was very "symmetric" and the nonlinearity in the QC used an above-band-edge state). Third, it appeared that using a nonlinearity within the QC laser was preferable to separate SLs to avoid domain formation in the SLs (which certainly breaks up level alignment) and limited current transport (and laser power). Alternatively, SLs designed specifically for large current transport would have to be used. Finally, the width of the ISBs provided for both SFG and SHG; therefore, just SHG would be simpler to use, allowing more of the waveguide core to be taken up by one type of nonlinearity and laser, thus strengthening the nonlinearity and increasing the laser power, and simplifying waveguide design as fewer wavelengths needed to be properly supported. As a result, subsequent work focused on optimization of the SHG signal, both in quantum design and later in phase-matching approaches.

5.4 Second Harmonic Generation in Quantum Cascade Lasers Using Resonant Intersubband Nonlinearities

After the first demonstration of nonlinear light emission from QC lasers, the primary focus was to increase the overall nonlinear optical power. Three different strategies are available to accomplish this: an improved quantum design of the nonlinear region, an increase in the pump laser power, and finally phase matching of nonlinear and pump light. All three strategies favor a move from SFG and separate (SL) nonlinear regions to SHG and nonlinearities implemented inside the laser active region. First, SHG provides a smaller number of wavelengths, thus simplifying phase matching (which is discussed in Sec. 5.7); second, not having to implement two pump laser wavelengths into one laser core enables more nonlinearity to be available for SHG; third, the integrated nonlinearity is more stable (as stable as QC lasers) under high current injection, and this added stability also allows for higher laser pump currents and pump power.

Even after the decision is made to implement the nonlinear transition into the active laser region, several options remain; the nonlinearity can use a separate set of energy levels from the laser (as shown in Sec. 5.3), or the laser transition can be one part of the nonlinear structure.

In implementing the latter, an automatic resonance of the pump light with one "leg" of the nonlinear optical transition cascade is achieved, hence maximizing one portion of the nonlinear susceptibility (e.g., by minimizing one denominator in Eq. 5.8). Furthermore, when one is implementing the nonlinearity in a cascade scheme, the electrons can be accumulated in an excited energy level, rather than the ground level, hence reducing resonant absorption for the nonlinear light. Nevertheless, such close integration of nonlinear and laser active regions may incur respective tradeoffs including bleaching of the laser transitions (however, such saturation has not yet been observed).

In this section, a "second-generation" SHG in QC lasers, in which the QWs of the active regions simultaneously function as nonlinear oscillators, is presented. These structures were originally described in Ref. 46. In this design the essential parameters for laser performance and SHG, such as resonance conditions, optical dipole matrix elements, and carrier density, have been optimized. As a result, up to ~1 W of pump power at ~7.5 μm was then converted up to ~1 μW of SHG at ~3.75-μm wavelength, which is a two-orders-of-magnitude improvement in the nonlinear power over that presented in Sec. 5.3. The linear-to-nonlinear power conversion efficiency was improved in devices operating at 9.1- (pump) and 4.55-μm (SHG) wavelength also by another factor ~100. Most of these improvements are the result of proper design of the electronic level structure; phase matching and improved waveguide design are the topics of Sec. 5.7.

At a minimum, high-performance QC lasers contain at least four energy levels, termed *g*, 3, 2, and 1, in their active regions, and can be envisioned as four-level laser schemes, with the laser transition set between levels 3 and 2. Examples of QC laser band structures are given in Fig. 5.8. For highest performance, the energy levels and wave functions, and the laser pumping mechanism (current), are designed such that under operation only levels 3 and *g* (closely coupled to 3) are occupied with electrons, and the optical dipole matrix element between levels 3 and 2 is as large as possible (usually 1 to 2 nm). These two conditions are also highly favorable for nonlinear ISB transitions, namely, a population concentrated in a single energy level (or a strongly coupled, close-level ensemble), and large optical dipole matrix elements.

In particular, as we discussed in Sec. 5.2.2, an ISB structure capable of efficient SHG contains a triplet of energy levels (i, ii, iii), in near resonance with the pump radiation at frequency ω and SHG at 2ω, and the product of the three optical dipole matrix elements $z_{\text{i-ii}}$, $z_{\text{i-iii}}$, and $z_{\text{ii-iii}}$ between all three levels is large. We adopt a simplified version of Eq. 5.8, the absolute value of the second-order nonlinear susceptibility $\chi^{(2)}$, as the figure of merit of our design; the latter is given as [12]

$$\chi^{(2)} \propto e^3 N_e \frac{z_{\text{i-ii}}\, z_{\text{ii-iii}}\, z_{\text{i-iii}}}{(\hbar\omega - E_{\text{i-ii}} - i\cdot\gamma_{\text{i-ii}})\cdot(2\hbar\omega - E_{\text{i-iii}} - i\cdot\gamma_{\text{i-iii}})} \tag{5.13}$$

where N_e is the electron density in the active region, assumed to be concentrated in energy level 3, and $E_{j\text{-}k}$ and $2\gamma_{j\text{-}k}$ are the energy difference and transition broadenings between levels j and k (j, k = i, ii, iii), respectively. Starting from the requirement that the upper QC laser level (levels 3 and *g*) be the only one significantly populated with electrons under laser operation, and hence must coincide with one energy level of the ISB nonlinearity, one can build the nonlinear cascade. Two separate workable design approaches are shown in the following.

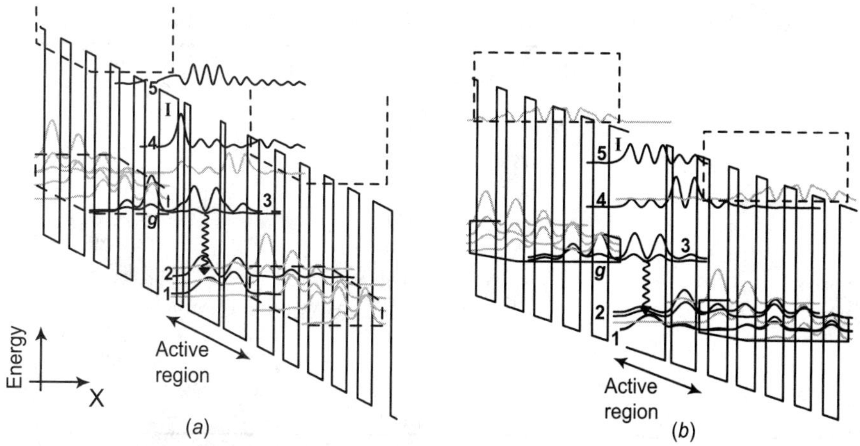

Figure 5.8 Portions of the conduction band diagram, one active region sandwiched between two injector regions, and the moduli squared of the essential wave functions of designs (*a*) D2886 and (*b*) D2912. The significant energy levels are labeled 1 to 5 inside the active region and *g* for the ground state of the injector. The dashed lines indicate the extent of the minibands inside the injector regions. The laser transition (3 → 2) is marked by a wavy arrow. The thicknesses in nanometers of the QWs and barriers of one period of active region and injector of D2886 are, from left to right (i.e., electron down-stream) and starting with the injection barrier *I*, **4.3**/1.4/**1.3**/6.8/**1.1**/5.1/**2.6**/3.6/**2.2**/3.4/**2.1**/3.3/**2.1**/3.3/**2.5**/3.2, the barriers are indicated by bold font, and the underlined layers are doped to 4.5 × 10^{17} cm^{-3}. The analogous layer sequence of D2912 is **4.1**/8.3/**1.3**/5.2/**2.6**/4.1/**2.1**/3.9/**2.3**/3.7/**2.5**/3.5/**2.6**/3.3; the doping density was 3.0 × 10^{17} cm^{-3}. (Source: Ref. 46.)

5.4.1 SHG in QC lasers with three-QW active regions

In Fig. 5.8*a* the conduction band structure of one active region with its preceding and subsequent injector regions is shown. This QC laser active region, labeled D2886 [46], contains three coupled QWs—a thin first QW, followed by two wider ones. The latter are the essential QWs sustaining the laser transition, between levels 3 and 2. The center QW is furthermore the main origin for energy level 5, and the thin first QW mainly supports level 4. The pump radiation at the fundamental frequency is generated between levels 3 and 2. Nonlinear cascades can be found for the level triplets 2–3–4 and 3–4–5. As the energy positions of levels 3 and 5 are mainly governed by the center QW, and that of level 4 by the thin first QW, variations in the relative thicknesses of the two QWs quickly allow optimization of the resonance condition; in particular, the thin well enables adjust one to level 4 practically independently of levels 3 and 5. Furthermore, the spatial asymmetry of the QWs—from localizing them in separate but adjacent wells—ensures sizable optical dipole matrix elements [5]. The specific design parameters for the SHG structure D2886 shown in Fig. 5.8*a* are given in Table 5.1.

TABLE 5.1 Design and Device Parameters of Samples D2886 and D2912

	D2886	D2912		D2886	D2912
E_{23} (meV)	167	131	γ_{45} (meV)	15	15
E_{34} (meV)	176	123	γ_{35} (meV)	25	20
E_{45} (meV)	143	105	τ_{32} (ps)	2.6	3.6
λ_1 (μm)	7.4	9.5	τ_3 (ps)	1.2	1.8
z_{23} (Å)	18	17	τ_2 (ps)	0.3	0.3
z_{24} (Å)	0.4	6	L_p (nm)	48.2	49.5
z_{34} (Å)	3.5	23	N_e (cm^{-3})	2.0×10^{17}	1.5×10^{17}
z_{45} (Å)	20	19	Waveguide Parameters		
z_{35} (Å)	2	8	$\mu_{\mathrm{eff},0}$	3.29	3.28
γ_{23} (meV)	12	10	$\mu_{\mathrm{eff},2\omega}$	3.35	3.35
γ_{24} (meV)	25	20	α_0 (cm^{-1})	10.5	15
γ_{34} (meV)	15	15	$\alpha_{2\omega}$ (cm^{-1})	2	3

Thirty-one periods of active regions (containing the laser active region and the nonlinearity) and injectors were grown at the core of a low-loss QC laser waveguide with ternary top cladding, which implies an about twofold increase in each active element (laser active regions and nonlinear regions) compared to the original first-generation designs shown in Sec. 5.3 [46, 47].

5.4.2 SHG in QC lasers with two-QW active regions

A closer examination of the structure of Fig. 5.8*a*, however, reveals some shortcomings. First, the very choice of a three-QW design also results in another energy level situated between levels 3 and 4. While out of resonance with any radiation inside the cavity, and thus not severely impeding the process, it still leads to smaller than possible dipole matrix elements for the other transitions. Such argument follows from a sum rule of the oscillator strength for intersubband transitions [57]. Therefore, we adopted a new design, labeled D2912 [46], of the QC laser active region with nonlinear cascade with only two QWs. The conduction band structure and moduli squared of the essential wave functions of one active region as well as of its preceding and following injector regions are shown in Fig. 5.8*b*. A detailed picture of the active region of this design is also shown in Fig. 5.2. The structure has been calculated for an applied electric field of 38 kV/cm; QC laser action again takes place between levels 3 and 2. As the first "leg" of the nonlinear cascade 2–3–4 coincides with the laser transition, i.e., the fundamental pump light, their remaining in resonance is trivial. Resonance of the second "leg" can be achieved by relative thickness variations of the two QWs and the barrier between them. It is worth noting that, in this design of the nonlinear cascade, resonant absorption of the SHG radiation is negligible, as the free-electron population is in level 3 rather than in level 2.

The full set of device parameters for this type of design (D2912) is also given in Table 5.1. Wafer D2912 was grown with a conventional QC laser waveguide [46, 49]. Fifty periods of active regions and injectors are sandwiched between two layers of InGaAs layers. The bottom cladding is provided by the InP substrate, and the top cladding is made from an inner 2.1-μm-thick AlInAs layer doped to $n_{Si} \sim 1 \times 10^{17}$ cm^{-3} and an outer AlInAs layer, 400 nm thick and doped to $n_{Si} \sim 2 \times 10^{17}$ cm^{-3}, which are also capped by a 350-nm-thick, highly doped ($n_{Si} \sim 5 \times 10^{16}$ cm^{-3}) layer of InGaAs. The calculated modal refractive indices and waveguide attenuation coefficients of all waveguides at both the fundamental ($\mu_{eff,0}$ and α_0, respectively) and SHG ($\mu_{eff,2\omega}$ and $\alpha_{2\omega}$, respectively) wavelengths are summarized in Table 5.1.

The estimations for $\chi^{(2)}$ may be uncertain within a factor of 2 or 3, mainly because of differences between the designed and actual transition frequencies and the doping levels. We obtain $|\chi^{(2)}| = 340$ and 2×10^4 pm/V, for structures D2886 and D2912, respectively. The nonlinear coefficient for D2912 is rather large, about two orders of magnitude higher than that in bulk III-V semiconductors.

In both cases D2886 and D2912, the contributions from 2–3–4 and 3–4–5 cascades turn out to be of comparable order of magnitude, but with leading contribution from the 2–3–4 cascade. In estimations of the linear-to-nonlinear power conversion efficiency η, the largest uncertainty comes from $|\chi^{(2)}|^2$, and there are additional sources of uncertainty: the phase mismatch Δk and overlap factor Σ. Variations of these parameters also cause the spread of η between different devices of the same structure. Nevertheless, values of $\eta = 0.8$ μW/W^2 for D2886 and $\eta = 300$ μW/W^2 for D2912 are calculated. These values are in reasonable agreement with the experimental values.

Shu et al. [58] recently reported improved designs of nonlinearities implemented in QC laser active regions only employing the 2–3–4 resonance triplet and maximizing the optical dipole matrix elements. These designs are shown in Fig. 5.9.

Design D3043 localizes and isolates the lower level of the laser transition, separating it from the injector to which it was coupled in previous designs. Decoupling the lower energy level from the injector states increases the values of the dipole matrix elements for transitions from and to this level. The coupling of the upper laser level with the injector ground level is taken into account by summing the oscillator strengths. The calculated dipole matrix elements are $z_{12} = 22.4$ Å, $z_{23} = 22.6$ Å, and $z_{31} = 6.88$ Å; and the optical nonlinear susceptibility is calculated to be $\chi^{(2)} = 8.8 \times 10^4$ pm/V, a factor-of-4 improvement in

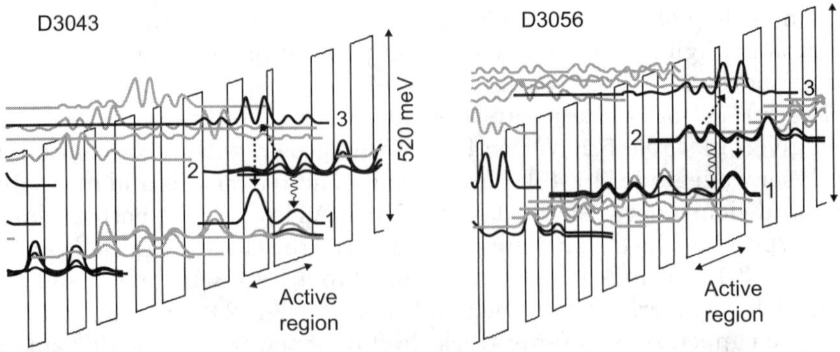

Figure 5.9 Schematics of the conduction band diagrams of two novel QC laser designs (D3043 and D3056) with integrated resonant optical nonlinearities [58]. The active region layer thicknesses in nanometers for D3043 and D3056 are 5.3/1.2/8.2 and 6.7/1.0/5.5, respectively, starting from the left wells. The respective laser transitions are indicated by the wavy arrows; the nonlinear transitions are indicated by dotted arrows. (Source: Ref. 58.)

optical nonlinearity over D2912, at the about same designed fundamental and SHG wavelengths of 9.2 and 4.6 μm, respectively.

In design D3056, optimization of the nonlinear susceptibility is achieved by isolating the uppermost level 3 of the optical nonlinearity. To accomplish this, the energy level was placed in the electronic "upstream" quantum well of the two-QW active region, away from the upper miniband where it had been located in previous designs. As a result, the upper laser level must be placed in the electronic "downstream" quantum well, a novel configuration for QC laser levels since such a design incurs a tunneling time penalty for electrons in the upper laser level and the injector ground level. The calculated dipole matrix elements are z_{12} = 24.1 Å, z_{23} = 29.3 Å, and z_{31} = 8.0 Å; and the optical nonlinear susceptibility is calculated to be $\chi^{(2)} = 1.3 \times 10^5$ pm/V, an almost one-order-of-magnitude improvement in nonlinear optical susceptibility over D2912. The designed fundamental and SHG wavelengths are 8.9 and 4.5 μm, respectively. Work on both structures, especially in light of improving their laser performance, is ongoing; both are examples in which the nonlinear design was the primary goal, and laser design came second, while earlier designs had been using successful laser designs as starting points and modified them to become nonlinear crystals.

Again the lasers were processed and characterized in conventional manner, as previously outlined. The cryostat was fitted with a doubly *antireflection* (AR) coated ZnSe window, yet control measurements have also been made using a quartz window; lenses with large numerical aperture, made from AR-coated ZnSe, Ge, or uncoated CaF, were used to collimate and refocus the highly divergent light emission from the cleaved facets. Quartz, silica glass, sapphire, and an undoped InP flat were used as optical filters to discriminate between fundamental pump and SHG light. When the laser light was measured at the fundamental frequency, attenuation screens were used to avoid saturation of the detector and detection electronics, which in itself would result in erroneous nonlinear signals. After measuring the linear and nonlinear light output versus current (*L-I*) characteristics, we deduce the external linear-to-nonlinear power conversion efficiency by graphing the nonlinear power versus the square value of the linear power. A linear fit to this curve (where suggested) results in the external power conversion efficiency.

Figure 5.10*a* shows the linear and nonlinear *L-I* characteristics of a representative device of D2886; approximately 0.5 W of fundamental peak power results in about 200 nW of SHG light. The inset reports the external linear-to-nonlinear power conversion efficiency, which is close to ~1 μW/W^2. The kink at about 2.4 A in the linear *L-I* curve is a frequent occurrence with QC lasers. It is mostly attributed to the onset of higher-order transverse modes; i.e., the laser ridges are wide enough to support two, sometimes even three transverse modes in lateral direction. The kink is also reproduced in the nonlinear *L-I* curve and the external conversion efficiency curve. As different transverse modes have different modal effective indices and attenuation coefficients, it is not surprising that the external efficiency can change noticeably at such a kink.

Figure 5.10*b* shows the linear and nonlinear *L-I* curves for a characteristic device of D2912. Here, about only 100 mW of fundamental peak output power is transformed to nearly 550 nW of

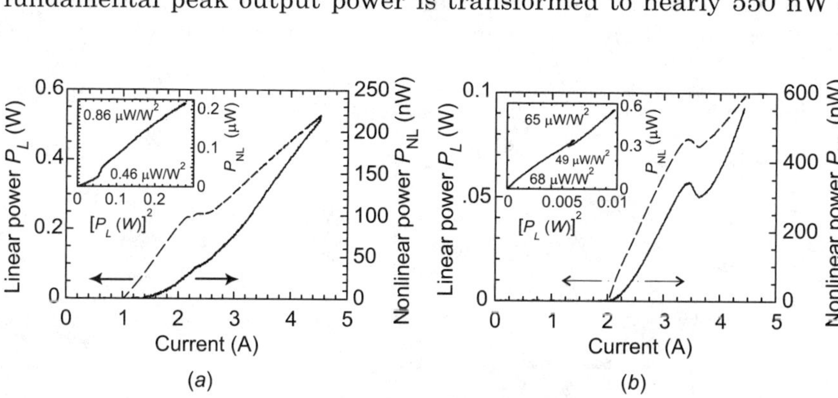

Figure 5.10 Linear (dashed) and nonlinear (solid) light output versus current characteristics of two representative devices (*a*) D2886 and (*b*) D2912. The insets show the respective external linear-to-nonlinear conversion efficiency. The numerical values inside the inset graphs have been extracted from least-squares fits to near-linear portions of these graphs. All data were taken in pulsed current mode at ~10 K heat sink temperature. The device of (*a*) was 12 μm wide and 3.04 mm long; the dimensions of the device of (*b*) were 13 μm and 2.25 mm, respectively. (Source: Ref. 46.)

SHG light. The external power conversion efficiency is accordingly much higher than that of the device of Fig. 5.3*a*, from a factor of 50 to 100 higher. Lasers of D2886 showed values up to ~1 μW/W^2. Devices of D2912 achieved greatly increased external conversion efficiencies up to 100 μW/W^2. A strong spread in the values of the external conversion efficiency, which can span almost a decade, can be detected.

The overall highest nonlinear power (of a nonlinear QC laser without a phase-matching waveguide) was measured from a laser of D2886, as shown in Fig. 5.11, which reports the linear and nonlinear light output and voltage versus current (*L-I-V*) characteristics of the one device of D2886. We obtained 2 μW of SHG light at 1 W of fundamental peak output power.

Figure 5.12*a* to *c* reports spectral data obtained from our lasers. Figure 5.12*a* displays large spectral range power versus wavelength data obtained with the InSb (top) and HgCdTe (bottom) detectors. These spectra demonstrate the absence of spurious signals in the power measurements of the two detectors, each targeted for its own specific (the fundamental or SHG) wavelength range. Figure 5.12*b* and *c* shows high-resolution spectra of representative devices D2886 and D2912, respectively. Spectra in the region of the fundamental as well as the SHG emission are shown. The expected direct correspondence between the respective spectra is observed in the data. It is necessary to note that what we term *SHG* here is actually a whole set of SHGs and SFGs as each of the individual Fabry-Perot modes of the free-running laser mixes with itself (SHG) and with all the other Fabry-Perot modes (SFG). This is so because the spectral width of the laser emission is narrower than the laser gain-band width, which is the same as or narrower (for SHG) than the nonlinear transition. Many of these SHG

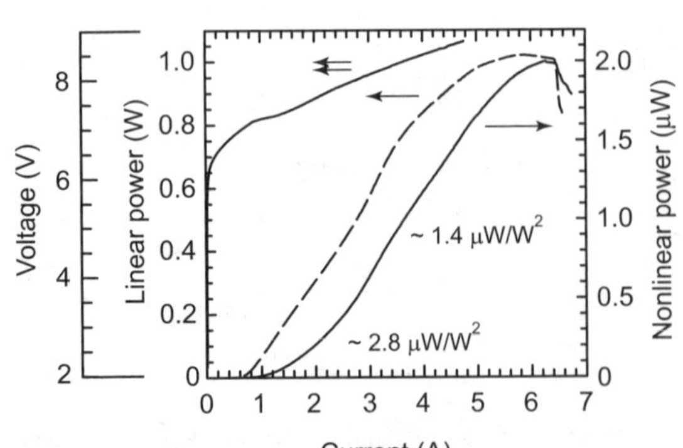

Figure 5.11 Linear (dashed) and nonlinear (solid) light output and voltage (solid line with separate *y* axis) versus current characteristics of a 10-μm-wide, 3.04-mm-long laser device of D2886. The laser was operated with 60-ns-long current pulses at 4.16-kHz repetition rate and at 10-K heat sink temperature. The external power conversion efficiencies are given in two regions. (Source: Ref. 46.)

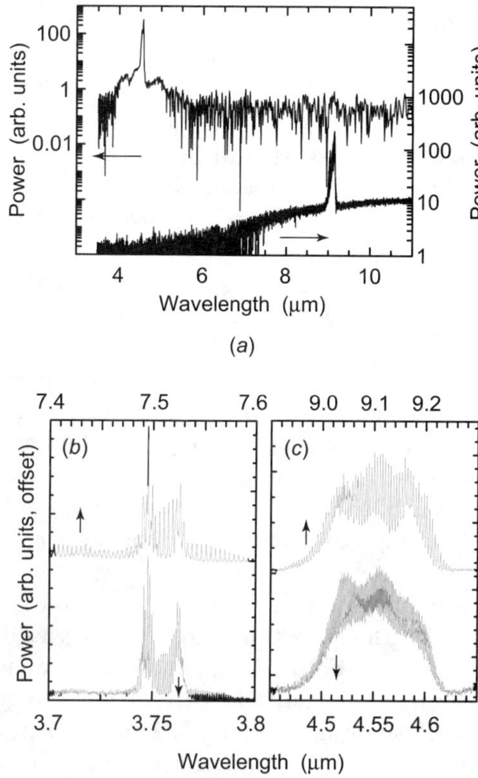

Figure 5.12 Emission spectra of QC lasers with integrated second-order nonlinear cascades. (*a*) Wide-range spectra of a characteristic laser of D2912. The spectrum on top has been taken in step-scan mode with an InSb detector fitted with a sapphire window and an additional quartz/silica glass filter. The bottom spectrum is taken with a HgCdTe detector; the laser radiation has been reduced with a metal screen to avoid saturation of the detector and detector electronics; no spurious electronic or optical signals are measured. (*b*) and (*c*) High-resolution spectra of representative laser devices of D2886 and D2912, respectively. The SHG (bottom) and fundamental (top) light emission is shown for both devices. (Source: Ref. 46.)

and SFG combinations are degenerate due to the low dispersion and hence the observed equidistant frequency comb results. Nevertheless, close inspection of the spectra allows one to count and attribute individual modes.

So far, in this section, we have demonstrated several strategies for improved quantum design of optical nonlinearities; a several-orders-of-magnitude improvement and continuing improvement have been shown. These values have been obtained without any phase matching, causing large spread in data quality and limited overall output power.

5.4.3 Temperature performance of nonlinear quantum cascade lasers

The early implementations of nonlinear QC lasers, as discussed in Secs. 5.4.1 and 5.4.2, and potential future implementations of more elaborate effects in parametric nonlinear lasers and light sources are limited to low temperatures, generally to keep a cold electron population. Nevertheless, for applications, such as trace gas sensing, two major questions arise: (1) How is performance at high temperatures? (2) What is the single-mode performance of the system? To answer these questions, the previously discussed SHG QC lasers were presented in Ref. 59.

In particular, design D2886 was used for the study of the temperature behavior, and again conventional laser processing and characterization (as briefly outlined above) were employed. In particular, the lasers were operated in pulsed mode at a very low duty cycle (< 1%) and pulse duration (50 to 100 ns), such that the measured heat sink temperature can be assumed to be practically identical to the laser core temperature.

Figure 5.13 shows the linear and nonlinear light output and voltage versus current characteristics of a 10-μm-wide and 3-mm-long device of D2886 at various heat sink temperatures. Figure 5.13*a* shows the linear optical power at ~7.5-μm wavelength measured with nearly unity collection efficiency; Fig. 5.13*b* shows the nonlinear SHG light output at ~3.75-μm wavelength measured with a cooled InSb photovoltaic detector and collected with 60% collection efficiency from the same facet. From the linear and quadratic portions of the *L-I* curves in the top and bottom panels, respectively, an external linear-to-nonlinear power conversion efficiency can be deduced. The latter is shown as a function of the heat sink temperature in the inset of Fig. 5.13*b*. Figure 5.14 reports spectra taken of the same device of Fig. 5.2 at cryogenic temperature as well as at a 250 K heat sink temperature.

As the heat sink temperature is raised, the nonlinear power drops quickly, a combined effect of the reduced power level of the fundamental pump laser and the reduced linear-to-nonlinear conversion efficiency. Nevertheless, at 250 K, a clear SHG signal can be measured, as shown in Fig. 5.14*b* top, with a peak power level of ~50 nW. Only at 300 K had the signal disappeared into a rising short-wavelength background.

The reduction of the fundamental optical power with increased heat sink and active region temperature is a well-known effect with QC lasers (as with most semiconductor lasers). Nevertheless, QC lasers can be obtained with high optical power even at room temperature [48]. Therefore, efficient SHG should be possible with QC lasers having (linear) peak output power levels in the watt range at room

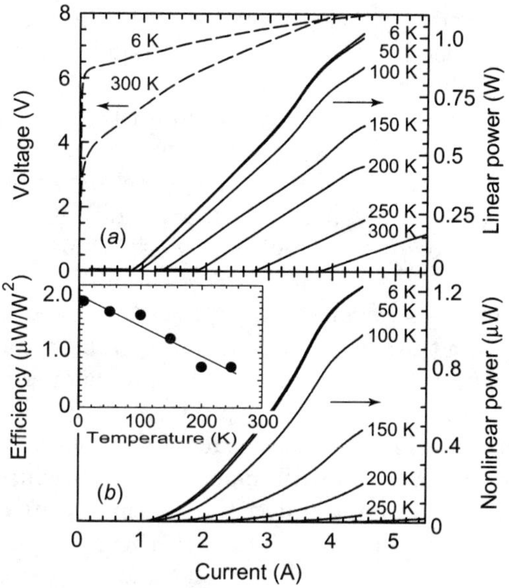

Figure 5.13 Linear (top, solid lines) and nonlinear (bottom) light output and voltage (top, dashed lines) versus current (L-I-V) characteristics of a 10-μm-wide, 3-mm-long QC laser of D2886 at various heat sink temperatures, as indicated. The inset shows the linear-to-nonlinear power conversion efficiency as extracted from the L-I-V curves versus the heat sink temperature. The solid line of the inset is a linear least-squares fit to the data, resulting in a slope of –5 nW/(W^2·K). (Source: Ref. 59.)

temperature. The deterioration of the external linear-to-nonlinear power conversion efficiency, conversely, can be understood from a discussion of the second-order nonlinear susceptibility $\chi^{(2)}_{2\omega}$, to which it

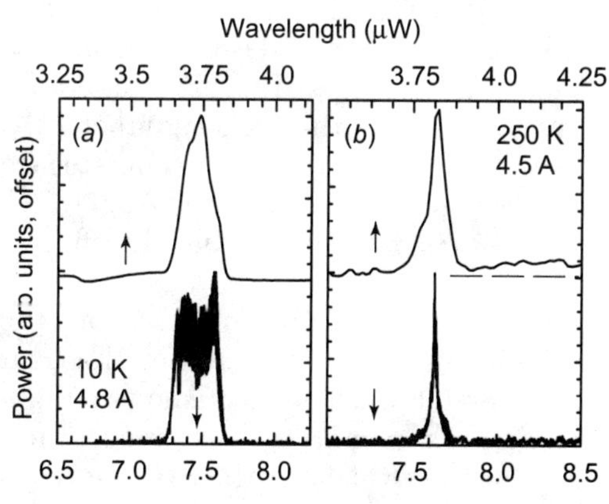

Figure 5.14 Linear (bottom) and nonlinear (top) emission spectra of QC lasers with integrated second-order nonlinear cascades obtained at (a) 10 K and (b) 250 K and at high peak current levels, as indicated. (Source: Ref. 59.)

is closely related through $W_2(2\omega) \sim |\chi^{(2)}_{2\omega}|^2 W_1(\omega)^2$, where $W_1(\omega)$ and $W_2(2\omega)$ are the linear and nonlinear (SHG) power at frequencies ω and 2ω, respectively. The second-order nonlinear susceptibility (Eqs. 5.8 and 5.13) contains several temperature-dependent quantities. First, the transition line widths of QC lasers are known to increase significantly, by ~ 50%, between 10 and 300 K [7]. We estimate values γ_{ij} for the line broadening (full width at half maximum) of the transitions between levels i and j of γ_{23} = 12 meV and γ_{24} = 20 meV for 10 K heat sink temperature and γ_{23} = 18 meV and γ_{24} = 30 meV for 300 K. The larger contribution to the drop of $|\chi^{(2)}_{2\omega}|$ with increasing temperature thereby results from the increase in γ_{24}, as the increase in γ_{23} is partly (except for excited state absorption 3 → 4) compensated by the increase in the population difference (population inversion) between the upper and lower laser levels, as a result of the increased laser threshold following the broadening of the laser gain bandwidth. Another source of reduced $|\chi^{(2)}_{2\omega}|$ can be found in the thermal shift of the various energy levels that may lead to a change in the resonance conditions.

5.4.4 Single-mode emission and tunability in nonlinear quantum cascade lasers

In Sec. 5.4.3 we referred to SHG of Fabry-Perot type spectra, which (as we pointed out) is somewhat of a misnomer, as a good fraction of the nonlinear signal is in fact SFG of many Farby-Perot modes mixing with one another. For precise determination of the nonlinearity, as well as for applications that require single-mode spectra, such single-mode performance needs to be assessed in the nonlinear domain.

It has long been known how mid-ir QC lasers can be made into single-mode and tunable lasers; one straightforward method is to fabricate the QC lasers as *distributed feedback* (DFB) lasers. In this section we discuss the single-mode emission and tuning of the SHG signal of a QC DFB laser. Quantum cascade DFB devices with first-order top-surface Bragg gratings (with grating periodicity Λ_B) were fabricated from wafer D2882, a design very similar to D2886, but with slightly shorter wavelength [46].

Figure 5.15 shows the single-mode emission wavelength at both the fundamental pump and SHG wavelengths as a function of the heat sink temperature for a device with $\Lambda_B = 1.1$ μm. The inset shows a high-resolution short-wavelength spectrum taken at cryogenic temperature. As we obtain single-mode emission for the QC lasers, the SHG signal is also single-mode. The fundamental emission wavelength λω of the QC DFB laser is given by $\lambda_\omega = 2n_{\mathrm{eff}}(T)\cdot\Lambda_B$, with $n_{\mathrm{eff}}(T)$ being the temperature-dependent effective refractive index of the waveguide.

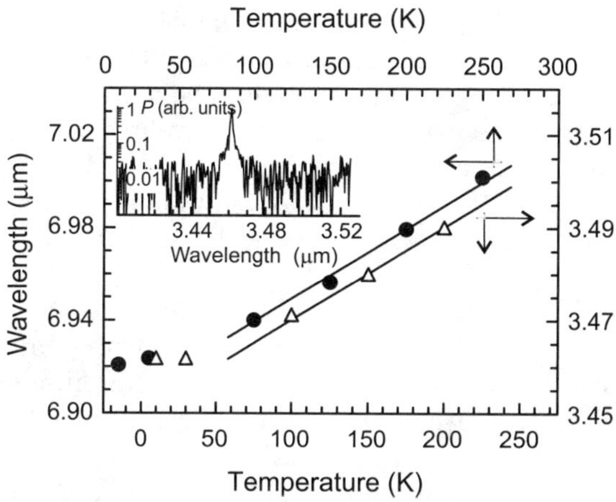

Figure 5.15 Single-mode emission wavelength versus temperature for a QC DFB laser with active regions that also include nonlinear optical cascades. The solid circles indicate the emission wavelength in the fundamental wavelength range, the open triangles in the SHG wavelength range. The solid lines are fitted to the data using slopes of 0.4 and 0.2 nm/K for the fundamental and SHG wavelength range, respectively. The inset shows a high-resolution spectrum of the single-mode SHG signal obtained in step-scan mode at cryogenic temperatures. (Source: Ref. 59.)

The single-mode wavelength $\lambda 2\omega$ of the SHG signal is given by $\lambda_{2\omega} = \lambda_{\omega}/2 = n_{\text{eff}}(T)\cdot\Lambda_B$. Therefore, we expect a SHG tuning rate of one-half the value of that of the fundamental mode. In fact, as shown in Fig. 5.15, we measure tuning rates of ~0.4 and ~0.2 nm/K for the fundamental pump and SHG light, respectively. The latter is very comparable to the tuning rate expected for a QC DFB laser with fundamental light emission at such short wavelength [6], which is not surprising where we take into account that the temperature coefficients of the refractive indices of the various materials are very little wavelength-dependent at energies far away from the fundamental energy gap. Nevertheless, there are technological and merely practical advantages, such as a larger grating periodicity, to obtaining a single-mode shortwavelength nonlinear emitter by fabricating the pump laser as a single-mode DFB laser.

5.5 Third Harmonic Generation In Quantum Cascade Lasers Using Resonant Intersubband Nonlinearities

In earlier sections we have motivated our choice of nonlinearity that we implemented first with QC lasers; SHG and SFG were the nonlinearities of choice. The next nonlinear effect to explore was THG. Originally, THG was a by-product of the earlier designs for SHG, which included two interleaved SHG level triplets, providing an automatic THG level cascade [46]. Aside from the auspicious presence of the THG nonlinearity, it was an intellectual challenge to test if it could be made strong enough to be of value as a means to extend the short-wavelength limit of nonlinear QC lasers. The results of this work indicated that technologically it will be unlikely that THG will result in useful power levels; nevertheless, the intellectual challenge of design and evaluation is of interest.

The first observation of THG was a by-product of the improved SHG designs of active regions D2912 which contained two strong and interleaved nonlinear cascades for SHG that added up to a nonlinear cascade 2–3–4–5, as shown in the inset of Fig. 5.16. At resonance, the ratio of the third-order to second-order nonlinear susceptibility $|\chi^{(3)}/\chi^{(2)}| \cong ez_{25}/(\hbar\gamma_{52})$ and the ratio of powers W_3/W_2 is of the order of $|ez_{25}Ex(\omega)/(\hbar\gamma_{52})|^2$. This ratio can be on the order of 0.01 if the pump field intensity is about 1 MW/cm^2 and the dipole moment z_{25} is on order of 10 Å. In fact, the design of D2912 has nonnegligible, though in no way optimized, $\chi^{(3)}$. Figure 5.16 thus reports the high-energy emission spectrum of a representative device of D2912; a small signal at 3ω (at $\lambda_{3\omega} = 3.05$ μm wavelength) is observed [46].

Nevertheless, after accidental observation of the THG signal in D2912, an attempt was made to improve on the THG signal with targeted design, and the results were reported in Mosely et al. [60]. Again, the design strategy was to use the active region of the laser simultaneously as the nonlinear region, and to have one leg of the nonlinearity (the lowest transition) coincide with the laser transition.

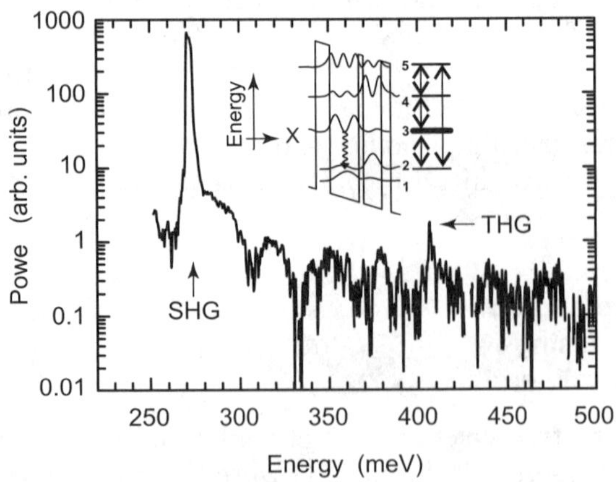

Figure 5.16 Emission spectrum of a representative device of D2912 in the very short-wavelength range. A small THG signal is measured. The energy of the fundamental light is 136 meV (9.1-μm wavelength); SHG and THG are at 271 meV (4.55 μm) and 407 meV (3.05 μm), respectively. (Source: Ref. 46.)

The design was made for pump light at ~11.1 μm wavelength and SHG and THG emission at 5.4 and 3.7 μm, respectively, where now the SHG signal is a by-product of a structure optimized for THG.

Fifty-two periods of QC laser active regions interleaved with injector regions have been grown with a ternary waveguide that is low-loss at the pump wavelength. The large wavelength step from the fundamental to the THG wavelength implies that the waveguide at the THG wavelength is minimally lossy (but also rather poorly confined; i.e., it is an oversized waveguide, in which a large number of transverse modes are allowed). Figure 5.17 shows the conduction band diagram for one active region sandwiched between two injector regions under an applied bias of 42 kV/cm. This field is considerably higher than the turn-on electric field calculated and measured as 30 kV/cm and is more appropriate to describe the structure under high pumping conditions, where we also measure the highest optical power. The structure is designed such that energy levels 1, 2, and 3 facilitate the laser transition at the fundamental frequency, and the nonlinear processes are supported by energy levels 2, 3, 4, and 5. The energy level separations ΔE_{23}, ΔE_{34}, and ΔE_{45} are 98, 104, and 90 meV, respectively. Since our target nonlinear light generation occurs as a result of the 2–3–4–5 cascade, we are most interested in the corresponding optical dipole matrix elements. They are calculated to be z_{23} = 14.6 Å, z_{34} = 21.5 Å, z_{45} = 33.2 Å, and z_{25} = 7.0 Å, yielding a significantly greater dipole matrix element product than that of previous asymmetric coupled QW designs intended for intersubband SHG or THG in the AlInAs/GaInAs material system [12, 46], which to some extent is attributed to the smaller laser and resonance energies ΔE_{ij}.

The lasers were processed and measured in conventional fashion. Figure 5.18 shows the emission spectra from the integrated device taken at cryogenic temperatures. Figure 5.18*a* shows a Fabry-Perot spectrum characteristic for the radiation at the pump wavelength; Fig. 5.18*b* shows the THG signal at $\lambda \sim 3.7$ μm from the same device that displays a fundamental emission at $\lambda \sim 11.1$ μm. The spectral response of the InSb detector fitted with a short-pass filter is given as a dashed line. The small broad background emission can be attributed to background luminescence from electrons excited to high energy levels; the SHG signal is present at 5.4-μm wavelength.

Using Eq. 5.9, the calculated dipole moments, and electron density $N_e \sim 5 \times 10^{15}$ cm^{-3}, assuming that all the population is in level 3, neglecting all detunings, and assuming all gammas (γ_{mn}) equal to 10 meV, except $\gamma_{52} \sim 20$ meV, we arrive at $\chi^{(3)} \sim 7 \times 10^{-8}$ esu, which is quite a large value. The ratio of third-order to second-order susceptibilities for the same structure scales as

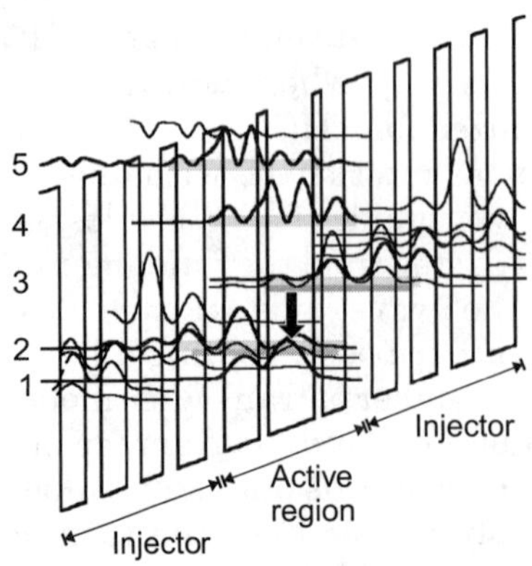

Figure 5.17 Conduction band energy diagram for one active region between two injector regions. The significant energy levels inside the active region are labeled 1 to 5. The thicknesses in nanometers of the InGaAs QWs and AlInAs barriers of one period of injector and active region are, from right to left, 3.7/**2.1**/3.0/**2.1**/3.5/**2.1**/3.4/**3.6**/3.1/**1.2**/6.4/**1.3**/4.7/**2.6**, the barriers are indicated in bold font, and the underlined layers are doped to $n_{\mathrm{Si}} = 2.5 \times 10^{16}$ cm^{-3}. The light gray bars indicate the levels involved in THG; the dark gray bars and arrow indicate the laser transition. (Source: Ref. 60.)

$$\left| \frac{\chi^{(3)}}{\chi^{(2)}} \right| \sim \frac{ez_{52}}{\hbar\gamma_{52}} \sim 10^{-3} \tag{5.14}$$

This gives the ratio of signal powers at second and third harmonics on the order of 0.01, which compares well with the experimental value of ~1.5×10^{-2}. Nevertheless, as outlined in Mosely et al. [60], by using this reasonably optimized THG process, even a fundamental power of 100 mW will only result in 30 to 50 nW of nonlinear THG power, a power level that was consistent with the measured detector signal.

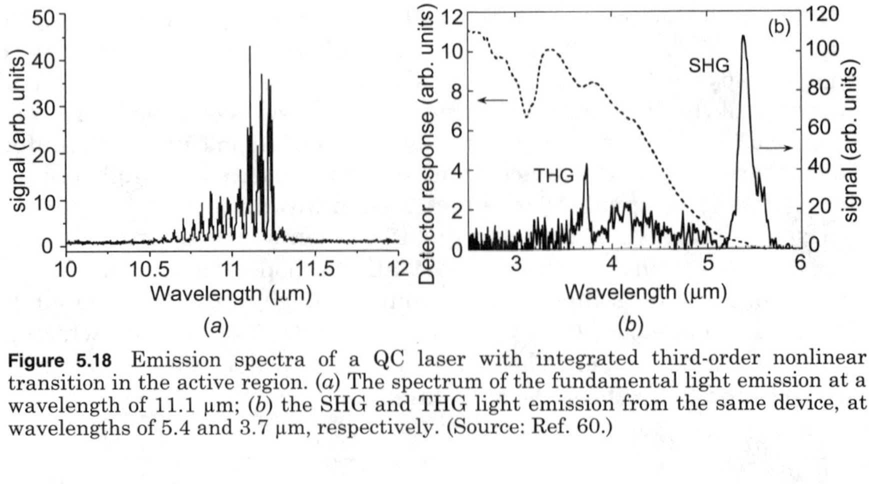

Figure 5.18 Emission spectra of a QC laser with integrated third-order nonlinear transition in the active region. (*a*) The spectrum of the fundamental light emission at a wavelength of 11.1 μm; (*b*) the SHG and THG light emission from the same device, at wavelengths of 5.4 and 3.7 μm, respectively. (Source: Ref. 60.)

5.6 Nonlinear Light Generation in Quantum Cascade Lasers Using Lattice Nonlinearities

So far, we have considered only resonant optical nonlinearities based on ISB transitions. These are advantageous, as they are much larger than the average bulk nonlinearities and thus result in high nonlinear optical power output and linear-to-nonlinear conversion efficiency. Nevertheless, they also have two drawbacks. First, the magnitude of the nonlinearity is about proportional to the free-carrier density, but a high carrier density also increases the free-carrier loss in the cavity. Second, the nonlinearities are resonant and must be tuned by design to the correct wavelengths. Given those shortcomings, it is worthwhile to consider nonlinear QC lasers based on the bulk, i.e., lattice, nonlinearities.

The GaAs- and InP-based III-V semiconductors have zinkblende lattice structure with two different atoms in the unit cell. The resulting asymmetry leads to a distinct second-order nonlinear susceptibility tensor, with nonzero components for electric pump fields that are oriented away from the main axes of symmetry. For the usual combination of transverse magnetic polarization, due to strict selection rules, and growth on (100)-oriented substrates, the second-order nonlinearity is vanishingly small in conventional InP- or GaAs-based QC lasers, much smaller than the one induced by the ISB transitions.

Neverthless, Bengloan et al. [61] succeeded in growing GaAs-based QC lasers on (111) substrate. In this orientation, the TM pump field is at an angle with all the major axes, and a significant nonlinear field is generated. This nonlinear process has the advantage that it is not resonant, i.e., is largely independent of wavelength, and that it is a property of the entire lattice; i.e., the overlap with the guided modes is unity, independent of the waveguide geometry.

In their work, Bengloan et al. [61] demonstrated GaAs/$Al_{0.45}Ga_{0.55}As$ QC lasers emitting at ~11.5-μm wavelength and SHG emission at ~5.8 μm. In a non-phase-matched waveguide, ~200 mW of pump light was converted to ~20 μW of SHG signal. The authors estimate a maximum power conversion efficiency of 1% to 10% for a phase-matched structure.

Giovannini et al. [62] reported SHG in nonlinear QC lasers based on a (111) lattice nonlinearity in InP-based lasers.

5.7 Phase-Matching Considerations in Nonlinear Quantum Cascade Lasers

In previous sections, emphasis was put on the proof of principle of nonlinear light generation in QC lasers, and on the quantum design.

Yet, practically no attempt was made to phase-match the fundamental and excited nonlinear guided modes, which would be essential to reap the full benefit of the nonlinear light generation. Phase matching is particularly important, because in the simplest geometry, the fundamental and harmonic modes copropagate within the same waveguide. Nevertheless, even for the non-phase-matched case, several microwatts of nonlinear light could be measured. This is partially the result of the low material dispersion for wavelengths far off the material bandgap and a particularly beneficial "side effect" of the QC laser which uses ISB instead of interband optical transitions [63]. Nevertheless, as the quantum design of the nonlinear QC laser improved, it became clear that the power was now limited by the lack of a phase-matching strategy. Hence, the next development step was dedicated to implementing phase matching into the nonlinear QC laser waveguide.

Before we review the various means of phase matching reported so far in the literature, a quick estimate of the magnitude of a typical phase mismatch in a conventional mid-ir QC laser waveguide is given. For the SHG structures presented above, the mismatch factor Δk in the structures under study is between 1000 and 1500 cm^{-1}, which is more than 100 times larger than the modal losses α_2. Therefore, we could obtain a three- to four-order-of-magnitude increase in nonlinear power efficiency by reducing the mismatch factor to the level of the optical losses.

The common techniques for phase matching, such as birefringence phase matching [64] or quasi-phase matching, are not practical for QC lasers due to the linear waveguide geometry. Moreover, the various schemes proposed for SHG phase matching based on refractive index matching in asymmetric double quantum well structures [65] cannot be easily applied to QC lasers because of the intrinsic waveguide dispersion, weak voltage tunability, and strict current transport requirements.

5.7.1 Modal phase matching

The technique of phase matching described here takes advantage of the flexibility in the design of the integrated QC laser waveguide and the fact that these are "large" waveguides capable of sustaining several transverse modes. These two properties allow for true matching of the fundamental (pump) and SHG modal refractive indices of different transverse order waveguide modes, i.e., true modal phase matching, while simultaneously preserving a large nonlinear overlap of the interacting modes with the nonlinear active region. While this phase-matching technique was first demonstrated using SHG nonlinear QC lasers, as reported by Malis et al. [66], we estimate that it can readily be extended to other nonlinear optical processes.

The QC laser structures used in this study [66] employed the fairly optimized active region D2912 (shown in Figs. 5.8 and 5.16 and discussed previously [46]) and a waveguide, as described in Fig. 5.19*a*. For this structure a maximum nonlinear susceptibility of $|\chi^{(2)}| \sim 2 \times 10^4$ pm/V is calculated, assuming exact resonance for all interacting fields, while the effective interaction cross section of the fundamental and nonlinear modes with the nonlinear region $\hat{A}$ varies between 600 and 1400 μm^2 as the ridge width varies between 6 and 12 μm. The power transfer from the fundamental to the nonlinear signal is maximized by minimizing the phase mismatch Δk and effective area $\hat{A}$, as outlined in Eqs. 5.11 and 5.12 and the related discussion in Sec. 5.2.2.

For QC lasers, the phase mismatch Δk is determined by the structure of the laser waveguide. Due to material dispersion at the two wavelengths (~9.1 and 4.55 μm) used in this study, phase matching is not possible between the zero-order transverse waveguide modes of the fundamental and SHG signals, or between the fundamental zero-order transverse magnetic (TM_{00}) mode and the SHG first-order transverse mode (TM_{01}). However, numerical analysis shows that phase matching is possible between the fundamental (TM_{00}) and the SHG (TM_{02}) modes.

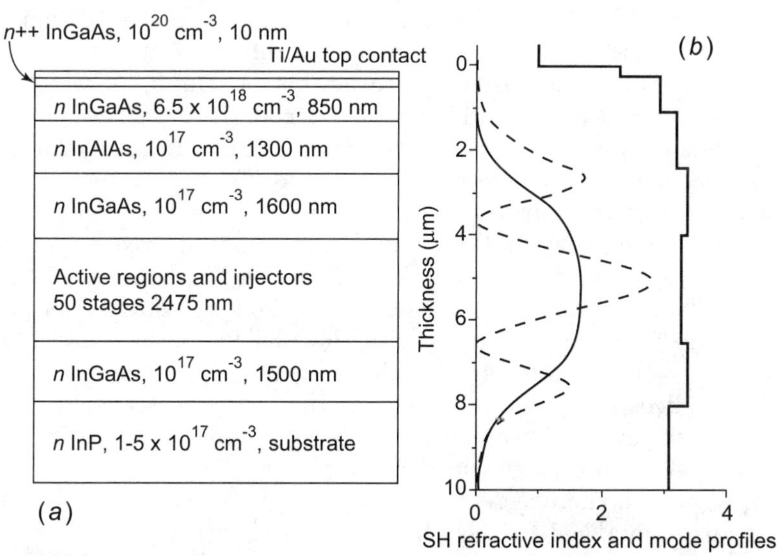

Figure 5.19 Waveguide structure of sample D2957, a SHG nonlinear QC laser with modal phase-matching waveguide. (*a*) The layer structure, materials, doping levels, and thicknesses; (*b*) refractive index profile at the SHG wavelength, and the intensity of the magnetic field profiles (arbitrary units) for the TM_{00} fundamental mode (solid line) and TM_{02} SH mode (dashed line) for this waveguide. (Source: Ref. 66.)

When modes of different mode orders are matched, one consideration is to maximize the overlap integral of these modes with the nonlinear mixing region. Modes of different transverse order and the same frequency are exactly orthogonal. Since the frequencies differ by a factor of 2 in the SHG case, this orthogonality is partially lifted; but still the overlap integral between these modes over the transverse cross section of the waveguide can be quite small. Hence, the nonlinear region (which doubles as the laser active material) must be placed in such a position in the waveguide so that the overlap is maximized (Fig. 5.19*b*).

The refractive indices of undoped InGaAs and InAlAs were derived by linear interpolation between the published values of the end alloys at 9.1 and 4.55 μm. The complex refractive indices of the variously doped layers were then calculated, using the Drude formalism. The refractive index profile has the shape plotted in Fig. 5.19*b*. This profile is similar to that of the so-called *M* waveguides that have been proposed and demonstrated for SHG phase matching in $LiNbO_3$ and AlGaAs in the near-ir [67].

The waveguide layer thicknesses were first optimized by minimizing the phase mismatch corresponding to an infinitely wide waveguide. The difficulty of designing a phase-matched waveguide lies in the uncertainty of the refractive indices for the various materials and in the inevitable differences between a designed structure and the real wafer and processed device. To keep the phase mismatch Δk below 100 cm^{-1}, e.g., one would need to know and control the refractive indices of the waveguide layers with an absolute accuracy better than 0.1%, which is not feasible. Therefore, we need a degree of freedom that allows one to tune the finished structure through the phase-matching point. Such a degree of freedom is provided by the dependence of the effective refractive indices of the various modes on the laser ridge width. The effective refractive index of the fundamental mode decreases faster with decreasing ridge width than the SHG refractive index (inset of Fig. 5.20*b*). This behavior can be understood from the slightly larger overlap of the fundamental mode with the SiN insulator and metal contact layers outside the semiconductor ridge, and the different refractive indices at the two wavelengths. The crossover point, corresponding to exact phase matching, was designed to lie within the available experimental ridge width range of 4 to 16 μm.

The waveguide structure detailed in Fig. 5.19*b* (sample D2957 [66]) provides an optimal calculated phase mismatch of 367 cm^{-1} for an infinitely wide waveguide, a value that is theoretically reduced to zero for an 8.8-μm-wide ridge laser. Figure 5.20*a* shows the *L-I* curves at the fundamental and SHG wavelengths for an 8-μm-wide, 1.5-mm-long laser. The laser displays a nonlinear power of 240 μW and a nonlinear

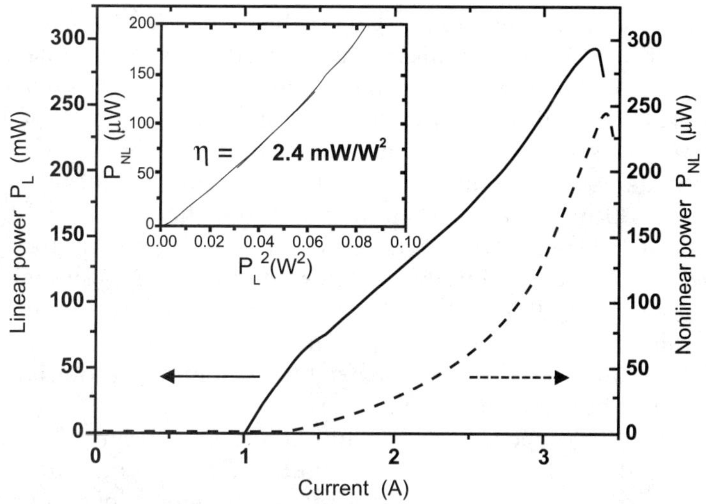

(a)

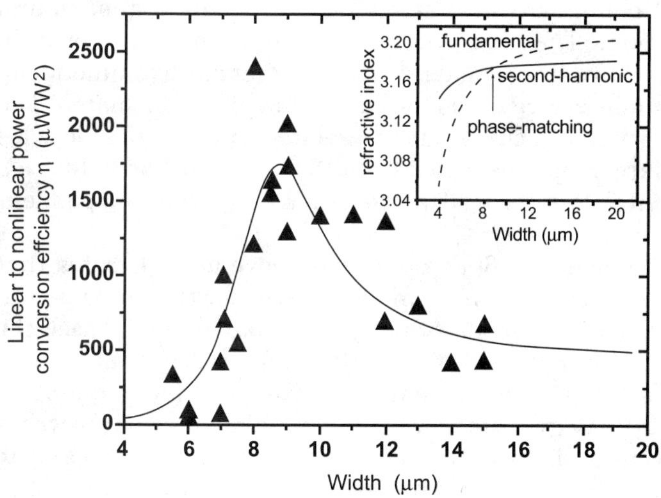

(b)

Figure 5.20 (*a*) Fundamental (solid) and SHG (dashed line) light output power versus current characteristics for an 8-μm-wide, 1.5-mm-long laser of wafer D2957 (see Fig. 5.19). The inset shows the SHG power as a function of the linear power squared and a fit of the curve with a straight line in the central region to obtain the external conversion efficiency. (*b*) Ridge width dependence of the nonlinear power conversion efficiency for wafer D2957. The solid line through the data points (triangles) represents a fit with a lorentzianlike function. The inset shows the theoretical dependence of the fundamental and SHG effective refractive indices on the ridge width. (Source: Ref. 66.)

efficiency of approximately 2.4 mW/W^2. Both values are significantly (by at least two to three orders of magnitude) higher than what was possible without phase matching. The highest nonlinear conversion efficiency of 36 mW/W^2 was achieved for a similar waveguide design, D2944, for a ridge width of 4 μm at the lowest end of what was feasible to process. The maximum theoretical nonlinear efficiency is 2 W/W^2, assuming $\Delta k = \alpha_2 = 10$ cm^{-1} and undepleted pumping. With this efficiency, pump depletion is reached when the fundamental laser power goes beyond ~200 mW. The discrepancy between the theoretical prediction and the experimental results can be explained by a 0.5- to 1-μm ridge width nonuniformity intrinsic to the wet-chemical mesa etching process; a deviation by 0.5 μm from the optimal ridge width would lead to the mismatch increase to $\Delta k \sim 50$ cm^{-1}. Another possible factor contributing to lower efficiency is the excitation of higher-order lateral laser modes that have much lower conversion efficiency.

The ridge width dependence of the nonlinear power efficiency was studied in detail for sample D2957, and the results are shown in Fig. 5.20*b*. The ridge width was measured at the height of the active region with a precision of ±0.5 μm. As expected from the theoretical calculations, the nonlinear conversion efficiency displays a maximum around 8 to 9 μm. The line through the data points shown in Fig. 5.20*b* represents a fit of the data with a lorentzianlike function of the form $1/(\Delta k^2 + \alpha_2{}^2)$, where Δk^2 is the calculated phase mismatch. The width of the experimental curve (closely related to the optical loss α_2) is considerably larger than expected from theory, most likely due to a larger experimental loss as well as to other ridge-width-dependent factors such as nonuniformity.

The work by Malis et al. [66] as outlined above demonstrates that true (near) modal phase matching can be achieved for the fundamental TM_{00} mode and SHG TM_{02} modes. Such phase matching increased the nonlinear output power by a factor of 100 compared to that of non-phase-matched structures with near optimized nonlinear quantum design. The third strategy to improve on the nonlinear output power is to reduce the waveguide loss, to improve the overlap factor, and to increase the overall laser power. All these requirements can be met or at least addressed by using a waveguide that employs InP for both the top and the bottom (substrate) claddings [68]. InP offers the advantage of better optical confinement, i.e., overlap, due to its lower refractive index, higher material quality, and lower optical losses at both wavelengths of interest. Hence, an effort was made to implement nonlinear QC lasers and phase-matching waveguides with InP top claddings; this work was reported in Ref. 69. The thicknesses of the waveguide layers are detailed in Fig. 5.21*a* and were chosen to reach the phase-matching condition between the fundamental TM_{00} and SHG TM_{02} modes for the waveguide ridge width of ~7.3 μm; and the phase mismatch was $\Delta k = 538$ cm^{-1} for an infinitely wide waveguide.

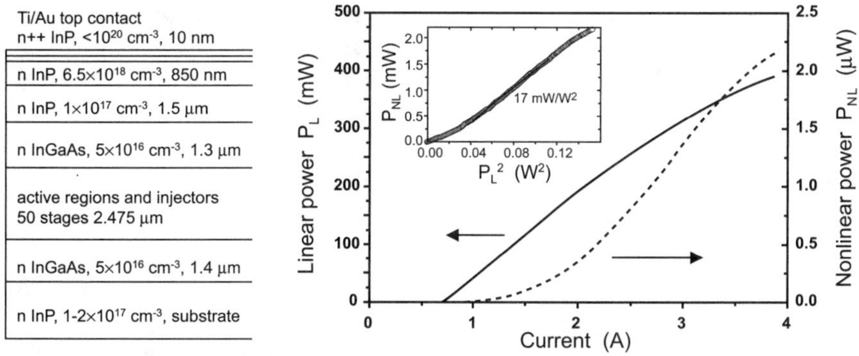

(*a*) (*b*)

Figure 5.21 (*a*) InP-based waveguide structure optimized for SHG phase matching (D3014). The calculated refractive indices of the fundamental TM_{00} and SH TM_{02} modes for an infinitely wide waveguide are 3.2079 and 3.1689, respectively. (*b*) Fundamental (solid) and SHG (dashed) light output power versus current characteristics at 10 K. The inset shows the SHG power as a function of the linear power squared. The fit of the curve with a straight line in the central region gives the nonlinear conversion efficiency. (Source: Ref. 69.)

The QC laser structure used in this study (D3014 [69]) was grown by *molecular beam epitaxy* (MBE) on InP substrates in two steps [68]. First, the ternary waveguide core was grown, and the structure was capped with an As layer. Then the wafer was transferred to a second MBE machine equipped with a solid phosphorus source, and the top waveguide layers were deposited. The devices were processed into 5- to 12-μm-wide, 0.5- to 1.5-mm-long deep-etched ridge waveguide lasers, and the laser back facets were *high-reflectance* (HR) coated, starting at the semiconductor, with SiO_2 2500 Å / Ti 160 Å / Au 1000 Å / SiO_2 1000 Å (reflectivity > 99%).

Devices with ridge widths between 5 and 12 μm were characterized, and the optimal width was found to be around 7 μm, as expected from theory. Figure 5.21*b* shows the *L-I* curves at the fundamental and SHG wavelengths for a 7-μm-wide, 1.5-mm-long laser measured at cryogenic temperatures. The device displays a maximum nonlinear power of 2 mW and a linear-to-nonlinear power conversion efficiency of approximately 17 mW/W^2. Both are exceptionally high values for nonlinear QC lasers; reaching the milliwatt power range is significant as such a power level is sufficient for trace gas point sensors using mid-ir sources.

The SHG far-field radiation pattern was found to have two unequal maxima located symmetrically with respect to the waveguide axis at angles of approximately 45 degrees from the normal to the laser facet in a plane perpendicular to the device layers. This is consistent with the expected far-field distribution for the TM_{02} mode.

The narrower optimal ridge width (7 μm as opposed to 9 μm in the previous study [6]) is also beneficial because it suppresses the turnon of higher-order in-plane transversal laser modes that are not phase-matched to the SHG modes.

5.7.2 Other methods of phase matching

The above-discussed method for true modal phase matching has the advantage of being a resonant effect with strong enhancement of several orders of magnitude at true resonance. The strategy is also straightforward, as the crossing point of the waveguide dispersion curves and hence the phase-matching point can be calculated and with some effort can even be made robust toward ridge width variations; i.e., the crossing of the dispersion curves (versus ridge width) can be made more oblique. It is nevertheless up to the experimenter to fabricate lasers at the precise ridge width. While this is a seemingly trivial point, a more robust technique for phase matching would be welcome.

Several techniques have been proposed, many of which adapt strategies of quasi-phase matching to the nonlinear QC lasers. Belkin et al. recently [44] reported on such quasi-phase-matched structures, where the laser waveguide was periodically modified, by removal (etching) of a significant portion of the top cladding and the metal contact layer, which periodically changes the current injection and modal refractive index along the laser ridge. The authors report an improvement in the SHG conversion efficiency for a D2912-type SHG structure by a factor of about 10.

Belyanin et al. [70] recently suggested a surface-emitting approach in which the nonlinear radiation is coupled out normal to the laser surface by using a deep-etched surface (or bulk) grating. The periodicity of the grating thereby corresponds to a second-order Bragg grating for the nonlinear light, while the fundamental pump light is still coupled through the facets of the edge emitter.

Berger and Sirtori [71] finally noticed that near-ir and terahertz radiation in waveguides can be directly phase-matched as a result of the strong materials dispersion including anomalous dispersion around the *reststrahlenband* and waveguide dispersion, which in fact results in similar (or at full matching, the same) refractive indices in GaAs-based waveguides for radiation around 1 and 100 μm.

5.8 Other Nonlinear Effects in Quantum Cascade Lasers

The above-discussed applications of resonant optical nonlinearities in QC lasers were chosen with the particular goal of nonlinear light generation (higher harmonics and SFG) in mind. In the following, we briefly discuss a wider spectrum of optical nonlinearities and their use in QC lasers. These include nonlinear light generation through electronic Raman scattering and the role of nonlinearities, especially the optical Kerr effect, in mode-locked QC lasers.

5.8.1 Resonant Raman up-conversion scattering

Raman scattering is—essentially by definition and origin—a means of wavelength conversion. While Raman scattering involving the absorption or emission of phonons is the most familiar form of Raman scattering, other means of scattering exist within this class as well. In the context of QC lasers, resonant electronic Raman scattering involving an excited ISB transition is the obvious choice. So far, Stokes Raman ISB lasing has been reported following two different approaches. Liu et al. [72] reported an optically pumped ISB laser, in which Raman lasing involving the GaAs phonon was observed. In the second report, Troccoli et al. [73] demonstrated electronic Stokes lasing based on a designed nonlinear structure; this work is the focus of Chap. 6. While Stokes lasing is an efficient means of frequency down-conversion, here we briefly discuss the complementary frequency up-conversion process, i.e., stimulated electronic *anti-Stokes* (AS) Raman emission in QC lasers. As nonlinear light generation from Raman processes does not require phase matching, this work [74] may constitute a more straightforward route to unstrained (i.e., lattice-matched) InP-based QC lasers operating at wavelengths around and below ~4 μm.

In this particular work [74], we chose to implement the nonlinear and QC laser regions as two separate, monolithically integrated stacks of active regions and injectors [50]. Figure 5.22 shows a schematic of the device, including the pump laser and the AS nonlinear region stacks. The ISB transitions in the nonlinear regions (also shown in Fig. 5.22 for two specific examples) were designed to include an energy level triplet 2–3–4 embedded in a cascade, such that the maximum population is again in the excited energy level 3. The optical dipole matrix elements between levels 2, 3 and 4, z_{43} and z_{42}, respectively, were made as large as possible by design, to enhance the relevant third-order AS nonlinearity. The electrons in level 3 undergo AS Raman scattering stimulated by the pump light (resulting from the adjacent stack of QC laser active regions and injectors), emitting photons with energy $h\upsilon_{AS} = \Delta E_{42} + \delta = \Delta E_{32} + h\upsilon_{pump}$, where ΔE_{ij} is the energy separation between levels i and j and δ is the detuning, defined as $\delta = h\upsilon_{pump} - \Delta E_{43}$. Fast depletion of the lower level (2) is achieved through LO phonon enhanced scattering into the lower energy levels.

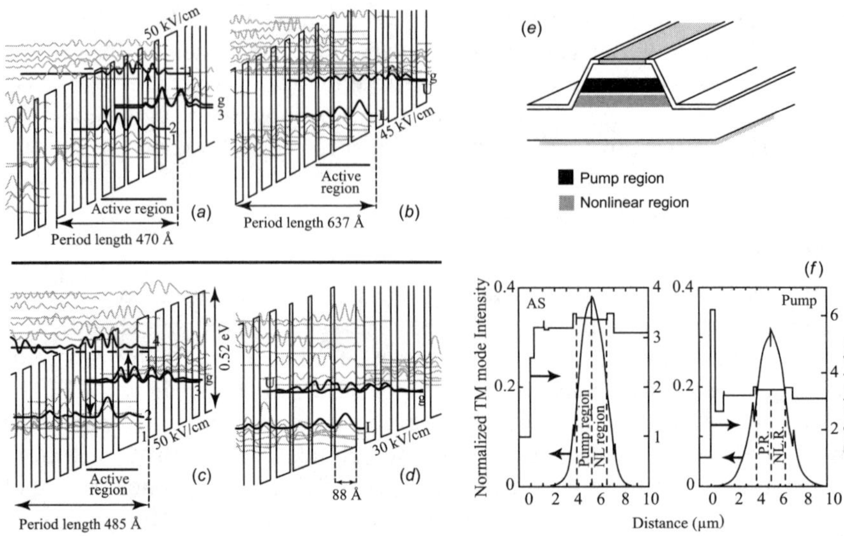

Figure 5.22 (*a*) and (*b*) Conduction band diagrams of the (*a*) nonlinear and (*b*) pump regions, respectively, in wafer D3015, a nonlinear QC laser structure for anti-stokes Raman up-conversion. The barrier/well widths of each of the 26 periods of the nonlinear stack, in nanometers, are 4.5/**3.1**/2.9/**3.1**/2.5/**3.2**/3.0/**3.3**/1.5/**3.7**/1.5/**4.2**/1.5/**4.8**/1.5/**2.7**, with the wells given in boldface. The underlined layers are doped to 2×10^{17} cm^{-3}. The pump laser is a conventional 20-stage QC stack emitting at $\lambda \sim 8.3$ μm. (*c*) and (*d*) Diagram of the conduction band of the (*c*) nonlinear region and (*d*) pump region, respectively, in wafer D2924. The pump is a conventional 30-period QC laser emitting at $\lambda \sim 10.5$ μm. The layer thicknesses of each of the 30 periods comprised in the nonlinear stack are 4.0/**2.6**/2.9/**2.9**/2.6/**2.7**/2.3/**2.7**/2.0/**2.7**/1.7/**3.8**/2.5/**4.2**/1.8/**7.1**, following the same notation as above. In (*a*) and (*c*) the black arrows indicate the Raman transition, and the levels involved are numbered. In (*b*) and (*d*), the letters *U* and *L* indicate the upper and lower pump laser levels, respectively. The ground level of the injector is labeled *g* in all four plots. (*e*) Schematic of the laser ridge showing the stacked configuration of QC laser and anti-Stokes Raman structure. (*f*) Normalized mode intensity profile and profile of the real part of the refractive index in the reverse growth direction for both anti-Stokes and pump fundamental modes in wafer D3015. (Source: Ref. 74.)

Stimulated AS emission—but so far no laser action—has been observed in a variety of samples, with both, positive and negative detuning. Two designs, D3015 (positive detuning) and D2924 (negative detuning), are shown in Fig. 5.22 and discussed in Ref. 74 and below.

The lasers are processed in conventional fashion, with the ridges etched deeply enough to provide strong current confinement for both active stacks, AS nonlinearity and QC laser, in the waveguide core. Figure 5.22 shows the normalized mode intensity profiles of both the AS and pump laser modes in sample D3015. The normalized mode overlap of the pump mode with the AS and pump active regions is 0.32 and 0.33, respectively; the confinement factor of the AS light is 0.38 for the nonlinear region. When one is implementing the nonlinear region and QC laser in a two- or potentially multistack approach, such as here, the relative overlap of the modes with the various stacks becomes a new degree of freedom, which must be carefully tailored to maximize energy transfer without depleting (or otherwise compromising) the pump field.

Figure 5.23*a* shows the spectra of sample D3015 measured at different peak currents above pump laser threshold. The inset shows the spectrum of the pump light for a peak current of 2 A. The main peak at 4.8 μm (258 meV) is attributed to stimulated AS Raman scattering [$h\upsilon_{\text{pump}}$ = 150 meV and ΔE_{32} = 110 meV, obtained from luminescence measurements below pump laser threshold, giving an expected λ_{AS} = 4.8 μm ($h\upsilon_{\text{AS}}$ = 260 meV)]. Electrons in level 3, when excited by the pump photons, can also easily scatter into energy level 4 and then return via spontaneous emission to level 2. This incoherent up-conversion process is the likely cause of the peak observed at 5.3 μm (235 meV), which is truncated by the wavelength cutoff at ~5.4 μm of the InSb detector. The detuning, obtained from the shift between the lines corresponding to AS Raman scattering and incoherent up-conversion peak, is δ = 258 meV – 235 meV = +23 meV.

The short-wavelength emission spectra of the sample with negative detuning, D2924, were measured at 7 K for different peak currents above the pump laser threshold and are shown in Fig. 5.23*b*. The main feature is a peak at around 4.5 μm (276 meV), which is again attributed to AS Raman scattering ($h\upsilon_{\text{pump}}$ = 118 meV and ΔE_{32} = 155 meV, from luminescence measurements below threshold, giving an expected λ_{AS} = 4.5 μm ($h\upsilon_{\text{AS}}$ = 273 meV).

Both examples present experimental evidence of stimulated electronic AS Raman emission in QC lasers. The most beneficial feature of this process is the large frequency shifts between the AS Raman and pump photons that can be achieved, namely, over 100 meV. Nevertheless, so far no narrowing of the AS emission peak or other evidence of gain or impending laser action has been observed. Work on improved designs of the nonlinear active regions, especially in light of an optimized (smaller) detuning δ, and on a better balance of the nonlinear and QC laser active regions in the waveguide, to achieve a larger pump laser power, is underway.

5.8.2 Intersubband nonlinearities in mode-locked QC lasers

All the nonlinearities discussed so far in connection with QC lasers involve specially designed multilevel ISB cascades. However, even before this work, a simple two-level ISB nonlinearity has been shown to be at work in QC lasers, namely, the Kerr nonlinearity, *or* intensity-dependent refractive index n around the laser transition $n = n_0 + I \cdot n_2$, where

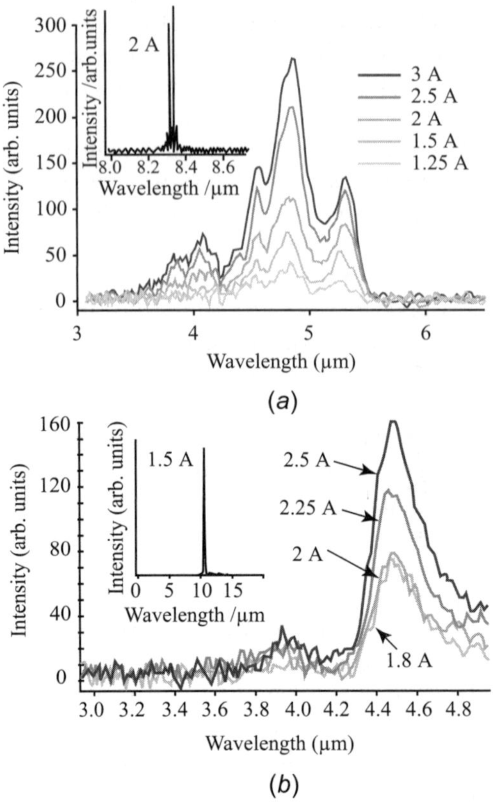

Figure 5.23 (*a*) Short-wavelength emission spectra of a laser of wafer D3015 above laser threshold. Peaks attributed to stimulated anti-Stokes emission (4.8 μm), and incoherent up-conversion (5.3 μm) can be distinguished. The inset shows the spectrum of the pump laser measured at a heat sink temperature of ~90 K and 2-A peak current. (*b*) Short-wavelength luminescence spectra, above threshold, of D2924 measured near liquid helium temperatures (~7 K). The main peak at 4.5 μm is attributed to anti-Stokes emission. The inset shows the pump spectrum measured at ~90 K and a peak current of 1.5 A. (Source: Ref. 74.)

n_0 and n_2 are the linear and nonlinear refractive indices, respectively, and I is the mode intensity in the cavity. When the QC laser field consists of ultrashort pulses, as in actively or passively mode-locked devices, such nonlinearity gives rise to self-phase modulation, which in turn introduces a characteristic oscillatory envelope in the emission spectrum. In addition, on the long-wavelength side of the laser transition n_2 is positive and hence leads to self-focusing of the mode in the cavity. Self-focusing in QC lasers also reduces the mode losses, as the overlap of the guided mode with the surrounding metal contact layers is reduced; therefore it provides a saturable loss mechanism which has been invoked to explain the observation of self-mode-locked QC lasers in Ref. 75.

In later work, Soibel et al. [76] combined the mode locking of QC lasers and the then-available knowledge on SHG in these devices to employ the latter as an analytical tool to estimate the time duration of the mode-locked pulses. In particular, from the increase in nonlinear power between cw and self-mode-locked operation, a pulse duration of ~12 ps was estimated in a QC laser capable of SHG, consistent with other measurements.

These examples illustrate cases in which ISB nonlinearities serve mainly supporting functions in QC lasers; they are excellent reminders that practically all ISB transitions are nonlinear to some degree, and that these properties need to be taken into account in a complete description of QC lasers and other ISB-derived optoelectronic components.

5.9 Conclusions and Outlook

In summary, we have reviewed here briefly ISB-based optical nonlinearities and then in greater detail the first implementation of ISB nonlinearities into QC lasers. The latter has been especially successful with the demonstration of frequency up-conversion such as SHG, SFG, and THG. Such nonlinear QC lasers are an alternative route to narrowband, short-wavelength ($\lambda \leq 4.5$ μm), coherent emitters, especially those grown with lattice-matched, unstrained heterostructures. Some steps of design optimization for nonlinear up-conversion lasers have been discussed; however, given the newness of the field, several more iterations of improved quantum design of nonlinearities are likely. Similarly, phase matching has been demonstrated in several attempts, most of which require some external tuning parameter that is still difficult to control, given current processing techniques. Thus, better implementations for phase matching are expected for future developments of nonlinear QC lasers.

Finally, a significant future development will be the demonstration of successful integration of down-conversion optical nonlinearities into QC lasers, especially DFG and photon-pair generation by parametric down-conversion, as well as monolithic parametric amplification in QC lasers.

5.10 Acknowledgments

The work reviewed here and originally presented in a range of publications as referenced above is the result of many fruitful and rewarding collaborations with wonderful individuals and teams with whom working has always been, and continues to be, a real pleasure. While we won't be able to list all those who merit thanks for the work performed, we would like to especially thank the following contributors: F. Capasso, J. Chen, A. Y. Cho, R. Colombelli, A. Gomez-Iglesias, V. Kocharovsky, H. C. Liu, T. S. Mosely, N. Owschimikow, R. Paiella, M. L. Peabody, G. Shu, A. M. Sergent, D. L. Sivco, A. Soibel, and D. Wasserman. The work performed at Princeton University was supported by the MRSEC grant at Princeton University under account NSF DMR-0213706 and by NSF grant ECS-0400615. Belyanin acknowledges support from NSF and AFOSR.

References

1. P. A. Franken, A. E. Hill, C. W. Peters, and G. Weinreich, *Phys. Rev. Lett.* **7**, 118 (1961); T. H. Maiman, *Nature* **187**, 493 (1960).
2. Ya. I. Khanin, *Principles of Laser Dynamics*, Elsevier, Amsterdam, 1995.
3. Y. R. Shen, *The Princeiples of Nonlinear Optics*, John Wiley & Sons, New York, 1984.
4. R. W. Boyd, *Nonlinear Optics*, Academic Press, New York, 2003.
5. H. C. Liu, and F. Capasso (eds.), *Intersubband Transitions in Quantum Wells: Physics and Device Applications I and II, Semiconductors and Semimetals*, vols. 62 and 66, Academic Press, New York, 2000.
6. J. Faist, F. Capasso, D. L. Sivco, C. Sirtori, A. L. Hutchinson, and A. Y. Cho, *Science* **264**, 553 (1994).
7. C. Gmachl, F. Capasso, D. L. Sivco, and A. Y. Cho, *Rep. on Progress in Physics* **64**, 1533 (2001) and references therein.
8. H. C. Liu, *Physica* **E8**, 170 (2000).
9. J. Allam and M. Wagner, *Appl. Phys. Lett.* **60**, 2726 (1992); S. Krishna, S. Raghavan, G. Von Winckel, A. Stintz, G. Ariyawansa, S. G. Matsik, and A. G. U. Perera, *Appl. Phys. Lett.* **83**, 2745 (2003); P. Bhattacharya, S. Gosh, and A. D. Stiff-Roberts, *Ann. Rev. Mat. Res.* **34**, 1 (2004).
10. J. Christen, M. Joschko, O. Stier, D. Bimberg, and E. Kapon, *Semicond. Sci. Techn.* **9**, 1939 (1994); A. Weber, O. Gauthier-Lafaye, F. H. Julien, J. Brault, M. Gendry, Y. Desieres, and T. Benyattou, *Appl. Phys. Lett.* **74**, 413 (1999); S. Schmult, I. Keck, T. Herrle, W. Wegscheider, M. Bichler, D. Schuh, and G. Abstreiter, *Appl. Phys. Lett.* **83**, 1909 (2003).
11. M. K. Gurnick and T. A. De Temple, *IEEE J. Quantum Electron.* **18**, 791 (1983).
12. F. Capasso, C. Sirtori, and A. Y. Cho, *IEEE J. Quantum Electron.* **30**, 1313 (1994); C. Sirtori, F. Capasso, D. L. Sivco, and A. Y. Cho, *Phys. Rev. Lett.* **68**, 1010 (1992).
13. E. Rosencher, A. Fiore, B. Vinter, V. Berger, Ph. Bois, and J. Nagle, *Science* **271**, 168 (1996).

14. L. Tsang, D. Ahn, and S. L. Chuang, *Appl. Phys. Lett.* **52**, 697 (1988); L. Tsang, S. L. Chuang, and S. M. Lee, *Phys. Rev.* **B41**, 5942 (1990).
15. P. Boucaud, F. H. Julien, D. D. Yang, J. M. Lourtioz, E. Rosencher, P. Bois, and J. Nagle, *Appl. Phys. Lett.* **57**, 215 (1990).
16. C. Sirtori, F. Capasso, D. L. Sivco, S. N. G. Chu, and A. Y. Cho, *Appl. Phys. Lett.* **59**, 2302 (1991).
17. J. Hamazaki, H. Kunugita, K. Ema, A. Kikuchi, and K. Kishino, *Phys. Rev.* **B71**, 165334 (2005); E. Baumann, F. R. Giorgetta, D. Hofstetter, H. Wu, W. J. Schaff, L. F. Eastman, and L. Kirste, *Appl. Phys. Lett.* **86**, 032110 (2005); I. Friel, K. Driscoll, E. Kulenica, M. Dutta, R. Paiella, and T. D. Moustakas, *J. Cryst. Growth* **278**, 387 (2005); C. Gmachl and H. M. Ng, *Electron. Lett.* **39**, 567 (2003).
18. H. Yoshida, T. Simoyama, A. V. Gopal, J. Kasai, T. Mozume, and H. Ishikawa, *IEICE Trans. Electron.* **E87C**, 1134 (2004); P. Cristea, Y. Fedoryshyn, and H. Jackel, *J. Cryst. Growth* **278**, 544 (2005).
19. M. Munoz, H. Lu, X. C. Zhou, M. C. Tamargo, and F. H. Pollak, *Appl. Phys. Lett.* **83**, 1995 (2003); B. S. Li, R. Akimoto, K. Akita, and H. Hasama, *J. Appl. Phys.* **95**, 5352 (2004).
20. J. Khurgin, *Phys. Rev.* **B38**, 4056 (1988).
21. M. M. Fejer, S. J. B. Yoo, R. L. Byer, A. Harwit, and J. S. Harris, *Phys. Rev Lett.* **62**, 1041 (1989).
22. E. Rosencher, P. Bois, J. Nagle, and S. Delaitre, *Electron. Lett.* **25**, 1063 (1989); E. Rosencher, P. Bois, J. Nagle, E. Costard, and S. Delaitre, *Appl. Phys. Lett.* **55**, 1597 (1989).
23. E. Rosencher, P. Bois, B. Vinter, J. Nagle, and D. Kaplan, *Appl. Phys. Lett.* **56**, 1822 (1990).
24. C. Sirtori, F. Capasso, D. L. Sivco, S. N. G. Chu, and A. Y. Cho, *Appl. Phys. Lett.* **59**, 2302 (1991).
25. M. Segev, I. Grave, and A. Yariv, *Appl. Phys. Lett.* **61**, 2403 (1992).
26. W. W. Bewley, C. L. Felix, J. J. Plombon, M. S. Sherwin, M. Sundaram, P. F. Hopkins, and A. C. Gossard, *Phys. Rev.* **B48**, 2376 (1993).
27. C. Sirtori, F. Capasso, J. Faist, L. N. Pfeiffer, and K. W. West, *Appl. Phys. Lett.* **65**, 445 (1994).
28. J. Heyman, K. Craig, B. Galdrikian, M. S. Sherwin, K. Campman, P. F. Hopkins, S. Fafard, and A. C. Gossard, *Phys. Rev. Lett.* **72**, 2183 (1994).
29. L. Tsang and S. L. Chuang, *Appl. Phys. Lett.* **60**, 2543 (1992); D. S. Citrin, *Opt. Lett.* **26**, 554 (2001); Z. Xu, P. M. Fauchet, C. W. Rella, B. A. Richman, H. A. Schwettman, and G. W. Wicks, *Solid State Commun.* **93**, 903 (1995).
30. S. Z. Li and J. Khurgin, *Appl. Phys. Lett.* **62**, 1727 (1993).
31. H. Xie, W. I. Wang, J. R. Meyer, and L. R. Ram-Mohan, *Appl. Phys. Lett.* **65**, 2048 (1994).

32. C. Fischer and M. W. Sigrist, in *Solid State Mid-Infrared Laser Sources; Topics in Applied Physics* **89**, 97 (2003); D. Richter, A. Fried, B. P. Wert, J. G. Walega, and F. K. Tittel, *Appl. Phys.* **B75**, 281 (2002); D. Richter, D. G. Lancaster, R. F. Curl, W. Neu, and F. K. Tittel, *Appl. Phys.* **B67**, 347 (1998).

33. A. A. Afonenko, V. Y. Aleshkin, and A. A. Dubinov, *Semicond.* **38**, 239 (2004); T. Kaing, J. J. Zondy, A. P. Yeliseyev, S. I. Lobanov, and L. I. Isaenko, *IEEE Trans. Instr. and Measurement* **48**, 592 (1999).

34. O. Wada, *New J. Phys.* **6**, 183 (2004); A. Harwit, R. Fernandez, W. D. Eades, D. K. Kinell, and R. L. Whitney, *Superlatt. and Microstruct.* **19**, 39 (1996).

35. I. Vurgaftman, J. R. Meyer, and L. R. Ram-Mohan, *IEEE J. Quantum Electron.* **32**, 1334 (1996); M. Zaluzny, *J. Appl. Phys.* **74**, 4716 (1993); R. Rapaport, G. Chen, O. Mitrofanov, C. Gmachl, H. M. Ng, and S. N. G. Chu, *Appl. Phys. Lett.* **83**, 263 (2003).

36. J. R. Meyer, I. Vurgaftman, and C. A. Hoffman, *J. Appl. Phys.* **81**, 7662 (1997).

37. X. H. Zhang, Z. H. Chen, D. F. Cui, G. Z. Yang, and R. Z. Wang, *Appl. Phys. Lett.* **69**, 3164 (1996).

38. N. Suzuki, *J. Opt. Soc. Am.* **B21**, 2017 (2004); N. Iizuka, K. Kaneko, and N. Suzuki, *Electron. Lett.* **40**, 962 (2004); Claire Gmachl, Sergey V. Frolov, Hock M. Ng, S.-N. George Chu, and A. Y. Cho, *Electron. Lett.* **37**, 378 (2001).

39. J. Hamazaki, S. Matsui, H. Kunugita, K. Ema, H. Kanazawa, T. Tachibana, A. Kikuchi, and K. Kishino, *Appl. Phys. Lett.* **84**, 1102 (2004); Claire Gmachl, Hock M. Ng, Sung-Nee G. Chu, and Alfred Y. Cho, *Appl. Phys.* Lett. **77**, 3722 (2000); V. D. Jovanovic, Z. Ikonic, D. Indjin, P. Harrison, V. Milanovic, and R. A. Soref, *J. Appl. Phys.* **93**, 3194 (2003); K. Kishino, A. Kikuchi, H. Kanazawa, and T. Tachibana, *Appl. Phys. Lett.* **81**, 1234 (2002).

40. H. C. Liu and E. Dupont, *J. Nonlin. Opt. Phys. & Mat.* **11**, 433 (2002).

41. H. Schmidt, K. L. Campman, A. C. Gossard, and A. Imamoglu, *Appl. Phys. Lett.* **70**, 3455 (1997); J. F. Dynes, M. D. Frogley, M. Beck, J. Faist, and C. C. Phillips, *Phys. Rev. Lett.* **94**, 157403 (2005); D. Dini, R. Koehler, A. Tredicucci, G. Biasiol, and L. Sorba, *Phys. Rev. Lett.* **90**, 116401 (2003).

42. H. Schmidt and R. J. Ram, *Appl. Phys. Lett.* **76**, 3173 (2000).

43. J. Faist, F. Capasso, C. Sirtori, D. L. Sivco, A. L. Hutchinson, M. S. Hybertson, and A. Y. Cho, *Phys. Rev. Lett.* **76**, 411 (1996); H. Willenberg, G. H. Dohler, and J. Faist, *Phys. Rev.* **B67**, 085315 (2003).

44. M. A. Belkin, M. Troccoli, L. Diehl, F. Capasso, A. A. Belyanin, and D. L. Sivco, *Conference on Lasers and Electro-Optics—Post-Deadline Session QPDA1,* Baltimore, MD, May 22–27, 2005.

45. A. A. Belyanin, F. Capasso, V. Kocharovsky, and M. O. Scully, *Phys. Rev.* ***A63***, 053803 (2001); A. A. Belyanin, F. Capasso, V. V. Kocharovsky, Vl. V.

Kocharovsky, and M. O. Scully, *Phys. Rev.* **A63**, 053803 (2001); ibid. **65**, 053824 (2002).

46. C. Gmachl, A. Belyanin, D. L. Sivco, M. L. Peabody, N. Owschimikow, A. M. Sergent, F. Capasso, and A. Y. Cho, *IEEE J. Quantum Electron.* **39**, 1345 (2003).
47. Nina Owschimikow, Claire Gmachl, Alexey Belyanin, Vitaly Kocharovsky, Deborah L. Sivco, Raffaele Colombelli, Federico Capasso, and Alfred Y. Cho, *Phys. Rev. Lett.* **90**, 043902 (2003).
48. S. Slivken, A. Evans, J. David, and M. Razeghi, *Appl. Phys. Lett.* **81**, 4321 (2002); J. S. Yu, A. Evans, J. David, L. Doris, S. Slivken, and M. Razeghi, *IEEE Photon. Techn. Lett.* **16**, 747 (2004); A. Evans, J. S. Yu, J. David, L. Doris, K. Mi, S. Slivken, and M. Razeghi, *Appl. Phys. Lett.* **84**, 314 (2004); S. Forget, C. Faugeras, J. Y. Bengloan, M. Calligaro, O. Parillaud, M. Giovannini, J. Faist, and C. Sirtori, *Electron. Lett.* **41**, 418 (2005).
49. C. Sirtori, J. Faist, F. Capasso, D. L. Sivco, A. L. Hutchinson, and A. Y. Cho, *Appl. Phys. Lett.* **66**, 3242 (1995).
50. Claire Gmachl, Deborah L. Sivco, James N. Baillargeon, Albert L. Hutchinson, Federico Capasso, and Alfred Y. Cho, *Appl. Phys. Lett.* **79**, 572 (2001).
51. A. Straub, T. S. Mosely, C. Gmachl, R. Colombelli, M. Troccoli, F. Capasso, D. L. Sivco, and A. Y. Cho, *Appl. Phys. Lett.* **80**, 2845 (2002); C. Gmachl, D. L. Sivco, R. Colombelli, F. Capasso, and A. Y. Cho, *Nature* **415**, 883 (2002).
52. H. T. Grahn, R. J. Haug, W. Müller, and K. Ploog, *Phys. Rev. Lett.* **67**, 1618 (1991).
53. S. Illek, W. Thulke, C. Schanen, H. Lang, and M. C. Amann, *Electron. Lett.* **26**, 46 (1990).
54. A. Belyanin, V. Kocharovsky, Vl. Kocharovsky, and M. Scully, *Phys. Rev.* **A65**, 053824 (2002); S. M. Wang, Y. H. Shen, J. X. Xu, L. G. Hu, J. Zhu, D. R. Yang, H. Zhang, Y. W. Zeng, and J. Q. Yao, *Appl. Phys. Lett.* **84**, 3007 (2004).
55. J. S. Yu, S. Slivken, S. R. Darvish, and M. Razeghi, *IEEE Photon. Techn.Lett.* **17**, 1154 (2005); M. P. Semtsiv, M. Ziegler, S. Dressler, W. T. Masselink, N. Georgiev, T. Dekorsy, and M. Helm, *Appl. Phys. Lett.* **85**, 1478 (2004); R. Teissier, D. Barate, A. Vicet, C. Alibert, A. N. Baranov, X. Marcadet, C. Renard, M. Garcia, C. Sirtori, D. Revin, and J. Cockburn, *Appl. Phys. Lett.* **85**, 167 (2004).
56. M. W. Sigrist (ed.), *Air Monitoring by Spectroscopic Techniques*, Wiley Interscience, New York, 1994.
57. C. Sirtori, F. Capasso, J. Faist, and S. Scandolo, *Phys. Rev.* **B50**, 8663–8674 (1994).

58. G. Shu, E. E. Narimanov, C. Gmachl, D. L. Sivco, and A. Y. Cho, *Conference on Lasers and Electro-Optics, CTuM5*, Baltimore, MD, May 22–27, 2005.

59. C. Gmachl, N. Owschimikow, A. Belyanin, A. M. Sergent, D. L. Sivco, M. L. Peabody, A. Y. Cho, and F. Capasso, *Appl. Phys. Lett.* **84**, 2751 (2004).

60. T. S. Mosely, A. Belyanin, C. Gmachl, D. L. Sivco, M. L. Peabody, and A. Y. Cho, *Opt. Express* **12**, 2972 (2004)

61. X. Marcadet, V. Ortiz, J.-Y. Bengloan, S. Dhillon, M. Calligaro, and C. Sirtori, *J. Vac. Sci. and Techn.* **22**, 1558 (2004); J.-Y. Bengloan, M. Calligaro, V. Ortiz, X. Mercadet, I. Maurin, and C. Sirtori, *Intersubband Transitions in Quantum Wells 2003, R65*, Evolene, Switzerland, September 1–5, 2003.

62. H. Giovannini, M. Beck, N. Hoyler, and J. Faist, *Intersubband Transitions in Quantum Wells 2003, R64*, Evolene, Switzerland, September 1–5, 2003.

63. C. Gmachl, A. Soibel, R. Colombelli, D. L. Sivco, F. Capasso, and A. Y. Cho, *IEEE Photon. Techn. Lett.* **14**, 1671 (2002).

64. A. Fiore, V. Berger, E. Rosencher, P. Bravetti, and J. Nagle, *Nature* **391**, 463 (1998).

65. J. R. Meyer, C. A. Hoffman, F. J. Bartoli, and L. R. Ram-Mohan, *Appl. Phys. Lett.* **67**, 608 (1995); K. L. Vodopyanov, K. O'Neill, G. B. Serapiglia, C. C. Phillips, M. Hopkinson, I. Vurgaftman, and J. R. Meyer, *Appl. Phys. Lett.* **72**, 2654 (1998).

66. Oana Malis, Alexey Belyanin, Claire Gmachl, Deborah L. Sivco, Milton L. Peabody, A. Michael Sergent, and Alfred Y. Cho, *Appl. Phys. Lett.* **84**, 2721 (2004).

67. A. Chowdhury and L. McCaughan, *IEEE Photon. Techn. Lett.* **12**, 486 (2000); K. Moutzouris, S. V. Rao, M. Ebrahimzadeh, A. De Rossi, M. Calligaro, V. Ortiz, and V. Berger, *Appl. Phys. Lett.* **83**, 620 (2003).

68. J. Faist, F. Capasso, C. Sirtori, D. L. Sivco, J. N. Baillargeon, A. L. Hutchinson, S. N. G. Chu, and A. Y. Cho, *Appl. Phys. Lett.* **68**, 3680 (1996).

69. O. Malis, A. Belyanin, D. L. Sivco, J. Chen, A. M. Sergent, C. Gmachl, and A. Y. Cho, *Electron. Lett.* **40**, 1586 (2004).

70. A. A. Belyanin, V. V. Kocharovsky, Vl. V. Kocharovsky, and M. O. Scully, "Surface-Emitting Grating-Outcoupled Lasers for Mid/Far-IR Difference-Frequency Generation," submitted to *Appl. Phys. Lett.* (2005).

71. V. Berger and C. Sirtori, *Semicon. Science and Techn.* **19**, 964 (2004).

72. H. C. Liu, C. Y. Song, Z. R. Wasilewski, A. J. Spring Thorpe, J. C. Cao, C. Dharma-wardana, G. C. Aers, D. J. Lockwood, and J. A. Gupta, *Phys. Rev. Lett.* **90**, 077402 (2003).

73. M. Troccoli, A. Belyanin, F. Capasso, E. Cubucku, D. L. Sivco, and A. Y. Cho, *Nature* **433**, 845 (2005).

74. A. Gomez-Iglesias, D. Wasserman, C. Gmachl, A. Belyanin, and D. L. Sivco, *Appl. Phys. Lett.* **87**, 261113 (2005).

75. R. Paiella, F. Capasso, C. Gmachl, D. L. Sivco, J. N. Baillargeon, A. L. Hutchinson, A. Y. Cho, and H. C. Liu, *Science* **290**, 1739 (2000).

76. A. Soibel, F. Capasso, C. Gmachl, M. L. Peabody, A. M. Sergent, R. Paiella, D. L. Sivco, A. Y. Cho, and H. C. Liu, *IEEE J. Quantum Electron.* **40**, 197 (2004).

Chapter

6

Raman Injection and Inversionless Intersubband Lasers

Alexey Belyanin

Department of Physics, Texas A&M University, College Station, Texas

Mariano Troccoli and Federico Capasso

Division of Engineering and Applied Sciences Harvard University, Cambridge, Massachusetts

6.1 Introduction

Quantum cascade lasers (QCLs) are only 11 years old [1]; however, they have already reached a remarkably high level of maturity. QCLs operate due to population inversion on the intersubband transition between quantum-confined electron states in semiconductor quantum well structures. They utilize a typical four-level lasing scheme [2] (Fig. 6.1), in which electrons are injected to the upper laser state 3 from the upstream injector 4 by resonant tunneling, make radiative or nonradiative transition to the lower laser state 2, and then rapidly depopulate this state via phonon emission and tunneling out of the active region to the injector section downstream. The latter serves as an upstream injector for the next active stage. The active region can be formed by several coupled *quantum wells* (QWs) or by a superlattice. Details of different active region and laser designs are presented in Chaps. 2 through 4 of this volume.

Although the scheme in Fig. 6.1 includes at least four states, there are only two states, 2 and 3, interacting resonantly with laser radiation. In this case, positive gain for the laser signal exists only in the presence

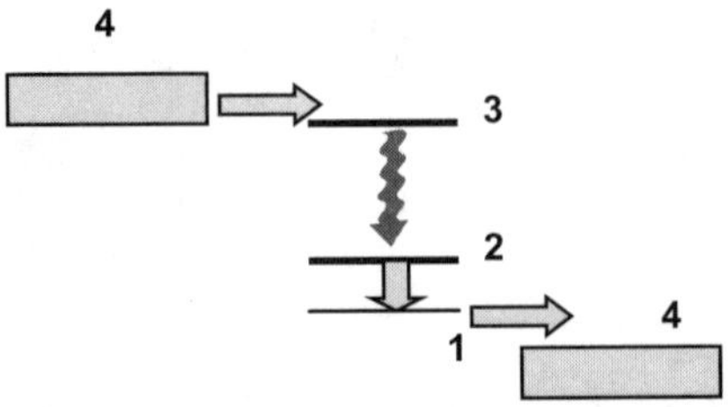

Figure 6.1 Sketch of the four-level lasing scheme implemented in one active stage of a conventional QCL. There are usually many stages in a device.

of population inversion between these states. Of course, population inversion should be defined by taking into account the density of states in the momentum space and the distribution function of electrons.

The situation changes if one of the laser states is strongly coupled to some other state(s) by the electromagnetic field, through coherent tunneling, or even through an incoherent relaxation process. Coupling means the existence of the coherence or the off-diagonal element of the density matrix on one of the transitions. It gives rise to an additional contribution to laser gain and frequency. Under favorable conditions this contribution adds a positive gain, and amplification becomes possible in the absence of population inversion. This phenomenon has been called *lasing without inversion* (LWI) and has been the subject of much debate and investigation since the 1980s; see, e.g. Refs. 3 and 4 for the review and references to original works.

In the "quantum" language, the origin of the effect is the quantum interference between multiple transition channels, which violates the reciprocity between the absorption and emission processes. Hereafter by *states* we will mean "bare" states of the hamiltonian of the electron subsystem. In most cases, the inversion exists in the basis of states "dressed" by the coupling, i.e., of the eigenstates of the total hamiltonian that includes the coherent coupling term. In some cases, the gain exists without population inversion in any meaningful basis.

In the language of macroscopic electrodynamics, the effect is described by resonant polarization of the multistate system at the signal frequency. The polarization is strongly affected by the resonant coupling, which gives rise to the change in the amplification condition for the signal, accompanied by peculiar changes in the refractive index and group velocity that have been the subject of active investigation and controversy as well; see Ref. 5 for the review. Note that although the polarization can be treated as linear with respect to the weak signal, it is highly nonlinear with respect to the electromagnetic field (the "drive") that is used to create coherent coupling. Therefore, with respect to the total field we are dealing with resonant nonlinear optical interaction.

The simplest system that already exhibits LWI consists of three states in the presence of a strong drive field. Four possible configurations are shown in Fig. 6.2.

Despite obvious differences, the physics of inversionless gain in these systems is remarkably similar. In the absence of an optical drive, the signal gain is proportional to the population difference on the signal transition, and population inversion is required for amplification. When the optical drive is on, there is an additional contribution to the gain coming from the nonlinear mixing of the drive electromagnetic field and the coherent polarization oscillations at frequency resonant to transition 1–2 in Fig. 6.2*a* and *b* and transition 1–3 in Fig. 6.2*c* and *d*. This polarization, in turn, is excited self-consistently by the interaction between the drive and signal field. Under certain conditions, this Raman-type contribution can exceed both resonant absorption on the signal transition and all other losses, leading to laser action. When all fields are close to resonance, Raman-type inversion on the two-photon transition (e.g., 1–3–2 in Fig. 6.2*a* and *b*) may or may not be present in a system. When the drive field is detuned very far from resonance with the drive transition 1–3, the gain exists only in the presence of the Raman inversion $n_1 - n_2 > 0$. Raman inversion in cascade schemes in Fig. 6.2*c* and *d* means $n_3 > n_1$. Similarly to the Λ scheme, gain on the signal transition in cascade schemes is possible in the absence of any inversion including Raman inversion. However, we found that for realistic parameters in QC structures, the gain exceeds losses when Raman inversion is already present in the system. A similar conclusion has been reached in studies of other systems [6, 7].

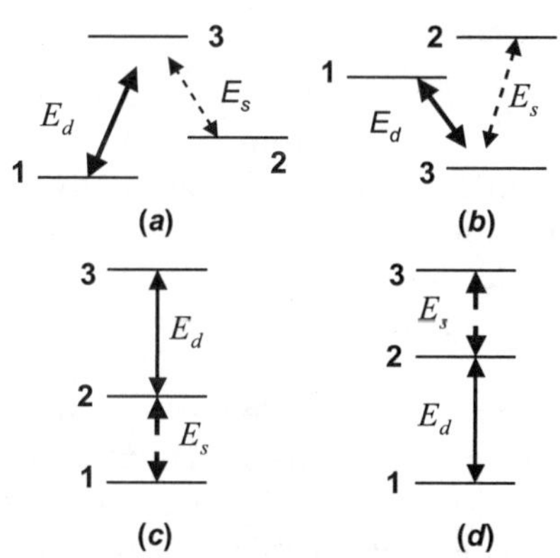

Figure 6.2 Simplest three-level systems exhibiting the effects of Raman and inversionless lasing of the signal field E_s in the presence of a coherent drive field E_d. (*a*) Λ scheme; (*b*) *V* scheme; (*c*) and (*d*) cascade, or ladder, schemes.

Although there have been many theoretical studies of LWI, experimental progress has not been that impressive; see Ref. 4 for the review. It has been limited to the proof-of-principle experiments with rarefied atomic vapors, in which the multilevel atomic system strongly driven by external laser field demonstrated amplification on the order of several percent at a signal frequency very close to the frequency of the drive laser.

In semiconductor quantum wells, it was suggested to reach LWI by employing Fano-type interference between two states induced by tunneling to the quasi-continuum; see Ref. 8 for the review. There was also a suggestion to use coherent laser coupling to achieve LWI in a system of excitons in quantum dots [9]. As far as we know, these ideas have not been realized yet.

Another issue with LWI in any material system, not only semiconductors, is the lack of a strong practical motivation. It would be very important to achieve inversionless lasing on the rapidly decaying transition when the population inversion is difficult to implement, e.g., on the X-ray transitions in atomic ions or intersubband transitions in a QW. However, note that in realistic situations LWI occurs under conditions that are actually harder to implement than usual lasing with inversion [6–8]. Selection rules in atomic systems place strong restriction on the type and efficiency of the particular LWI scheme. In addition, the necessity of a strong external laser drive limits practical applications of such schemes and gives rise to additional problems related to the drive absorption, scattering, and overlap with a signal beam. For example, in many cases the drive field experiences resonant absorption at a characteristic length, which is shorter than the amplification length of the signal. The latter is especially true for semiconductor QWs in which the drive absorption on the interband or intersubband transition makes the realization of the LWI schemes with an external drive unfeasible.

In Ref. 5 we argued that the above problems can be overcome if the drive field is generated in the same multilevel system that is used for inversionless amplification of the signal, as shown schematically in Fig. 6.3. This idea of resonant intracavity optical pumping is applicable not only to Raman or inversionless lasing, but also to any nonlinear optical process. Such schemes and devices can demonstrate new physics and offer significant practical benefits.

QCL structures seem to be an ideal system for implementing such integration of lasing and nonlinear optical generation. Intracavity intensities of the laser field in QCLs can be quite high—many megawatts per square centimeter, which corresponds to Rabi frequencies exceeding the spectral line width of intersubband transitions. Intersubband transitions in coupled asymmetric QWs can

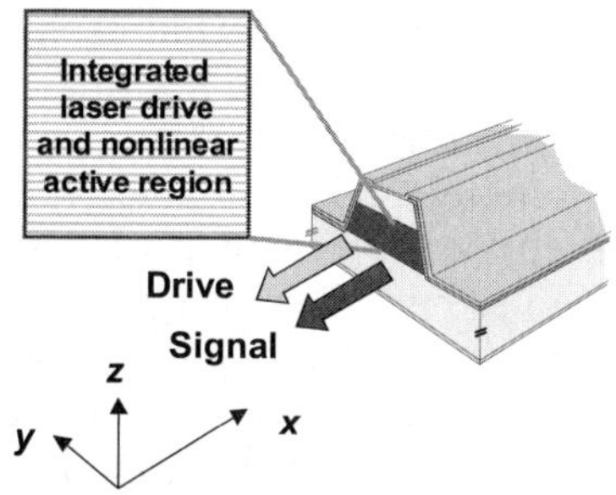

Figure 6.3 Integration of the conventional injection laser and the nonlinear region supporting resonant nonlinear conversion of the laser (drive) field into radiation at a signal frequency. Coordinate frame used in the equations below is indicated.

be manipulated to control their dipole moments and relaxation rates. These dipole moments are usually very large, between 1 and 10 nm, giving rise to giant resonant nonlinearities. For example, the values of $\chi^{(2)} \sim 10^6$ pm/V have been demonstrated; see, e.g., Refs. 10 through 13. Note that in conventional nonlinear optical crystals $\chi^{(2)}$ is on the order of several picometers per volt. Electron transport through the structure and populations of electron states can be efficiently controlled by resonant tunneling and phonon emission. Finally, QC structures offer a great flexibility in combining several laser transitions or heterogeneous active stages designed to emit or absorb at different wavelengths [14]. In particular, lasing in QCLs at two or more frequencies has been demonstrated [15, 16].

We have recently shown [17–23] that the standard active region of a QCL can be integrated with a cascade of intersubband transitions designed for the nonlinear conversion of the laser field into coherent radiation at a different frequency. Generation of the second and third harmonics and the sum frequency has been demonstrated. Milliwatt power level for second harmonic generation has been achieved [22]. Recently the first successful experiment on Raman injection laser has been reported [23].

Here we consider the possibility to integrate a QC laser with a multiple-QW nonlinear section implementing one of the schemes shown in Fig. 6.2, in which states 1, 2, and 3 are electron subbands in the conduction band. Besides discussing the physics of Raman and inversionless lasing with application to QC lasers, our goal is to identify the particular schemes that could bring meaningful practical benefits.

From a practical point of view, the schemes in Fig. 6.2 could be used to implement a nonlinear conversion of the optical drive into the signal at wavelengths that can be hard to reach for standard QC lasers and that can vary in a very broad range from the terahertz region to short mid-infrared and even near-infrared region. The physical reasons underlying this advantage of nonlinear optical schemes are that (1) no resonant tunneling and even no current are required in the nonlinear section of the laser and (2) no population inversion is necessary on the signal transition. Of course, population inversion and electron transport are required in the active region supporting the drive laser. However, the drive wavelength can be chosen to lie in the "sweet spot" of QC lasers (7 to 9 μm for InGaAs/AlInAs structures) where population inversion is easy to provide and where they are most powerful, efficient, and capable of continuous-wave room-temperature operation [24, 25]. In addition, one may expect new or enhanced functionalities stemming from the nature of the nonlinear optical process, e.g., electric tuning by Stark effect and wider tuning range as compared to conventional QC lasers. To summarize, the primary practical appeal of integrated devices is the potential to combine the advantages of semiconductor injection lasers (compactness, robustness, and handling convenience) with the benefits of nonlinear optical sources.

We give a review of opportunities and limitations related to the simplest three-level Raman and LWI-type schemes in QCL structures, in which the laser field serves as an internally generated coherent optical drive.

6.2 Theoretical Framework

6.2.1 Electromagnetic modes

Throughout the chapter we assume that the electromagnetic field is classical and is described by the Maxwell equations. An interesting problem of the statistics of light generated in our resonant nonlinear systems is beyond the scope of this review and will be considered elsewhere.

In QCL structures we are dealing with interaction of tightly confined waveguide modes. Although the nonlinear optics of waveguide modes has been considered many times in the past (see, e.g., Ref. 26), for the reader's convenience we present the general equations in the form suitable for QC structures.

For intersubband transitions close to the Γ point in the conduction band, the dipole moments have only z components. Therefore, transverse magnetic (TM) polarized modes with a large z component of the electric field are preferentially excited. In Maxwell's equations it is convenient to separate resonant intersubband polarization **P** from the nonresonant dielectric response described by the complex dielectric function $\varepsilon(x, y, z)$ as follows:

$$\nabla \times \mathbf{H} = \frac{\varepsilon}{c}\frac{\partial \mathbf{E}}{\partial t} + \frac{4\pi}{c}\frac{\partial \mathbf{P}}{\partial t} \tag{6.1}$$

$$\nabla \times \mathbf{E} = -\frac{1}{c}\frac{\partial \mathbf{H}}{\partial t} \tag{6.2}$$

In the simplest case of a plane-parallel semiconductor waveguide, the dielectric function has a piecewise-constant profile in the z direction and is constant in the y direction, with jumps on the lateral walls of the waveguide.

In deriving a single wave equation for the magnetic field in the TM mode, it is convenient to introduce a new operator $\tilde{\nabla} = (1/\varepsilon)\nabla$, which is equivalent to the transformation to new variables

$$\tilde{x} = \int_0^x \varepsilon(x, y, z)\, dx \qquad \tilde{y} = \int_0^y \varepsilon(x, y, z)\, dy \qquad \tilde{z} = \int_0^z \varepsilon(x, y, z)\, dz$$

Then Eqs. 6.1 and 6.2 can be combined as

$$\tilde{\nabla}^2 \mathbf{H} - \frac{1}{\varepsilon c^2}\frac{\partial^2 \mathbf{H}}{\partial t^2} = -\frac{4\pi}{c}\tilde{\nabla} \times \frac{\partial}{\partial t}\left(\frac{\mathbf{P}}{\varepsilon}\right) \tag{6.3}$$

Let us further assume for simplicity that $\varepsilon = \varepsilon\,(y, z)$ is independent of x, and let us expand the magnetic field over the quasi-orthogonal set of the transverse waveguide modes $\mathbf{F}_j(y, z)$:

$$\mathbf{H} = \sum_j h_j(x)\mathbf{F}_j(y, z)e^{i\beta_j x - i\omega_j t} + c.c. \tag{6.4}$$

Here the functions $h_j(x)$ are slowly varying with x, and we can neglect their second derivatives. In general, h_j are also the functions of time, but here we consider the continuous-wave operation or pulses that are much longer than all relaxation times in a QCL. The waveguide modes are eigenfunctions of the equation

$$\varepsilon\left(\frac{\partial}{\partial y}\frac{\partial}{\varepsilon\,\partial y} + \frac{\partial}{\partial z}\frac{\partial}{\varepsilon\,\partial z}\right)\mathbf{F}_j + \left(\frac{\omega_j^2 \varepsilon}{c^2} - \beta_j^2\right)\mathbf{F}_j = 0 \tag{6.5}$$

The solution of Eq. 6.5 defines both the transverse profile of a given waveguide mode $\mathbf{F}_j(y, z)$ and its dispersion, i.e., the complex propagation constant $\beta_j(\omega_j)$ as a function of the real frequency of the mode. The modes with different index j are weakly coupled due to waveguide losses, but we neglect this coupling here, assuming the orthogonality condition

$$\int \frac{1}{\varepsilon} F_i F_j^* dA = \delta_{ij} \tag{6.6}$$

where the integration is over the waveguide cross section and the asterisk denotes complex conjugation.

After substituting the expansion 6.4 into Eq. 6.3 and using Eq. 6.5, we arrive at

$$\sum_j \frac{2i\beta_j}{\varepsilon} \mathbf{F}_j e^{i\beta_j x - i\omega_j t} \frac{\partial h_j}{\partial x} = -\frac{4\pi\varepsilon}{c} \tilde{\nabla} \times \frac{\partial}{\partial t} \left(\frac{\mathbf{P}}{\varepsilon} \right) \tag{6.7}$$

Multiplying Eq. 6.7 by $\mathbf{F}_k^*$, integrating over the waveguide cross section, and using the condition 6.6, we finally obtain the equation describing the excitation of a given waveguide mode:

$$\frac{\partial h_k}{\partial x} = \frac{2\pi i}{c\beta_k} e^{-i\beta_k x + i\omega_k t} \int \varepsilon \tilde{\nabla} \times \frac{\partial}{\partial t} \left(\frac{\mathbf{P}}{\varepsilon} \right) \mathbf{F}_k^* \, dA \tag{6.8}$$

The polarization that excites a given mode can be due to nonlinear optical processes involving other modes, in which case we should write the equation similar to Eq. 6.8 for each participating mode and solve the whole coupled system together with material equations for the polarization.

Further simplification occurs if we recall that the intersubband electronic polarization is directed along the z axis perpendicular to the QW plane: $\mathbf{P} = P\mathbf{z}^0$. For the waveguide that is sufficiently broad in the y direction, we may also assume that the magnetic field has only a y component. Then Eq. 6.8 can be rewritten as

$$\frac{\partial h_k}{\partial x} = -\frac{2\pi i}{c\beta_k} e^{-i\beta_k x + i\omega_k t} \int \frac{1}{\varepsilon} \frac{\partial^2 P}{\partial t \, \partial x} F_k^* \, dA \tag{6.9}$$

This equation has to be solved with appropriate boundary conditions on the cavity facets $x = 0$ and $x = L$, taking into account the reflection and transmission of the optical power. For a weak signal field the polarization can be expanded in series in powers of the field. For example, the terms of zeroth order with respect to the signal field describe intracavity generation of harmonics by a strong laser drive field [18, 21]. Here we restrict ourselves to the processes of true lasing, in which the weak-signal polarization is proportional to the z component of the electric field of a given mode: $P = \chi E_z$. This would lead to the exponential weak-signal gain.

Note that the polarization may be linear with respect to the weak-signal field but is highly nonlinear with respect to the drive field. In fact, it *has* to be highly nonlinear. Indeed, the main nonlinearity present in our system is resonant saturation nonlinearity. The expansion of the polarization in powers of the drive field breaks down when the drive field intensity approaches the saturation value (between 0.1 and 1 MW/cm^2 for a strong intersubband transition). However, since the drive field is an internally generated laser field, it is inevitably close to saturation as soon as the lasing threshold is surpassed by a factor of 2. Therefore, we are in the nonperturbative regime, and in calculating the polarization, the drive field needs to be taken into account exactly.

The resonant susceptibility χ is a function of all three coordinates x, y, z as it depends on the position of the active layers in a waveguide and the distribution of the drive field intensity across the waveguide and along the cavity. We will, however, assume that the x dependence of χ is quite smooth and neglect its x derivative in the right-hand side of Eq. 6.9. Then, from Eqs. 6.1 and 6.2 one can derive the relation

$$\frac{\partial^2 P}{\partial t\,\partial x} \approx -\frac{c\chi}{\varepsilon}\frac{\partial^2 H}{\partial x^2} \approx \frac{c\chi}{\varepsilon}\sum_j h_j \beta_j^2 F_j e^{i\beta_j x - i\omega_j t} \tag{6.10}$$

If only one waveguide mode is excited at a signal frequency (because, e.g., it has the excitation threshold considerably lower than that of all other modes), we can obtain from Eqs. 6.9 and 6.10 the equation for the amplitude of this mode

$$\frac{\partial h}{\partial x} \approx \frac{2\pi i\omega\mu h}{c}\int \frac{\chi}{\varepsilon^2}\,|F|^2\,dA = \frac{2\pi i\omega\bar{\chi}}{c\mu}\Gamma h \tag{6.11}$$

where $\mu = c\,\mathrm{Re}(\beta)/\omega$ is the effective refractive index of the given waveguide mode and

$$\Gamma = \mu^2 \int_{\chi\neq 0} \frac{1}{\varepsilon^2}\,|F|^2\,dA \tag{6.12}$$

is the effective optical confinement factor proportional to the overlap of the Poynting vector of the mode with the active region. The integral in Eq. 6.12 is over the active layers where $\chi \neq 0$ and $\bar{\chi}$ is the averaged value of χ in the active region; its definition is evident from comparing Eqs. 6.11 and 6.12. We will omit the overbar hereafter. The integral in Eq. 6.12 extended over the whole waveguide cross section would give the total flux of the Poynting vector of the mode in the x direction.

The real part of the coefficient before h on the right-hand side of Eq. 6.11 defines the modal gain of the field

$$g_M = \frac{2\pi\omega \text{ Im } \chi}{c\mu}\Gamma \tag{6.13}$$

while its imaginary part determines the resonant frequency shift.

Solving Eq. 6.3 with boundary conditions on the cavity facets leads to a set of discrete longitudinal modes of the Fabry-Perot cavity, spaced by $\Delta\nu = m\pi c/L$, where L is the cavity length and m is an integer. This spacing is usually much smaller than the spacing between neighboring transverse modes. For a weak signal field below saturation, when χ is independent of the signal intensity, the x distribution of the total field amplitude in a cavity is the sum of two counterpropagating exponents

$$H_{\text{tot}} = \frac{H_{\text{out}}}{1 - R}\left(\text{R} e^{(g_M - \alpha_w)x} + e^{-(g_M - \alpha_w)x}\right) \tag{6.14}$$

where $\alpha_w = \text{Im } \beta$ is waveguide losses, $R = (\mu - 1)/(\mu + 1)$ is the reflection factor of the facets for the magnetic field in the TM mode, and H_{out} is the amplitude of the magnetic field in the outgoing radiation from one facet. We assumed the reflection factors for the two facets to be equal for simplicity. The excitation condition for a given mode has the usual form

$$g_M > \alpha_w + \frac{1}{2L} \ln \frac{1}{R} \tag{6.15}$$

Note that both gain and losses are defined here with respect to the field amplitude. The same quantities for the field intensity are 2 times larger.

6.2.2 Material equations

Since we are dealing with resonant optical nonlinearities, the natural way to treat the interaction of light with an electron subsystem is through the density matrix equations. It allows one to easily incorporate many interacting fields and various incoherent relaxation and scattering processes. In this approach the polarization **P** of the medium that serves as a source of the field in Maxwell's equations is calculated as a trace of the density matrix

$$\mathbf{P} = \frac{1}{\Delta V}\sum \text{Tr}\,(\mathbf{d}\rho) \tag{6.16}$$

where the summation is performed over all electron states in the volume ΔV and $\mathbf{d}_{ij} = e\mathbf{z}_{ij}$ are the dipole moments of the intersubband transitions i–j. The elements of the density matrix are found from the master equation of the type

$$\frac{\partial \rho_k}{\partial t} = -\frac{i}{\hbar}(H,\ \rho_k) + \frac{\partial \rho_k}{\partial t}\bigg|_{\text{col}} \tag{6.17}$$

where the hamiltonian includes a system of electron states and its coherent interaction with the electromagnetic field, while all relaxation and scattering processes are included in the collision integral, which is the second term on the right-hand side.

We proceed with several important approximations as specified below.

1. We neglect most many-body effects due to Coulomb interaction of electrons. The only Coulomb effect that we include is the electric field due to charge separation in a doped structure. For average electron densities in the active region at the level of several 10^{16} cm^{-3} typical for QCLs, the neglected Hartree-Fock and higher-order corrections lead to the energy shift of the *intersubband* (ISB) transition of the order of 1 meV or less, which is insignificant compared to the line broadening and experimental uncertainty. Of greater potential importance is the effect of Coulomb interactions on the decoherence rates. There have been arguments that Coulomb interactions may actually decrease the decoherence rate of the optical polarization [27], which could be important for the coherent phenomena considered here. However, the line broadening typically observed in InGaAs/AlInAs structures in luminescence and absorption experiments is at the level of 10 to 15 meV *full width at half maximum* (FWHM). This value is much larger than any positive or negative effect associated with Coulomb interactions. The origin of the observed broadening is uncertain at this point. It could be homogeneous broadening related to the interface roughness [28], which is supported by an approximately lorentzian profile of the absorption lines. Another option is the inhomogeneous broadening that could be associated, e.g., with inhomogeneity of the QW structure. In any case, we treat the decoherence rates of the off-diagonal matrix elements as external parameters coming from experiment.
2. We neglect the effect of the laser field on the electron wave functions and dipole moments of the intersubband transitions. In principle, the intracavity laser intensities in QCLs can reach the level of several megawatts per square centimeter. This corresponds to the electric field of tens of kilovolts per centimeter, which is comparable to the external electric field created by a voltage bias across the contacts. Such fields could modify the electron wave functions, leading to a change in the dipole moments and scattering rates, although we are not aware of any studies of this effect for intersubband transitions.

This could be an interesting problem by itself, which is beyond the scope of this chapter.

3. In evaluating the dynamics of populations (diagonal terms in the density matrix), we replace the realistic spatial kinetics of electron transport through the structure with rate equations for a set of electron states, with the transition times between different states defined by phonon emission and resonant tunneling.

   Then the analysis can be split in two sequential steps. At the first step, the electron states are found in the absence of the optical field by solving the coupled Schrödinger and Poisson equations. The dipole moments and scattering rates are calculated. At the second step, the resonant interaction of the multifrequency laser field with a given set of electron states is studied on the basis of Maxwell's equations coupled with density matrix equations, with the dipole moments and relaxation rates provided from the previous step. As a result, we find the electron populations, drive and signal thresholds, and drive and signal intensities as functions of the single external parameter of injection current density.

Equations for the intersubband off-diagonal density matrix elements $\rho_{mn}(k)$ and the diagonal matrix elements $\rho_{mm}(k)$ can be written in the following general form [29]:

$$\frac{d\rho_{mn}}{dt} + (\gamma_{mn} + i\omega_{mn})\rho_{mn} = \frac{ie}{\hbar} E_z \sum_q (z_{mq}\rho_{qn} - z_{qn}\rho_{mq}) \tag{6.18}$$

$$\frac{d\rho_{mm}}{dt} + R_m = \frac{ie}{\hbar} E_z \sum_q (z_{mq}\rho_{qm} - z_{qm}\rho_{mq}) \tag{6.19}$$

Here $\gamma_{mn}(k)$ is the relaxation rate of the off-diagonal element $\rho_{mn}(k)$ of the density matrix, $\omega_{mn}(k)$ is the transition frequency and z_{mn} the dipole moment of the ISB transition between mth and nth subband states with a given in-plane momentum $\mathbf{k}$, and R_m denotes all relaxation and pumping terms that determine the electron distribution $\rho_{mm}(k)$ in the mth subband in the absence of radiation field.

Since we deal with resonant interactions, the electric component $E_z(\mathbf{r}, t)$ of the radiation field inside the laser cavity can be represented as a sum over quasi-monochromatic components with slowly varying amplitudes E_{mn} and frequencies ν_{mn} nearly resonant to the transition frequencies ω_{mn}:

$$E_z(\mathbf{r}, t) = \sum E_{mn}(\mathbf{r})\, e^{i\beta_{mn}x - i\nu_{mn}t} + c.c. \tag{6.20}$$

The function E_{mn} is related to the magnetic field amplitude $h_{mn}(x)$ $F_{mn}(y, z)$ for a given quasi-monochromatic field component through Maxwell's equations 6.1 and 6.2. In principle, there can be several waveguide modes for each frequency ν_{mn}. However, we will assume for simplicity that there is only one transverse waveguide mode excited for each frequency. Then one can write $E_{mn} \approx -h(x)[\mu/\varepsilon(y, z)]F(x, z)$. Here all terms on the right-hand side should also have the index mn, which we dropped for simplicity. The generalization to the case of many transverse modes of different order is straightforward.

Off-diagonal elements of the density matrix can be written as

$$\rho_{mn}(t) = \sigma_{mn}(t)e^{-i\nu_{mn}t} \tag{6.21}$$

where the amplitude σ_{mn} is a slowly varying function of time. For the diagonal elements, $\rho_{mm} = \sigma_{mm}$.

Using the above expressions in Eqs. 6.18 and 6.19, we arrive at the truncated density matrix equations

$$\frac{d\sigma_{mn}}{dt} + \Gamma_{mn}\sigma_{mn} = \frac{ie}{\hbar}\sum_q \left(z_{mq}E_{mq}\sigma_{qn} - z_{qn}E_{qn}\sigma_{mq}\right) \tag{6.22}$$

$$\frac{d\sigma_{mm}}{dt} + R_m = \frac{ie}{\hbar}\sum_q \left(z_{mq}E_{mq}\sigma_{qm} - z_{qm}E_{qm}\sigma_{mq}\right) \tag{6.23}$$

Here $\Gamma_{mn} = \gamma_{mn} + i\left(\omega_{mn} - \nu_{mn}\right)$. Note that $\sigma_{mn} = \sigma^*_{mn}$, $z_{mn} = z^*_{mn}$, and $E_{mn} = E^*_{nm}$.

6.3 Raman QC Laser

Let us apply the above general treatment to the generic Raman process of Fig. 6.2*a*.

Stimulated Raman scattering is a third-order nonlinear process that allows one, in specific materials, to generate gain at a wavelength different from the one of the incident radiation and can lead to laser emission at the peak of the Raman gain. The underlying physical mechanism is the scattering of the incident fundamental radiation by a certain eigenmode of oscillations in a material, which is self-consistently excited by the parametric interaction between the incident and the scattered light. The wavelength shift between the fundamental and the Raman emission is determined by a resonance frequency of the internal oscillations and can be due to vibrational (phonon), rotational, plasmon, or electronic resonances.

In a general scheme of the Raman Stokes process sketched in Fig. 6.4, the internal oscillations in a medium correspond to the transition

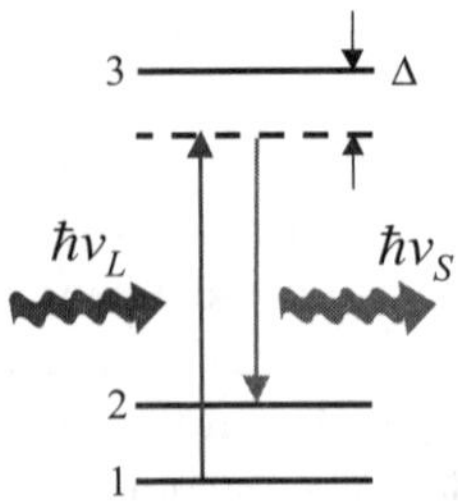

Figure 6.4 A schematic of Raman scattering of laser light (left arrow) into a lower-frequency Stokes wave (right arrow).

between electron or molecular states 1 and 2. The incident photon of energy $\hbar\omega_d$ is converted into a Stokes signal photon of energy $\hbar\omega_s$, where both the incident and scattered beams are usually far-detuned in frequency from state 3 and other higher-lying states to avoid strong absorption. In this case, only two-photon transitions between state 1 and 2 mediated by an intermediate, very short-lived virtual state may occur. As a result, the interaction between the fundamental and scattered light is of purely parametric origin and is ultimately constrained by the Manley-Rowe relation: conservation of a total number of photons in both electromagnetic modes.

Stimulated Raman scattering has been observed in all kinds of media: gases, liquids, solids, and plasmas. Until now, a distinctive feature of all Raman amplifiers has been the necessity of an external optical pump. Also, existing solid-state Raman sources universally employ scattering off a vibrational (phonon) resonance in a crystal or fiber [30]. A powerful fundamental laser radiation is usually required to offset a small value of the Raman gain (several 10^{-9} cm/W).

The possibility of Raman amplification using intersubband transitions has been theoretically discussed in Ref. 31. In Ref. 32, Raman lasing on intersubband transition in a GaAs/AlGaAs double quantum well structure optically pumped by CO_2 laser has been demonstrated, although the Raman shift was still determined by a phonon resonance. In recent work [23] we demonstrated the first *injection-pumped* Raman laser, in which the fundamental and the Raman radiation are both generated by intersubband electronic transitions in the very same active region of a quantum cascade laser. The stimulated scattering and lasing is due to the excitation of coherent electronic polarization on the mid-infrared intersubband transition 2–1. Its frequency $(E_2 - E_1) / \hbar$ defines the Raman shift; it has nothing to do with phonons and can vary in a very broad range. It could be also efficiently tuned by a voltage bias since the transition 2–1 is diagonal in real space.

6.3.1 Gain in the resonant $\wedge$ scheme

Let us first consider the generic properties of resonant gain at the Stokes frequency in a system of Fig. 6.4 without specifying its implementation in a particular semiconductor heterostructure and the way the electron transport is organized.

To get transparent semianalytic results, we assume for simplicity that all relevant ISB transitions are homogeneously broadened; i.e., the transition frequency between two subbands is the same for all electrons in these subbands. There are experimental indications and theoretical arguments that this may indeed be the case for QCL structures [28]. In doing this, we assume that the effects of subband nonparabolicity and inhomogeneity of the structure lead to smaller broadening than the homogeneous line width. The effects leading to inhomogeneous broadening can be easily taken into account if necessary. Then Eqs. 6.18 and 6.19 or 6.22 and 6.23 can be summed over the electron **k**-states in each subband, which amounts to dropping the argument **k** and replacing the diagonal matrix elements by the total populations of the subbands $n_m = \sum_k \rho_{mm}$. The result is

$$\frac{d\sigma_{21}}{dt} + \Gamma_{21}\sigma_{21} = -i\Omega_d\sigma_{32}^* + i\Omega_s^*\sigma_{31} \tag{6.24}$$

$$\frac{d\sigma_{31}}{dt} + \Gamma_{31}\sigma_{31} = i\Omega_d n_{13} + i\Omega_s\sigma_{21} \tag{6.25}$$

$$\frac{d\sigma_{32}}{dt} + \Gamma_{32}\sigma_{32} = i\Omega_s n_{23} + i\Omega_d\sigma_{21}^* \tag{6.26}$$

where

$$\Gamma_{32} = \gamma_{32} + i(\omega_{32} - \omega_s)$$

$$\Gamma_{31} = \gamma_{31} + i(\omega_{31} - \omega_d)$$

$$\Gamma_{21} = \gamma_{21} + i(\omega_{21} - \omega_d + \omega_s)$$

Here $\Omega_d = ez_{13}E_d/\hbar$ and $\Omega_s = ez_{23}E_s/\hbar$ are the Rabi frequency of the drive and the Stokes mode, respectively; $n_{12} = n_1 - n_2$; z_{mn} is the dipole moment of the transition $m - n$; and ω_d and ω_s are frequencies of the drive and the Stokes field modes, respectively.

The polarization on the Stokes transition 2–3 that should be substituted in the right-hand side of Eq. 6.9 or 6.11 is given by

$$P = e\,(z_{23}\,\rho_{32} + z_{32}\,\rho_{23}) \tag{6.27}$$

The system of Eqs. 6.24 through 6.26 is incomplete because the drive field and the population densities $n_{1,2,3}$ of electrons in states 1, 2, 3 are not free parameters. They must be calculated self-consistently from the full set of coupled density matrix and Maxwell equations including the QCL section in which the drive field is generated; see the modeling of the specific QCL structure below. The only external parameters are the injection current density and the dephasing rates γ_{mn} that are equal to the half-width at half-maximum of the corresponding transition line widths. Typical values of γ_{mn} are 5 to 10 meV for intersubband transitions in InGaAs/AlInAs quantum wells in the mid-infrared range, which exceeds the lifetime broadening, estimated to be on the order of 1 meV. Lifetime-broadened intersubband transitions have been reported so far only in GaAs QWs at low temperature [28]. In the terahertz range the broadening is usually smaller, on the order of 2 to 6 meV FWHM.

For the continuous-wave operation or pulses that are much longer than all relaxation times, one can neglect the time derivatives in Eqs. 6.24 to 6.26. Then the calculation of σ_{32} is reduced to algebra. According to Eq. 6.13, the gain is proportional to $\mathrm{Im}\,\chi = \mathrm{Im}(ez_{23}\sigma_{32}/E_s)$. Retaining only linear terms with respect to the weak field E_s, we obtain for the intensity gain (which is double the modal gain in Eq. 6.13)

$$g_1 = 2g_M = \frac{4\pi\,\omega_s e^2 z_{32}^2 \Gamma_s}{\hbar c \mu_s} \times \mathrm{Re}\left\{\frac{1}{\Gamma_{32} + |\Omega_d|^2/\Gamma_{21}^*}\left[\frac{|\Omega_d|^2 (n_1 - n_3)}{\Gamma_{21}^*\Gamma_{31}^*} - (n_2 - n_3)\right]\right\} \tag{6.28}$$

where Γ_s is the confinement factor for the Stokes mode.

An expression similar to Eq. 6.28 has been obtained before in theoretical papers on LWI [33]. Note several peculiar features in it, compared to the "usual" Raman physics in existing Raman lasers and amplifiers.

First, the gain is *not* proportional to the Raman inversion $n_1 - n_2$ between the initial and final states of the two-photon transition. Instead, there are two terms. Their origin can be seen from inspecting the general structure of Eqs. 6.24 to 6.26. It is clear that in the absence of the drive field, the density matrix element σ_{32} can be excited only by the first term on the right-hand side of Eq. 6.26, which is proportional to the Stokes field and the population difference $n_2 - n_3$ on the Stokes transition. This term describes linear absorption of the Stokes field on the transition 2–3 (the second term in brackets in Eq. 6.28).

When the drive field is on, there is a second term $i\Omega_d\sigma_{21}^*$ on the right-hand side of Eq. 6.26 that originates from the coherent beating between the drive field and the polarization σ_{12} on the transition 1–2. The latter, in turn, is excited by the coherent mixing of the Stokes field and the polarization σ_{31} on the drive transition 1–3, as described by the second term $i\Omega_s^*\sigma_{31}$ on the right-hand side of Eq. 6.24. In turn, according to Eq. 6.25 the polarization σ_{31} is excited by the strong drive field: $\sigma_{31} \approx in_{13}\Omega_d/\Gamma_{31}$. As a result, there is a contribution to the polarization σ_{32}, which scales as $\sigma_{32} \propto \Omega_d\sigma_{21}^* \propto \Omega_d\Omega_s\sigma_{31}^* \propto |\Omega_d|^2\Omega_s n_{13}$. It is proportional to the drive field intensity and the population difference on the drive transition n_{13}. This contribution gives rise to the first term in brackets in Eq. 6.28.

Second, note the first term $i\Omega_d\sigma_{32}^*$ in Eq. 6.24 that describes the shift of the resonances for the Stokes field due to the resonant ac Stark effect. It gives rise to the term $|\Omega_d|^2/\Gamma_{21}^*$ in the denominator of Eq. 6.28.

The net effect of the strong drive field on the propagation of a weak signal nearly resonant to the transition 2–3 can be described in terms of the splitting of the "bare" state 3 into two new states, "dressed" by the drive (Fig. 6.5). Instead of one resonance at frequency $\omega_s = \omega_{32}$, the weak Stokes signal has two resonances with these states. If all broadenings γ_{ij} are equal, the dressed states are shifted from the bare state 3 by

$$\omega_{31} - \frac{E_\pm}{\hbar} = \frac{\Delta}{2} \pm \sqrt{\left(\frac{\Delta}{2}\right)^2 + |\Omega_d|^2} \approx \begin{cases} \Delta + \dfrac{|\Omega_d|^2}{\Delta} \\ -\dfrac{|\Omega_d|^2}{\Delta} \end{cases} \tag{6.29}$$

where $\Delta = \omega_{31} - \omega_d$ and the energies of the dressed states $E_\pm$ are counted from the energy of state 1. The last approximate equality in Eq. 6.29 is in the limit of $|\Omega_d| << |\Delta|$. In this limit one of the dressed states (E_-) is close to the position of bare state 3: $E_- \approx \hbar\omega_{31}$, while another one is close to the two-photon resonance: $E_+ \approx \hbar\omega_d$. For arbitrary broadenings γ_{ij} the dressed states can be found by exact diagonalization of the hamiltonian, which leads to a cubic equation for the eigen energies.

The gain spectrum of Eq. 6.28, evaluated for the injection current density at threshold for the Stokes lasing for the structure reported in Ref. 23, is shown in Fig. 6.6.

As expected, it has a peak close to the position of the two-photon resonance $\delta = \Delta$, or $\omega_d - \omega_s = \omega_{21}$. The Stark shift is small for our experimental conditions. Figure 6.6 also shows the first and second term in Eq. 6.28 evaluated separately. One can see that at the point of

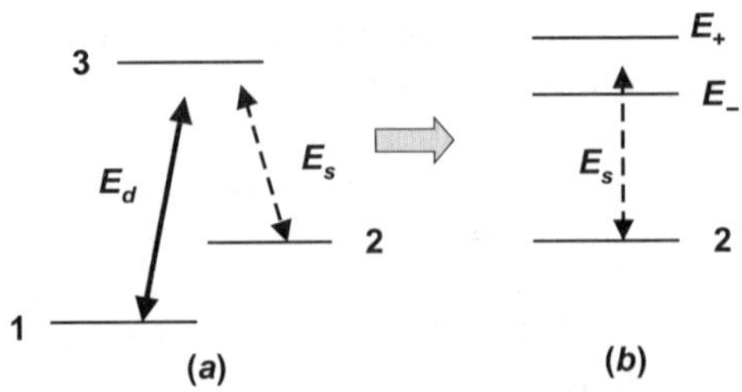

Figure 6.5 Interaction of a weak signal field with a strongly driven Λ system in the dressed-state picture. (*a*) Original system of bare states; (*b*) dressed states as probed by a weak signal field.

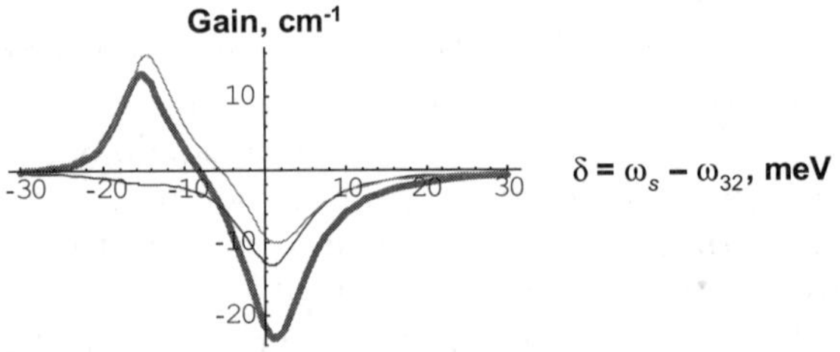

Figure 6.6 Gain spectrum for the Stokes field according to Eq. 6.28, evaluated for the Stokes laser threshold for the structure reported in Ref. 23. Bold line: total gain; thin solid line: first term in Eq. 6.28; dashed line: second term.

maximum gain, the first term dominates over the linear absorption by a factor of order 7.

The "usual" Raman physics is recovered in the limit when Δ is much larger than all linewidths γ_{ij} and the Rabi frequency $|\Omega_d|$. In this case the maximum gain is still reached close to the two-photon resonance (state E_+); its value is equal to

$$g_I = \frac{4\pi\omega_s e^2 z_{32}^2 \Gamma_s}{\hbar c \mu_s} \frac{|\Omega_d|^2 (n_1 - n_2)}{\Delta^2 \gamma_{21}} \tag{6.30}$$

i.e., it becomes proportional to the Raman inversion $n_1 - n_2$ and inversely proportional to the relaxation rate γ_{21} of the Raman coherence. In this case, real one-photon transitions to the excited state 3 with subsequent emission of a photon due to deexcitation to state 2 are much less probable than the two-photon transitions from state 1 to state 2 through an intermediate virtual state. Note that this limit is not identical to just dropping the linear absorption term from Eq. 6.28. The plot of the gain spectrum of Eq. 6.28 in the case of a large detuning $\Delta = 50$ meV is shown in Fig. 6.7.

In the opposite limit of a strong coherent drive, when $|\Omega_d| \gg \sqrt{\gamma_{21}\gamma_{31} + \Delta^2}$, it becomes impossible to distinguish a sequence

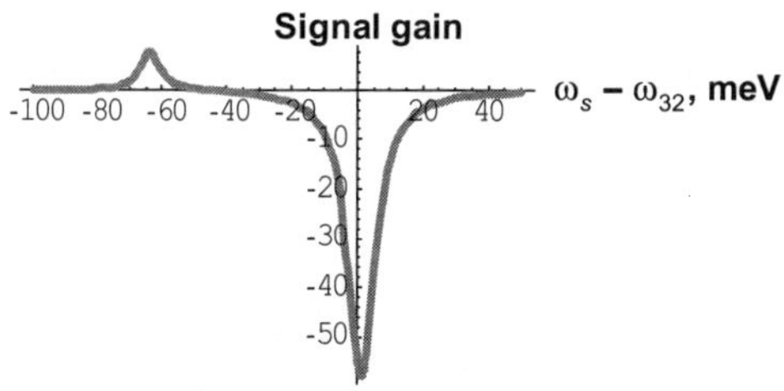

Figure 6.7 Gain spectrum for the Stokes field in the case of large detuning Δ of the drive.

of two consecutive one-photon processes accompanied by real transfer of electrons to the excited state 3 (e.g., absorption of the drive photon and subsequent emission of the Stokes photon) from a two-photon Raman process. The transition occurs simultaneously through both dressed states that are shifted by approximately $\pm|\Omega_d|\ /\ 2$ from state 3. It has been shown (see, e.g., Ref. 4 for review) that under certain conditions most of the population of the two states 1 and 2 becomes trapped in a nonabsorbing, "dark" state determined by the superposition of bare states |1> and |2>:

$$|D\rangle = \frac{\Omega_s\ |\ 1\rangle - \Omega_d\ |\ 2\rangle}{\sqrt{|\ \Omega_s\ |^2 + |\ \Omega_d\ |^2}} \tag{6.31}$$

where $\Omega_s = ez_{23}E_s/\hbar$. This phenomenon has been called *coherent population trapping* (CPT) [34]. As a result, one can get transparency for both propagating fields, and the inversionless amplification for one of them if incoherent pumping is present. In principle, this amplification may occur in the absence of any inversion in the system including Raman inversion. However, in practice there is a very narrow range of parameters in which the absence of Raman inversion and the presence of gain for the signal field coexist. In our lasers the Raman inversion is certainly present, because state 1 is close to the ground state of a doped injector and state 2 is relatively short-lived by design.

6.3.2 General properties of the nonlinear dynamics of Raman lasing

Equation 6.28 has been derived by assuming the Stokes signal to be weak and the drive field to be constant along the cavity. In reality, one has to consider the coupled propagation of both fields along the waveguide, taking into account their absorption and nonlinear interaction. Some general features of this process can be seen without solving the equations. Using Eqs. 6.11, 6.24 to 6.26, and the relation between H_y and E_z in the TM mode, one can write the propagation equations for the normalized drive and Stokes field intensities as

$$\frac{dI_s}{dx} = -\eta_s \,\mathrm{Re}\left(\frac{\sigma_{32}}{i\Omega_s}\right) I_s \tag{6.32}$$

$$\frac{dI_d}{dx} = -\eta_d \,\mathrm{Re}\left(\frac{\sigma_{31}}{i\Omega_d}\right) I_d + g_d I_d \tag{6.33}$$

where

$$\eta_s = \frac{4\pi\omega_s e^2 z_{23}^2 \Gamma_s}{\hbar\mu_s c} \qquad \eta_d = \frac{4\pi\omega_d e^2 z_{13}^2 \Gamma_d}{\hbar\mu_d c}$$

and Γ_s and Γ_d are the overlap factors of the Stokes and drive modes with the Raman active region, and g_d is the modal gain for the drive field due to the presence of the drive active region (see Fig. 6.8). Off-diagonal density matrix elements for arbitrarily strong fields are given by

$$\frac{\sigma_{32}}{i\Omega_s} = \frac{n_{23}\left(I_s + \dfrac{\Gamma_{21}^*\Gamma_{31}^*}{\gamma_{21}\gamma_{31}}\right) - n_{13}I_d}{\Gamma_{32}\left(I_s + \dfrac{\Gamma_{21}^*\Gamma_{31}^*}{\gamma_{21}\gamma_{31}}\right) + \Gamma_{31}^* I_d} \tag{6.34}$$

$$\frac{\sigma_{31}}{i\Omega_d} = \frac{n_{13}\left(I_d + \dfrac{\Gamma_{21}\Gamma_{32}^*}{\gamma_{21}\gamma_{31}}\right) - n_{23}I_s}{\Gamma_{31}\left(I_d + \dfrac{\Gamma_{21}\Gamma_{32}^*}{\gamma_{21}\gamma_{31}}\right) + \Gamma_{32}^* I_s} \tag{6.35}$$

The normalized intensities are defined by $I_{s,d} = |\Omega_{s,d}|^2/(\gamma_{21}\gamma_{31})$. In atomic systems it is possible to have very long-lived Raman coherence: $\gamma_{21} << \gamma_{31}$. Then I_s and I_d can be much greater than 1 and can excite maximum amplitude of coherence $|\sigma_{21}| = 1/2$ when the intensities are well below the values saturating corresponding transitions. In QC lasers operating above threshold, there are no long-lived coherences, so I_s and I_d become on the order of 1 when the intensities approach saturation.

Equations 6.32 through 6.35 should be solved together with equations for populations of states 1, 2, and 3. In the limit of a weak Stokes field, one can put $I_s = 0$ in Eqs. 6.34 and 6.35. Then the Stokes field experiences exponential amplification with the gain given in Eq. 6.28, while the drive field experiences resonant linear absorption on the transition 1–3 that should be compensated by gain g_d to ensure laser generation. In a general case of arbitrary fields, one needs to solve nonlinearly coupled Eqs. 6.32 through 6.35 to find the steady-state intensities.

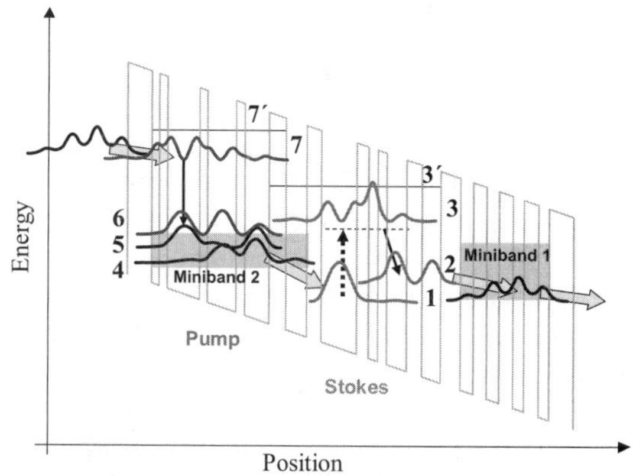

Figure 6.8 Active region of a resonant Raman laser that integrates fundamental laser cascade 7–6–5 and resonant Λ scheme of transitions 1–3–2–1 for Raman Stokes lasing. Calculated conduction band structure of one period of the 30-stage Raman laser. The plot represents the potential profile along the growth direction, where the square moduli of only the most significant wave functions are indicated for clarity. The energy barriers (0.52 eV) are made of $Al_{0.48}In_{0.52}As$ and the quantum wells of $Ga_{0.47}In_{0.53}As$. Shown are the quantum cascade laser states (4, 5, 6, 7) and Raman region states (1, 2, 3). Two upper-lying states are indicated by straight lines (3′ and 7′), while the gray boxes indicate manifolds of closely spaced states (minibands 1 and 2). The layer thicknesses are (starting from the left, in nanometers): **4.2**, 1.3, **1.4,** 5.6, **1.4,** 4.9, **1.5**, 4.3, **3.0**, 3.6, **2.5**, 6.1, **2.0**, 1.6, **1.5**, 3.2, **2.6**, 3.2, **3.4**, 2.2, **2.3**, 2.1, **2.4**, 1.9, **2.5**, 1.8, **4.2**, where the barriers are indicated in boldface and the underlined layers are doped to $n = 4 \times 10^{17}$ cm^{-3}. The thick arrows indicate the direction of electron transport. Electrons in the ground state of miniband 1 are injected by resonant tunneling into the upper laser level (state 7) of the following period. The solid and dashed vertical arrows represent the internally generated pump laser radiation. The relevant calculated transition energies, dipole matrix elements, and lifetimes are $E_{76} = 186$ meV, $E_{65} = 31$ meV, $E_{54} = 39$ meV, $E_{32} = 150$ meV, $E_{31} = 199$ meV, $E_{21} = 49$ meV; $z_{32} = 1.31$ nm, $z_{31} = 1.23$ nm, $z_{21} = 0.7$ nm; $\tau_{32} = 4.5$ ps, $\tau_{31} = 1.9$ ps, $\tau_{21} = 5.4$ ps, $\tau_2 = 0.2$ ps, $\tau_3 = 0.4$ ps. The total broadenings of the transitions of the Raman section are estimated from absorption and electroluminescence data in the literature as $\gamma_{32} \sim 4$ to 5 meV and $\gamma_{31} \sim 4$ to 5 meV; the value of γ_{21} is expected to be lower.

In the limit of very large detunings, and at the frequency corresponding to the two-photon resonance $\omega_d - \omega_s = \omega_{21}$ where the Stokes gain is maximized, we obtain

$$\frac{dI_s}{dx} \approx \eta_s \frac{\gamma_{31} n_{12}}{\Delta^2} I_d I_s \tag{6.36}$$

$$\frac{dI_d}{dx} \approx -\eta_d \frac{\gamma_{31} n_{12}}{\Delta^2} I_d I_s + g_d I_d \tag{6.37}$$

where $\Delta = \omega_{31} - \omega_d$.

If the gain term on the right-hand side of Eq. 6.37 is absent or small compared to the nonlinear coupling term proportional to $I_s I_d$, then Eqs. 6.36 and 6.37 have the first integral

$$\frac{I_s}{\eta_s} + \frac{I_d}{\eta_d} = \text{const.} \tag{6.38}$$

which expresses conservation of the total number of quanta (the Manley-Rowe relation). This means that in the limit of large detunings, the nonlinear interaction between the drive and Stokes modes is of purely parametric origin: The Stokes field takes energy from the drive field. However, since in our design the drive and Raman lasers are integrated, the drive gain can compensate for this nonlinear depletion. As a result, the combined number of photons in both modes will grow rather than stay constant; the efficiency of the transformation of the drive to Stokes power is not constrained, except for the limitations imposed by the material and current injection. The energy is of course drawn from the injection current source.

At exact resonance, when all detunings are equal to zero, one gets from Eqs. 6.32 to 6.35

$$\frac{dI_s}{dx} = \eta_s \frac{n_{13}I_d - n_{23}(I_s+1)}{\gamma_{32}(I_s+1) + \gamma_{31}I_d} I_s \tag{6.39}$$

$$\frac{dI_d}{dx} = -\eta_d \frac{n_{13}(I_d + \gamma_{32}/\gamma_{31}) - n_{23}I_s}{\gamma_{31}(I_d + \gamma_{32}/\gamma_{31}) + \gamma_{32}I_s} I_d + g_d I_d \tag{6.40}$$

In this case, when both intensities become very large, $I_{s,d} >> 1$, the above equations yield

$$\frac{dI_s}{dx} = \eta_s \, \Phi \, I_s \tag{6.41}$$

$$\begin{aligned} \frac{dI_d}{dx} &= -\eta_d \, \Phi \, I_d + g_d I_d \\ \Phi &\equiv \frac{n_{13}I_d - n_{23}I_s}{\gamma_{32}I_s + \gamma_{31}I_d} \end{aligned} \tag{6.42}$$

Population differences n_{13} and n_{23} are to be evaluated from the equations for diagonal density matrix elements. There is no conservation law of the Manley-Rowe type. In the absence of the drive gain term $g_d I_d$, Eqs. 6.41 and 6.42 have the first integral, which is different from Eqs. 6.38 but still corresponds to the nonlinear transfer of energy from the drive to the Stokes mode. However, the numerical analysis shows that in this limit $\Phi \to 0$ due to coherent population trapping and saturation effects, so that the coupling between the two fields becomes suppressed. In our QCL system, this condition of strong-field *electromagnetically induced transparency* (EIT) can occur only when both fields are well above saturation, with intensities greater than 1 MW/cm^2. In atomic systems the relaxation rate γ_{21} of the Raman coherence can be very small, so that the strong-coupling regime $I_{s,d} >> 1$ and the EIT can occur well below saturation.

6.3.3 Experimental realization of Raman QC laser

Here we consider the first realization of a Raman laser monolithically integrated with a QC laser [23]. To the best of our knowledge, this is the first Raman laser with injection pumping.

One period of the Raman laser structure is shown in Fig. 6.8.

In our injection-pumped resonant Raman laser, the fundamental radiation and the Stokes radiation are both generated by electronic transitions between confined states in the conduction band of the very same active region of a quantum cascade laser. The typical layout of a QC laser has been modified to include within the same band structure the Λ scheme of three intersubband transitions designed to produce stimulated Raman scattering of laser radiation. Each period of the laser multistage structure consists of an injector, a drive laser region, and a Raman active section where lasing at the Stokes frequency takes place (Fig. 6.8).

The drive section consists of a coupled well's vertical transition active region, used in the design of state-of-the-art mid-infrared QC lasers [35], suitably modified for optimum coupling to the Stokes section. The drive is generated across the E_{76} transition, where state 6 is depleted by resonant LO-phonon emission to level 5. The energy of the E_{13} transition is designed to be detuned by 13 meV from the intersubband resonance ω_{31}. Although the detuning lowers the gain, as shown in Eq. 6.28, we kept it on the order of the 1–3 transition full width at half maximum $2\gamma_{31}$ = 10 meV, in order to decrease resonant absorption for the drive, to minimize the threshold current for the drive and Stokes lasers, and to be able to distinguish between Raman lasing and usual lasing due to population inversion on the transition 3–2.

Laser light at 6.7 μm generated on the transition 7–6 serves as a resonant optical pump for lasing at the Stokes wavelength of 9 μm, corresponding to the two-photon resonance $\omega_s = \omega_d - \omega_{21}$. Resonant absorption of the drive on the transition 1–3 is overcome by amplification in the drive laser section on the transition 7–6, which has 2 times larger dipole moment and larger overall cross section of the stimulated emission. The triply resonant nature of the process makes stimulated Raman scattering very efficient: peak *intracavity* fundamental power is only 60 mW at the threshold for Raman lasing, which implies very large material gain coefficient on the order of 2×10^{-5} cm/W per period (6×10^{-4} cm/W for the whole stack of 30 stages) and high efficiency of the nonlinear interaction. Nonlinear conversion efficiency of around 30% has been measured.

The devices were based on the InGaAs/InAlAs heterostructure, grown by molecular beam epitaxy lattice-matched to the InP substrate. Typical emission spectra from ridge waveguide devices are shown in Fig. 6.9*a*. The pump wavelength (inset) at a heat sink temperature of 80 K is measured to be $\lambda_L = 6.7$ μm, corresponding very well to the predicted value of $\hbar\omega_d = E_{76} = 186$ meV. The Stokes spectrum is peaked

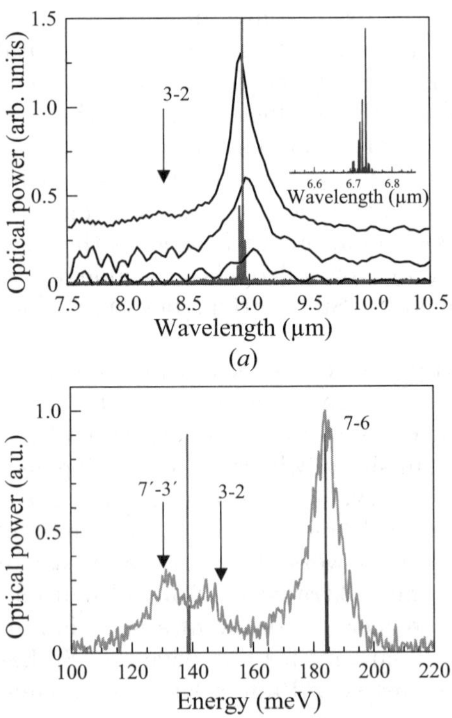

Figure 6.9 (*a*) Measured subthreshold emission spectra (black lines) of quantum cascade Raman laser at currents of (from bottom to top) $I = 2.43$, 2.45, and 2.5 A, offset for clarity. The gray curve is recorded above threshold for Raman lasing (I = 2.6 A). Inset: Fundamental laser emission spectrum at I = 0.8 A. All measurements shown are performed at a heat sink temperature of $T = 80$ K. The vertical arrow marks the position where the emission from the transition 3–2 would be expected if level 3 were populated by optical pumping. (*b*) Comparison of electroluminescence (gray curve) and fundamental and Raman laser (right and left peaks, respectively) emission spectra. The vertical arrows indicate the energy of the transitions 3–2, 3–1 and 7′–3′ as calculated from the band structure in Fig. 6.1*c*. All measurements were performed in pulsed operation with a repetition rate of 80 kHz and a pulse width of 100 ns.

at a wavelength λ_s = 8.9 μm corresponding to the expected value $\hbar\omega_s = \hbar\omega_d - E_{21} = 137$ meV, and its narrowing with increased current demonstrates that it is a stimulated process. The onset of Stokes lasing is observed above a threshold current of 2.6 A, as indicated by the appearance of narrow cavity modes centered nears λ_s. Figure 6.9*b* shows the electroluminescence spectrum of the same material processed into round mesa devices to prevent feedback from the facets and laser action at the same current density ($J \cong 4.5$ kA/cm^2) as the one in laser devices at threshold for Raman lasing. We can clearly identify the main peak as the pump transition E_{76}, while two low-intensity peaks appear at lower energies, none of which corresponds to the Stokes emission.

Raman lasing has been demonstrated in all 15 tested devices. The power output characteristics of one of the devices are displayed in Fig. 6.10. The measured samples exhibit typically two thresholds, the first one around 1 kA/cm^2 for the fundamental laser emission and the second one around 4.3 kA/cm^2 for the Stokes emission. The three curves shown in Fig. 6.10*a* represent the *I-V* curve (top curve) and the laser emission turnon of the fundamental radiation (middle curve) and Stokes (bottom curve) radiation. The Stokes laser turnon occurs when the output laser power reaches about 40 mW, and at higher currents the Stokes power reaches about 26% of the fundamental power, where the fundamental emission starts to saturate, as commonly observed in QC lasers at injection levels of several times the threshold current. This should be compared to conversion efficiencies for solid-state Raman lasers that are usually very low ($\leq 10^{-3}$), with the exception of ultrahigh Q spherical microcavities (~ 20%) [36] and spin-flip Raman lasers in InSb (~ 50%) [37]. Also shown in Fig. 6.10*b* are the emitted Stokes radiation and fundamental output power as a function of current on a logarithmic scale, to highlight the exponential dependence of the emission below threshold compared with the linear behavior above threshold for both laser lines. Given that the laser facet reflectivity does not change significantly in the wavelength range λ = 6 to 9 μm, we can consider the ratio of the emitted powers to be equal to the ratio of the internal power densities, measured by comparing the relative intensities of the laser peaks in spectra acquired over the entire mid-ir range without any optical filtering. From these measurements, the conversion efficiency is estimated to be about 30% at currents of 4.5 A, in good agreement with the value obtained from the power-current curves.

In response to suggestions by Faist and Colombelli [38], we conducted a detailed investigation to verify that the observed lasing at 9-μm wavelength is indeed due to coherent optical driving and not due to an accidental electrically pumped transition in the structure.

First, we compared devices with identical stripe width very close to each other on the wafer with somewhat different pump thresholds and

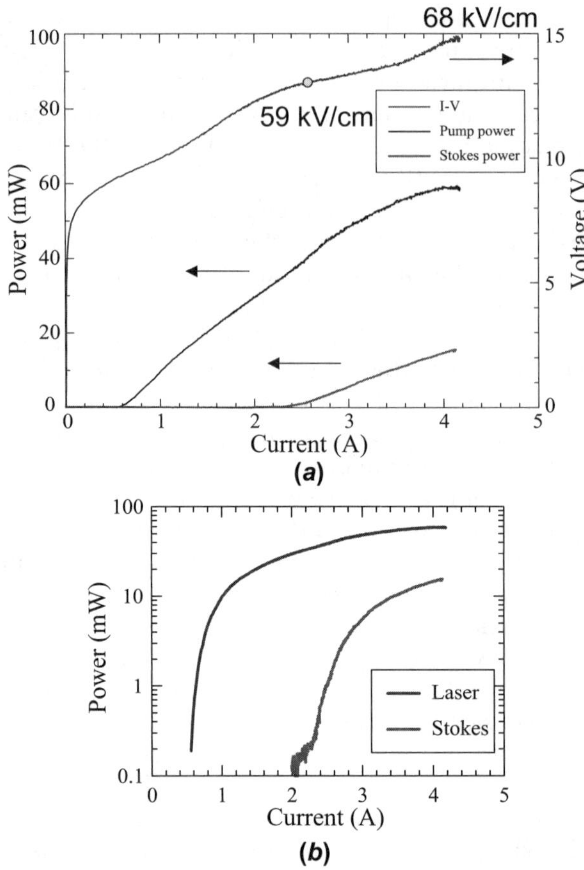

Figure 6.10 (*a*) Peak output power versus current characteristics measured for the fundamental (middle curve) and Stokes (lower curve) emissions at a temperature of 80 K. The upper curve is the *I-V* characteristics. Measurements were performed with a 5-kHz repetition rate at a 0.05% duty cycle. (*b*) Experimental power-current characteristics on a logarithmic scale, showing the exponential behavior of the emission close to threshold.

slope efficiencies. The data for two representative devices (Fig. 6.11) show that the longer-wavelength lasing turns on at the same pump power (which also corresponds to the same intracavity powers for the two devices). This is direct evidence that the 9-μm lasing results from internal pumping by the shorter-wavelength radiation rather than from direct electrical pumping.

Second, high-reflectivity SiO_2/Au back-facet coating was deposited on part of the samples. The resulting change in the threshold between coated and uncoated devices for the pump lasing (6.7 μm) and the Stokes lasing at 9 μm has been measured. The threshold current for the pump laser is simply proportional to the sum of the waveguide and

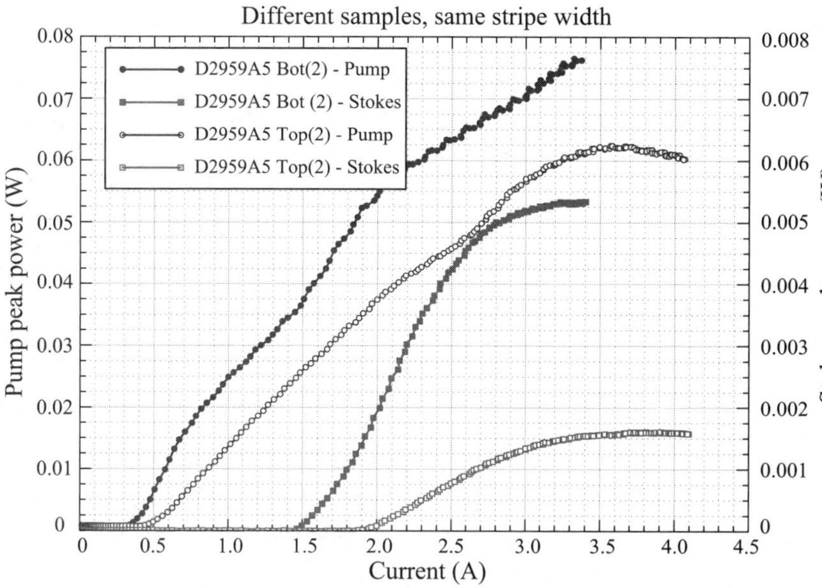

Figure 6.11 Power-current curves for two representative devices 1 mm apart in the wafer, denoted by filled and open squares, respectively. There are two curves for each device, corresponding to the pump (upper curve) and Stokes (lower curve) emission. Pump threshold difference is ~80 mA, Stokes threshold difference is ~0.5 A. Both Stokes lasers turn on at the same output power in both devices: ~35 mW.

mirror losses: $J_p \propto \alpha_w^p + \alpha_m^p$. If the longer-wavelength laser at 9 μm is indeed optically pumped, its gain must be proportional to the intracavity energy density of the pump field W_{pump}

$$g_s \propto W_{\text{pump}} \propto J\tau_p \propto \frac{J}{\alpha_w^p + \alpha_m^p}$$

which is proportional to the injection current times the pump photon lifetime τ_p. This lifetime is inversely proportional to the total pump losses. The superscript p in the losses in the equation above refers to the pump. From this relation we obtain that the current at Raman threshold scales as

$$J_s \propto g_s(\alpha_w^p + \alpha_m^p) \sim (\alpha_w^s + \alpha_m^s)(\alpha_w^p + \alpha_m^p)$$

where the superscript s refers to the Stokes field.

Both α_m^p and α_m^s decrease in the coated samples approximately by one-half, according to

$$\alpha_m = \frac{1}{L} \ln \frac{1}{\sqrt{R_1 R_2}}$$

where the cavity length $L \sim 1.25$ mm, reflection factors are $R_1 = R_2 = 0.27$ in the uncoated sample, and $R_1 = 0.96$ in the coated one for both wavelengths. Therefore, one should expect a greater decrease in the threshold current for the optically pumped Raman laser compared to an electrically pumped laser emitting at the Stokes wavelength $\lambda_s = 9$ μm. The measurements show that the reduction in threshold is ~12% to 15% for the pump radiation and ~25% to 28% for the Stokes radiation. These data are consistent with the optical pumping hypothesis and tend to rule out electrical pumping as the origin of the long-wavelength lasing.

We have also performed measurements of the corresponding changes of the slope efficiency and power. These data clearly show that after application of the coating, the slope efficiency of the Stokes lasing increases by a factor of 5.5 to 6, significantly more than that of the pump laser, which increases by a factor of 1.6. These results also point in the direction that the 9-μm lasing arises from internal optical pumping rather than electrical. However, a quantitative analysis predicts the lower magnitude of the effect.

Third, the roll-off of the pump and Stokes lasers occurs at the same current, which also suggests a strong correlation between them and contradicts the hypothesis of unrelated electrically pumped lasers.

Finally, analysis of the band structure of Raman devices does not reveal any transition that could have a significant population inversion in the whole range of applied voltages and could be a potential candidate for the electrically pumped laser at 9 μm.

6.3.4 Theoretical modeling of Raman QC laser

The simplest model of a system shown in Fig. 6.8 should include states 1, 2, 3, 6, and 7 coupled by two laser fields. We assume that the injector states remain undepleted and do not need to be included directly. This assumption is believed to work reasonably well for QC lasers; it can be easily dropped if necessary. Then the density matrix equations for the populations read

$$\frac{\partial n_1}{\partial t} = j_1 + r_{31}n_3 + r_{21}n_2 - r_{12}n_1 - r_1(n_1 - n_{1T}) - 2\,\mathrm{Im}\left(\sigma_{31}\Omega_d^*\right) \tag{6.43a}$$

$$\frac{\partial n_2}{\partial t} = j_2 + r_{32}n_3 + r_{12}n_1 - r_{21}n_2 - r_2(n_2 - n_{2T}) - 2\,\mathrm{Im}\left(\sigma_{32}\Omega_s^*\right) \tag{6.43b}$$

$$\frac{\partial n_3}{\partial t} = -r_3 n_3 + 2\,\mathrm{Im}\left(\sigma_{32}\Omega_s^*\right) + 2\,\mathrm{Im}\left(\sigma_{31}\Omega_d^*\right) \tag{6.43c}$$

$$\frac{\partial n_6}{\partial t} = j(1 - \eta) + r_{76}n_6 - r_6(n_6 - n_{6T}) - 2\,\mathrm{Im}\left(\sigma_{76}\Omega_L^*\right) \tag{6.43d}$$

$$\frac{\partial n_7}{\partial t} = j\,\eta - r_7 n_7 + 2\,\mathrm{Im}\left(\sigma_{76}\Omega_L^*\right) \tag{6.43e}$$

Here j is the total current density and $j_{1,2}$ are the densities of current flowing through states 1 and 2. We will assume that $j_1 + j_2 = j$, although this may not necessarily be the case. The factor η is the efficiency of injection to the upper drive laser state 7; we again assume that the rest of the current goes to state 6, although this may not be the case. Quantities n_{iT} are equilibrium populations in the absence of current injection and the optical fields. Equilibrium populations of states 3 and 7 are neglected.

The relaxation rate r_i is the inverse time during which state i reaches equilibrium with the injector; we assume that the injector remains undepleted and has Fermi distribution of electrons at temperature T. The relaxation rate r_{ij} is the inverse time of the transition from state i to state j.

Then we need Eqs. 6.24 through 6.26 for the off-diagonal density matrix elements that have to be supplemented by the equation

$$\frac{d\sigma_{76}}{dt} + \Gamma_{76}\sigma_{76} = i\Omega_L n_{67} \tag{6.44}$$

for the optical polarization on the laser drive transition. Note that since the dipole moments of the transitions 7–6 and 3–1 are different, the Rabi frequencies $\Omega_L = ez_{67}E_d/\hbar$ and $\Omega_d = ez_{13}E_d/\hbar$ are also different despite the fact that the drive field is the same. It is the Rabi frequency—not the amplitude of the field itself—that determines the strength of the coupling between the field and the transition of interest. This provides an important flexibility in designing the integrated devices: The same field interacts differently with the drive laser transition 7–6 and with the drive transition 3–1 in the Raman cascade.

Finally, we need two coupled equations for the drive and Stokes field amplitudes that can be written as

$$\frac{\partial \Omega_s}{\partial x} = \frac{2\pi i\omega_s c^z z_{23}^2}{c\mu_s\hbar}\Gamma_s\sigma_{32} \tag{6.45}$$

$$\frac{\partial \Omega_d}{\partial x} = \frac{2\pi i\omega_d e^2 z_{13}^2}{c\mu_d\hbar}\Gamma_d\sigma_{31} + \frac{2\pi i\omega_d e^2 z_{67}^2}{c\mu_d\hbar}\Gamma_L\sigma_{76} \tag{6.46}$$

The factor Γ_L is the transverse overlap of the drive mode with the drive laser active region; for our structure its value is very close to Γ_d. However, in the alternative design where the stacks of the Raman and drive laser active regions are separated, these two factors can be very different. The density matrix elements σ_{ij} on the right-hand sides of Eqs. 6.45 and 6.46 are averaged with the transverse mode distribution over the cross section of the active layer, according to Eq. 6.11.

The above set of equations 6.24 to 6.26 and 6.43 to 6.46 can also be used to describe the short-pulse regime after replacing $\partial/\partial x \rightarrow \partial/\partial x + (\mu/c)(\partial/\partial t)$. In this chapter, we are interested in the continuous-wave or long-pulse regime when the pulse duration is much longer than all relaxation times; therefore, we may put all time derivatives equal to zero in Eqs. 6.24 to 6.26 and 6.43.

An important feature of our system compared to many previous studies of LWI is that it is open: The sum of populations of states 1, 2, and 3 is not conserved. One can control them quite efficiently by a combination of scattering and tunneling processes, i.e., through the control of electron transport. This adds flexibility and significantly broadens the range of parameters under which the lasing is possible, compared to simple three-level schemes.

There are several parameters in the above equations that we know with a large degree of uncertainty or do not know at all: the electron temperature, relaxation rates γ_{ij} of the off-diagonal density matrix elements, and the efficiencies of current injection. We checked the sensitivity of the results to these parameters. It was found that the drive laser threshold is quite sensitive to the thermal population n_{6T} of the lower laser state 6 and the efficiency η of injection to the upper state 7. As expected, it increases with increasing n_{6T} and increases sharply with decreasing η—by a factor of 3 when η changed from 1 to 0.8. The Stokes threshold is not sensitive to n_{6T}, but increases significantly (by 50%) with the above decrease in η. The Stokes threshold is also not very sensitive to the change in the injection efficiency to states 1 and 2: It increases by only 20% when j_2/j increases from 0 to 1. The reason is that in our structure state 1 is very close in energy to the ground state of the doped injector; therefore it always stays filled. State 2 is elevated above the Fermi level of the injector by only 34 meV, so its thermal population is also nonnegligible. Therefore, the variation in the distribution of the injection current over these states does not lead to strong changes in the Stokes threshold. At the same time, the latter shows a significant dependence from the thermal population of state 2 and the line widths of all transitions. This is clear from the expression for the gain spectrum in Eq. 6.28.

It is also clear from Eq. 6.28 that having a large population of state 1 and narrow line widths is beneficial for the Stokes laser.

Figure 6.12 shows the output powers of the drive and Stokes lasers calculated for $\eta = 1$, $j_2/j = 0.2$, and electron temperature $T = 10$ meV. The total nonresonant losses for the drive and the Stokes modes were taken to be 12 and 13 cm^{-1}, respectively, according to the measurements and in agreement with calculations of the waveguide and cavity losses for TM_{00} modes. The line widths (FWHM) were assumed to be 8 meV for transitions 1–3 and 2–3, and 5 meV for transition 1–2. For this choice of parameters the agreement with experimental values of both thresholds is very good. Increasing γ_{21} by a factor of 2 increased the Stokes threshold by 30%. Decreasing the electron temperature by 50% decreases the Stokes threshold by 25%, while the same relative increase in temperature boosts the Stokes threshold by more than a factor of 2.

Figure 6.6 shows the gain spectrum for the Stokes field. The peak gain is achieved close to the two-photon resonance, because the Stark shift is very small: The Rabi frequency of the drive is only about 2 meV. Figure 6.13 shows this peak value of the gain as a function of current.

Experimental *L-I* curves (Fig. 6.10) show the roll-off of the drive laser power at a current several times above the drive threshold. Such rollover is not seen in our simulations. It is unlikely to result from the nonlinear interaction between the drive and Stokes fields. Such interactions are included in the model, but they become important at much higher Stokes powers. A plausible hypothesis is that the drive laser deteriorates with current due to decreasing efficiency of injection to state 7, which can originate from increasing misalignment between this state and the upstream injector. We can model this effect by including the dependence of efficiency on current $\eta = 1 - aj$, where a is a free parameter. The result for $a = 1$ indeed shows the behavior similar to that of the experimental curve, although there is not much predictive power in such empirical modeling.

Another feature of the measured light-current curves that we were not able to reproduce accurately is their slope. In general, the slope is much more sensitive to the uncertainties in the above input parameters

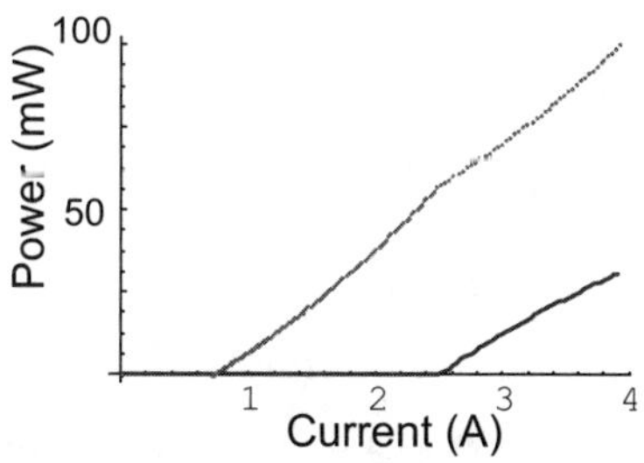

Figure 6.12 Calculated drive (upper curve) and Stokes (lower curve) powers as functions of injection current.

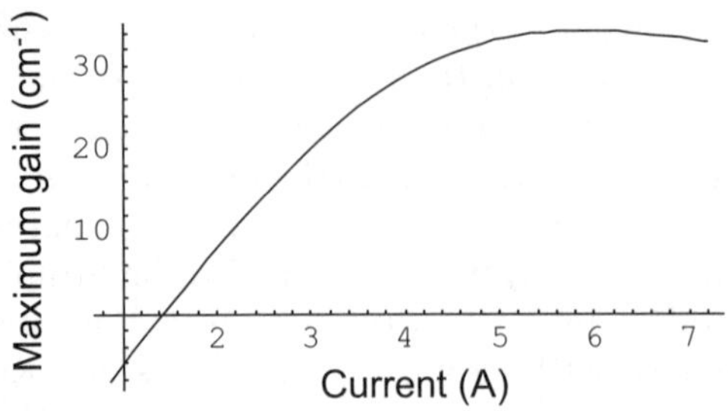

Figure 6.13 Maximum gain of the Stokes mode as a function of injection current.

than are the laser thresholds. The modeling predicts considerably higher power for both lasers compared to the measurements. Also, the drive laser curve in Fig. 6.12 demonstrates an increase in the slope when the Stokes laser turns on, while we observed only a decrease in the slope for all our simulations.

An interesting question involves the choice of the optimal detuning Δ of the drive field. The denominator in Eq. 6.28 is minimized when all detunings are equal to zero. At the same time, at zero detuning the resonant absorption of the drive field on the transition 1–3 is maximized, leading to an increase in both the drive and Stokes threshold. Figure 6.14 shows the dependence of both thresholds on the drive field detuning. It is clear that the designed value of $\Delta = 13$ meV is close to the optimum.

Finally, both Fig. 6.6 and our experimental results in Fig. 6.9 indicate that there is no net gain or conventional lasing on the transition 2–3 at the current corresponding to the Stokes threshold. However, for the drive field intensities about 2 times higher than those reached in these experimental devices, the optical pumping should lead to population

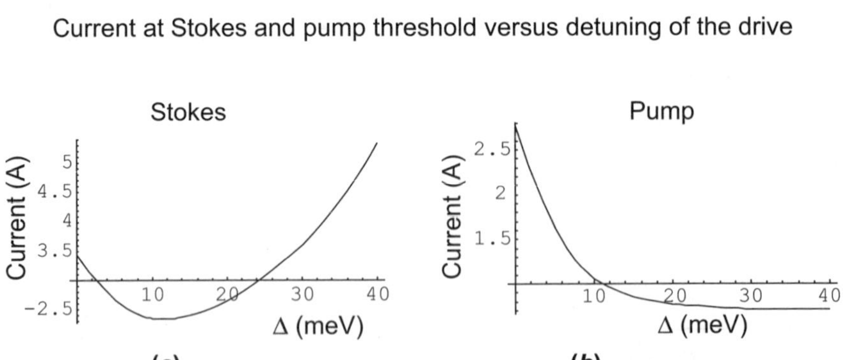

Figure 6.14 Threshold current for (*a*) Stokes and (*b*) drive field as a function of the drive field detuning.

inversion $n_3 > n_2$. In this case one could expect lasing to start around the frequency $\omega_s = \omega_{32}$. The laser modes will experience strong electrical and optical coupling, resulting in complex multimode dynamics.

To summarize, our calculations demonstrate reasonable robustness and good agreement with experiment with respect to laser thresholds. The details of the power-current curves turned out to be more difficult to reproduce, and their modeling has a low predictive power.

6.3.5 Anti-Stokes Raman laser

The physics of resonant anti-Stokes generation is similar to that of the Stokes case, but the specific implementation in the QCL structure is of course very different. Figure 6.15 shows one example of such a design.

Here the electrons need to be injected to a relatively long-lived initial state 2 and drained from the final state 1 by, e.g., resonant optical phonon emission, in order to maintain a large population difference on the drive transition 2–3 and decrease the resonant absorption of the signal on transition 1–3. The expression for the gain spectrum is similar to Eq. 6.28, after permutation of indices and changing of the phases of complex detunings:

$$g_1 = 2g_M = \frac{4\pi\omega_s e^z z_{13}^2 \Gamma_s}{\hbar c \mu_s} \times \mathrm{Re}\left\{ \frac{1}{\Gamma_{31} + |\Omega_d|^2/\Gamma_{21}} \left[\frac{|\Omega_d|^2 (n_2 - n_3)}{\Gamma_{21}\Gamma_{32}^*} - (n_1 - n_3) \right] \right\} \tag{6.28a}$$

Here $\Omega_d = ez_{23} E_d/\hbar$ and $\Omega_s = ez_{13} E_s/\hbar$ are the Rabi frequency of the drive and signal mode, respectively; the subscript s now relates to the anti-Stokes mode and means "signal" in order to avoid introducing new notation.

The above analysis of the gain optimization and the nonlinear dynamics of the Stokes lasing is directly applicable here. Some practical benefits are discussed below.

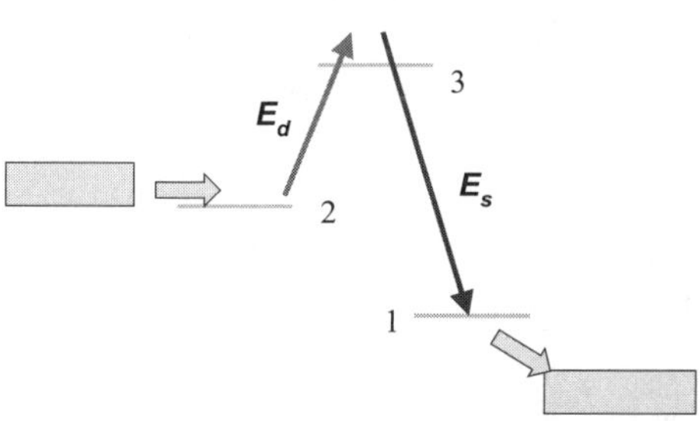

Figure 6.15 A schematic of the anti-Stokes laser.

One feature of the anti-Stokes laser is that initial state 2 of the Raman transition is elevated, and its population is mainly controlled by injection current. State 1 should stay as empty as possible, which means that it should also be elevated above the ground state of the injector. This puts some additional constraint on the design, but also adds flexibility in manipulating the populations by current.

6.3.6 Optically pumped laser

In the Λ scheme of Figs. 6.8 and 6.15, and in all other schemes in Fig. 6.2, the role of the drive field is twofold. On one hand, it helps to excite the Raman coherence between the initial and final states of the two-photon transition (states 1 and 2 in the Λ scheme) which creates maximum gain at the two-photon resonance $\omega_s = \omega_d - \omega_{21}$. (We will discuss the Raman Stokes scheme for definiteness, although the same reasoning applies to all other systems.) On the other hand, it transfers populations to states 3 and 2 by one-photon absorption on the drive transition with subsequent incoherent relaxation. Both effects are included in the expression 6.28 for the gain and similar equation 6.51 for the cascade scheme below. Under certain conditions, this optical pumping creates population inversion on the Stokes transition 3–2, and the second term in brackets in Eq. 6.28 may become dominant, while all effects related to the two-photon transitions and Raman coherence become less significant. This gives rise to the gain peaked at the frequency $\omega_s = \omega_{32}$ and conventional lasing with inversion on this transition. Of course, an intermediate case is also possible, when both contributions are comparable.

A natural question to ask is, What are the optimal conditions for one of these contributions to become dominant? This question has been addressed in Ref. 39 on the basis of simplified rate equations and an approximate expression for the Raman gain, valid for large detunings of the drive field from the transition 1–3 or very low drive powers, when the depletion of state 1 is neglected and only the two-photon Raman transitions through a virtual state are taken into account.

By using Eqs. 6.24 to 6.26, 6.28, and 6.43, it is easy to consider the general case of arbitrary detuning and field strength. The population differences n_{13} and n_{23} that enter the gain expression 6.28 can be expressed as a function of the optical drive intensity as

$$n_{13} = \frac{n_{1T} + j_1/r_1}{1 + I\left[1 + (r_3 - r_{31}/r_1)\right]}$$

$$n_{23} = n_{2T} + \frac{j_2}{r_2} - In_{13}\left(1 - \frac{r_{32}}{r_2}\right)$$

where $I = 2 \mid \Omega_d \mid^2 \gamma_{31} / r_3(\gamma_{31}^2 + \Delta^2)$ is the drive intensity normalized by saturation intensity, and we neglected the term $r_{21}n_2 - r_{12}n_1$ in Eqs. 6.43 for simplicity. This term is small in our case; it can be easily included if necessary.

In the experimental device described in Ref. 23, state 2 was only 34 meV above the Fermi level of the injector. Also, there was inevitable current injection into this state, although its rate was difficult to quantify. Even with small injection, we had $n_{23} \approx n_{2T} > In_{13}$ at the signal threshold for electron temperature above 10 meV. As a result, the gain was primarily due to the excitation of Raman coherence, and lasing started at the Stokes frequency.

Theoretically, we can consider a hypothetical structure optimized for the optical pumping, in which state 2 has negligible equilibrium population and current injection, and very short lifetime: $r_2 >> r_{32}$. In this case the transition 3–2 can be inverted by the optical pumping: $n_{32} \approx In_{13}$. Exactly at resonance, when $\Delta = 0$, the gain in Eq. 6.28 has a single maximum at $\omega_s = \omega_{32}$ where both terms in Eq. 6.28 are positive and the ratio of the second to the first term is equal to $|\Omega_d|^2 n_{13} / (\gamma_{21}\gamma_{31}n_{23}) = r_3/\gamma_{21}$. This ratio is generally less than 1, although the transition 2–1 can be forbidden and have a relatively narrow line width.

Very far from resonance, when Δ is much larger than all line widths and the Rabi frequency of the drive, the gain at the Stokes frequency is equal to

$$g(\omega_s = \omega_d - \omega_{21}) \approx \eta \frac{|\Omega_d|^2 n_{12}}{\gamma_{21}\Delta^2}$$

The gain at the frequency of the transition 3–2 is

$$g(\omega_s = \omega_{32}) = \eta \frac{|\Omega_d|^2 n_{13}}{\gamma_{32}\Delta^2}\left(\frac{\gamma_{31}}{r_3} - 1\right)$$

Note that the first term in Eq. (6.28) is negative at this frequency, contributing the −1 term in the above expression. Here the factor $\eta = 4\pi\omega_{32}e^2 z_{32}^2 \Gamma_s / (\hbar\mu_s c)$.

The ratio of the two peak gains is

$$\frac{g(\omega_s = \omega_d - \omega_{21})}{g(\omega_s = -\omega_{32})} = \frac{\gamma_{32}}{\gamma_{21}(\gamma_{31}/r_3 - 1)}$$

When all line broadenings are equal, this ratio is about 0.3 to 0.5 for realistic laser structures. Therefore, in a structure optimized for the optical pumping, Raman gain is always lower by a factor of 2 to 3, and lasing is expected to start at the frequency of the transition 3–2. Note that this conclusion depends sensitively on the relaxation rate of the Raman coherence γ_{21}, as should be expected. If this rate is low, Raman gain can win.

6.4 Vertical Cascade Schemes

Now we turn to other three-level laser schemes shown in Fig. 6.2. Among these, the cascade, or ladder, schemes in Fig. 6.2*c* and *d* turned out to be of the greatest potential importance for integration with QC lasers. The most interesting and potentially useful feature of these cascade schemes is that when the drive field is tuned to resonance to the drive transition 2–3, the coherent two-photon contribution to the signal gain is positive when there is population inversion on the drive transition and is negative otherwise. This pattern is reversed for large detuning of the drive, when, however, the gain is small. Therefore, it has been suggested in Ref. 6 that one can use the three-level cascade scheme for simultaneous generation of the drive field on the inverted transition 3–2 and inversionless lasing on the transition 2–1. Then one does not need a separate active region for generation of the drive field. This saves space in the laser core, maximizes the modal overlap, eliminates drive absorption in the active region for the signal, and eliminates the problem of tuning the drive field to resonance with the drive transition. Of course, the design similar to the one in Fig. 6.8, when these two active regions are separate, is always a possibility.

Two possible arrangements for a completely integrated design are sketched in Fig. 6.16*a* and *b*.

In the scheme of Fig. 6.16*a* there is population inversion between states 2 and 1 due to a very fast depletion of state 1 and the relatively long lifetime of state 2. The laser field generated on the transition 2–1 serves as a coherent drive for the inversionless lasing on a fast-decaying transition 3–2. Block arrows show the injection current of electrons. In the scheme of Fig. 6.16*b* the roles of transitions 1–2 and 2–3 are interchanged. Now the transition 3–2 is inverted and supports the drive laser generation, while the signal is generated on the fast-decaying transition 1–2 in the absence of population inversion on this transition. Note that an attempt to have lasing with inversion on both transitions 3–2 and 2–1 reported in Ref. 14 resulted in lasing on only one of these transitions.

The physics of the two schemes is similar. In both the drive field is generated automatically at resonance with the drive transition. As a result, the peak gain is also reached at resonance with the signal transition. Obviously, there is no problem with resonant absorption of the drive that has been a serious factor for the Λ scheme. This is a very

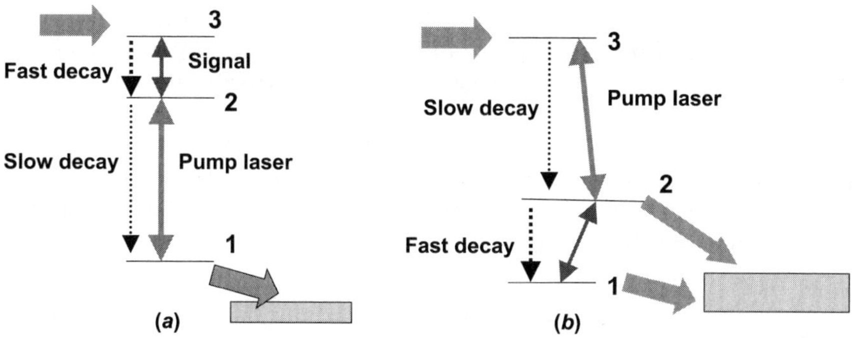

Figure 6.16 Two cascade schemes showing gain without inversion on the signal transition and lasing with inversion on the drive transition.

important advantage of the cascade configuration. Below we concentrate on the scheme shown in Fig. 6.16*b*.

The set of the density matrix equations for this scheme is given by

$$\frac{\partial \sigma_{21}}{dt} + \Gamma_{21}\sigma_{21} = i\Omega_s n_{12} + i\Omega_d^* \sigma_{31} \tag{6.47a}$$

$$\frac{\partial \sigma_{31}}{dt} + \Gamma_{31}\sigma_{31} = i\Omega_d \sigma_{21} - i\Omega_s \sigma_{32} \tag{6.47b}$$

$$\frac{\partial \sigma_{32}}{dt} + \Gamma_{32}\sigma_{32} = i\Omega_d n_{23} - i\Omega_s^* \sigma_{31} \tag{6.47c}$$

$$\frac{\partial n_1}{\partial t} = r_{31}n_3 + r_{21}n_2 - r_{21}n_1 - r_1(n_1 - n_{1T}) - 2\,\mathrm{Im}\,(\sigma_{21}\Omega_s^*) \tag{6.48a}$$

$$\begin{aligned}\frac{\partial n_2}{\partial t} &= j(1-\eta) + r_{32}n_3 + r_{12}n_1 - r_2(n_2 - n_{2T}) \\ &\quad + 2\,\mathrm{Im}\,(\sigma_{21}\Omega_s^*) - 2\,\mathrm{Im}\,(\sigma_{32}\Omega_d^*)\end{aligned} \tag{6.48b}$$

$$\frac{\partial n_3}{\partial t} = j\eta - r_3 n_3 + 2\,\mathrm{Im}\,(\sigma_{32}\Omega_d^*) \tag{6.48c}$$

where

$$\Gamma_{32} = \gamma_{32} + i(\omega_{32} - \omega_d)$$
$$\Gamma_{31} = \gamma_{31} + i(\omega_{31} - \omega_d - \omega_s)$$
$$\Gamma_{21} = \gamma_{21} + i(\omega_{21} - \omega_s)$$

Here $\Omega_d = ez_{23} E_d / \hbar$ and $\Omega_s = ez_{12} E_s / \hbar$ are the Rabi frequency of the drive and signal mode, respectively, and ω_d and ω_s are frequencies of the drive and signal modes, respectively. All other notations are the same as in Eqs. 6.43.

Wave equations for the field amplitudes are simpler than in the case of Raman generation, because there is no separate active region for the drive field. Most importantly, there is no absorption of the drive in the active region for the signal:

$$\frac{\partial \Omega_s}{\partial x} = \frac{2\pi i \omega_s e^2 z_{12}^2}{c\mu_s \hbar} \Gamma_s \sigma_{21} \tag{6.49}$$

$$\frac{\partial \Omega_d}{\partial x} = \frac{2\pi i \omega_d e^2 z_{23}^2}{c\mu_d \hbar} \Gamma_d \sigma_{32} \tag{6.50}$$

The small-signal gain is given by the expression generally similar to the one in Eq. 6.28:

$$g_1 = 2g_M = \frac{4\pi\omega_s e^2 z_{12}^2 \Gamma_s}{\hbar c \mu_s} \times \mathrm{Re}\left\{ \frac{1}{\Gamma_{21} + |\Omega_d|^2 / \Gamma_{31}} \left[\frac{|\Omega_d|^2 (n_3 - n_2)}{\Gamma_{31}\Gamma_{32}} - (n_1 - n_2) \right] \right\} \tag{6.51}$$

An example of the gain spectrum is plotted in Fig. 6.17 for the parameters corresponding to a real QCL structure, currently under investigation. The maximum gain is reached exactly at resonance, when all Γ_{ij} are real and equal to γ_{ij}.

Similarly to the Stokes generation, resonant absorption of the signal can be overcome by a two-photon coherence term proportional to the population difference on the drive transition and the drive intensity. However, in the present case, population inversion of the drive transition is required. There is not much flexibility in controlling the population difference on the drive transition and the resulting gain. In the linear approximation with respect to the signal field, the population inversion n_{32} is fixed at its value on the drive laser threshold. For a homogeneously broadened transition

$$n_{32} = \frac{c\mu_d \hbar \gamma_{32} \alpha_d}{4\pi\omega_d e^2 z_{23}^2 \Gamma_d} \tag{6.52}$$

where α_d is the total cavity losses for the drive mode. The drive field intensity is given by

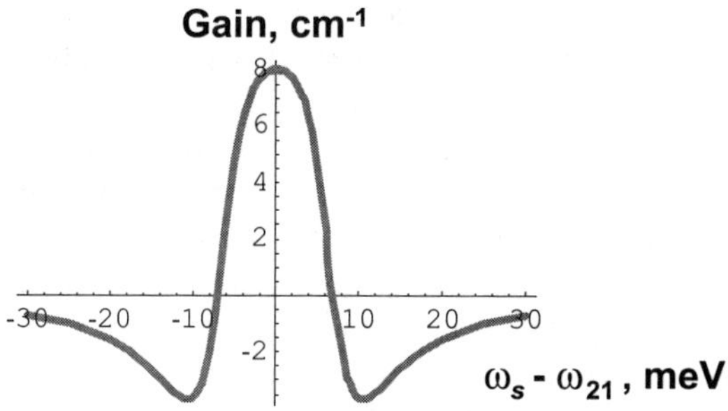

Figure 6.17 Gain in the scheme of Fig. 6.16*b*.

$$| \Omega_d |^2 = | \Omega_d^{sat} |^2 \left(\frac{\bar{n}_{32}}{n_{32}} - 1 \right) \tag{6.53}$$

where $|\Omega_d^{sat}| \sim r_2\gamma_{32}$ is the value of the Rabi frequency that saturates the transition 2–3. It is a cumbersome function of relaxation rates. In Eq. 6.53, $\bar{n}_{32}$ is the population inversion supported by injection pumping in the absence of both drive and signal field.

One can see from the above expressions that as long as $| \Omega_d | \leq \gamma_{21}\gamma_{31}$ and one-photon absorption due to n_{12} is smaller than the two-photon term, the signal gain increases with current proportionally to $\bar{n}_{32}$. The factor $|\Omega_d|^2/\gamma_{31}\gamma_{32}$ before n_{32} in brackets in Eq. 6.51 is smaller than or on the order of 1. Therefore, to overcome the resonant absorption on the transition 1–2 one needs $n_{32} > n_{12}$. This means (1) that state 1 cannot be the ground state and (2) that the threshold for the drive laser should not be too low, so there is a tradeoff between the drive and signal lasers.

When the opposite inequality is satisfied and the drive field is very strong, the factors $|\Omega_d|^2$ in the numerator and denominator of Eq. 6.51 cancel each other, and the ratio of the signal gain to signal losses becomes equal to

$$\frac{g_1}{\alpha_s} \approx \frac{\mu_d}{\mu_s} \frac{\omega_s}{\omega_d} \frac{z_{12}^2}{z_{23}^2} \frac{\Gamma_s}{\Gamma_d} \frac{\alpha_d}{\alpha_s} > 1 \tag{6.54}$$

where the last inequality is required for lasing. This inequality seems to be easy to satisfy when $\omega_s > \omega_d$. In the opposite case of the frequency down-conversion, all ratios in Eq. 6.54 except the ratio of the dipole moments are smaller than or on the order of 1; therefore one needs z_{12} to be significantly larger than z_{23} in order to reach the lasing threshold for the signal.

The above semiqualitative conclusions are confirmed by numerical results. We have also found that by the time the signal laser threshold is reached, the population of state 3 is greater than or very close to the population of state 1. This means that the two-photon Raman-type inversion is usually required for lasing. This is not surprising, since the condition $n_{32} > n_{12}$ mentioned above corresponds to Raman inversion $n_3 > n_1$. If the transition 1–3 has a strong dipole moment, one can get lasing on this transition as well. Two signal modes resonant to the transitions 1–2 and 1–3, respectively, can actually help each other to turn on. The analysis of the amplification of such a bichromatic signal presented in Ref. 6 shows that in this case the threshold can be significantly lower than that for a monochromatic signal. However, one needs to put extra effort into the design to provide fast depopulation of both state 1 and state 2 simultaneously and strong enough dipole moment on the transition 1–3.

6.5 Practical Considerations and Benefits

The experiment reported in Ref. 23 was mainly a proof of principle. The natural question to ask is, Does this kind of device have any potential benefits compared to standard QC lasers? One can envision at least several benefits that are usually associated with nonlinear optical sources:

1. Generation at wavelengths that are not easily accessible for QC lasers in general, or within a given material system: for example, in the terahertz range or at short wavelengths below 3 to 4 μm.
2. Significantly better tunability and wavelength agility compared to standard QC lasers.
3. Higher operating temperature, which is essentially defined in our devices by the operating conditions for the drive laser. The drive wavelength can be chosen to fit within the “sweet spot” of the QC lasers (7 to 9 μm for InGaAs/AlInAs structures) where they are most powerful, efficient, and capable of the continuous-wave room temperature operation.
4. Other benefits, such as small beam divergence or generation of light with interesting statistical properties, e.g., squeezed or entangled light.

Below we consider some examples from the above list.

6.5.1 Frequency down-conversion to the terahertz region

Inversionless or Raman-type lasing could be especially valuable in the situations in which the population inversion is difficult to reach. One example of such a situation is the lasing in the terahertz spectral region. With decreasing energy separation between subbands it becomes increasingly difficult to provide selective injection to the upper laser state, and selective depopulation of the lower state and to prevent backfilling of the lower state. These factors limit operation of existing terahertz QC lasers to cryogenic temperatures. The nonlinear optics may provide an alternative to usual lasing.

Two possible Λ-type schemes for terahertz generation are shown in Fig. 6.18. In the scheme of Fig. 6.18*a*, the separation between states 2 and 3 is below the optical phonon energy, and the drive field is generated in a separate active region. If these states are high enough above the ground state, their population is only due to the optical pumping. They are likely to have comparable lifetimes; therefore, the transition 2–3 has the population difference close to zero or even inverted due to the selective optical pumping to state 3. The two-photon term in the gain expression in Eq. 6.28 is proportional to the population difference $n_1 - n_3$ that can be quite large if state 1 is the ground state or close to it. Therefore, potentially the Stokes gain in the terahertz range can be much larger than the gain for standard lasing with inversion on the transition 2–3 and much larger than the gain in the mid-ir QC laser. Plugging the typical numbers into Eq. 6.28, we obtain the following scaling formula for the gain:

$$g = \frac{(z_{32}/5\,\text{nm})^2(n_{13}/3\times10^{16}\text{cm}^{-3})(\Gamma_s/0.3)}{(\lambda/100\,\mu\text{m})(\gamma_{32}/2\text{ meV})}\left(\frac{1}{\gamma_{21}\tau_3}\right)130\text{ cm}^{-1}$$

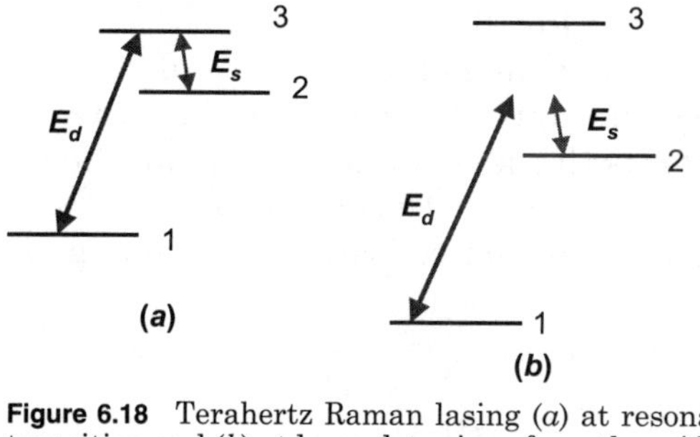

Figure 6.18 Terahertz Raman lasing (*a*) at resonance with the terahertz intersubband transition and (*b*) at large detunings from the mid-infrared intersubband transition.

This estimate is obtained for the optimal value of the average drive intensity in the Raman active region $|\Omega_d|^2 \tau_3/\gamma_{31} \sim 1$ corresponding to the drive power of ~50 to 100 mW.

Of course, care should be taken to avoid too strong absorption of the drive on the transition 1–3, especially when this field is close to resonance. In particular, the dipole moment of the laser transition in the drive active region should be larger than z_{13} by at least a factor of 2.

In the scheme of Fig. 6.18*b*, the transition 2–3 may be in the mid-ir, but the drive field is far detuned from state 3 to shift the Stokes wavelength to the terahertz region. An advantage of this scheme is negligible resonant absorption of the drive field. However, the gain is smaller in this case, which imposes more stringent limitations on the maximum terahertz losses in the waveguide.

For the cascade schemes in Fig. 6.16, the requirement imposed by Eq. 6.54 makes it very difficult to reach the threshold when the ratio of the signal to drive field frequencies is on the order of 0.1 or less. Therefore, it makes sense to have separate active regions for the drive and signal lasers. Population inversion on the drive transition in the cascade scheme is still required; however, its magnitude and the drive field amplitude are not controlled by the drive lasing threshold. As a result, the gain can be significantly enhanced, although extra space is needed for the drive active region.

The main challenge with realization of the Raman-type terahertz laser is a more complicated waveguide design. The waveguide needs to accommodate both mid-ir and terahertz modes and to provide a good overlap between them. Probably the simplest way to achieve it is to have a double-metal or surface-plasmon terahertz waveguide [40–43] and group the active region for the drive and Stokes fields in two separate stacks. Then the whole Raman stack may be an injectorless doped multiple-QW region or a superlattice, since we do not need any current injection to excited states (and actually no current at all) to obtain Raman Stokes lasing. The Raman active region needs only optical pumping. The injectorless design saves space in the waveguide and helps to increase the optical confinement factors; see also Ref. 44. One example of such a design is shown in Fig. 6.19.

The terahertz losses are inevitably higher than in terahertz QC lasers, because both active regions must be doped in the range of 10^{16} cm^{-3} to provide high gain for both the drive and the Stokes fields. Typical numbers in the devices we modeled range between 40 and 100 cm^{-1}. These losses are expected to be compensated by a higher gain.

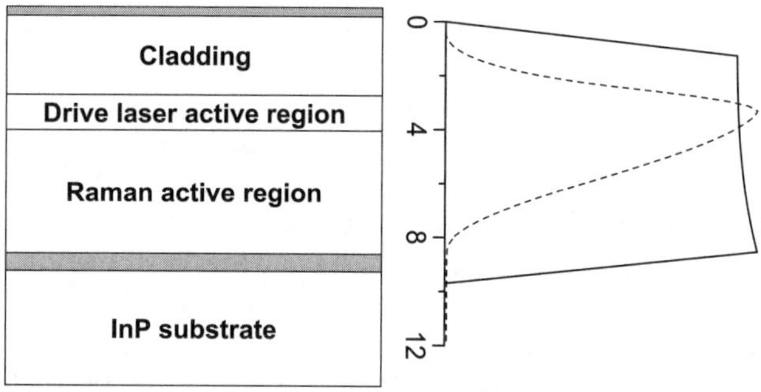

Figure 6.19 A sketch of the InGaAs/AlInAs Raman terahertz laser with a double-metal waveguide and separately stacked active regions for the drive and Stokes laser. Transverse profiles of the magnetic fields squared of the mid-ir drive (λ_d = 7 μm) and terahertz Stokes (λ_s = 70 μm) laser modes are shown with dashed and solid lines, respectively. Metal layers are shaded.

6.5.2 Frequency up-conversion

Frequency up-conversion is in general easier to realize than down-conversion, because free-carrier losses decrease with frequency, and there is no need in a special waveguide design. The practical motivation is the possibility to shift the operating frequency to the region close to the band offset of a given material system, when the usual lasing becomes difficult to achieve. For the InGaAs/AlInAs material system, this means lasing at wavelengths shorter than about 4 to 5 μm, although lasing at wavelengths as short as 3.5 μm [45] and room temperature pulsed operation at ~3.8 μm [46] has been demonstrated in strain-compensated heterostructures.

The anti-Stokes scheme shown in Fig. 6.15 is suitable for this purpose. As is clear from the figure and Eq. 6.28*a*, the upper state 3 can be arbitrarily close to the band offset or even in the continuum, because its population is unimportant. There is no need to provide resonant tunneling to the high-lying state 3; therefore, almost full band offset can be utilized. Moreover, one can implement this scheme for the frequency up-conversion into the near-infrared range by using heterostructures with large conduction band discontinuity, such as InGaAs/AlAsSb, in which the presence of low-lying L and X valleys creates a fundamental problem for conventional injection-pumped QC lasers. Evidently, lateral valleys are not a problem for the optically pumped Raman process that does not require any electron current through the excited state 3.

6.5.3 Tunability issues

Conventional QC lasers are based on resonant tunneling structures. They "resist" any tuning by Stark effect above laser threshold, after the alignment between the injector and the upper laser state is reached. In practice, QC lasers can only be tuned over a limited range by changing the temperature or using an external cavity.

Tunability in conventional Raman lasers or amplifiers has been associated with tuning the external optical pump, while the amount of frequency shift has remained fixed and determined by the phonon frequency. Here we have an opposite situation: The drive laser serving as an optical pump for the Raman laser is internally generated, and its tunability is limited; at the same time, the Raman shift is not fixed. It is determined by the energy separation between subbands denoted by 1 and 2 in Fig. 6.8 or 6.15. If the transition 1–2 is diagonal in space (which is a desirable feature for the Raman laser anyway), its energy can be efficiently tuned by the electric field. To increase the electric tuning range and flexibility, the Raman active region should be grouped in a separate stack from the drive laser, so that it can be independently contacted and biased. This design is shown in Fig. 6.20. Note that the device is expected to generate TM_1 modes that have zero field at the position of a strongly doped side contact layer, thus minimizing the losses. Also, the TM_1 mode provides better confinement factors and the overlap of the drive field with the Raman region.

The Raman active region of the device in Fig. 6.20 can be an injectorless stack of coupled QWs, such as shown in Fig. 6.21. This structure has large dipole elements $z_{13} = 15$ nm and $z_{23} = 14$ nm, while the transition 2–1 is diagonal in space and partially suppressed. Changing the electric field from 0 to 15 kV/cm shifts the Stokes frequency by as much as 15 meV, or by 120 cm^{-1}. At the same time, the frequency detunings of both drive and Stokes fields from their respective vertical transitions 1–3 and 2–3 remain practically unchanged: Their shift is only 3 meV. The dipole moments of these transitions are also changed very little — by 0.2 nm. Moreover, these

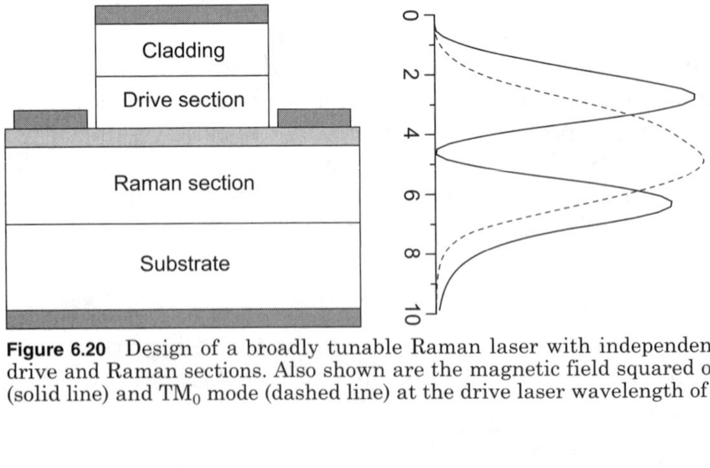

Figure 6.20 Design of a broadly tunable Raman laser with independent biasing of the drive and Raman sections. Also shown are the magnetic field squared of the TM_1 mode (solid line) and TM_0 mode (dashed line) at the drive laser wavelength of 6 μm.

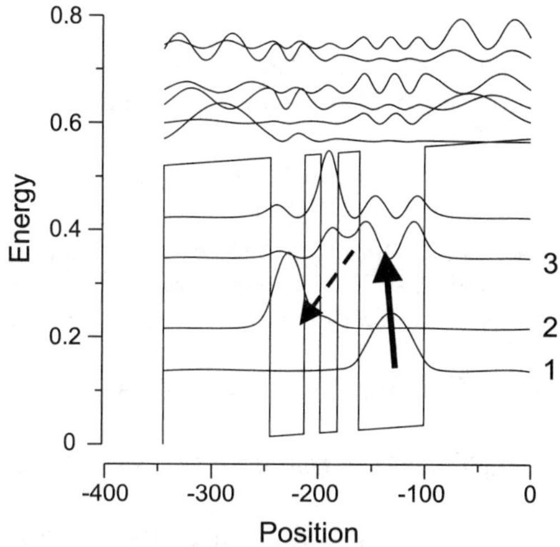

Figure 6.21 One stage of the Raman active region in an InGaAs/AlInAs structure. The drive field (solid arrow) is nearly resonant to the transition 1–3, while the Stokes field (dashed arrow) is generated at frequency $\omega_s = \omega_d - \omega_{21}$ and close to resonance with the transition 2–3.

small changes in the detunings and dipole moments can partially compensate one another in expression 6.28 for the Raman gain. Therefore, the electric tuning of the Stokes field is not accompanied by a significant modulation of the output intensity. This is a very important advantage of the proposed broadly tunable laser.

6.6 Acknowledgments

It is our pleasure to acknowledge numerous helpful discussions with our colleagues and collaborators M. Belkin, R. Colombelli, J. Faist, C. Gmachl, O. Kocharovskaya, V. Kocharovsky, O. Malis, Y. Rostovtsev, M. Scully, C. Sirtori, A. Tredicucci, and B. Williams. This work has been supported by the Air Force Office for Scientific Research and the National Science Foundation.

References

1. J. Faist, F. Capasso, D. L. Sivco, C. Sirtori, A. L. Hutchinson, and A. Y. Cho, *Science* **264,** 553 (1994).
2. O. Svelto, *Principles of Lasers*, Plenum Press, New York, 1998.
3. M. O. Scully and M. S. Zubairy, *Quantum Optics*, Cambridge Univ. Press, U.K., 1997.
4. J. Mompart and R. Corbalán, *Quant. & Semiclass. Opt.* **2,** R7 (2000).

5. A. B. Matsko, O. Kocharovskaya, Y. Rostovtsev, G. R. Welch, A. S. Zibrov, and M. O. Scully, *Adv. in Atom., Mol. and Opt. Phys.* **46,** 191–242 (2001).
6. A. A. Belyanin, C. Bentley, F. Capasso, O. Kocharovskaya, and M. O. Scully, *Phys. Rev.* **A64**, 013814 (2001).
7. J. Khurgin and E. Rosencher, *IEEE J. Quantum Electron.* **32,** 1882–1896 (1996).
8. J. Faist, C. Sirtori, F. Capasso, et al., Quantum interference effects in intersubband transitions, in *Intersubband Transitions in Quantum Wells: Physics and Device Applications I*, H. C. Liu, and F. Capasso (eds.), Academic Press, San Diego, 2000.
9. W. W. Chow, H. C. Schneider, and M. C. Phillips, *Phys. Rev.* **A68,** 053802 (2003).
10. M. K. Gurnick and T. A. De Temple, *IEEE J. Quantum Electron.* **19,** 791 (1983).
11. M. M. Fejer, S. J. B. Yoo, R. L. Byer, A. Harwit, and J. S. Harris, *Phys. Rev. Lett.* **62,** 1041 (1989).
12. C. Sirtori, F. Capasso, D. Sivco, S. N. G. Chu, and A. Y. Cho, *Appl. Phys. Lett.* **59,** 2302 (1991).
13. F. Capasso, C. Sirtori, D. L. Sivco, and A. Y. Cho, Nonlinear optics in coupled-quantum-well quasi-molecules, in *Intersubband Transitions in Quantum Wells: Physics and Device Applications II*, H. C. Liu, and F. Capasso (eds.), Academic Press, San Diego, 2000.
14. C. Sirtori, A. Tredicucci, F. Capasso, et al., *Opt. Lett.* **23,** 463 (1998).
15. C. Claire, D. L. Sivco, and J. N. Baillargeon, *Appl. Phys. Lett.* **79,** 572 (2001).
16. C. Gmachl, D. L. Sivco, R. Colombelli, et al., *Nature* **415,** 883–887 (2002).
17. N. Owschimikow, C. Gmachl, A. Belyanin, V. Kocharovsky, D. L. Sivco, R. Colombelli, F. Capasso, and A. Y. Cho, *Phys. Rev. Lett.* **90,** 043902 (2003).
18. C. Gmachl, A. Belyanin, D. L. Sivco, Milton L. Peabody, N. Owschimikow, A. M. Sergent, F. Capasso, and A. Y. Cho, *IEEE J. Quantum Electron.* **39,** 1345 (2003).
19. C. Gmachl, N. Owschimikow, A. Belyanin, A. M. Sergent, D. L. Sivco, M. L. Peabody, F. Capasso, and A. Y. Cho, *Appl. Phys. Lett.* **84,** 2751 (2004).
20. O. Malis, A. Belyanin, C. Gmachl, D. L. Sivco, M. L. Peabody, A. M. Sergent, and A. Y. Cho, *Appl. Phys. Lett.* **84,** 2721 (2004).
21. T. S. Mosely, A. Belyanin, C. Gmachl, D. L. Sivco, M. L. Peabody, and A. Y. Cho, *Opt. Express,* **12,** 2972 (2004).
22. O. Malis, A. Belyanin, D. L. Sivco, J. Chen, A. M. Sergent, C. Gmachl, and A. Y. Cho, *Electron. Lett.* **40,** 1586 (2004).
23. M. Troccoli, A. Belyanin, F. Capasso, E. Cubukcu, D. L. Sivco, and A. Y. Cho, *Nature* **433,** 845 (2005).

24. A. Evans, J. S. Yu, S. Slivken, and M. Razeghi, *Appl. Phys. Lett.* **85,** 2166 (2004).

25. S. Blaser, D. A. Yarekha, L. Hvozdara, Y. Bonetti, A. Muller, M. Giovannini, and J. Faist, *Appl. Phys. Lett.* **86,** 041109 (2005).

26. A. Yariv, The coupled-mode formalism in guided-wave optics, in *Proc. of the NATO Advanced Study Institute on Fiber and Integrated Optics*, D. B. Ostrowsky, ed., Plenum Press, New York, 1978.

27. D. Nikonov, A. Imamoglu, and M. Scully, *Phys. Rev.* **B59,** 12,212 (1999).

28. J. Faist, C. Sirtori, F. Capasso, L. Pfeiffer, and K.W. West, *Appl. Phys. Lett.* **64,** 872 (1994).

29. Ya. I. Khanin, *Principles of Laser Dynamics*, Elsevier, Amsterdam, 1995.

30. H. M. Pask, *Progress in Quantum Electron.* **27,** 3–56 (2003).

31. G. Sun, J. Khurgin, L. Friedman, and R. A. Soref, *J. Opt. Soc. Am.* **B15,** 648 (1998).

32. H. C. Liu et al., *Appl. Phys. Lett.* **78,** 3580 (2001).

33. O. Kocharovskaya, Yu. V. Rostovtsev, and A. Imamoglu, *Phys. Rev.* **A58,** 649–654 (1998).

34. E. Arimondo and G. Orriols, *Nuovo Cim. Lett.* **17,** 333 (1976).

35. F. Capasso, C. Gmachl, R. Paiella, A. Tredicucci, A. L. Hutchinson, D. L. Sivco, J. N. Baillargeon, A. Y. Cho, and H. C. Liu, *IEEE J. Select. Topics Quantum Electron.* **6,** 931–947 (2000).

36. S. M. Spillane, T. J. Kippenberg, and L. J. Vahala, *Nature* **415,** 621–623 (2002).

37. Y. R. Shen, *The Principles of Nonlinear Optics,* John Wiley & Sons, New York, 1984.

38. J. Faist and R. Colombelli, private communication.

39. J. Khurgin, G. Sun, L. R. Friedman, and R. A. Soref, *J. Appl. Phys.* **78,** 7398 (1995).

40. R. Koehler, A. Tredicucci, F. Beltram, H. E. Beere, E. H. Linfield, A. G. Davies, D. A. Ritchie, R. C. Iotti, and F. Rossi, *Nature* **417,** 157 (2002).

41. B. S. Williams, S. Kumar, Q. Hu, and J. L. Reno, *Opt. Express* **13,** 3331 (2005).

42. B. S. Williams, S. Kumar, H. Callebaut, Q. Hu, and J. L. Reno, *Appl. Phys. Lett.* **83,** 2124 (2003).

43. G. Scalari, N. Hoyler, M. Giovannini, and J. Faist, *Appl. Phys. Lett.* **86,** 181, 101 (2005).

44. M. C. Wanke, F. Capasso, C. Gmachl, et al., Injectorless quantum cascade lasers, *Appl. Phys. Lett.* **78,** 3950 (2001).

45. J. Faist, F. Capasso, D. L. Sivco, et al., *Appl. Phys. Lett.* **72,** 680 (1998).

46. M. P. Semtsiv, M. Ziegler, S. Dressler, W. T. Masselink, N. Georgiev, T. Dekorsy, and M. Helm, *Appl. Phys. Lett.* **85,** 1478 (2004).

Chapter

7

Quantum Well Infrared Photodetector: High-Absorption and High-Speed Properties, and Two-Photon Response

H. C. Liu[a]

Institute for Microstructural Sciences
National Research Council
Ottawa, Ontario, Canada

Harald Schneider[b]

Fraunhofer-Institute for Applied Solid-State Physics
D-79108 Freiburg, Germany

7.1 Introduction

Similar to all other devices discussed in this book, the *quantum well infrared photodetector* (QWIP) makes use of intersubband transitions between quantized states within (commonly) the conduction band. The most popular materials system is GaAs/AlGaAs because of its maturity in fabrication techniques, e.g., grown by *molecular beam epitaxy* (MBE). For a given materials system, the conduction band offset or bandgap discontinuity is limited to a given range. For example, the wavelengths covered by GaAs/AlGaAs QWIPs are longer than about 3 μm. Nearly 20

[a]Electronic mail: h.c.liu@nrc.ca.

[b]Present address: Institute of Ion-Beam Physics and Materials Research, Forschungszentrum Rossendorf, D-01314 Dresden, Germany.

years of research and development activities have placed QWIP arrays into the market for thermal imaging applications in the two atmospheric transmission windows of 3 to 5 and 8 to 12 μm in wavelength.

The physics [1] and applications [2–4] of GaAs/AlGaAs QWIPs are well established. Thermal imaging is the main existing application of QWIPs fabricated into arrays [2]. For detecting weak signals, the device parameters are chosen to obtain the highest possible sensitivity and *background limited infrared performance* (BLIP) temperature. To reduce and in most cases eliminate the noise caused by dark current, the devices are cooled to cryogenic temperatures. For example, for a 10-μm cutoff QWIP viewing the usual room temperature background, the operating temperature is about 77 K. The design principles for this existing application are reviewed in Ref. 1. Presently QWIP arrays are on the verge of infiltrating existing applications and *more importantly* finding new ones [5, 6].

A schematic illustration of the QWIP operation is shown in Fig. 7.1. The intrinsic detector response time is limited by either the photoelectron lifetime or the total transit time, whichever is smaller. The transit time is usually much longer than the lifetime, leading to an intrinsic response time of about 5 ps due to the fast intersubband scattering processes by electron-phonon, electron-impurity, or electron-electron interactions.

To turn to the subject of this chapter, QWIPs are also well suited for high-frequency and high-speed applications. A distinct advantage of QWIPs over standard detectors made of HgCdTe is their high intrinsic speed. This is related to the inherent short carrier lifetime $\tau \sim 5$ ps as inferred both from heterodyne experiments [7, 8] showing a cutoff frequency of about 30 GHz and from time-resolved photocurrent measurements [9–11].

In this chapter, we review a simple way to achieve high absorption and demonstrate the high-speed capability. For characterizing short laser pulses, an autocorrelation technique is commonly used. This can be done either by a second harmonic generation crystal plus a linear detector or by a two-photon detector. We investigate the nonlinear behavior of QWIPs having three energetically equidistant energy levels. Resonantly enhanced nonlinear absorption, six orders of magnitude higher compared to typical bulk semiconductors, leads to a threshold power density for quadratic detection as low as 0.1 W/cm^2 and femtosecond time resolution. The approach enables dynamic characterization of the optical light field of infrared (ir) emitters with un-precedented sensitivity. We demonstrate that second-order auto-correlation measurements of ultrashort mid-infrared laser pulses in the picojoule regime can be performed by using these ultrasensitive two-photon detectors.

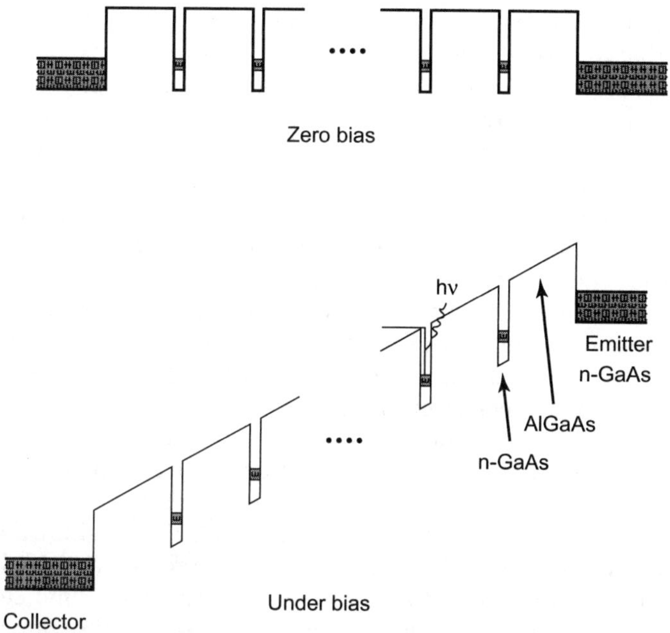

Figure 7.1 Schematic conduction band edge profile of a GaAs/AlGaAs QWIP under zero (above) and finite (below) bias. The electron population in the n-type wells is provided by doping using silicon. The emitter and collector contact layers are doped with silicon. Electrons occupy the ground state subband. Photons ($h\nu$) excite electrons from quantum wells, causing a photocurrent.

7.2 Pictures of Intersubband Transition and Photoconductivity

The success of QWIPs has been made possible by the semiconductor crystal growth technology such as MBE [12] and by the device concepts using bandgap engineering [13]. Semiconductor structures with atomic layer control are now routinely fabricated by MBEs. Several introductory textbooks on quantum well physics have been written, e.g., by Bastard [14], Weisbuch and Vinter [15], Shik [16], and Harrison [17].

The term *intersubband transition*, abbreviated as ISBT hereafter, refers to the electronic transition between the confined states in quantum wells. The physics related to the optical ISBT in quantum wells is treated in great detail by Helm [18]. Other materials of specific relevance to QWIPs can be found in Refs. 1, 19, and 20. Here, we discuss only ISBT in the conduction band. Examples presented here are all on samples made by GaAs-based MBE. For this system, a polarization selection rule was established in the early days [21]: Only the light polarized in the growth direction can cause ISBT. The selection rule is valid for quantum wells where the single isotropic effective mass approximation holds. Since the argument is based on the effective mass approximation, the selection rule is naturally not rigorous. As band mixing is the cause of the breakdown of the selection rule, physical intuition tells us that the accuracy of the selection rule should be related to the ratio of the energy scales involved. The closest band to the conduction band (at least for the GaAs case) is the valence band. The relevant energy ratio is then E_n/E_g, where E_n are the eigenenergies of the confined states in the conduction band quantum well and E_g is the bandgap. (Note that the reference point—zero energy—for E_n is chosen at the conduction band edge of the well.) For common quantum wells used in QWIPs (see, e.g., Ref. 1) the ratio is small, and therefore the selection rule is expected to be quite accurate, with a deviation of at most 10% level, as shown experimentally [22].

Because of the selection rule, a normal-incidence geometry (i.e., light incident normal to the wafer and along the growth direction) is not suited. A commonly employed 45-degree edge facet geometry is shown in Fig. 7.2, as first used in Levine et al. [23]. For unpolarized light, this geometry "throws away" one-half of the power, but is simple and convenient and is usually used to obtain a detector performance benchmark. This geometry is also suited for high-speed applications involving lasers. The majority of applications of QWIPs require large *two-dimensional* (2D) arrays where the facet geometry is not suited. For imaging arrays where the incident infrared is normal to the wafer, a diffraction grating is fabricated on each pixel to bend the light.

Within the single-band effective mass approximation, the Schrödinger equation reads

$$-\frac{\hbar^2}{2m}\nabla^2 \psi + V \psi = E \psi \tag{7.1}$$

where m is the effective mass and V is the potential. If we choose the quantum well direction as the z axis, the potential V depends only on z, and the wave function is separable into lateral xy plane and z parts. The lateral part is simply a plane wave.

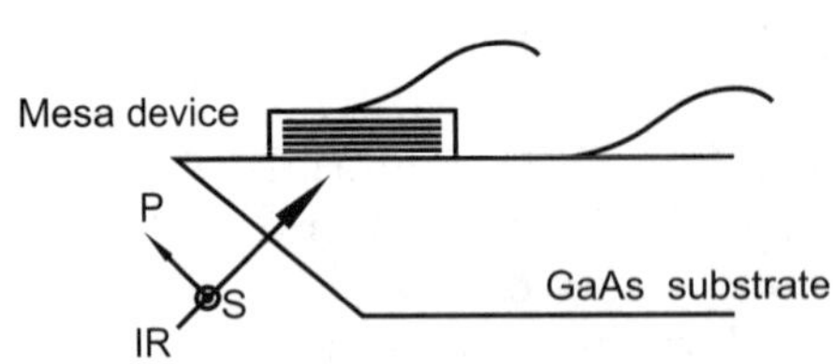

Figure 7.2 A 45-degree edge facet light coupling geometry. The figure is not drawn to scale. The semi-insulating GaAs substrate thickness is usually in the range of 400 to 700 μm. For testing individual detector performance, mesa devices of areas from about 100^2 to 1000^2 μm^2 are used. The infrared (IR) light is shone normal to the facet surface. The *P* and *S* polarizations are defined with respect to the light incidence on the quantum wells.

To obtain a simple picture of ISBT, much of the physics can be illustrated by using the simplest model of an infinitely high-barrier square quantum well [24]; that is, $V = 0$ for $0 \le z \le L_w$ and $V = \infty$ for $z < 0$ and $z > L_w$, where L_w is the well width. In this case the eigenstate wave function and energy are trivial:

$$\psi_n(\vec{k}_{xy}) = \sqrt{\frac{2}{L_w A}} \sin \frac{\pi n z}{L_w} \exp(i\vec{k}_{xy} \cdot \vec{x}) \tag{7.2}$$

$$E_n(\vec{k}_{xy}) = \frac{\hbar^2}{2m}\left(\frac{\pi^2 n^2}{L_w^2} + k_{xy}^2\right) \tag{7.3}$$

where A is the normalization area in the xy plane, n is a positive integer, $\vec{k}_{xy}$ is the in-plane wave vector, and m is the effective mass in the well. Equation 7.3 explains the reason for the term *subband* instead of a single state. For a given quantized state, one can put many electrons occupying different in-plane momenta. For the ground state in equilibrium, the occupation of electrons leads to a Fermi energy determined by $n_{2D} = (m/\pi\hbar^2)E_f$, where n_{2D} is the 2D quantum well carrier density and $m/\pi\hbar^2$ is the 2D density of states.

The dipole matrix moment between any two states (say, n and n') with opposite parity is

$$e\langle z\rangle_{n,n'} = eL_w \frac{8}{\pi^2} \frac{nn'}{(n^2 - n'^2)^2} \tag{7.4}$$

Note that the in-plane momentum remains the same between initial and final states.

The oscillator strength is

$$f \equiv \frac{2m\omega}{\hbar} \langle z\rangle_{n,n'}^2 = \frac{64}{\pi^2} \frac{n^2 n'^2}{(n^2 - n'^2)^3} \tag{7.5}$$

The absorption probability for an ir beam polarized in the plane of incidence and propagating at an angle of θ with respect to the growth axis is

$$\eta = \frac{e^2 h}{4\varepsilon_0 n_r mc} \frac{\sin^2\theta}{\cos\theta} n_{2D} f \frac{1}{\pi} \frac{\Delta E}{(E_{n,n'} - \hbar\omega)^2 + (\Delta E)^2} \tag{7.6}$$

where ε_0 is the vacuum permittivity, n_r is the index of refraction, c is the speed of light, $E_{n,n'} \equiv E_n - E_{n'}$, and ΔE is the broadening half-width. The line shape associated with broadening is modeled by a lorenzian. At the peak ($E_{n,n'} = \hbar\omega$) the absorption is inversely proportional to ΔE.

For a given n_{2D}, the absorption is inversely proportional to m; that is, the smaller the effective mass, the larger the absorption. Note that for a given E_f, however, the absorption is independent of m since $n_{2D} = [m/(\pi\hbar^2)]E_f$. Note also that the integrated absorption is independent of ΔE. The derivation of Eq. 7.6 is straightforward by using a dipole interaction hamiltonian and Fermi's golden rule (see, e.g., Helm [18]).

The factor $\sin^2\theta$ in Eq. 7.6 comes from the polarization selection rule discussed before. The factor $\cos\theta$ seems to give an unphysical result when $\theta \to 90°$. However, since the meaning of η is the absorption probability of light passing through the well, in this extreme the passing length becomes infinitely long, resulting in an infinitely large absorption. If one had considered a quantity of absorption constant α defined by $\eta = \alpha \times$ length, where length $= L_w/\cos\theta$ is the propagation length, this quantity would have been always finite. For a real quantum well with finite barrier height, it would be more physical to choose the length using the quantum well structure thickness (including barriers) instead of only L_w taken here.

Let us put some typical numbers into Eq. 7.6 to get a feeling for how strong the ISBT absorption is. For a typical 8- to 12-µm peaked QWIP, the half-width is about $\Delta E = 0.01$ eV. For ground state to first excited state transition, the oscillator strength is $f = 0.961$ (see Eq. 7.5 with $n' = 1$ and $n = 2$). For 77 K operation, the carrier density is set to about $n_{2D} = 5 \times 10^{11}$ cm^{-2}. For GaAs well, the reduced effective mass is $m^* = 0.067$ ($m = m^* \times m_e$, where m_e is the free electron mass), and the refractive index is about $n_r = 3.3$. With these values and for a 45-degree angle ($\theta = 45°$), the absorption for a single quantum well is $\eta = 0.54\%$ (for polarized light). Let us also evaluate the absorption constant for the case of $\theta = 90°$, with other parameters the same as above. Taking a quantum well structure thickness of 50 nm (including barriers), we find the peak absorption constant is $\alpha = 1520$ cm^{-1}.

For a standard QWIP, the optimum well design is the one having the first excited state in resonance with the top of the barrier. This configuration gives *at the same time* both a large peak absorption (similar to the bound state to bound state transition discussed above) and a rapid escape for the excited electrons. The optimum design configuration has been experimentally proved [25, 26]. To design a quantum well for a given QWIP wavelength, one needs to know how the barrier height (conduction band offset) relates to heterosystem parameters, i.e., the Al fraction x in the GaAs/Al$_x$Ga$_{1-x}$As case. Surveying many samples by comparing the calculated transition energy with the experimental peak absorption, we find a range of values for the conduction band offset $\Delta E_c = (0.87 \pm 0.04) \times x$ eV, where x is the Al fraction. The calculation is a simple eigenenergy calculation of a square quantum well. Higher-order effects, such as band nonparabolicity and many-body effects, influence the precise values of the calculation (as discussed by Helm [18]).

To describe the photocurrent from a QWIP, the standard photoconductivity formulas can be used. The current responsivity (in amperes per watt) is written as

$$R_i = \frac{e}{h\nu}\eta g \tag{7.7}$$

where η now is the total absorption quantum efficiency and g is the photoconductive gain given by $g = \tau_{\text{life}}/\tau_{\text{trans}}$, where τ_{trans} is the total transit time and τ_{life} is the excited carrier lifetime or more specifically for QWIP the capture time into the quantum well. The noise in a common QWIP is of a generation recombination nature and is therefore given by

$$i_{\text{noise}}^2 = 4egIB \tag{7.8}$$

where I is the device current (including dark current and photocurrent) and B is the measurement bandwidth. Given the incident photon power, the photocurrent is calculated by Eq. 7.7, while the dark current can be estimated by the simple expression

$$J_{\text{dark}} = eN_{\text{3D}}\upsilon \tag{7.9}$$

where υ is the drift velocity for carriers in the barrier region and the *three-dimensional* (3D) mobile carrier density in the barrier is estimated by

$$N_{\text{3D}} \approx 2\left(\frac{m_b k_B T}{2\pi\hbar^2}\right)^{3/2} \exp\left(\frac{-E_{\text{act}}}{k_B T}\right) \tag{7.10}$$

where m_b is the barrier effective mass. E_{act} is the thermal activation energy, which equals the energy difference between the top of the barrier and the top of the Fermi sea in the well. We have assumed that $E_{\text{act}}/(k_B T) \gg 1$, appropriate for most practical cases.

7.3 High-Absorption QWIPs

As shown later, especially for the case of heterodyne detection, a high absorption efficiency directly translates to a high sensitivity and is therefore of critical importance. In this section, we show experimental results to demonstrate the high-absorption capability. We also show that these devices can work at elevated temperatures either within the thermoelectric cooling range (> 200 K) or near room temperature.

7.3.1 Absorption measurements

Starting from a standard QWIP structure, the simplest way to enhance absorption is to dope the wells more heavily and to grow more wells [27, 28]. At 45-degree incidence and for polarized light, the absorption per quantum well per pass is about 0.54% per well for a standard GaAs/AlGaAs QWIP with 5×10^{11} cm^{-2} doping. For a doping density of 1.5×10^{12} cm^{-2}, the one-well, one-pass absorption is expected to be $\eta_1 \approx 1.6\%$. If a 90% QWIP absorption is desired, the number of wells needed is determined by $\exp(-2N_{\eta 1}) = 10\%$, which gives $N = 72$. (The factor of 2 in the exponential accounts for the double passes in the 45-degree facet detector geometry.) In the following experiments, we have chosen $N = 100$ to ensure high absorption.

All QWIP wafers were grown in a MBE system. Here we show three examples with different detection wavelengths centered at about 10, 8, and 5 μm. The period of the 100-repeat multiple-quantum-well structure consists of a GaAs well and $Al_xGa_{1-x}As$ barriers. The GaAs well center region is doped with Si to give an equivalent 2D density of 1.5×10^{12} cm^{-2}. The top and bottom GaAs contact layers are 400 and 800 nm thick, doped with Si to 2×10^{18} cm^{-3}. Other device parameters are listed in Table 7.1. Mesa devices were fabricated by standard GaAs processing techniques. All devices were packaged into the 45-degree edge facet geometry for optical coupling.

With the high well doping densities used in these QWIPs, we expect a large increase in dark current [29]. Using a set of samples having similar cutoff wavelengths but different doping densities of 5×10^{11}, 1.0×10^{12}, 1.5×10^{12}, and 2.0×10^{12} cm^{-2}, measurements on current-voltage (*I-V*) characteristics at 77 K show approximately a factor-of-10 increase between consecutive samples. The increase is mainly caused by the increase in the well Fermi energy ΔE_f which raises the dark current by $\exp\left[\Delta E_f / (k_B T)\right]$. Measurements of the BLIP temperature show that T_{BLIP} is degraded with increasing doping and $T_{\mathrm{BLIP}} \approx 60$, 45, and 30 K for 1.0×10^{12}, 1.5×10^{12}, and 2.0×10^{12} cm^{-2} doping samples, respectively. An optimized QWIP (with about 4×10^{11} cm^{-2} doping) covering the same wavelength range would have a T_{BLIP} of about 77 K

TABLE 7.1 Sample Parameters

The doping in the well is Si, δ-doped at the center to 1.5×10^{12} cm^{-2}. The number of periods is 100. The symbols are λ_p for peak detection wavelength, x for Al fraction, L_w for well width, and L_b for barrier thickness.

Sample	λ_p (μm)	x	L_w (nm)	L_b (nm)
QWIP10	10.0	0.19	6.6	25
QWIP8	8.4	0.24	5.9	24
QWIP5	5.0	0.48	5.0	22

for a 2π-solid-angle field of view. High-doping devices are therefore far from optimum for the detection of weak signals.

To show the anticipated high absorption, Fig. 7.3 presents the measured double-pass polarized 45-degree incident transmission spectra for the three samples. For samples QWIP10 and QWIP8, the high absorption is clearly demonstrated. Ideally, QWIP5 should have a similar absorption, and it is not clear why a somewhat lower absorption is obtained. Nevertheless, a 70% absorption is still quite high. In a separate experiment on a set of 10-μm QWIPs [27], it was noted that if the doping was further increased from 1.5×10^{12} to 2.0×10^{12} cm^{-2}, then the peak absorption did not increase and only a broadening in the spectral width was seen. The same experiment also indicated that a doping of 1.0×10^{12} cm^{-2} is sufficient for the 10-μm QWIP to have 90% absorption.

7.3.2 Detector characteristics

Figure 7.4 shows the spectral response curves for the three samples, covering the designed wavelength regions. A further goal of these samples is to attempt operation on them at elevated temperatures. To maximize the dark current limited detectivity, it has been established [1] that the well doping density should be such that the Fermi energy is $E_f = 2k_BT$, where T is the desired operating temperature. The doping in the well is assumed to be completely ionized; i.e., doping density equals electron density (n_{2D}). The Fermi energy relates to n_{2D} by $n_{2D} = \left[m/\left(\pi\hbar^2\right)\right]E_f$, where m is the well effective mass. For $T = 80$ K, the required density is about 4×10^{11} cm^{-2} for GaAs wells. It is then

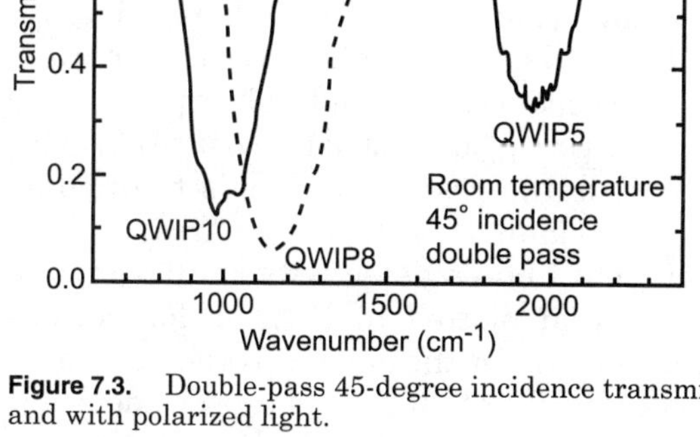

Figure 7.3. Double-pass 45-degree incidence transmission spectra at room temperature and with polarized light.

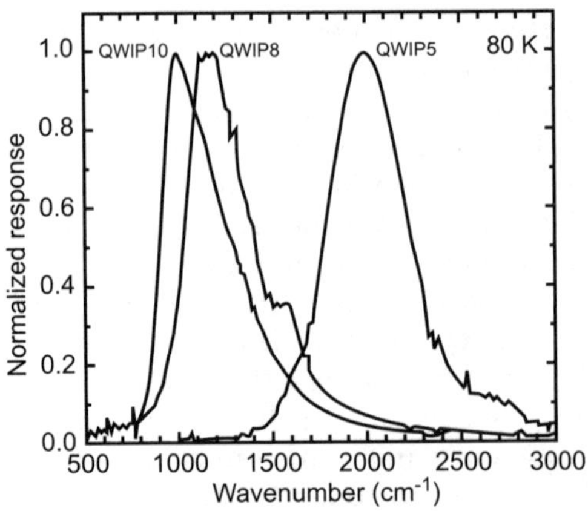

Figure 7.4. Normalized spectral response curves at 80 K and 3 V. The device geometry is equivalent to that of Fig. 7.3 so that the absorption efficiency is represented by Fig. 7.3.

expected that the doping range of 1 to 2 × 10^{12} cm^{-2} is where QWIPs may operate near room temperature, albeit with a reduced sensitivity. Measured results at various temperatures using a CO_2 laser tuned to 10.6 μm are shown in Fig. 7.5 for sample QWIP10. It is clear that the device does work up to room temperature.

It is interesting to note that with increasing temperature, the responsivity first goes up and then comes down for a given bias voltage. Since the responsivity is proportional to the mobility, this behavior is attributed (in a large part) to the temperature dependence of the mobility. It is well known that the impurity scattering limited mobility increases with temperature whereas the phonon scattering limited mobility decreases. The two scattering effects result in the observed behavior [29]. For elevated-temperature operation where the QWIP resistance is low, a high responsivity is highly desirable to overcome the large noise associated with the low resistance and the 50-Ω termination. It is difficult to accurately measure the responsivity when the device resistance is low and comparable to contact and/or series resistance. The low resistance causes a "short circuit" for the photocurrent, leading to an apparent low responsivity. The values in Fig. 7.5 for high temperatures (> 200 K) could be somewhat lower than reality.

Given the measured *I-V*, responsivity, and absorption, the detectivity D^* is easily evaluated. Using the relations $R_i = [e/(h\nu)]g\eta$ and $D^* = R_i\sqrt{A}/\sqrt{4egI}$, where R_i is the responsivity, ν is the photon frequency, g is the photoconductive gain, A is the device area, and I is the device current, the dark current limited detectivity at 10.6 μm and for polarized light is calculated and plotted in Fig. 7.6 for various temperatures. A state-of-the-art QWIP for the same wavelength would have a D^* value of over 10^{10} cm·Hz$^{1/2}$/W at 80 K. The present device is

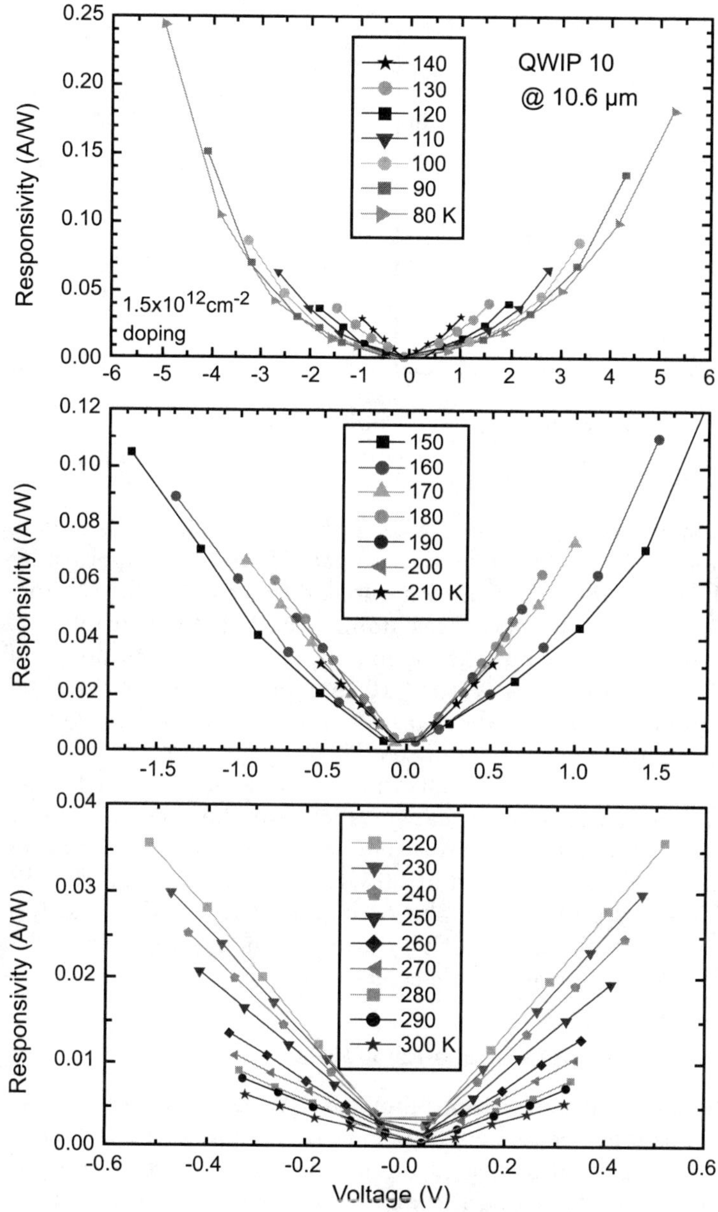

Figure 7.5. Responsivity versus applied voltage under a CO_2 laser (10.6 μm) illumination and at various temperatures for sample QWIP10.

(again) then far from optimum for low-signal and low-temperature use. Comparing with a room temperature thermal detector having a D^* value of about 10^8 cm·Hz$^{1/2}$/W, which is commonly used in a

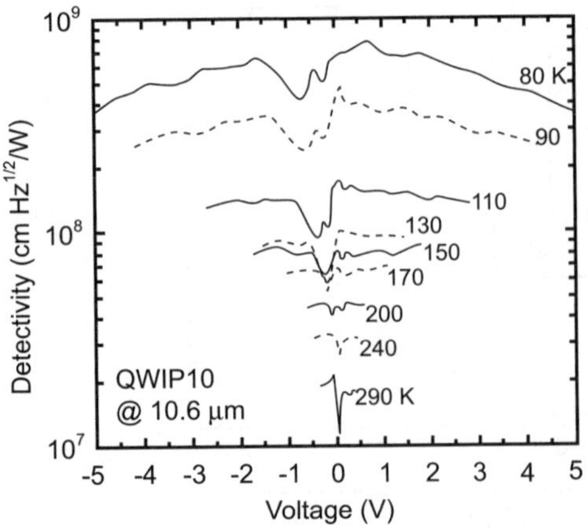

Figure 7.6 Detectivity at 10.6 μm for sample QWIP10 and at various temperatures.

mid-infrared spectrometer system, we see the present device needs to be cooled to about 150 K to achieve the same D^*. Note that temperatures about 200 K and higher are attainable by thermoelectric cooling. Operation at elevated temperatures implies high dark currents. To overcome the dark current noise at room temperature, a laser power of about 10 mW is required for a 10 × 10 μm^2 active-area device; and similarly at 200 K a power of about 5 mW is needed. High-speed operation requires small device capacitances. The present approach employing a large number of quantum wells makes the device thicker than a typical QWIP and hence of lower capacitance. For the present devices with a thickness of about 3 μm, an active area of about 50 × 50 μm^2 or smaller is sufficient for operation at 30 GHz or higher. Similarly, the detectivity for sample QWIP5 is shown in Fig. 7.7. For this shorter-wavelength device, a higher detectivity is obtained, making this device well operable with a thermoelectric cooler.

7.4 High-Speed and High-Frequency QWIPs

Thanks to the intrinsic short carrier lifetime, QWIPs are well suited for high-speed and high-frequency applications. High-speed detectors may create new applications, e.g., environmental remote sensing of molecules [30] and CO_2 or quantum cascade laser-based communication [31], as well as laboratory and instrumentation [32–36]. For these applications, there is commonly a strong signal or a powerful local oscillator, in most cases employing lasers. Under such circumstances, a high dark current can be tolerated to a large degree, and a high absorption and high operating temperature are desirable. Note also that

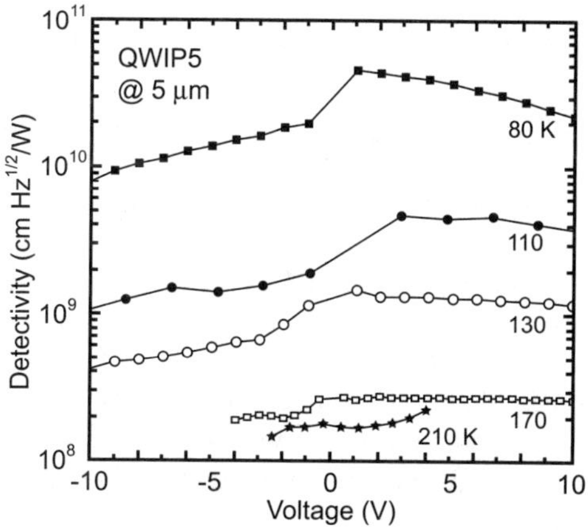

Figure 7.7 Detectivity at 5 μm for sample QWIP5 and at various temperatures.

the polarization sensitivity of the n-type QWIP is no longer a disadvantage for (polarized) laser-based systems.

The intrinsic detector response time is limited by either the photoelectron lifetime or the transit time, whichever is smaller. For a large number of QWs, the total transit time is usually much longer than the lifetime, leading to a lifetime-limited intrinsic response time of about 5 ps. If, on the other hand, a QWIP has a small number of QWs (for example, 10 or less), photoexcited carriers will be swept out before capture, resulting in a transit time-limited situation. At present, QWIPs hold the unique position of having high-speed, high-frequency capability and high absorption for the thermal infrared region. There are no competitive alternatives.

The measurement of high-speed and high-frequency characteristics can be done in either the frequency [7, 8, 37] or the time [9–11] domain. In this section, we discuss the physics and show results in the frequency domain.

7.4.1 Microwave rectification

As a simple way to measure the QWIP high-frequency behavior, we first discuss the microwave rectification technique [37]. We apply a microwave signal to the QWIP and measure the change in its direct biasing current. This is complementary to the optical heterodyne technique, which involves generating a microwave signal within a QWIP at the difference frequency of two optical beams [7]. The rectification in the QWIP relies on its inherent nonlinear I-V characteristic and therefore probes its transport properties.

The small-signal rectified direct current is given by

$$I_{\text{rect}} = \frac{1}{4} I'' V_{\mu}^{2} \tag{7.11}$$

where I'' is the second derivative of the *I-V* curve and V_μ is the amplitude of the microwave voltage applied to the device. Both I'' and V_μ depend on the microwave frequency ω. The dependence of I'' on ω reflects the frequency roll-off behavior of the intrinsic transport mechanism and therefore is expected to behave as $1/[1 + (\omega\tau)^2]$, where τ is some characteristic time. This τ is expected to be approximately the excited electron lifetime or the photoconductive lifetime. Given a constant output power from a microwave source, V_μ varies as a function of frequency because of the circuit limited by the device capacitance and differential resistance, and other parasitics. We then rewrite Eq. 7.11, separating out the frequency dependencies:

$$I_{\text{rect}} = \frac{1}{4} I_0'' V_{\mu 0}^{2}\, \alpha(\omega)\, \beta(\omega) \tag{7.12}$$

where

$$\alpha(\omega) = \frac{1}{1 + (\omega\tau)^2} \tag{7.13}$$

Here $\beta(\omega)$ is the circuit dependence, and I_0'' and $V_{\mu 0}$ are the low-frequency limiting values of I'' and V_μ, respectively.

The experiment is schematically shown in Fig. 7.8. QWIPs were connected to the end of a 50-Ω coplanar transmission line by a short wire bond. The dc bias is applied through a bias-T. The microwave power is supplied by a microwave source capable of frequencies up to 40 GHz. The QWIP is modeled by a parallel *resistance-capacitance*

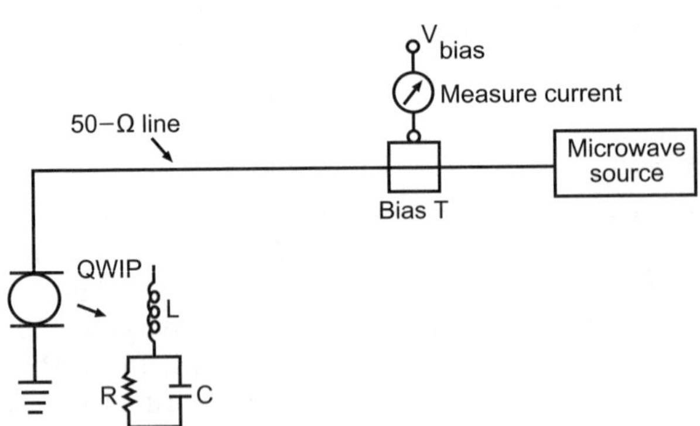

Figure 7.8 Schematic of the microwave rectification experiment. The QWIP is mounted at the end of a 50-Ω transmission line. The QWIP is modeled by a parallel resistance-capacitance equivalent circuit, and the parasitic inductance is caused by the short wire bond.

(RC) equivalent circuit, and the parasitic inductance L is caused by the wire bond. Given the equivalent circuit of the device and its parasitic inductance (Fig. 7.8), the microwave voltage across the device is straightforward

$$V_\mu^2 = \frac{8R_L}{(1-\omega^2 LC)^2 + \omega^2(R_L C + L/R)^2} P_{\text{out}} \tag{7.14}$$

where $R_L = 50\ \Omega$ is the line impedance, $R = 1/I'$ is the device differential resistance, I' is the derivative of the I-V curve, and P_{out} is the output power from the microwave source. We have made the approximation $R + R_L \approx R$ because $R_L \ll R$ for a typical QWIP. For $\omega \to 0$ we get

$$V_{\mu 0}^2 = 8R_L P_{\text{out}} \tag{7.15}$$

and by comparison with Eq. 7.12

$$\beta(\omega) = \frac{1}{(1-\omega^2 LC)^2 + \omega^2(R_L C + L/R)^2} \tag{7.16}$$

The $L \to 0$ limit of the above gives the usual RC roll-off: $1/[1 + (\omega R_L C)^2]$.

We show experimental results on three samples listed in Table 7.2. The samples differ mainly by the number of quantum wells. All experiments were carried out with the sample at 77 K. The relevant device parameters of these samples are listed in Table 7.3. The expected RC characteristic frequency is $f_{RC} = 1/(2\pi R_L C)$ for R_L = 50-Ω

TABLE 7.2 Sample Structural Parameters

The Si doping density in the two contact layers was 1.5×10^{18} cm^{-3}, and the center Si δ-doping density in the wells was 9×10^{11} cm^{-2} for all samples.

Sample	x	L_w (nm)	L_b (nm)	Repeats
16W	0.265	5.7	23.7	16
8W	0.260	5.9	24.6	8
4W	0.276	6.0	24.6	4

TABLE 7.3 Sample Device Parameters

The device size is $10 \times 10\ \mu m^2$. Here L_{tot} is the total active device thickness, C is the device capacitance, $f_{RC} = 1/(2\pi R_L C)$, $R_L = 50\ \Omega$, f_{max} is the lifetime-limited cutoff frequency, and $\tau = 1/(2\pi f_{max})$.

Sample	L_{tot} (µm)	C (fF)	f_{RC} (GHz)	f_{max} (GHz)	τ (ps)
16W	0.49	21.5	148	33	4.8
8W	0.27	39.4	81	≈33	≈4.8
4W	0.15	72.2	44	≈33	≈4.8

load resistance. The estimated carrier lifetime τ and the intrinsic cutoff frequency f_{max} [defined by $\tau = 1/(2\pi f_{max})$] are discussed next.

Figure 7.9 shows the measured (dots) and calculated (lines) rectified current versus frequency for the three samples. The 3-dB point from the maximum f_{max} for the 16-well sample (sample 16W) is about 33 GHz. For this sample, the RC-limited frequency is much larger than 33 GHz, and hence the roll-off here directly relates to τ, resulting in a value of 4.8 ps. However, for the 4- and 8-well samples, the RC effect is clearly seen. The rectified current starts to roll off at a lower frequency and with a much slower rate. The calculated curves in Fig. 7.9 used the expressions given above with the inductance L determined by the length of the wire bond. Further details are given in Ref. 37.

7.4.2 Heterodyne detection

Although very well suited, heterodyne detection using QWIPs has received limited attention [38, 39, 7, 8, 32]. Heterodyne detection involves a *local oscillator* (LO), commonly a laser, at a slightly different wavelength from that of the signal. The difference-frequency signal (often referred to as the IF—the intermediate-frequency signal) is measured.

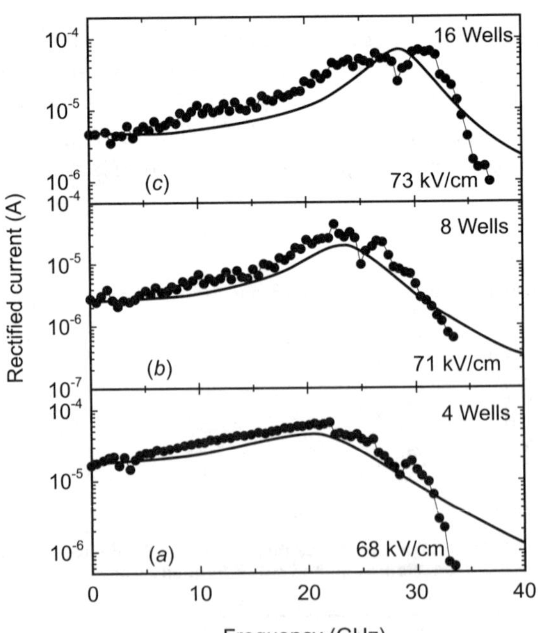

Figure 7.9 Rectified current versus frequency for (*a*) the 4-well, (*b*) the 8-well, and (*c*) the 16-well samples. The solid lines are calculated. The bias voltages were chosen to give approximately the same electric field of 70 kV/cm. All device areas were 10 × 10 μm^2. The device temperature was 77 K. The microwave power was 1.0 mW.

This method is also used to characterize the high-frequency behavior by beating two lasers and measuring the IF as a function of the frequency. The advantage of heterodyne detection is well known and described [40]. The key point is that with a sufficiently high LO power, the ideal detection limit can be reached, and the sensitivity [e.g., the *noise equivalent power* (NEP)] depends only on the absorption quantum efficiency for a given wavelength. The argument can be easily constructed for a photoconductive QWIP as follows. As established, the detector current responsivity is written as

$$R_i = \frac{e}{h\nu}\eta g \tag{7.17}$$

where η is the absorption quantum efficiency and g is the photoconductive gain. Under a strong LO power P_{LO}, i.e., LO-induced current I_{LO} dominates over dark current, the g-r noise current power spectral density is

$$S_i = 4egI_{\mathrm{LO}} = 4egR_iP_{\mathrm{LO}} \tag{7.18}$$

By using Eq. 7.17, Eq. 7.18 becomes

$$S_i = (2eg)^2\frac{\eta}{h\nu}P_{\mathrm{LO}} \tag{7.19}$$

Hence the noise current is

$$i_{\mathrm{noise}} = \sqrt{S_iB} = 2eg\sqrt{\frac{\eta}{h\nu}P_{\mathrm{LO}}B} \tag{7.20}$$

where B is the measurement bandwidth.

The heterodyne current for a signal power of P_{sig} is

$$i_{\mathrm{het}} = 2R_i\sqrt{P_{\mathrm{LO}}P_{\mathrm{sig}}} = 2eg\frac{\eta}{h\nu}\sqrt{P_{\mathrm{LO}}P_{\mathrm{sig}}} \tag{7.21}$$

The last step used Eq. 7.17.

The minimum detectable signal occurs when $i_{\mathrm{noise}} = i_{\mathrm{het}}$; from Eqs. 7.20 and 7.21, we then have

$$\frac{P_{\mathrm{sig,min}}}{B} = \frac{h\nu}{\eta} \tag{7.22}$$

which leads to the following definition [38, 39]:

$$(\mathrm{NEP})_{\mathrm{het}} = \frac{h\nu}{\eta} \tag{7.23}$$

The NEP is measured in watts per hertz (not in the common unit of W/ $Hz^{1/2}$ for the usual detection scheme). It is shown from Eq. (7.23) that the heterodyne NEP depends only on the absorption η and photon quantum $h\nu$, and is independent of gain g. This of course is in the regime where LO-induced current is dominant, referred to as photon noise limited.

The direct measurement of IF signal is limited to frequencies up to the capability of the available spectrum analyzer. To achieve higher frequencies, we can employ mixing in the QWIP to down-convert the IF signal generated by the optical heterodyne [8]. This again (as in the rectification case) relies on the nonlinear *I-V* characteristic as in rectification. We apply to the QWIP not only the two ir beams, as in a conventional heterodyne experiment, but an additional microwave excitation. Specifically, let f_{ir1} and f_{ir2} denote the two ir frequencies, and $f_{\mu wave}$ be the microwave frequency. Then the ir heterodyne frequency is $f_{het} = |f_{ir1} - f_{ir2}|$, while the down-converted signal frequency is $|f_{het} - f_{\mu wave}|$. In this scheme, one can reach very high f_{het} frequencies by using a standard spectrum analyzer at the output of the QWIP. Of course, when f_{het} and $f_{\mu wave}$ are both sufficiently small, we also observe the up-converted signal at $f_{het} + f_{\mu wave}$ on the spectrum analyzer.

The device used in this experiment was a 100-well QWIP with a GaAs well width of 4.5 nm and an $Al_{0.21}Ga_{0.79}As$ barrier width of 40 nm. The center 2.5 nm of each well was doped with Si to 2.5×10^{18} cm^{-3}. The ir sources were either two CO_2 lasers [41] or a CO_2 laser and a lead-salt temperature *tunable diode laser* (TDL) [42]. The microwave or millimeter-wave radiation was generated by either a microwave source tunable up to 40 GHz or a Gunn oscillator mechanically tunable in the range of 91 to 94 GHz. The CO_2 lasers were operating in the neighborhood of 10.3 μm on individual lines selected by gratings with a separation between adjacent lines of about 41 GHz. The TDL was used only for $f_{het} < 26.5$ GHz, which was the limit of our spectrum analyzer. A schematic of the experimental techniques is shown in Fig. 7.10.

The experimental results covering the frequency range of about 1 to 100 GHz are shown in Fig. 7.11. The curves are the expected roll-off behavior due to the device *RC* time constant and the photocarrier lifetime, again using $\tau \sim 5$ ps. The optical heterodyne data were taken with a CO_2 laser and the TDL laser [7]. We have normalized the signal for a constant incident power of about 0.2 mW from each of the ir lasers and 0.3 mW from the microwave source. The five data points [8] shown in Fig. 7.11 for the mixing experiment were taken using different sources and different microwave coupling schemes. For the first three points at $f_{het} = 1.83$, 5.37, and 15.5 GHz, we used one CO_2 laser, the TDL, and the microwave source at frequencies of $f_{\mu wave} = 9$, 15, and 10 GHz, respectively, and measured the sum signals of f_{het} and $f_{\mu wave}$.

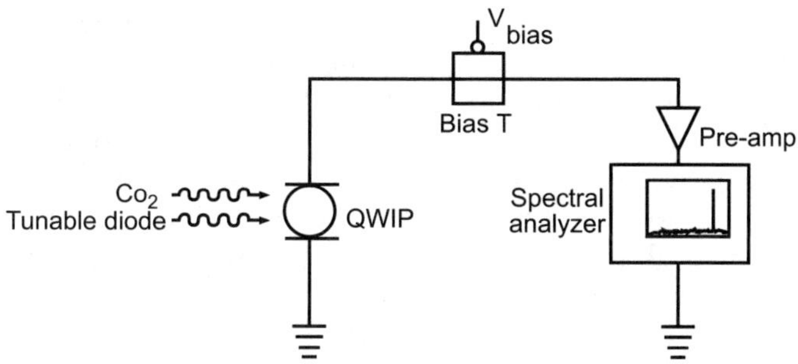

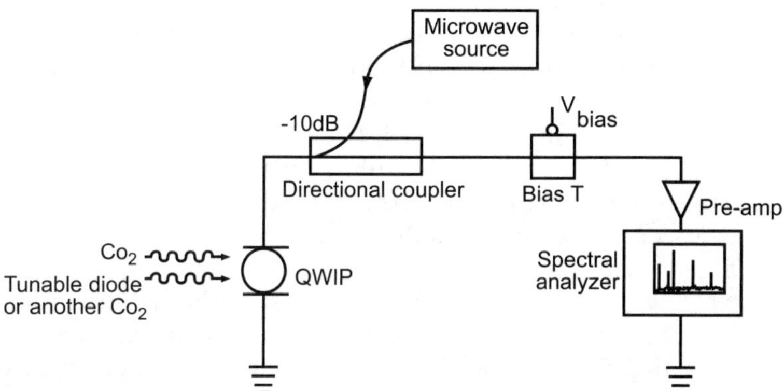

Figure 7.10 A schematic of heterodyne detection experiment: (*above*) direct measurement of the IF and (*below*) mixing on QWIP for extending the measurement frequencies.

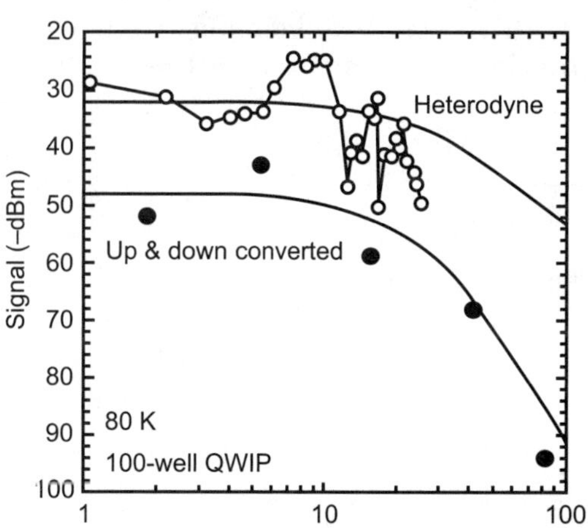

Figure 7.11 Direct infrared heterodyne and mixed heterodyne frequency with microwave frequency signal versus heterodyne frequency for a bias voltage of 2 V. The curves show the expected roll-off behaviors. The heterodyne frequency is defined as the difference between the two infrared frequencies. The incident powers from the two infrared sources and the microwave source are normalized to about 0.2 and 0.3 mW, respectively. The device temperature was about 80 K.

The point at 41.42 GHz was obtained by using two CO_2 lasers separated by f_{het} = 41.42 GHz and the microwave source at $f_{\mu wave}$ = 20 GHz, and measuring the difference signal at 41.42 – 20 = 21.42 GHz. The point at 82.16 GHz was obtained by using two CO_2 lasers separated by this frequency and a Gunn oscillator mechanically tuned in the range of 91 to 94 GHz. The difference signal in the range of 9 to 12 GHz was measured. For this data point, the millimeter-wave radiation was coupled into the device through free space.

To cross-check the consistency between results measured in frequency and time domains, both microwave rectification and time-resolved photocurrent measurements were made on one of our QWIPs. The Fourier transform of the time-resolved photocurrent is compared with the microwave rectification curve in Fig. 7.12. For this experiment, we used a QWIP very similar to the sample QWIP10 in Sec. 7.3, but with a doping concentration of 10^{12} cm^{-2}. The details of this sample are given in Ref. 27. The plot in Fig. 7.12 exhibits an excellent quantitative agreement between the two experimental methods. There was a defect in the packaging of this device, resulting in a drop in the frequency response at about 8 GHz.

Recently we have made substantial advances in both fabrication and packaging of high-frequency devices and in measurement techniques [43]. The air bridge and coplanar waveguide are monolithically integrated with QWIP, eliminating the wire bond. Direct measurements of heterodyne signal over coaxial cables have been performed up to 110 GHz and from cryogenic to room temperatures. Data measured on a 100-well QWIP with 10^{12} cm^{-2} well doping are shown in Fig. 7.13, representing a substantial improvement over the results in Fig. 7.11.

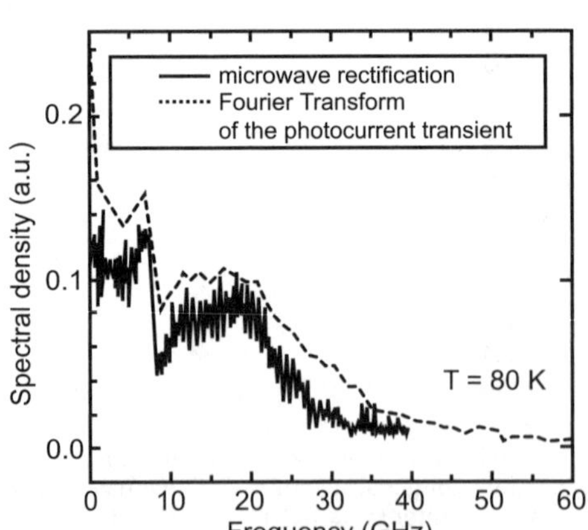

Figure 7.12 Comparison of the Fourier transform of the photocurrent transient and the frequency dependence obtained from the microwave rectification method.

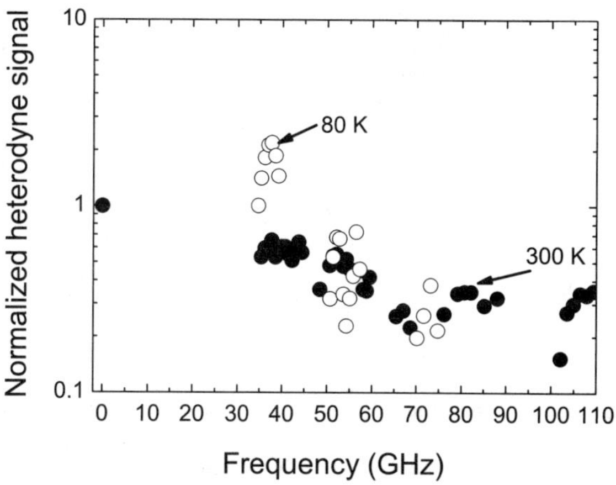

Figure 7.13 Normalized heterodyne signal taken at 80 K (circles) and 300 K (dots).

7.5 Two-Photon QWIPs

7.5.1 Equidistant three-level system for quadratic detection

While the high-speed detection methods discussed so far need to take into account the residual capacitance and inductance of the QWIP device and its packaging, much faster signals can be studied by exploiting the intrinsic nonlinearity. Since such nonlinear spectroscopy usually suffers from a lack of detection sensitivity, an artificial three-level system has been realized by energetically equidistant subbands [44, 45], as indicated in Fig. 7.14*a*. Subbands 1 and 2 are bound in the QW, whereas state 3 is a continuum resonance located close to the barrier energy. In an external electric field, the carriers excited into the continuum are swept out of the QW and give rise to a photocurrent. According to numerical simulations, a 7.6-nm-thick GaAs QW, sandwiched between $Al_{0.33}Ga_{0.67}As$ barriers, is optimized for a transition wavelength of 10.2 μm, whereas operation at 7.9 μm is achieved by using a 6.8-nm $In_{0.10}Ga_{0.90}As$ QW and $Al_{0.38}Ga_{0.62}As$ barriers.

Figure 7.14*b* summarizes the spectral characteristics of a two-photon QWIP comprising 20 GaAs QWs of 7.6-nm width, Si-doped to an electron concentration of $n_{2D} = 4 \times 10^{11}$ cm^{-2}, and $Al_{0.33}Ga_{0.67}As$ barriers (device 1). The spectral dependence of the optical transition from the first to the second subband (1 → 2) is obtained through intersubband absorption measurements in a Brewster-angle geometry at 77 K. The 2 → 3 transition was studied through photocurrent measurements at 130 K, where the thermal population of the second subband causes a signal. The spectral dependence of the two-photon photocurrent at

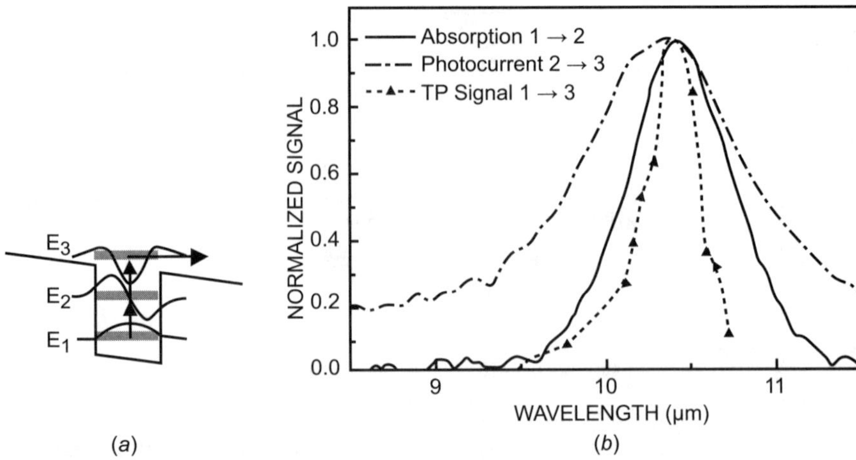

Figure 7.14 (*a*) Schematics of the two-photon QWIP. (*b*) Normalized absorption at 77 K in Brewster-angle geometry, photocurrent at 130 K due to excitation from thermally populated subband 2 into subband 3, and two-photon photocurrent at 77 K under excitation by a continuous-wave CO_2 laser versus excitation wavelength. (Source: Ref. 44.)

77 K has been measured with a wavelength-tunable CO_2 laser. Remarkably, all three experimental curves point to identical peak wavelengths (10.4 μm), indicating perfectly equidistant subband energies ($E_2 - E_1 = E_3 - E_2 = 119$ meV). The spectrum associated with CO_2 laser illumination is narrower than the absorption spectrum, which reflects the quadratic nature of the two-photon transition [46].

Figure 7.15 shows the photocurrent versus the power density P upon CO_2 laser illumination at 10.2 μm. The photocurrent exhibits a linear increase with P up to about 0.1 W/cm^2 at 70 K (up to 1 W/cm^2 at 90 K) followed by a quadratic increase. Closer investigation at different temperatures up to 160 K reveals thermally activated behavior of the

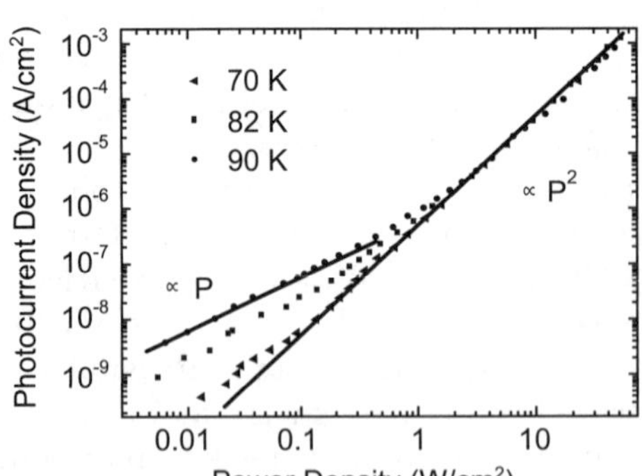

Figure 7.15 Photocurrent density under continuous-wave illumination at a wavelength of 10.3 μm and 1.5-V detector bias versus power density at different temperatures, as indicated. (Source: Ref. 44.)

photocurrent in the linear regime, with an activation energy of 104 meV. This result allows us to conclude that the linear regime is caused by thermal population of E_2. Quadratic behavior can be achieved even below 0.1 W/cm^2 when the device is operated at lower temperatures.

From the data of Fig. 7.15, the two-photon absorption coefficient β can be readily determined. In a two-photon detector with an absorbing region of thickness L, the associated quantum efficiency η_{2P} is given by $\eta_{2P} = \beta PLf_\theta$. Here the parameter f_θ relates to the selection rule for intersubband transitions. If θ is the angle between light propagation and the sample normal, we have $f_\theta = \sin^4\theta/\cos\theta$ for two-photon absorption, whereas the corresponding factor in the linear case is $\sin^2\theta/\cos\theta$. Using the usual definition of responsivity, the two-photon photocurrent density j_{2P} is thus expressed as

$$j_{2P} = \frac{eg}{h\nu}\beta L f_\theta P^2 \tag{7.24}$$

For a double-pass at $\theta = 45°$, the light traverses $2N = 40$ absorbing QWs with a total thickness of $L = NL_w \approx 0.3$ μm. The gain as determined from noise measurements amounts to $g = 0.3$ at 1.5 V. Equation 7.24 thus yields a two-photon absorption coefficient of 1.3×10^7 cm/GW. This nonlinearity is six orders of magnitude higher than that in typical bulk semiconductors (Si, ZnSe, GaAs), where values of $\beta \approx 10$ cm/GW are observed. These results also indicate a huge increase of β compared to the detuned two-photon QWIP [47], where β takes a value of about 5×10^3 cm/GW.

7.5.2 Autocorrelation of sub-picosecond optical pulses

We now demonstrate the use of this detection scheme for quadratic autocorrelation measurements of ultrashort optical pulses. With a Michelson interferometer, the pulses are split into two parts separated by a variable time delay. While quadratic *intensity* autocorrelation measurements have been studied in Ref. 45, we concentrate here on *interferometric* autocorrelation. The interference between the pulses contains additional information associated with the relative phase between the pulses. The optical pulse source used for the experiment is generated by the difference frequencies between the signal and idler pulses of an optical parametric oscillator pumped by a femtosecond Ti:sapphire laser [48]. The pulse source is wavelength-tunable from 6 to 18 μm, with 10-pJ pulse intensity, 160-fs pulse width, and 76-MHz repetition rate. Due to the high sensitivity of the present two-photon

QWIP, interferometric autocorrelation measurements can be performed in spite of the low optical power.

The upper part of Fig. 7.16 shows an autocorrelation trace measured at 77 K on a device with 6.8-nm $In_{0.10}Ga_{0.90}As$ quantum wells and $Al_{0.38}Ga_{0.62}As$ barriers. From these data, important information both on the temporal width of the optical pulses and on the dynamic properties of the two-photon QWIP can be obtained. While the interference fringes around zero time delay correspond to those of a nearly ideal autocorrelation of the optical field, two additional signatures are associated with intrinsic time constants of the detector. First, the amplitude of the interference fringes toward increasing positive and negative delay times shows an exponential rather than gaussianlike decay. This behavior is attributable to the decay time T_2 (dephasing time) of the coherent intersubband polarization between the first and second subbands. Second, after the disappearance of the fringes, the signal still decreases exponentially with delay time toward its asymptotic value. The associated time constant T_1 is due to intersubband relaxation of electrons from the second back to the first subband.

Using the model function of Nessler et al. [49], we have carried out a numerical least-squares fit of the measured autocorrelation (lower part of Fig. 7.16). The procedure is facilitated because the dephasing time associated with the continuum resonance 3 is negligibly small. The fit yields $T_1 = 0.46$ ps and $T_2 = 0.10$ ps, which agrees with typical values obtained from degenerate four-wave-mixing experiments [50]. The value of T_2 corresponds to a broadening with a full-width at half

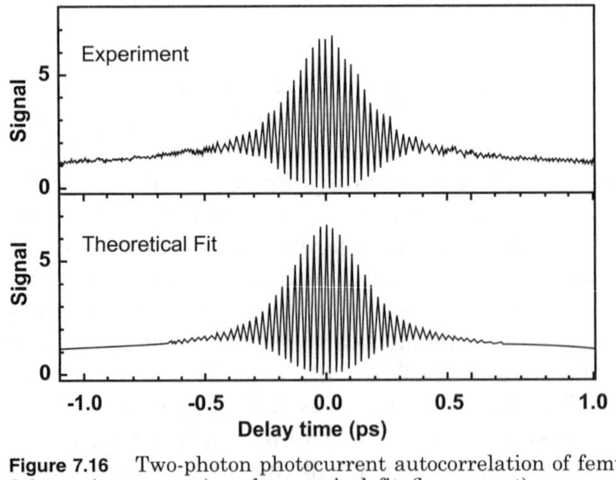

Figure 7.16 Two-photon photocurrent autocorrelation of femtosecond optical pulses at 8.0 μm (upper part) and numerical fit (lower part) versus delay time. The signal is normalized to its value at a large delay time above 10 ps. (Source: Ref. 44.)

maximum of $\Gamma_{12} = h/(\pi T_2) = 13$ meV. Since the absorption line width of this sample has a similar value (11 meV), we conclude that the $1 \rightarrow 2$ transition is homogeneously broadened. We note that similar time constants $T_1 = 0.53$ ps and $T_2 = 0.13$ ps have been obtained for the sample of Fig. 7.14*b* [45].

Theoretically, two different mechanisms give rise to photoemission involving two absorbed photons. One mechanism is a coherent two-photon transition which, like degenerate four-wave mixing, is associated with the third-order nonlinear susceptibility χ^3. Within second-order perturbation theory, it gives rise to the two-photon absorption coefficient

$$\beta_{\chi^3} = \left(\frac{e^2}{4\varepsilon_0 n_r mc}\right)^2 \frac{n_{2D}}{L_w h\nu} f_{12} f_{23} 2T_2^2 T_e \tag{7.25}$$

with the well width L_w, the refractive index n_r, and the dimensionless oscillator strengths f_{12} and f_{23} associated with the $1 \rightarrow 2$ and $2 \rightarrow 3$ transitions, respectively. Here T_e is the time constant associated with the broadening of state 3, which is attributed to the escape time. Equations 7.25 (together with Eq. 7.24) can be seen as a generalization of the two-photon quantum efficiency $\eta^{(2)}$ to the case of N quantum wells.

The other mechanism is a *sequential* two-step absorption

$$\beta_{2\text{step}} = \left(\frac{e^2}{4\varepsilon_0 n_r mc}\right)^2 \frac{n_{2\text{D}}}{L_w h\nu} f_{12} f_{23} 4T_1 T_2 T_e \tag{7.26}$$

where the population of subband 2 either is excited into 3 or decays with the time constant T_1.

While the two processes cannot be distinguished in a continuous-wave experiment, they have different signatures in the autocorrelation traces in Fig. 7.16. In fact, the χ_3 process manifests itself by interference fringes, limited with increasing time delay by the decay time T_2 of the coherent intersubband polarization. Sequential absorption produces the "slow" decrease in the autocorrelation trace, prominent at larger time delays when the coherence has been lost.

Interestingly, expressions 7.25 and 7.26 take the same value if the decoherence of the $1 \rightarrow 2$ transition is entirely due to relaxation (i.e., absence of pure dephasing [50]), and the relation $T_2 = 2T_1$ holds. In the present experiment, the coherence loss is mainly caused by pure dephasing, such that $\beta_{2\text{step}}$ is almost an order of magnitude larger than β^3_χ. Using an estimated value of $T_e = 50$ fs (consistent with the line width of the measured $2 \rightarrow 3$ absorption), Eq. 7.26 in fact yields a value of 1×10^7 cm/GW, which is in excellent agreement with the experimentally observed two-photon absorption coefficient. Even though $\beta_{\chi^3} \ll \beta_{2step}$, the χ^3 process is still dominant at short delay times, as Eqs. 7.25 and 7.26 refer to the time-integrated response.

7.6 Conclusion

In this chapter, we summarized the high-speed and nonlinear properties of QWIPs. We have shown that QWIPs can have high absorption, contrary to popular belief. Our study indicates that a 100-well QWIP with 1 to 1.5×10^{12} cm^{-2} well doping is close to the optimum. Decreasing the doping leads to a reduction in absorption, while increasing the doping seems to only result in a spectral broadening. A further advantageous point of high doping is that it shortens the carrier lifetime, but this also leads to a disadvantageous point of lower photoconductive gain and lower responsivity. We have also shown that these high-absorption QWIPs can operate at elevated temperatures, even near room temperature. Furthermore, we have shown by various experimental techniques that QWIPs have high intrinsic speeds. It is also possible to go even higher (> 30 GHz) by, e.g., using an even shorter lifetime system such as the *p*-type valence band quantum well. Since limited work on QWIPs for high-absorption and near-room-temperature operation has been carried out so far [51, 27, 28], further work will surely lead to better performance and will help to create new applications.

We have also shown that by properly designing a double-resonance quantum well, nonlinear absorption ($\beta = 1.3 \times 10^7$ cm/GW) is enhanced by more than three orders of magnitude higher than in previous, nonresonant devices. In this equidistant three-level system, both coherent and sequential two-photon absorptions are necessarily present, and they manifest themselves through different experimental signatures, which also allow us to determine the dephasing time T_2 and the intersubband relaxation time T_1. This approach enables quadratic detection at much lower radiation density, and it is thus expected to make a substantial impact on the characterization and development of novel infrared and terahertz radiation sources. Two-photon QWIPs based on resonant optical transitions between three energetically equidistant subbands allow us to study the autocorrelation of weak infrared sources with unprecedented sensitivity and femtosecond temporal resolution.

7.7 Acknowledgments

We thank many of our coworkers and colleagues at both the National Research Council of Canada and Fraunhofer-Institute for Applied Solid-State Physics in Freiburg, Germany.

References

1. H. C. Liu, "Quantum well infrared photodetector physics and novel devices," in *Intersubband Transition in Quantum Wells: Physics and Device Applications I*, H. C. Liu and F. Capasso, eds., Academic Press, San Diego, 2000, *Semiconductors and Semimetals*, vol. 62, chap. 3, pp. 126–196.
2. S. D. Gunapala and S. V. Bandara, "Quantum well infrared photodetector focal plane arrays," in *Intersubband Transition in Quantum Wells: Physics and Device Applications I*, H. C. Liu and F. Capasso, eds., Academic Press, San Diego, 2000, *Semiconductors and Semimetals*, vol. 62, chap. 4, pp. 197–282.
3. H. Schneider, T. Maier, J. Fleissner, M. Walther, P. Koidl, G. Weimann, W. Cabanski, M. Finck, P. Menger, W. Rode and J. Ziegler, *Electron. Lett.* **40**, 831 (2004).
4. H. Schneider, T. Maier, J. Fleissner, M. Walther, P. Koidl, G. Weimann, W. Cabanski, M. Finck, P. Menger, W. Rode, and J. Ziegler, *SPIE* **5726**, 35 (2005).
5. M. A. Fauci, R. Breiter, W. Cabanski, W. Fick, R. Koch, J. Ziegler, and S. D. Gunapala, *Infrared Phys. Technol.* **42,** 337 (2001).
6. W. A. Cabanski, R. Breiter, W. Rode, J. Ziegler, H. Schneider, M. Walther, and M. A. Fauci, *SPIE* **4721,** 165 (2002); in E.L. Dereniak and R.E. Sampson, eds., *Infrared Detectors and Focal Plane Arrays VII.*
7. H. C. Liu, G. E. Jenkins, E. R. Brown, K. A. McIntosh, K. B. Nichols, and M. J. Manfra, *IEEE Elect. Dev. Lett.* **16,** 253 (1995).
8. H. C. Liu, J. Li, E. R. Brown, K. A. McIntosh, K. B. Nichols, and M. J. Manfra, *Appl. Phys. Lett.* **67**, 1594 (1995).
9. S. Ehret, H. Schneider, J. Fleissner, P. Koidl, and G. Böhm, *Appl. Phys. Lett.* **71**, 641 (1997).
10. S. Steinkogler, H. Schneider, R. Rehm, M. Walther, and P. Koidl, *Appl. Phys. Lett.* **81**, 3401 (2002).
11. S. Steinkogler, H. Schneider, M. Walther, and P. Koidl, *Appl. Phys. Lett.* **82**, 3925 (2003).
12. A. Y. Cho, ed., *Molecular Beam Epitaxy,* American Institute of Physics, New York, 1994.
13. F. Capasso, *Science* **235,** 172 (1987).
14. G. Bastard, *Wave Mechanics Applied to Semiconductor Heterostructures*, Les Éditions de Physique, Les Ulis, France, 1988.
15. C. Weisbuch and B. Vinter, *Quantum Semiconductor Structures*, Academic Press, Boston, 1991.
16. A. Shik, *Quantum Wells*, World Scientific, Singapore, 1997.
17. P. Harrison, *Quantum Wells, Wires and Dots: Theoretical and Computational Physics*, Wiley, New York, 1999.

18. M. Helm, "The basic physics of intersubband transitions," in *Intersubband Transition in Quantum Wells: Physics and Device Applications I*, H. C. Liu and F. Capasso, eds., Academic Press, San Diego, 2000, *Semiconductors and Semimetals*, vol. 62, chap. 1, pp. 1–99.
19. B. F. Levine, *J. Appl. Phys.* **78**, R1 (1993).
20. K. K. Choi, *The Physics of Quantum Well Infrared Photodetectors*, World Scientific, Singapore, 1997.
21. D. D. Coon and R. P. G. Karunasiri, *Appl. Phys. Lett.* **45**, 649 (1984).
22. H. C. Liu, M. Buchanan, and Z. R. Wasilewski, *Appl. Phys. Lett.* **72**, 1682 (1998).
23. B. F. Levine, K. K. Choi, C. G. Bethea, J. Walker, and R. J. Malik, *Appl. Phys. Lett.* **50**, 1092 (1987).
24. L. C. West and S. J. Eglash, *Appl. Phys. Lett.* **46**, 1156 (1985).
25. A. G. Steele, H. C. Liu, M. Buchanan, and Z. R. Wasilewski, *Appl. Phys. Lett.* **59**, 3625 (1991).
26. H. C. Liu, *J. Appl. Phys.* **73**, 3062 (1993).
27. H. C. Liu, R. Dudek, A. Shen, E. Dupont, C.-Y. Song, Z. R. Wasilewski, and M. Buchanan, *Appl. Phys. Lett.* **79**, 4273 (2001).
28. T. Oogarah, H. C. Liu, E. Dupont, Z. R. Wasilewski, M. Byloos, and M. Buchanan, *Semicond. Sci. Technol.* **17**, L41 (2002).
29. H. C. Liu, R. Dudek, A. Shen, E. Dupont, C.-Y. Song, Z. R. Wasilewski, and M. Buchanan, *J. High Speed Electronics and Systems* **12,** 803 (2003).
30. E. R. Brown and K. A. McIntosh, U.S. Patent 5,304,805, Apr. 19, 1994.
31. R. Martini, R. Paiella, C. Gmachl, F. Capasso, E. A. Whittaker, H. C. Liu, H. Y. Hwang, D. L. Sivco, J. N. Baillargeon, and A. Y. Cho, *Electron. Lett.* **37**, 1290 (2001).
32. R. K. Richards, D. P. Hutchinson, C. A. Bennett, M. L. Simpson, H. C. Liu, and M. Buchanan, *ECS Proc.* **98-21**, 95 (1998); in S. S. Li, H. C. Liu, M. Z. Tidrow, and S. D. Gunapala, eds., *Long Wavelength Infrared Detectors and Arrays: Physics and Applications VI.*
33. R. Paiella, F. Capasso, C. Gmachl, D. L. Sivco, J. N. Baillargeon, A. L. Hutchinson, A. Y. Cho, and H. C. Liu, *Science* **290,** 1739 (2000).
34. R. Paiella, F. Capasso, C. Gmachl, H. Y. Hwang, D. L. Sivco, A. L. Hutchinson, A. Y. Cho, and H. C. Liu, *Appl. Phys. Lett.* **77**, 169 (2000).
35. T. Kondoh, Y. Miura, S. Lee, R. K. Richards, D. P. Hutchinson, and C. A. Bennett, *Rev. Sci. Instrum.* **74**, 1642 (2003).
36. R. K. Richards, D. P. Hutchinson, C. A. Bennett, T. Kondoh, Y. Miura, and S. Lee, *Rev. Sci. Instrum.* **74**, 1646 (2003).
37. H. C. Liu, J. Li, M. Buchanan, and Z. R. Wasilewski, *IEEE J. Quantum Electron.* **32**, 1024 (1996).
38. E. R. Brown, K. A. McIntosh, F. W. Smith, and M. J. Manfra, *Appl. Phys. Lett.* **62**, 1513 (1993)

39. E. R. Brown, K. A. McIntosh, K. B. Nichols, F. W. Smith, and M. J. Manfra, "CO_2-laser heterodyne detection with GaAs/AlGaAs MQW structures", in *Quantum Well Intersubband Transition Physics and Devices*, H. C. Liu, B. F. Levine, and J. Y. Andersson, eds., Kluwer, Dordrecht, Netherlands, 1994, pp. 207– 220.

40. R. H. Kingston, *Detection of Optical and Infrared Radiation,* Springer-Verlag, New York, 1978.

41. K. J. Siemsen and H. D. Riccius, *Appl. Phys.* **A35**, 177 (1984).

42. T. Koizumi, H. Tashiro, K. Nagasaka, and S. Namba, *Jpn. J. Appl. Phys.* **23**, L809 (1984).

43. P. D. Grant, R. Dudek, L. Wolfson, M. Buchanan, and H. C. Liu, *Electron. Lett.* **41**, 69 (2005).

44. H. Schneider, T. Maier, H. C. Liu, M. Walther, and P. Koidl, *Opt. Lett.* **30**, 287 (2005).

45. T. Maier, H. Schneider, M. Walther, P. Koidl, and H. C. Liu, *Appl. Phys. Lett.* **84**, 5162 (2004).

46. H. C. Liu, E. Dupont, and M. Ershov, *J. Nonlin. Opt. Phys. Mat.* **11**, 433 (2002).

47. A. Zavriyev, E. Dupont, P. B. Corkum, H. C. Liu, and Z. Biglov, *Opt. Lett.* **20**, 1886 (1995).

48. S. Ehret and H. Schneider, *Appl. Phys.* **B66**, 27 (1998).

49. W. Nessler , S. Ogawa, H. Nagano, H. Petek, J. Shimoyama, Y. Nakayama, and K. Kisho, *J. Electron. Spectrosc. Rel. Phenom.* **88**, 495 (1998).

50. R. A. Kaindl, K. Reimann, M. Woerner, T. Elsaesser, R. Hey, and K. Ploog, *Phys. Rev.* **B24**, 161308 (2001).

51. H. Schneider, C. Schönbein, G. Bihlmann, P. van Son, and H. Sigg, *Appl. Phys. Lett.* **70**, 1602 (1997).

Chapter

8

Intersubband Transitions in Quantum Dots

Pallab Bhattacharya, Xiaohua Su, and Subhananda Chakrabarti

Department of Electrical Engineering and Computer Science
University of Michigan
Ann Arbor, Michigan

Adrienne D. Stiff-Roberts

Department of Electrical and Computer Engineering
Duke University
Durham, North Carolina

Carl H. Fischer

M.I.T. Lincoln Laboratory
Lexington, Massachusetts

8.1 Introduction

Bulk semiconductor materials are optically sensitive to a large portion of the electromagnetic spectrum, with their bandgap energy E_g ranging from the ultraviolet to the *infrared* (ir). However, the low-bandgap materials appropriate for detection and emission in the long- and far-ir ranges (≥ 8 μm) are inherently soft and brittle, and it can be difficult to epitaxially grow and fabricate ir devices with bulk active regions based on these materials. In addition, intrinsic long/far-ir photodetectors have limited applicability. However, many more options for extending the ir detection and emission ranges of compound semiconductors exist if the advantages of an extrinsic scheme are realized. As a result, large-bandgap materials that are more amenable to various fabrication techniques become an option. An important extrinsic scheme is the reduced dimensionality of semiconductor active regions so that *intersubband* transitions are possible. Through epitaxial growth techniques, active region heterostructures with *three-dimensional* (3D) quantum confinement, called *quantum dots* (QDs), can be realized. In this chapter, the strained-layer epitaxy of a typical QD material system is described, the optoelectronic properties of intersubband transitions in QDs relevant to ir absorption and emission are discussed, and current research related to QD ir detectors and emitters is reviewed.

8.2 Molecular Beam Epitaxy of Self-Assembled Quantum Dots

The rapid improvement of epitaxial growth techniques, such as *molecular beam epitaxy* (MBE) and *metal-organic chemical vapor deposition* (MOCVD), has enabled the routine growth of excellent compound semiconductor materials [1]. In particular, the epitaxial growth of quantum-confined heterostructures has been pursued for decades due to enhanced performance of electronic and optoelectronic devices using such materials in the active region. While MBE is well suited for the growth of quantum wells, in which the active region experiences quantum confinement in the growth direction only, achieving the 3D quantum confinement required for device-quality QDs is much more challenging. Currently, the most successful approach to achieve defect-free, multiple-layer, high-density QD ensembles is the heteroepitaxy of coherently strained, 3D, self-assembled islands [2–9], a process referred to as the *Stranski-Krastanow* (S-K) *growth mode* [10, 11].

8.2.1 Stranski-Krastanow growth mode

Strained-layer epitaxy is controlled by the strain energy between the substrate and the epitaxially grown overlayer, or epilayer. Typically, the epilayer is biaxially strained in the plane of the substrate and uniaxially strained in the perpendicular direction. For a thick substrate, the biaxial strain is determined from the bulk lattice constants of the substrate material a_s and the epilayer material a_l [12]:

$$\varepsilon \| = \frac{a_s}{a_l} - 1 \tag{8.1}$$

When strain is incorporated into the epitaxial crystal coherently, the lattice constant of the epilayer parallel to the interface is forced to be equal to the lattice constant of the substrate. Consequently, the lattice constant of the epilayer perpendicular to the substrate is changed by the Poisson effect [12]

$$\varepsilon \perp = -\frac{\varepsilon \|}{\sigma} \tag{8.2}$$

where σ is the Poisson ratio. Therefore, compressive strain ($a_l > a_s$) reduces the parallel lattice constant of the epilayer and increases the perpendicular lattice constant, while tensile strain ($a_l < a_s$) increases the parallel lattice constant and reduces the perpendicular lattice constant.

The strained-layer epitaxy of the S-K growth mode is limited by both the strain energy and the dislocation formation energy. In other words, for a lattice mismatch in the range from 1.8% to 10%, it is possible to grow a pseudomorphic, 2D epilayer, or wetting layer, that is lattice-matched to the substrate for very small thicknesses. Beyond some critical thickness, the strain energy in the system is relaxed by the formation of misfit dislocations at the epilayer/substrate interface. However, as long as the epilayer is below the critical thickness, the strain is absorbed coherently and minimal dislocations are produced by the formation of coherently strained 3D islands on the epilayer surface. Thermodynamic considerations have demonstrated that the minimum Helmholtz free-energy surface configuration in a coherently strained system is not atomically flat, but has a 3D form [13]. Thus, self-assembled QDs result from the minimization of the Helmholtz free energy during strained-layer epitaxy such that thermodynamic equilibrium is achieved in the heterostructure. Note that thermodynamic equilibrium includes the effect of microscopic, kinetic processes of adatoms on the crystal surface, such as diffusion. Thermodynamic equilibrium is achieved when the total free energy is minimized, given a fixed amount of deposited material. The driving forces for achieving equilibrium in these self-assembled QDs are elastic relaxation at the facet edges, strain-induced renormalization of the surface energy at the facets, and interaction between neighboring islands via the substrate.

8.2.2 Standard MBE growth technique for InAs/GaAs QDs

The lattice mismatch at the interface of an InAs epilayer and a GaAs substrate is

$$f = \frac{a_{\text{InAs}} - a_{\text{GaAs}}}{a_{\text{GaAs}}} \approx 7\% \tag{8.3}$$

where the lattice constant in InAs is 6.06 Å and the lattice constant in GaAs is 5.65 Å. Therefore, the S-K growth mode for the self-assembly of 3D islands will occur under appropriate growth conditions. The typical InAs growth rate is 0.1 ML/s, and the GaAs substrate temperature is usually cooled to 500°C. This slow rate, coupled with the low substrate temperature, enables the self-assembly of QDs since surface kinetics are minimized and thermodynamic equilibrium by the coherent incorporation of strain can be achieved more easily. The wetting layer thickness in the InAs/GaAs QD system is 1.7 *monolayers* (ML), where 1 ML ~ 2.83 Å for III-V materials. The transition from the 2D wetting layer to 3D islands is observed by using in situ *reflection high-energy electron diffraction* (RHEED), in which a streaky pattern gives way to a spotty pattern when QDs begin to form. Once this transition is observed, InAs overgrowth occurs for 2 to 5 s to provide enough InAs charge (2.2 ML) for pyramidal QDs, and a 30-s growth-interrupt pause is used to allow the complete formation of QDs. After this pause, an intrinsic GaAs cap layer (~25 to 60 nm) is grown on top of the InAs QDs, thereby completing the QD potential barrier. This sequence of growth can be repeated to achieve a specified number of QD layers.

For a given volume of a coherently strained island, there exists an equilibrium shape. For growth along the [100] direction in the InAs/GaAs QD system, the self-assembled QDs can be lens- or pyramid-shaped with a base in the (100) plane [14]. For typical growth parameters used in MBE or MOCVD, an ensemble of InAs/GaAs QDs has lateral sizes ranging from 15 to 25 nm and heights ranging from 5 to 8 nm. The surface density of InAs/GaAs QDs ranges from 10^9 to 10^{11} cm^{-2}. An AFM image of an ensemble of InAs/GaAs QDs grown by MBE at 500°C at a rate of 0.1 ML/s is shown in Fig. 8.1. From this image, the dot density is estimated to be 1×10^{11} cm^{-2}.

While the S-K growth mode is well established, having been reported first in 1938, there is still much to learn about the controlled growth of self-assembled QDs, particularly those with device-quality electrical and optical properties. In particular, since the S-K growth mode is a random process, QD ensembles suffer from large inhomogeneous line width broadening, which adversely affects photonic device properties. In addition, the nonuniform size distribution of the QD ensemble makes it difficult to control the intersubband spacing for ir transitions. The S-K growth mode also introduces challenges due to the buildup of strain for repeated dot layers. Thus, the self-assembly of QDs is a complicated process, subject to many parameters for controlling epitaxial growth.

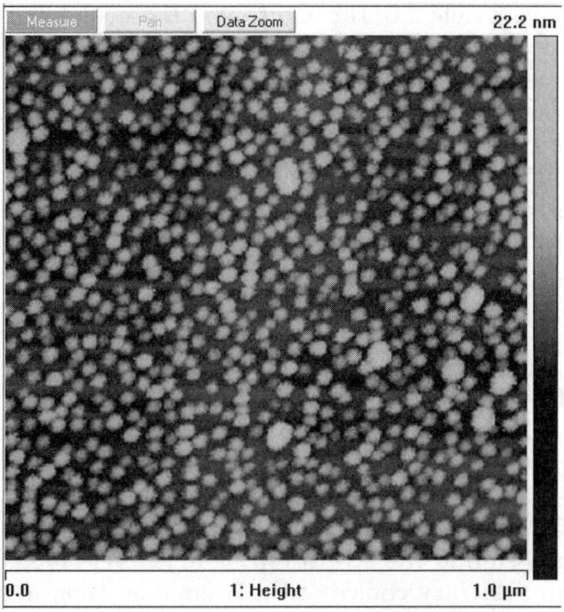

Figure 8.1 AFM image of 2.2-ML InAs quantum dots on GaAs grown at 500°C (1-µm scale).

8.3 Quantum Dot Intersubband Transitions

In this section, the optoelectronic properties of intersubband transitions in QDs are discussed, including bound-state energy levels, density of states, intersubband transition rates, and the phonon bottleneck. The determination of these quantities is complicated since the strain field at the heterojunction heavily influences the potential barrier and the resulting electronic band structure. However, it is important to understand these quantities as they are responsible for important differences in the performance of intersubband photonic devices featuring quantum well and quantum dot active regions.

8.3.1 Optoelectronic properties

Bound-state energy levels. The eight-band $\mathbf{k}\cdot\mathbf{p}$ theory is used to determine the electronic band structure in semiconductor heterostructures near the band edge only. For devices that operate using carrier phenomena near the band edge, the $\mathbf{k}\cdot\mathbf{p}$ perturbation theory not only is simple, but also yields very accurate results. An important advantage of this approach is the inclusion of the influence of remote bands. This is particularly important for strained systems such as QD heterostructures since the conduction band and valence band cannot be decoupled in the presence of large strain fields. In the eight-band **k**·**p** model, only the nearest eight bands with sufficiently small energy differences from the given state are assumed to contribute to the solution [12].

In applying the **k**·**p** theory to self-assembled QDs, it is necessary to take into account the unique strain distribution of these coherently strained, quasi-zero-dimensional heterostructures. As has been demonstrated experimentally, the large built-in strain of the QDs causes a significant change in the properties of the electronic band structure. For example, the bandgap of InAs/GaAs QDs has been measured to be ~1.05 eV, even though the intrinsic bandgap of bulk InAs is 0.4 eV [14]. Thus the effect of strain is incorporated into the energy band structure of a semiconductor by (1) determining the elastic strain energy of the system and (2) including this energy in the hamiltonian as an additional perturbation through deformation potential theory. Jiang and Singh [14] have used the *valence force field* (VFF) model [15, 16] to determine the strain distribution in self-assembled InAs/GaAs QDs for the eight-band **k**·**p** calculation of the electronic band structure. To find the strain tensor in self-assembled QDs, they begin with an arbitrary choice of the atomic position and minimize the system energy, using the hamiltonian for the elastic strain energy. This determines the appropriate configuration of the atoms for calculating the strain distribution, yielding a very detailed strain tensor for QDs and the surrounding barrier material. They then use the finite-difference method to solve the matrix equation for the perturbed, **k**·**p** hamiltonian.

By using the eight-band **k**·**p** model developed by Jiang and Singh, the first four bound states in the conduction band of pyramidal InAs/GaAs QDs with dot height of 6.8 nm and base width of 13.5 nm have been calculated. The spatial dependence of the electron wave functions for the four double-degenerate energy levels is shown in Fig. 8.2*a*. The listed energy levels are assumed to be from the InAs valence band edge. In Fig. 8.2*b*, the QD intersubband energy levels in the conduction band are shown schematically, assuming that the bandgap in strained InAs is 1.05 eV. According to this calculation, bound-to-bound transitions from the QD ground state to the top excited state correspond to 9.25 μm. Thus, the electron intersubband energies can vary greatly (from ~50 to 500 meV) depending on the QD size, composition, and heterostructure band offsets.

Density of states. Another unique property of QDs is the zero-dimensional density of states, which leads to favorable carrier dynamics in photonic devices. In the case of QDs, **k**-space is zero-dimensional and can be represented by a point. As a result, the density of states in **k**-space is infinite for each state of constant energy. Thus, unlike bulk

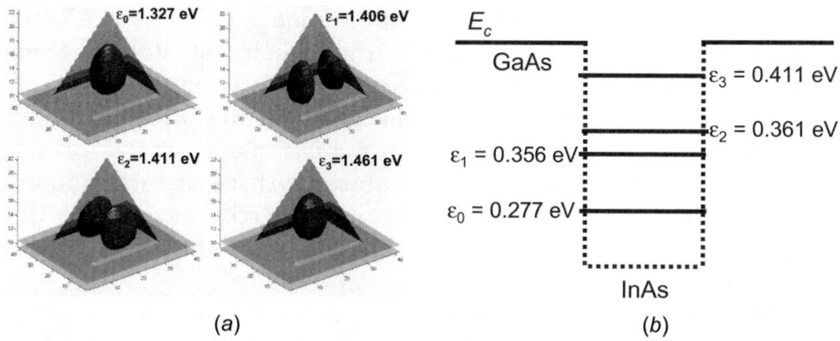

Figure 8.2 Calculated energy levels using eight-band $\mathbf{k}\cdot\mathbf{p}$ model for InAs/GaAs QD with height of 6.8 nm and base width of 13.5 nm. (*a*) Spatial variation of the electron wave function for the first four bound-state energy levels (with respect to the InAs valence band edge); (*b*) schematics of InAs QD intersubband energy levels in the conduction band assuming an InAs bandgap of 1.05 eV in strained QDs.

material or quantum-confined structures in one and two dimensions, the QD *does not* have a continuous density of states for any direction. However, this infinite density of states is for an ideal QD. In reality, the size of QDs varies spatially such that the density of states is actually represented by inhomogeneously broadened δ functions in a gaussian distribution. For instance, the joint density of states can be written as

$$N(\hbar\omega) = \frac{1}{\sqrt{2\pi}\sigma} \exp \frac{(E_{fi} - \hbar\omega)^2}{2\sigma^2} \tag{8.4}$$

where E_{fi} is the energy separation between states f and i and σ is the gaussian line width of the transition. The QD electronic density of states and corresponding electron distribution are shown schematically in Fig. 8.3*a* and *b*, respectively.

Despite such inhomogeneous broadening, the discrete nature of the density of states leads to significant results in terms of the carrier

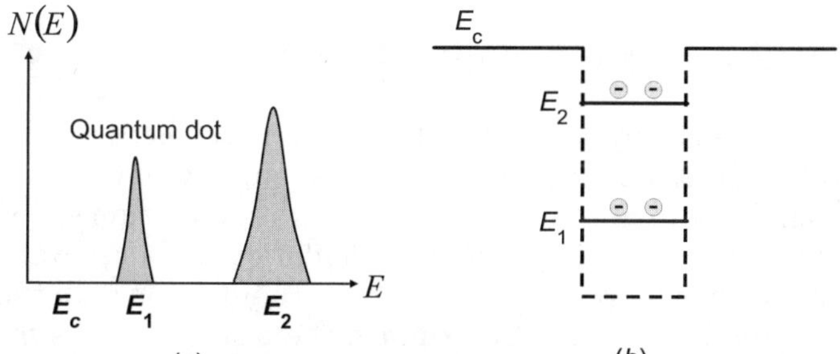

Figure 8.3 Schematic diagrams of (*a*) the density of states in QDs and (*b*) the corresponding carrier distribution in QDs.

dynamics in QDs. It is clear that as the dimensionality of the active region is reduced, the energy levels and density of states become quantized. In addition, the electron distribution is maximized near the conduction band edge, and it is independent of temperature to a large extent. However, at room temperature, high-energy states of QDs become occupied, where the density of states is three-dimensional and continuous. Thus, as the temperature increases, the advantages that result from quantum confinement become less pronounced.

Fermi golden rule. The Fermi golden rule, derived by using first-order, time-dependent perturbation theory, determines the transition rate of electrons from some initial state $|i\rangle$ to some final state $|f\rangle$. In the case of electron interactions with an electromagnetic field, this transition occurs due to the absorption or emission of a photon. The *Fermi golden rule* for these processes is

$$W(i) = \frac{2\pi}{\hbar} \sum_f |\langle f \mid H_1 \mid i \rangle|^2 \delta(E_f - E_i \mp \hbar\omega) \tag{8.5}$$

where the minus sign indicates absorption and the plus sign indicates emission. Here H_1 represents the first-order perturbation to the hamiltonian in the time-dependent Schrödinger equation

$$H_1 = \frac{ie\hbar}{mc} \mathbf{A} \cdot \nabla = \frac{-e}{mc} \mathbf{A} \cdot \hat{\mathbf{p}} \tag{8.6}$$

where e is the electron charge, c is the velocity of light, and $\mathbf{A}$ is the vector potential whose direction is equal to that of the electric field. Thus, the momentum matrix element between the initial and final states $\mathbf{p}_{fi} \equiv \langle f \mid \mathbf{p} \mid i \rangle$ must be evaluated in the direction of the vector potential $\mathbf{a}$ or, equivalently, the polarization of the electric field. The momentum matrix element for intersubband transitions is calculated by using the electron wave functions determined by the eight-band $\mathbf{k}{\cdot}\mathbf{p}$ model in pyramidal QDs

$$\mathbf{p}_{fi} = \langle f \mid \mathbf{p} \mid i \rangle = \sum_{j,k} \langle \varphi_{fj} \mid \varphi_{ik} \rangle \langle u_j \mid \mathbf{p} \mid u_k \rangle + \sum_j \langle \varphi_{fj} \mid \mathbf{p} \mid \varphi_{ij} \rangle \tag{8.7}$$

where φ and u are the envelope and central-cell functions, respectively. Due to considerable band mixing in QDs, the first summation term is nonzero, as is often the case in intraband transitions in quantum wells, and therefore cannot be excluded from the calculation. An important aspect of the momentum matrix element is that, in conjunction with the polarization of the electric field, it determines the selection rules for permitted transitions in which photons are absorbed (or emitted). In contrast to quantum wells, quantum dots are sensitive to both normal and lateral incident radiation, which is an important advantage for photonic devices.

Once the rate of photon absorption W_{abs} is known, it is relatively straightforward to determine the absorption coefficient α by considering the continuity equation for the photon density n_{ph} under steady-state conditions:

$$\alpha = \frac{W_{abs}}{cn_{ph}} \tag{8.8}$$

In the case of a QD ensemble, the absorption coefficient for photons with energy $\hbar\omega$ is

$$\alpha(\hbar\omega) = \frac{\pi e^2 \hbar}{\varepsilon_0 n_0 c m_0^2 V_{av}} \sum_{fi} \frac{1}{\hbar\omega} |\mathbf{a}\cdot\mathbf{p}_{fi}|^2 N(\hbar\omega) \tag{8.9}$$

where ε_0 is the permittivity of free space, n_0 is the index of refraction, V_{av} is the average dot volume, and $N(\hbar\omega)$ is the intersubband joint density of states for QDs. Given the gaussian distribution of this density of states, the absorption coefficient can be calculated as a function of the photon energy or, equivalently, the incident wavelength. Therefore, typical absorption spectra have a peak absorption coefficient at the wavelength corresponding to the average intersubband transition and a line width that depends on the uniformity of the QD ensemble.

Phonon bottleneck. The carrier dynamics in 3D, quantum-confined heterostructures differ significantly from those of bulk materials and quantum wells. The most significant scattering processes for electrons occupying the excited energy levels in QDs have been identified as [17] electron-phonon scattering due to a single-phonon process, electron-multiphonon scattering, electron-hole scattering, and electron-electron scattering. In addition to these carrier relaxation processes, transport mechanisms play integral roles in determining: (1) the carrier relaxation time from a QD excited state to the ground state, (2) the carrier capture time from the continuum to a QD bound state, and (3) the effective carrier lifetime, which describes the time elapsed before a photoexcited carrier falls into the ground state of any QD.

Note that while first-order, electron-phonon interactions cause rapid relaxation in bulk and quantum well semiconductor materials, electron-phonon coupling is suppressed in QDs due to the magnitude of the intraband energy spacing between confined levels in QDs [18, 19]. More specifically, the intraband energy spacing is > 50 meV for practical QD sizes, while the *longitudinal optical* (LO) phonon energy is < 40 meV [20]. Thus conservation of energy cannot be satisfied with the emission of a single LO phonon. Similarly, single *longitudinal-acoustic* (LA) phonon emission (achieved by coupling with the deformation potential) is reduced for intraband energy spacings greater than a few millielectronvolts [21]. This reduction of electro-phonon scattering for QDs, which leads to longer electron lifetimes in QDs (and increased photoconductive gain in detectors), is known as the *phonon bottleneck*. Several experiments have been conducted to confirm the existence of the phonon bottleneck, including electroluminescence and time-resolved photoluminescence measurements of InGaAs/GaAs QDs [22], high-frequency electrical impedance measurements in $In_{0.4}Ga_{0.6}As$/GaAs QD lasers [23], and time-resolved *differential transmission spectroscopy* (DTS) measurements of $In_{0.4}Ga_{0.6}As$/GaAs QDs [24].

Thus, the suppression of phonon scattering in QDs suggests that electron relaxation lifetimes are significantly longer than those in *quantum wells* (QWs) (~1 to 10 ps) [25], which are dominated by scattering due to the emission of single optical phonons. The phonon bottleneck also affects the effective carrier lifetime in QDs, resulting in unique carrier dynamics that are beneficial to the operation of both intersubband detectors and emitters. For detectors, the long lifetime of the carriers in the higher-energy dot states ensures that photoexcited carriers will contribute to the photocurrent. In intersubband light emitters, a long effective lifetime in the upper states implies that the intersubband stimulated emission time can also be large, and therefore the threshold current can be low. In addition, the phonon bottleneck is also promising for temperature-independent performance since the electrons are energetically decoupled from the optical phonon, whose properties depend very heavily on temperature.

Note that the carrier configuration in QDs, i.e., geminate versus nongeminate, significantly influences the effect of the phonon bottleneck. For a geminate QD configuration in which electrons and holes are present simultaneously, electron-hole scattering increases due to the strong overlap between the two carriers. However, for a nongeminate QD configuration in which only electrons or only holes are present, electron-hole scattering is reduced. As a result, the phonon bottleneck contributes to increased gain in QD intersubband detectors since the carrier density is low and nongeminate configurations dominate. In the case of bipolar QD intersubband emitters, the carrier density is much higher, leading to geminate configurations and increased electron-hole scattering that can dominate the effect of the phonon bottleneck.

8.3.2 Intersubband absorption

Intersubband absorption in QDs spans a broad spectrum of ir wavelengths, including the *near-ir* (NIR) from 1 to 3 μm, the *mid-ir* (MIR) from 3 to 5 μm, the *long-ir* (LIR) from 8 to 14 μm, and the *far-ir* (FIR) for wavelengths greater than 14 μm. The first observation of intersubband absorption in QDs [26] reported absorption in the 13- to 15-μm spectral range. Since then, intersubband absorption has been reported throughout the ir spectrum by varying the QD heterostructure through epitaxial growth conditions, QD material composition, and/or modification of the QD barrier layer [27–34]. In general, there are two types of intersubband transitions that can originate in the QD. The first type of transition, in which ground state electrons are photoexcited to excited states of the QD, is called a *bound-to-bound* (B-B) transition. The second type of transition is called *bound-to-continuum* (B-C), and in this case an electron in a ground or excited state is photoexcited out of the QD into the continuum of energy levels near the barrier bulk material band edge. For either case, the intersubband absorption process requires the presence of electrons in a bound state of the QD. Electrons can be introduced in the QD by direct doping [35], modulation doping [36], or capture of injected electrons from the device contact [37]. It has been shown that *quantum dot infrared photodetector* (QDIP) performance has a distinct tradeoff between, and dependence on, QD density and dopant concentration [38]. For this reason, it is customary to design QDIPs with doping such that there are approximately two electrons per QD. The dark current increases for higher dopant densities, and the absorption efficiency decreases for lower dopant densities.

The absorption coefficient is an important quantity for determining the applicability of a given QD active region to the detection of ir light in that it determines, in part, both the quantum efficiency of the device and the spectral response. The QD absorption spectrum is generally broad due to the size variation of QDs comprising the active region of a device. Therefore, QD ir photodetectors experience a tradeoff between absorption strength and absorption line width. For example, a perfectly uniform QD ensemble will have large peak absorption and a narrow absorption line width since all the dots experience peak absorption at the same wavelength. However, a nonuniform QD ensemble will have reduced peak absorption and a broad absorption line width since many dots with varying sizes experience peak absorption at different wavelengths. Thus, even though each QD individually has sharp atomlike electronic states, for an ensemble of QDs the inhomogeneous gaussian line width of the density of states σ (measured in millielectronvolts) significantly influences the absorption. Specifically, the peak absorption coefficient α (measured in cm^{-1}) is inversely proportional to the line width [39]

$$\alpha \approx \frac{3.5 \times 10^5}{\sigma} \tag{8.10}$$

8.3.3 Intersubband emission

The greatest challenge to controlling intersubband transitions in QDs for ir emitters is the achievement of population inversion between ground and excited states. For population inversion to occur, the energy relaxation time between the upper-level excited states and the ground state τ_{u1} should be as long as possible, and the lifetime of electrons in the ground states τ_1 should be very short. Providing a high density of coherent photons in the cavity, which can greatly reduce the interband electron-hole recombination time τ_{stim}, can decrease the lifetime of the electron in the ground state. By using a bipolar scheme, a high density of coherent photons can be made available in the intersubband laser cavity by simultaneous interband lasing due to current injection.

To examine population inversion between the ground state and the excited dot states, the following rate equations are solved self-consistently:

$$\begin{aligned}
\frac{dn_u}{dt} &= \eta_{in}\frac{J}{e} - \frac{n_u g_u(1-f_1)}{\tau_{u1}} + \frac{n_1 g_1(1-f_u)}{\tau_{1u}} \\
&\quad - \frac{n_u g_u f_u f_h}{\tau_u} + \frac{n_{phu} n_u g_u (1-f_u-f_h)}{\tau_u} = 0 \\
\frac{dn_1}{dt} &= (1-\eta_{in})\frac{J}{e} + \frac{n_u g_u(1-f_1)}{\tau_{u1}} - \frac{n_1 g_1(1-f_u)}{\tau_{1u}} \\
&\quad - \frac{n_1 g_1 f_1 f_h}{\tau_1} + \frac{n_{ph1} n_1 g_1 (1-f_1-f_h)}{\tau_1} = 0
\end{aligned} \tag{8.11}$$

where η_{in} is the injection efficiency; n_{ph1} and n_{phu} are the densities of photons that are involved in the lasing process from the ground states and the excited states, respectively; n_1 and n_u are the densities of electrons in the ground states and excited states, respectively; and f_1 and f_u are the occupational probabilities of the ground states and excited states, respectively. The degeneracy of the ground state ($g_1 = 2$) and the excited states ($g_u = 4$) [14] is taken into account in the rate equation calculations. Interband recombination times for the ground state τ_1 (= 700 ps) and the excited state τ_u (= 250 ps) have been derived from time-resolved photoluminescence measurements [40], and an intersubband relaxation time of τ_{u1} = 60 ps is assumed. A thermal distribution of holes is used ($f_h = 0.45$). When n_{ph1} is zero, $f_1 > f_u$ for all values of the injection current, and no population inversion is possible. However, when n_{ph1} is increased to 50, f_1 is pinned at a value of

~0.5, whereas f_u increases linearly with the current. Similarly, when n_{phu} is increased to 50, $f_1 > f_u$ for all values of injection. This is consistent with the fact that if $n_{phu} = 50$, interband lasing occurs from the excited state in the dot and no MIR emission results. The intersubband population inversion and lasing processes, together with the bipolar recombination, are illustrated in Fig. 8.4.

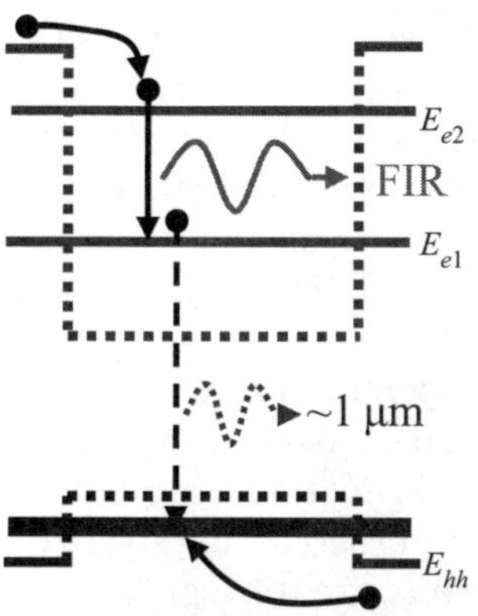

Figure 8.4 Schematic illustration of electronic bound states, carrier relaxation processes, and intersubband population inversion in QDs.

8.4 Quantum Dot Intersubband Photonic Devices

Quasi-zero-dimensional QD heterostructures with intersubband transitions in the ir are important for numerous military, commercial, and scientific photonic device applications. The ir detectors and the resultant imaging capability are useful for military targeting and tracking, law enforcement, medical diagnoses, and space science, while ir emitters can be used for optical spectroscopy, point-to-point atmospheric communication, fiber-optic telecommunication, and optical radar. In particular, intersubband photonic devices comprising QDs offer several favorable attributes due to their three-dimensional quantum confinement. Quantum dot infrared photodetectors are inherently sensitive to normal-incidence light, can achieve high-temperature operation (> 150 K), and have low dark currents. Quantum dot infrared emitters benefit from advantages such as high-temperature operation, low threshold current, and high modulation bandwidth.

8.4.1 Quantum dot infrared detectors

QD active region heterostructures result in several advantages that should enhance ir detection in QDIPs [30, 35, 41–50]. The 3D quantum confinement of electrons leads to low dark currents in QDIPs. The bound-state, atomlike energy levels that depend on QD size and shape provide tunability of the ir absorption wavelengths and of the spectral responsivity. The carrier distribution that results from the zero-dimensional density of states in QDs is beneficial in that it reduces the temperature dependence of the QDIP characteristics. The phonon bottleneck in QD carrier relaxation leads to increased gain, quantum efficiency, and responsivity. In addition, QDIPs are inherently sensitive to normal-incidence ir radiation due to the polarization selection rules. Finally, since electrons and holes are localized due to quantum confinement, in general photonic devices with QD active regions exhibit better defect and radiation tolerance. Both the increased electron lifetime and the reduced dark current indicate that QDIPs should be able to provide high-temperature operation ($\geq$ 150 K).

Several QDIP heterostructure designs have been investigated for use as ir photodetectors in addition to the standard InAs/GaAs QDIP. As an example, InAs QDs embedded in a strain-relieving InGaAs quantum well are known as *dot-in-a-well* (DWELL) heterostructures. QDIPs using a DWELL heterostructure not only permit greater control over wavelength tunability, but also have demonstrated excellent device performance [51–54]. QDIPs using an undoped active region of InAs quantum dots grown directly on AlGaAs with a GaAs cap layer have also demonstrated promising device performance [37]. Other III-V material systems under investigation include InAs/InAlAs QDs grown on InP [55] and InGaAs/InGaP QDs grown on GaAs [56–58]. Intersubband transitions in the ir have also been observed in the valence band of Ge/Si QDs [59–64]. Three QDIP designs demonstrating state-of-the-art performance are discussed in greater detail in the following sections.

The 70-layer InAs/GaAs quantum dot infrared photodetector. Some 70 layers of InAs/GaAs QDs were grown by MBE in an attempt to increase the absorption, and thereby the quantum efficiency, in QDIPs [65]. Including 70 QD layers in the InAs/GaAs QDIP increased the amount of absorbed ir light and contributed to better detector performance. In addition, 50-nm GaAs barriers required for reducing dislocation propagation in the large QD stack contribute to extremely low dark currents in the 70-layer QDIP. Schematic diagrams of the device heterostructure and conduction band profile under bias are shown in Fig. 8.5*a* and *b*, respectively.

Figure 8.6*a* shows the dark current densities for $T = 140$, 160, and 200 K (measured at the Jet Propulsion Laboratory), demonstrating extremely low dark currents at elevated temperatures ($J_{\text{dark}} = 2.3 \times 10^{-6}$ A/cm^2, $V_{\text{bias}} = -2.0$ V, $T = 160$ K). The spectral responsivity, as

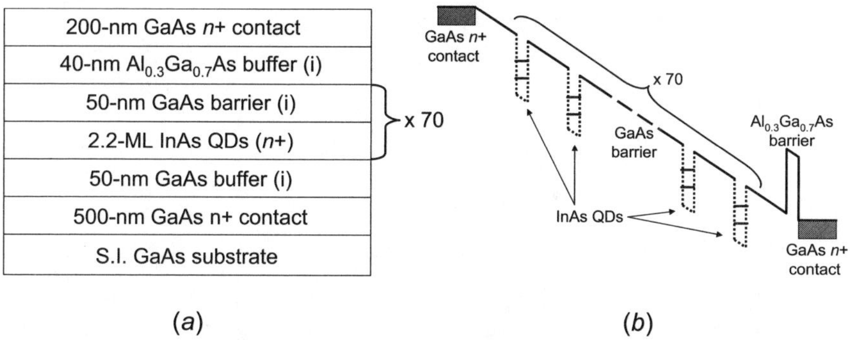

Figure 8.5 Schematic diagrams of (*a*) the MBE device heterostructure and (*b*) the conduction band profile under bias in the 70-layer InAs/GaAs QDIP.

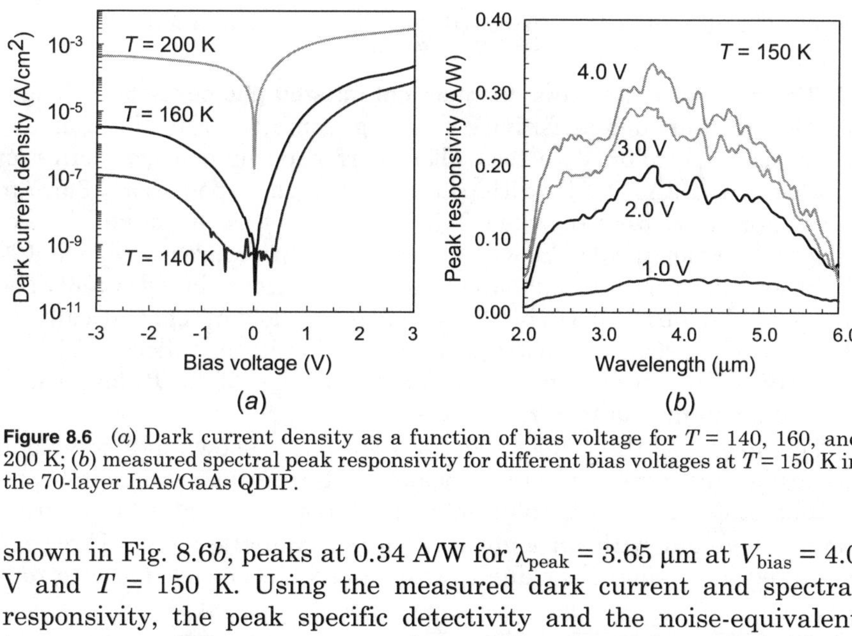

Figure 8.6 (*a*) Dark current density as a function of bias voltage for $T = 140$, 160, and 200 K; (*b*) measured spectral peak responsivity for different bias voltages at $T = 150$ K in the 70-layer InAs/GaAs QDIP.

shown in Fig. 8.6*b*, peaks at 0.34 A/W for $\lambda_{peak} = 3.65$ μm at $V_{bias} = 4.0$ V and $T = 150$ K. Using the measured dark current and spectral responsivity, the peak specific detectivity and the noise-equivalent temperature difference were estimated for the 70-layer InAs/GaAs QDIP. The peak detectivity, shown in Fig. 8.7 as a function of temperature, is $\sim 10^{11}$ cm·Hz$^{1/2}$/W at $T = 100$ K and $\sim 6 \times 10^9$ cm·Hz$^{1/2}$/W at 200 K. These expected detectivities indicate that the 70-layer QDIP will provide high-temperature performance, currently unattainable in ir photodetectors, due to the simultaneous reduction of dark current by the GaAs barrier and increase of responsivity by the 70-layer absorption region. Therefore, by carefully designing the 70-layer InAs/GaAs QDIP heterostructure to minimize dark current and maximize absorption, this device provided low dark current and relatively high responsivity at elevated operating temperatures.

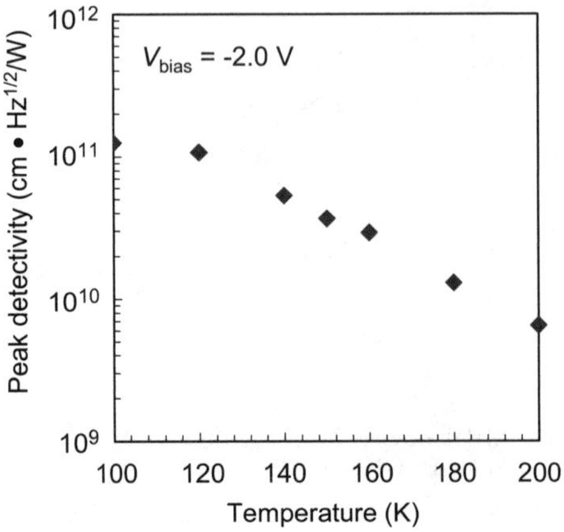

Figure 8.7 Estimated peak specific detectivity in the 70-layer InAs/GaAs QDIP.

Multicolor $In_{0.4}Ga_{0.6}As$/GaAs quantum dot infrared photodetector. High-resolution and high-sensitivity imaging using ir focal plane arrays requires multicolor detectors. The performance characteristics of $In_{0.4}Ga_{0.6}As$/GaAs QDIPs with peak wavelength responses at ~3.5, 7.5, and 22 μm have been reported. This was the first report of multiwavelength detection in QDIPS, with operating temperatures as high as 120 k, based solely on intersubband transitions within the QDs [66]. The $In_{0.4}Ga_{0.6}As$/GaAs QDIP heterostructure is shown schematically in Fig. 8.8. The use of n^+ substrates minimized multiple reflections in the substrate. The active region of the device consists of 20 QD layers separated by 50-nm GaAs barrier layers.

The measured three-color spectral response is shown in Fig. 8.9 and its inset. The origin of the three peaks, centered at 3.5, 7.5, and 22 μm, can be understood from the electronic states of the InGaAs quantum dots calculated with an eight-band **k·p** formulation [12]. The peak centered at 22 μm arises due to transitions of photoexcited electrons

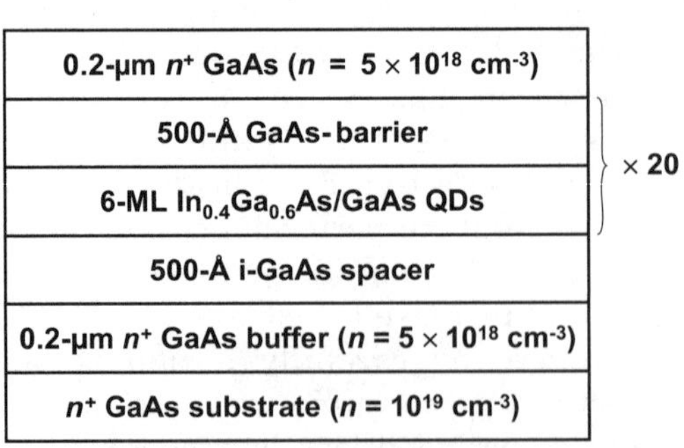

Figure 8.8 Schematic of the 20-layer $In_{0.4}Ga_{0.6}As$/GaAs QDIP heterostructure grown by MBE.

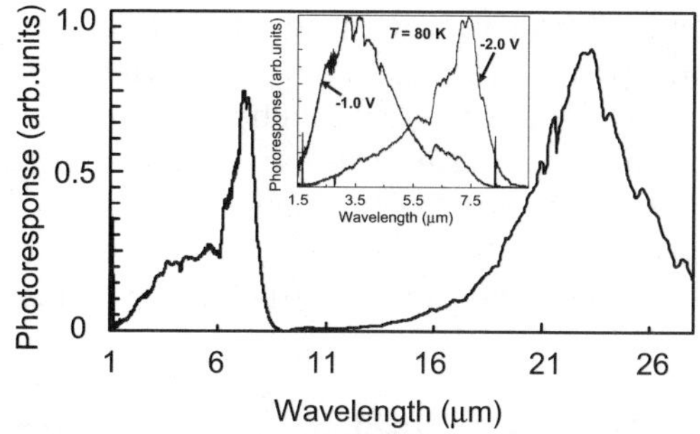

Figure 8.9 Spectral response of the 20-layer InGaAs/GaAs QDIP at $T = 20$ K. The inset shows the normalized two-color response at 80 K.

from the ground states to the first excited states ($\Delta E \sim 50$ to 60 meV) while the one at 7.5 μm (155 meV) is believed to be due to transitions from the dot ground states to the GaAs barrier states or the wetting layer states. The broad response with peak at 3.5 μm is probably due to bound-to-continuum transitions from the dot ground states to dot levels in the continuum. The peaks at 7.5 and 22 μm, resulting from transitions between quasi-bound states, are narrower ($\Delta\lambda/\lambda \sim 0.2$) than the one at 3.5 μm resulting from bound-to-continuum transitions ($\Delta\lambda/\lambda \sim 0.79$). The 22-μm transition is observed distinctly at very low temperatures (~20 K) and disappears with an increase of temperature as the excited states become more populated and thereby the rate of transitions into these states decreases.

The peak responsivity of the 20 dot-layer device for the 7.5-μm photoresponse, plotted in Fig. 8.10*a* as a function of bias for $T = 80$ K, is adequately high for application in focal plane arrays. The specific detectivity D^* was obtained from the measured peak responsivity and

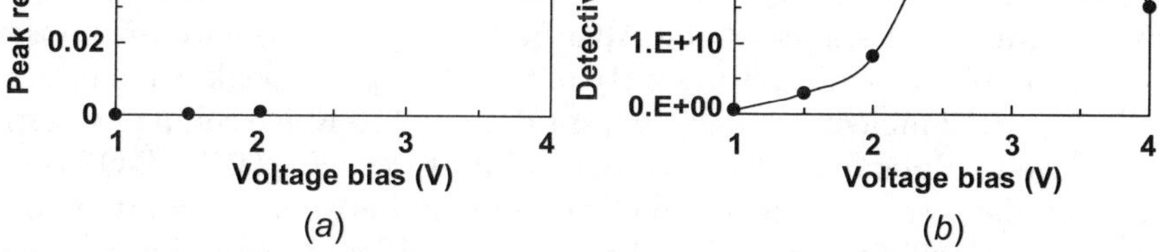

Figure 8.10 Measured (*a*) peak responsivity and (*b*) peak detectivity as a function of applied bias for the 7.5-μm response at $T = 80$ K in multicolor $In_{0.4}Ga_{0.6}As$/GaAs QDIPs.

noise density spectra at different temperatures and applied biases. These measured values of D^* are plotted in Fig. 8.10*b* as a function of bias for the 7.5-μm response at 80 K. Detectivity D^* reaches a maximum of 4.8×10^{10} cm·Hz$^{1/2}$/W at 3 V and decreases thereafter, due to the monotonic increase of the dark current with bias. These values of D^* are among the highest measured for QDIPs in the LWIR range and are attributed to the extremely low dark currents measured in these devices.

Resonant tunneling $In_{0.4}Ga_{0.6}As$/GaAs quantum dot infrared photodetector. In the commonly investigated vertical QDIPs, with or without confined dot layers, the dark current is determined by thermionic emission and field-assisted tunneling. The dark current can be effectively reduced by increasing the dot confinement potential, by increasing the thickness of the active region, or by incorporating additional energy barriers. Unfortunately, these measures also reduce the photocurrent, since the transport paths of the carriers are identical. It is therefore essential to explore device designs by which the dark current can be decoupled from the thermal background which enhances thermionic emission at high operating temperatures, but the device photocurrent is not affected.

The properties of a novel QDIP design in which a resonant tunneling heterostructure is incorporated with each QD layer have been reported recently [67–69]. The heterostructure and conduction band diagram of a single period of the active region are schematically shown in Fig. 8.11*a* and *b* respectively. The resonant tunneling double barrier is designed so that the electron tunneling probability is unity at the energy of the final state of the photoexcited electrons. At energies sufficiently removed from this resonance, the tunneling probability is small; thus the transport of the carriers contributing to the dark current, which have a broad energy distribution at high temperatures, will be reduced. A single $Al_{0.1}Ga_{0.9}As$ barrier is also included on the side of the dots opposite to the tunnel barriers. The inclusion of this layer creates a quantum well and quasi-bound final states for the photoexcited electrons from the QDs. These states are designed to resonate with the tunnel states in the double-barrier heterostructure. Furthermore, the energy position of the states in the well can be tuned by varying the distance of the AlGaAs barrier from the quantum dot layer, thereby providing tunability of the absorption peak wavelength.

In Fig. 8.12 measured dark current data at 140 K are compared with calculated values for both conventional and *tunnel QDIPs* (T-QDIPs). It is evident that there is good agreement between calculated and measured dark currents, and there is a significant reduction in dark currents with the incorporation of the resonant tunnel filters. The

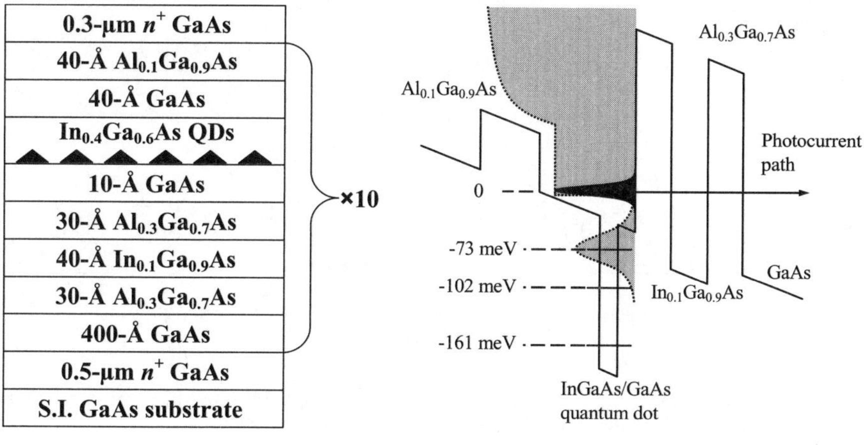

Figure 8.11 (*a*) Schematic heterostructure and (*b*) conduction band profile of an InAs/GaAs T-QDIP with a GaAs/$Al_{0.3}Ga_{0.7}As$ resonant tunneling heterostructure under an applied bias. The calculated energy levels in the well and dot are indicated.

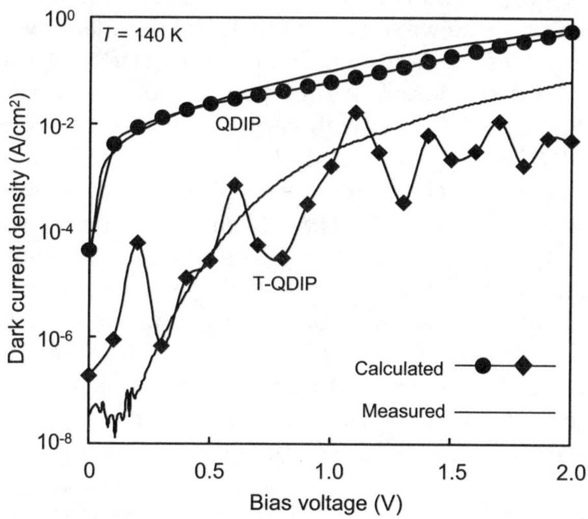

Figure 8.12 Comparison of calculated and measured dark current densities as a function of bias voltage at $T = 140$ K for conventional and tunneling QDIPs using In0.4Ga0.6As/GaAs quantum dots.

disagreement at large bias values is due to nonequilibrium conditions which were neglected in our calculation.

Figure 8.13*a* and *b* depicts the mid-ir spectral response of the T-QDIP for different bias values and different ambient temperatures (up to 300 K), respectively. The peak wavelength is in excellent agreement

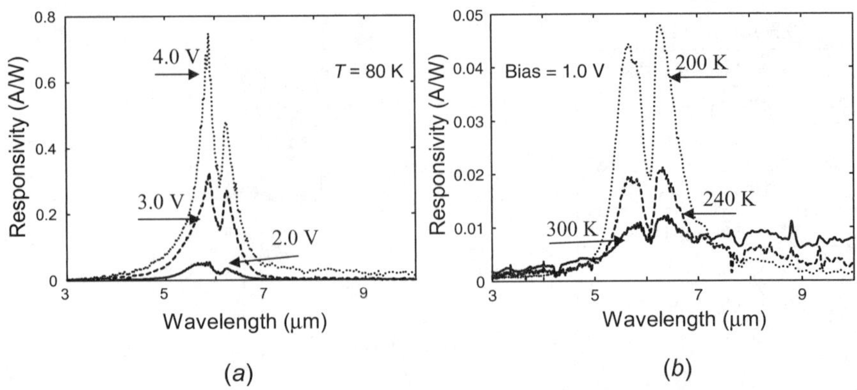

Figure 8.13 Measured spectral responsivity of the T-QDIP: (*a*) in the bias range 2–4 V at 80 K and (*b*) in the temperature range 200–300 K under 1-V bias.

with the designed and calculated transition energy of 161 meV for the photoexcited electrons from the ground state in the QD to the quasi-bound state in the well. The spectral response exhibits closely spaced dual peaks with an energy separation of ~15 meV. The dual peaks arise from coupling and splitting of the wave functions of the quantum well states and the bound states of the double-barrier heterostructure and provide evidence of resonant tunneling in the operation of the device. From the data of Fig. 8.13*a* and *b*, the peak responsivity is 0.75 A/W (4-V bias at 80 K) and ~0.01 A/W (1-V bias at 300 K).

Long-ir response at 17 μm is observed in the T-QDIPs at higher temperatures, as shown in Fig. 8.14. This 17-μm response peak intensity increases with temperature, and at 300 K its peak responsivity is higher than that of the 6-μm peak. With reference to the calculated energies of the bound and quasi-bound states in the dot and well, shown in Fig. 8.11*b*, this peak results from transitions from the second excited state of the dot to the well state (ΔE = 73 meV). The transition line width of ~26 meV is close to the inhomogenenous broadening of the QD states at 300 K. We believe that the long-ir transition is dominant at high temperatures because the probability of occupation of the dot excited states increases with temperature. Due to the symmetry of the dot geometry, the excited states have a higher degeneracy than the ground state. Additionally, for the same incident power, there are more photons at 17 μm than at 6 μm. It is extremely important to note that it is the low dark current in the T-QDIP that makes detection possible at these high temperatures.

The measured values of D^* for mid-ir peaks at T = 80 K reach a maximum value of 2.4×10^{10}cm $\cdot$Hz$^{1/2}$/W at 2 V and decrease again due to the monotonic increase of the dark current with bias. The values of

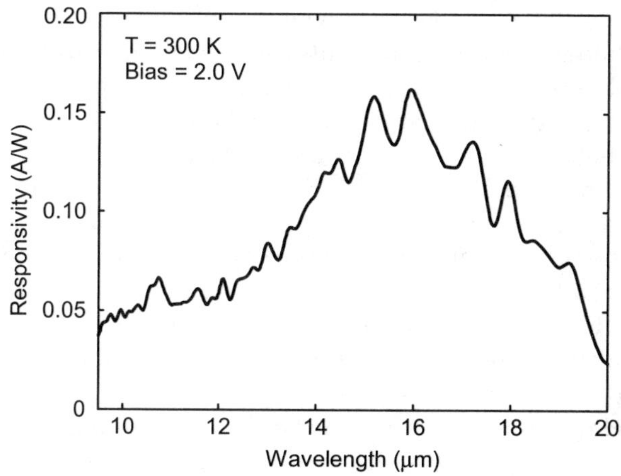

Figure 8.14 Measured spectral responsivity of the T-QDIP in the long wavelength range at 300 K under a bias of 2 V.

D^* for the 17-μm response at a bias of 1 V are in the 10^7 cm·Hz$^{1/2}$/W range and are reasonable at these high temperatures.

8.4.2 Quantum dot infrared emitters

The use of self-assembled QDs in the active region of intersubband emitters is expected to improve ir sources in two fundamental ways: (1) Three-dimensional electron confinement enables surface-normal light emission due to polarization selection rules, and (2) the inhomogeneously broadened, deltalike density of states in QDs introduces strict energy conservation rules that significantly reduce phonon- or carrier-mediated scattering. These scattering processes are a significant limiting factor in the efficiency of modern long-wavelength sources, especially for efficient CW laser operation at room temperature. Yet, favorable relaxation times in the excited states of QDs invoke the possibility of intersubband lasing. This was first suggested by Singh [70], who proposed the use of an external interband laser to rapidly depopulate the ground state electrons by stimulated emission, thus creating a favorable nonequilibrium carrier distribution between the ground and excited states for MIR emission. Kastalsky et al. have theoretically analyzed a similar dual color laser by using a three-level carrier rate equation [71]. Also, Krishna et al. have demonstrated intersubband stimulated emission in interband QD lasers [72].

Bipolar quantum dot intersubband emitters. Spontaneous emission at 12 μm, resulting from intersubband transitions in $In_{0.4}Ga_{0.6}As/GaAs$ QDs [72], has been reported. The output from the surface-emitting device was measured at 80 and 300 K. The device was essentially a four-dot layer interband laser heterostructure. The interband lasing occurring in the device at 1 μm due to ground state electron-hole recombination helped to deplete the electron ground state in the dots and to establish a population inversion between the excited states and the ground state. The far-ir signal was enhanced after the interband transition reached threshold at 300 K. To determine the possibility of stimulated emission and lasing with this scheme, the condition for intersubband population inversion was determined from the carrier and photon rate equations. In addition, the intersubband gain was calculated as a function of the inhomogeneous line width broadening of the transition, demonstrating that gains as high as 150 cm^{-1} can be achieved.

For observing stimulated emission, plasmon-enhanced waveguides were designed and grown by MBE. Edge-emitting devices were fabricated using standard photolithography, lift-off, and a combination of dry and wet etching, followed by ohmic contact formation. The width of the waveguide was varied from 20 to 60 μm, and the length was varied from 800 to 1200 μm. Therefore, the devices are multimode laterally. The output spectrum is broad and multimode, with a more distinct peak at 13 μm which, in all probability, originates from the electron transitions from higher QD excited levels to the ground state, separated by ~90 to 100 meV. Theoretical calculations have confirmed the presence of these excited states [20]. The threshold in the FIR output occurs at 1.6 times the interband laser threshold. The additional carriers injected after the interband laser reaches threshold recombine to provide for high coherent photon density required for intersubband gain. In essence, the device converts the more readily available near-ir photons to the more difficult to obtain mid-ir photons. Unfortunately, device heating prevents measurements at higher injection currents. However, these devices demonstrate intersubband gain and stimulated emission. The polarization dependence of the far-ir emission, measured by using a far-ir polarizer, is found to be strongly TE polarized, as shown in Fig. 8.15.

Unipolar quantum dot cascade lasers. While quantum cascade devices have hitherto been designed with quantum well active regions, it is of interest to investigate QD active regions for the two reasons just mentioned: (1) polarization selection rules allow surface-normal emitting devices, and (2) reduced phonon coupling in self-assembled QDs, compared to bulk material or quantum wells, promises high-temperature operation. Note that the bipolar device described in the previous section, while adequate to demonstrate intersubband light emission, is not suitable for achieving significant stimulated emission and lasing. However, a unipolar cascade design, in which the ground state of the

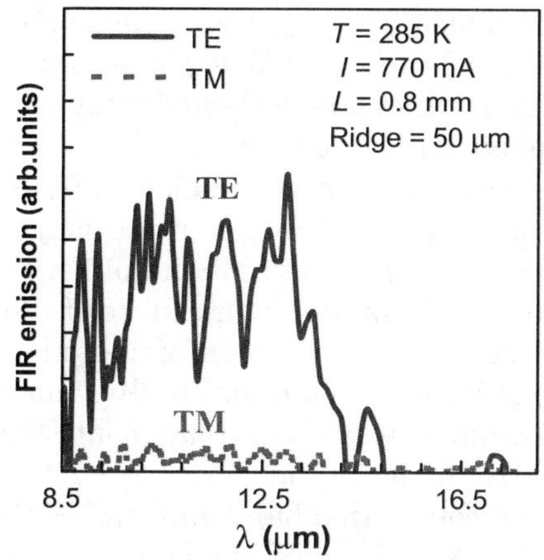

Figure 8.15 Far-ir (FIR) emission from the bipolar QD emitter, showing the intersubband transition to be strongly TE polarized.

QD is depopulated by tunneling to the excited state of the next dot (as shown in Fig. 8.16*a*) should provide stimulated emission [73]. This approach allows multiple photons to be emitted for each injected electron, and it eliminates the need for interband transitions. To facilitate tunneling between ground and excited states of adjacent QD layers, a *chirped superlattice* (CSL) structure must be designed to create a narrow miniband connecting these states. It has been established that injected electrons preferentially occupy the excited dot states and wetting layer states at high temperatures due to the higher density of states in these regions, suggesting that population inversion can be readily obtained. Measured and calculated carrier lifetimes in dot excited states are in the range of tens to hundreds of picoseconds,

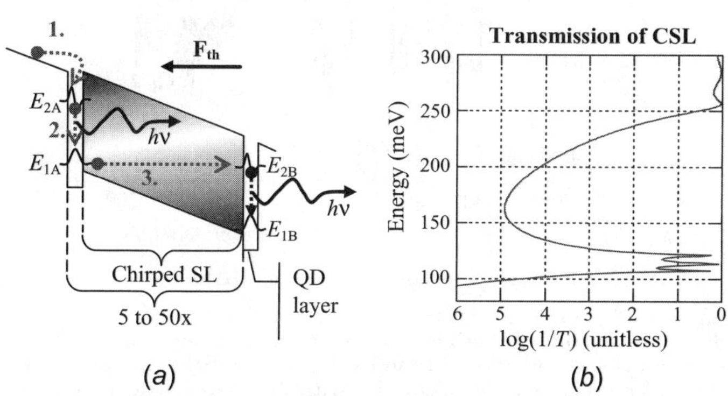

Figure 8.16 (*a*) Conduction band schematic of a QD cascade emitter. Dashed lines show the steps an electron follows through the device. (*b*) Transmission characteristics of a 4-period $GaAs/GaAs_{0.96}N_{0.04}$ chirped superlattice. The miniband is centered at 114 meV and the minigap extends to roughly 225 meV.

depending on excitation and bias conditions [74]. In contrast, the upper-state lifetime in quantum well cascade lasers is a few picoseconds, determined by phonon scattering. It is therefore expected that the threshold current will be low in unipolar QD cascade lasers.

If $In_{0.4}Ga_{0.6}As/GaAs$ QDs are to be used for the active region, GaAsN is an excellent choice for quantum wells in the CSL for three major reasons. First, GaAsN has a large negative conduction band offset to GaAs, thus allowing the electron to tunnel from the ground state of the QD. Second, advances in the growth of low-nitrogen-composition GaAsN have led to its use as laser-quality material [75]. Finally, since GaAsN is in tensile strain on a GaAs substrate, it can compensate for the compressively strained InGaAs QDs and permit more periods in the cascade structure. Using the measured conduction band shift in MBE-grown $GaAs_{0.96}N_{0.04}$, the design for a $GaAs/GaAs_{0.96}N_{0.04}$ CSL with over 40 dB of contrast between ground and excited state transmission is shown in Fig. 8.16*b*.

Figure 8.17*a* shows light-current (*L-I*) measurements made on edge-emitting devices at 90 K. Since the injector region is designed to operate under a bias of ~60 mV per period (660 mV over the entire device structure), the "turn-on" slightly above this bias, as illustrated by the dashed lines, is expected. The polarization-resolved electroluminescent spectra for the same device at 18 K, as measured by FTIR, are shown in Fig. 8.17*b* along with the predicted quantum dot energy transitions

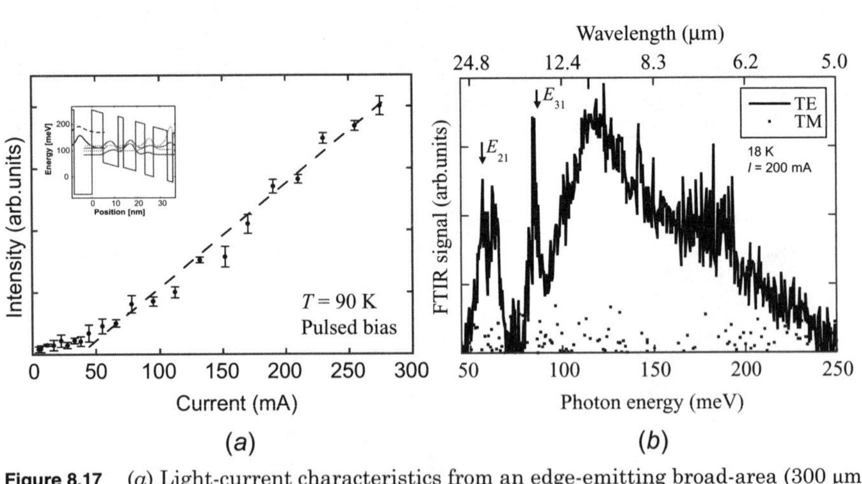

Figure 8.17 (*a*) Light-current characteristics from an edge-emitting broad-area (300 μm × 2 mm) QD cascade device at 90 K. A distinct injector "turn-on" is evident, illustrated by the dashed lines. The inset shows the solution to the coupled Schrödinger–Poisson equations for the chirped superlattice at 60 mV/period with the quantum dot states from from an 8-band **k·p** model superimposed on left. (*b*) Polarization-dependent spectral measurements for the broad-area device at 18 K showing dominant TE polarization. Theoretically predicted energies for first- and second-excited to ground-state transitions are indicated (as E_{21} and E_{31}) for comparison.

described above. Background spectra were collected with the device unbiased and subtracted from the biased data. All TE polarized outputs show the same three emission peaks, while none of the TM polarized outputs exhibit these features. Two peaks (57 and 90 meV) agree well with first and second excited to ground state transitions, while a third peak (115 meV) arises, most likely, from wetting layer to ground state transitions. The dominant TE polarization agrees with theory [74]. Unlike in quantum wells, the oscillator strength in quantum dots arises from the central-cell part of the wave function. Strong biaxial strain in dots leads to band intermixing, which results in a larger momentum matrix element for TE polarized light. Total integrated output power for all three peaks is estimated to be 90 nW.

8.5 Conclusions

Intersubband transitions in QDs for ir detectors and emitters are positioned to become an important technology for the type of high-temperature, low-cost, high-yield photonic devices required for highly sophisticated sensing and imaging applications. The 3D confinement of QD heterostructures provides important advantages in ir optoelectronic properties compared to quantum wells, such as surface-normal detection and emission, extended carrier lifetimes due to the phonon bottleneck, and temperature invariance due to the deltalike density of states. Significant challenges exist in reaching the full potential of such devices, due in large part to the nonuniformity of self-assembled QDs. However, continued advances in epitaxial growth and device simulation should enable the desired device improvements.

References

1. E. H. C. Parker, in *The Technology and Physics of Molecular Beam Epitaxy*. New York: Plenum Press, 1985.
2. L. Goldstein, F. Glas, J. Y. Marzin, M. N. Charasse, and G. LeRoux, "Growth by molecular beam epitaxy and characterization of InAs/GaAs strained-layer superlattices," *Applied Physics Letters*, vol. 47, pp. 1099–101, 1985.
3. S. Fujita, Y. Matsuda, and A. Sasaki, "Blue luminescence of a ZnSe-$ZnS_{0.1}Se_{0.9}$ strained-layer superlattice on a GaAs substrate grown by low-pressure organometallic vapor phase epitaxy," *Applied Physics Letters*, vol. 47, pp. 955–957, 1985.
4. P. R. Berger, K. Chang, P. K. Bhattacharya, and J. Singh, "A study of strain-related effects in the molecular-beam epitaxy growth of $In_xGa_{1-x}As$ on GaAs using reflection high-energy electron diffraction," *Journal of Vacuum Science and Technology B*, vol. 5, pp. 1162–1166, 1987.

5. M. Tabuchi, S. Noda, and A. Sasaki, “Strain energy and critical thickness of heteroepitaxial InGaAs layers on GaAs substrate,” *Journal of Crystal Growth*, vol. 115, pp. 169–173, 1991.
6. D. Leonard, M. Krishnamurthy, C. M. Reaves, S. P. Denbaars, and P. M. Petroff, “Direct formation of quantum-sized dots from uniform coherent islands of InGaAs on GaAs surfaces,” *Applied Physics Letters*, vol. 63, pp. 3203–3205, 1993.
7. Q. Xie, P. Chen, A. Kalburge, T. R. Ramachandran, A. Nayfonov, A. Konkar, and A. Madhukar, “Realization of optically active strained InAs island quantum boxes on GaAs (100) via molecular beam epitaxy and the role of island induced strain fields,” *Journal of Crystal Growth*, vol. 150, pp. 357–363, 1995.
8. D. Bimberg, M. Grundmann, and N. N. Ledentsov, “Growth, spectroscopy, and laser application of self-ordered III-V quantum dots,” *Materials Research Society Bulletin*, vol. 23, pp. 31–34, 1998.
9. P. Bhattacharya, K. Kamath, J. Phillips, and D. Klotzkin, “Self-organized growth of In(Ga)As/GaAs quantum dots and their opto-electronic device applications,” *Bulletin of Material Science*, vol. 22, pp. 519–529, 1999.
10. I. N. Stranski and L. Krastanow, “Theory of orientation separation of ionic crystals,” *Sitzungsberichte*, vol. Abteilung IIb. 146, pp. 797–810, 1938.
11. D. Bimberg, M. Grundmann, and N. N. Ledentsov, *Quantum Dot Heterostructures*. Chichester: John Wiley & Sons, 1999.
12. J. Singh, *Physics of Semiconductors and Their Heterostructures*. New York: McGraw-Hill, 1993.
13. P. R. Berger, K. Chang, P. Bhattacharya, J. Singh, and K. K. Bajaj, “Role of strain and growth conditions on the growth front profile of $In_xGa_{1-x}As$ on GaAs during the pseudomorphic growth regime,” *Applied Physics Letters*, vol. 53, pp. 684–686, 1988.
14. H. Jiang and J. Singh, “Strain distribution and electronic spectra of InAs/GaAs self-assembled dots: an eight-band study,” *Physical Review B*, vol. 56, pp. 4696–4701, 1997.
15. P. N. Keating, “Effect of invariance requirements on the elastic strain energy of crystals with application to the diamond structure,” *Physical Review*, vol. 145, pp. 637–645, 1966.
16. R. M. Martin, “Elastic properties of ZnS structure semiconductors,” *Physical Review B*, vol. 1, pp. 4005–4011, 1970.
17. H. Jiang and J. Singh, “Self-assembled semiconductor structures: electronic and optoelectronic properties,” *IEEE Journal of Quantum Electronics*, vol. 34, pp. 1188–1196, 1998.
18. U. Bockelmann and G. Bastard, “Phonon scattering and energy relaxation in two-, one-, and zero-dimensional electron gases,” *Physical Review B*, vol. 42, pp. 8947–8951, 1990.

19. H. Benisty, C. M. Sotomayor-Torres, and C. Weisbuch, "Intrinsic mechanism for the poor luminescence properties of quantum-box systems," *Physical Review B*, vol. 44, pp. 10,945–10,948, 1991.

20. P. Bhattacharya, K. K. Kamath, J. Singh, D. Klotzkin, J. Phillips, H. Jiang, N. Chervela, T. B. Norris, T. Sosnowski, J. Laskar, and M. R. Murty, "In(Ga)As/GaAs self-organized quantum dot lasers: DC and small-signal modulation properties," *IEEE Transactions on Electron Devices*, vol. 46, pp. 871–883, 1999.

21. B. Ohnesorge, M. Albrecht, J. Oshinowo, and A. Forchel, "Rapid carrier relaxation in self-assembled $In_xGa_{1-x}As$/GaAs quantum dots," *Physical Review B*, vol. 54, pp. 11,532– 11,538, 1996.

22. K. Mukai, N. Ohtsuka, H. Shoji, and M. Sugawara, "Phonon bottleneck in self-formed $In_xGa_{1-x}As$/GaAs quantum dots by electroluminescence and time-resolved photoluminescence," *Physical Review B*, vol. 54, pp. R5243–R5246, 1996.

23. D. Klotzkin, K. Kamath, and P. Bhattacharya, "Quantum capture times at room temperature in high-speed $In_{0.4}Ga_{0.6}As$-GaAs self-organised quantum-dot lasers," *IEEE Photonics Technology Letters*, vol. 9, pp. 1301–1303, 1997.

24. J. Urayama, T. B. Norris, J. Singh, and P. Bhattacharya, "Observation of phonon bottleneck in quantum dot electronic relaxation," *Physical Review Letters*, vol. 86, pp. 4930–4933, 2001.

25. D. Klotzkin and P. Bhattacharya, "Temperature dependence of dynamic and DC characteristics of quantum-well and quantum-dot lasers: a comparative study," *Journal of Lightwave Technology*, vol. 17, pp. 1634–1642, 1999.

26. D. Pan, Y. P. Zeng, M. Y. Kong, J. Wu, Y. Q. Zhu, C. H. Zhang, J. M. Li, and C. Y. Wang, "Normal incident infrared absorption from InGaAs/GaAs quantum dot superlattice," *Electronics Letters*, vol. 32, pp. 1726–1727, 1996.

27. S. Sauvage, P. Boucaud, F. H. Julien, J.-M. Gérard, and J.-Y. Marzin, "Infrared spectroscopy of intraband transitions in self-organized InAs/GaAs quantum dots," *Journal of Applied Physics*, vol. 82, pp. 3396–3401, 1997.

28. J. Phillips, K. Kamath, X. Zhou, N. Chervela, and P. Bhattacharya, "Photoluminescence and far-infrared absorption in Si-doped self-organized InAs quantum dots," *Applied Physics Letters*, vol. 71, pp. 2079–2081, 1997.

29. K. W. Berryman, S. A. Lyon, and M. Segev, "Electronic structure and optical behavior of self-assembled InAs quantum dots," *Journal of Vacuum Science and Technology*, vol. 15, pp. 1045–1050, 1997.

30. S. Maimon, E. Finkman, G. Bahir, S. E. Schacham, J. M. Garcia, and P. M. Petroff, "Intersublevel transitions in InAs/GaAs quantum dots

infrared photodetectors," *Applied Physics Letters*, vol. 73, pp. 2003–2005, 1998.

31. Q. D. Zhuang, J. M. Li, H. X. Li, Y. P. Zeng, and L. Pan, "Intraband absorption in the 8-12 μm band from Si-doped vertically aligned InGaAs/GaAs quantum-dot superlattice," *Applied Physics Letters*, vol. 73, pp. 3706–3708, 1998.
32. L. Chu, A. Zrenner, G. Böhm, and G. Abstreiter, *Applied Physics Letters*, vol. 75, p. 3599, 1999.
33. I. Mukhametzhanov, Z. H. Chen, O. Baklenov, E. T. Kim, and A. Madhukar, *Physica Status Solidi B*, vol. 224, p. 697, 2001.
34. Z. Chen, E.-T. Kim, and A. Madhukar, *Applied Physics Letters*, vol. 80, p. 2490, 2002.
35. J. Phillips, K. Kamath, and P. Bhattacharya, "Far-infrared photoconductivity in self-organized InAs quantum dots," *Applied Physics Letters*, vol. 72, pp. 2020–2022, 1998.
36. N. Horiguchi, T. Futatsugi, Y. Nakata, N. Yokoyama, T. Mankad, and P. M. Petroff, "Quantum dot infrared photodetector using modulation doped InAs self-assembled quantum dots," *Japanese Journal of Applied Physics*, vol. 38, pp. 2559–2561, 1999.
37. Z. Chen, O. Baklenov, E. T. Kim, I. Mukhametzhanov, J. Tie, A. Madhukar, Z. Ye, and J. C. Campbell, "Normal incidence InAs/$Al_xGa_{1-x}As$ quantum dot infrared photodetectors with undoped active region," *Journal of Applied Physics*, vol. 89, pp. 4558–4563, 2001.
38. J. Phillips, "Evaluation of the fundamental properties of quantum dot infrared detectors," *Journal of Applied Physics*, vol. 91, pp. 4590–4594, 2002.
39. J. Singh, *Electronic and Optoelectronic Properties of Semiconductor Structures*. New York: Cambridge University Press, 2003.
40. K. Kamath, N. Chervela, K. K. Linder, T. Sosnowski, H. T. Jiang, T. Norris, J. Singh, and P. Bhattacharya, "Photoluminescence and time-resolved photoluminescence characteristics of $In_xGa_{1-x}As$/GaAs self-organized single- and multiple-layer quantum dot laser structures," *Applied Physics Letters*, vol. 71, p. 927, 1997.
41. V. Ryzhii, "The theory of quantum-dot infrared phototransistors," *Semiconductor Science and Technology*, vol. 11, pp. 759–765, 1996.
42. K. W. Berryman, S. A. Lyon, and M. Segev, "Mid-infrared photoconductivity in InAs quantum dots," *Applied Physics Letters*, vol. 70, pp. 1861–1863, 1997.
43. D. Pan, E. Towe, and S. Kennerly, "Normal-incidence intersubband (In, Ga)As/GaAs quantum dot infrared photodetectors," *Applied Physics Letters*, vol. 73, pp. 1937–1939, 1998.
44. S. J. Xu, S. J. Chua, T. Mei, X. C. Wang, X. H. Zhang, G. Karunasiri, W. J. Fan, C. H. Wang, J. Jiang, S. Wang, and X. G. Xie, "Characteristics

of InGaAs quantum dot infrared photodetectors," *Applied Physics Letters*, vol. 73, pp. 3153–3155, 1998.

45. J. Phillips, P. Bhattacharya, S. W. Kennerly, D. W. Beekman, and M. Dutta, "Self-assembled InAs-GaAs quantum-dot intersubband detectors," *IEEE Journal of Quantum Electronics*, vol. 35, pp. 936–943, 1999.
46. D. Pan, E. Towe, and S. Kennerly, "Photovoltaic quantum-dot infrared detectors," *Applied Physics Letters*, vol. 76, pp. 3301–3303, 2000.
47. H. C. Liu, M. Gao, J. McCafferey, Z. R. Wasilewski, and S. Fafard, "Quantum dot infrared photodetectors," *Applied Physics Letters*, vol. 78, pp. 79–81, 2001.
48. S. Y. Wang, S. D. Lin, H. W. Wu, and C. P. Lee, "Low dark current quantum-dot infrared photodetectors with an AlGaAs current blocking layer," *Applied Physics Letters*, vol. 78, pp. 1023–1025, 2001.
49. A. D. Stiff, S. Krishna, P. Bhattacharya, and S. Kennerly, "Normal-incidence, high-temperature, mid-infrared InAs-GaAs vertical quantum-dot infrared photodetector," *IEEE Journal of Quantum Electronics*, vol. 37, pp. 1412–1419, 2001.
50. S. Krishna, A. D. Stiff-Roberts, J. D. Phillips, P. Bhattacharya, and S. W. Kennerly, "Hot dot detectors: infrared quantum dot intersubband photodetectors are a promising technology for multiwavelength IR detection," *IEEE Circuits & Devices*, vol. 18, pp. 14–24, 2002.
51. E.-T. Kim, Z. Chen, and A. Madhukar, "Tailoring detection bands of InAs quantum-dot infrared photodetectors using $In_xGa_{1-x}As$ strain-relieving quantum wells," *Applied Physics Letters*, vol. 79, pp. 3341–3343, 2001.
52. S. Raghavan, P. Rotella, A. Stintz, B. Fuchs, S. Krishna, C. Morath, D. A. Cardimona, and S. W. Kennerly, "High-responsivity, normal-incidence long-wave infrared (λ = 7.2 μm) InAs/$In_{0.15}Ga_{0.85}As$ dots-in-a-well detector," *Applied Physics Letters,* vol. 81, pp. 1369–1371, 2002.
53. S. Krishna, S. Raghavan, G. V. Winckel, P. Rotella, A. Stintz, C. P. Morath, D. Le, and S. W. Kennerly, "Two color InAs/InGaAs dots-in-a-well detector with background-limited performance at 91 K," *Applied Physics Letters*, vol. 82, p. 2574, 2003.
54. U. Sakoglu, J. S. Tyo, M. M. Hayat, S. Raghavan, and S. Krishna, "Spectrally adaptive infrared photodetectors with bias-tunable quantum dots," *Journal of Optical Society of America B*, vol. 21, pp. 7–17, 2004.
55. A. Weber, O. Gauthier-Lafaye, F. H. Julien, J. Brault, M. Gendry, Y. Désieres, and T. Benyattou, "Strong normal-incidence infrared absorption in self-organized InAs/InAlAs quantum dots grown on InP(001)," *Applied Physics Letters*, vol. 74, pp. 413–415, 1999.
56. S. Kim, H. Mohseni, M. Erdtmann, E. Michel, C. Jelen, and M. Razeghi, "Growth and characterization of InGaAs/InGaP quantum dots for mid-infrared photoconductive detector," *Applied Physics Letters*, vol. 73, pp. 963–965, 1998.

57. J. Jiang, K. Mi, S. Tsao, W. Zhang, H. Lim, T. O'Sullivan, T. Sills, M. Razeghi, G. J. Brown, and M. Z. Tidrow, "Demonstration of a 256 × 256 middle-wavelength infrared focal plane array based on InGaAs/InGaP quantum dot infrared photodetectors," *Applied Physics Letters*, vol. 84, pp. 2232–2234, 2004.

58. J. Jiang, S. Tsao, T. O'Sullivan, W. Zhang, H. Lim, T. Sills, K. Mi, M. Razeghi, G. J. Brown, and M. Z. Tidrow, "High detectivity InGaAs/InGaP quantum-dot infrared photodetectors grown by low pressure metalorganic chemical vapor deposition," *Applied Physics Letters*, vol. 84, pp. 2166–2168, 2004.

59. J. L. Liu, W. G. Wu, A. Balandin, G. L. Jin, and K. L. Wang, "Intersubband absorption in boron-doped multiple Ge quantum dots," *Applied Physics Letters*, vol. 74, pp. 185–187, 1999.

60. P. Boucaud, V. L. Thanh, S. Sauvage, D. DéBarre, and D. Bouchier, "Intraband absorption in Ge/Si self-assembled quantum dots," *Applied Physics Letters*, vol. 74, pp. 401–403, 1999.

61. J. L. Liu, W. G. Wu, A. Balandin, G. Jin, Y. H. Luo, S. G. Thomas, Y. Lu, and K. L. Wang, "Observation of inter-sub-level transitions in modulation-doped Ge quantum dots," *Applied Physics Letters*, vol. 75, pp. 1745–1747, 1999.

62. A. I. Yakimov, A. V. Dvurechenskii, Y. Y. Proskuryakov, A. I. Nikiforov, O. P. Pchelyakov, S. A. Teys, and A. K. Gutakovskii, "Normal-incidence infrared photoconductivity in Si p-i-n diode with embedded Ge self-assembled quantum dot," *Applied Physics Letters*, vol. 75, pp. 1413–1415, 1999.

63. C. Miesner, O. Röthig, K. Brunner, and G. Abstreiter, "Mid-infrared photocurrent measurements on self-assembled Ge dots in Si," *Physica E*, vol. 7, pp. 146–150, 2000.

64. A. I. Yakimov, A. V. Dvurechenskii, A. I. Nikiforov, and Y. Y. Proskuryakov, "Interlevel Ge/Si quantum dot infrared photodetector," *Journal of Applied Physics*, vol. 89, pp. 5676–5681, 2001.

65. S. Chakrabarti, A. D. Stiff-Roberts, P. Bhattacharya, S. Gunapala, S. Bandara, S. B. Rafol, and S. W. Kennerly, "High-temperature operation of InAs/GaAs quantum dot infrared photodetectors with large responsivity and detectivity," *IEEE Photonics Technology Letters*, vol. 16, pp. 1361–1363, 2004.

66. S. Chakrabarti, X. H. Su, P. Bhattacharya, G. Ariyawansa, and A. G. U. Perera, "Characteristics of a multi-color InGaAs/GaAs quantum dot infrared photodetector," *IEEE Photonics Technology Letters*, vol. 17, pp. 178–180, 2005.

67. X. H. Su, S. Chakrabarti, A. D. Stiff-Roberts, J. Singh, and P. Bhattacharya, "Quantum dot infrared photodetector design based on double-barrier resonant tunneling," *Electronics Letters*, vol. 40, pp. 1082–1083, 2004.

68. P. Bhattacharya, X. H. Su, S. Chakrabarti, G. Ariyawansa, and A. G. U. Perera, "Characteristics of a tunneling quantum dot infrared photodetector operating at room temperature," *Applied Physics Letters*, vol. 86, pp. 191,106–191,108, 2005.

69. X. H. Su, S. Chakrabarti, P. Bhattacharya, G. Ariyawansa, and A. G. U. Perera, "A resonant tunneling quantum dot infrared photodetector," *IEEE Journal of Quantum Electronics*, vol. 41, pp. 974–979, 2005.

70. J. Singh, "Possibility of room temperature intra-band lasing in quantum dot structures placed in high-photon density cavities," *IEEE Photonics Technology Letters*, vol. 8, p. 488, 1996.

71. A. Kastalsky, L. E. Vorobjev, D. A. Firsov, V. L. Zerova, and E. Towe, "A dual-color injection laser based on intra- and inter-band carrier transitions in semiconductor quantum wells or quantum dots," *IEEE Journal of Quantum Electronics*, vol. 37, p. 1356, 2001.

72. S. Krishna, P. Bhattacharya, J. Singh, T. Norris, J. Urayama, P. J. McCann, and K. Namjou, "Intersubband gain and stimulated emission in long wavelength (l = 13 mm) intersubband quantum dot emitters," *IEEE Journal of Quantum Electronics*, vol. 37, pp. 1066–1074, 2001.

73. C.-F. Hsu, J.-S. O, P. Zory, and D. Botez, "Intersubband quantum-box semiconductor lasers," *IEEE Journal of Selected Topics in Quantum Electronics*, vol. 6, pp. 491–503, 2000.

74. B. Kochman, A. D. Stiff-Roberts, S. Chakrabarti, J. D. Phillips, S. Krishna, J. Singh, and P. Bhattacharya, "Absorption, carrier lifetime, and gain in InAs-GaAs quantum-dot infrared photodetectors," *IEEE Journal of Quantum Electronics*, vol. 39, pp. 459–467, 2003.

75. A. Jackson, *Electronics Letters*, vol. 37, pp. 355–356, 2001.

Chapter

9

Intersubband Transitions in Si/SiGe Heterojunctions, Quantum Dots, and Quantum Wells

Hans Sigg

Paul Scherrer Institut, Villigen PSI, Switzerland

9.1 Introduction

The investigation of intersubband transitions in silicon- and germanium-based quantum wells has attracted considerable interest over the past 15 years. Such investigations have helped to improve the atomic layer growth of group IV materials on silicon, and were crucial in the development of the detailed knowledge we have now of the complex band structure of strained SiGe quantum structures, particularly for the valence bands. Additionally, Si-based intersubband systems have generated interest because of their potential applications in infrared detection and emission.

In particular, the integration of such systems into the powerful silicon microelectronics platform promises the development of optical devices with complex functionalities. For example, monolithic integration of a chip containing large arrays of intersubband detectors with associated electric drivers and logic circuitries for signal analysis may allow the production of sophisticated infrared *focal plane arrays* (FPAs) at low cost. Furthermore, if the quantum cascade laser concept is successfully transferred from III-V to group IV semiconductors, silicon, in spite of its unfavorable indirect band structure, could become a competitive player in optoelectronics. Si-based quantum cascade lasers could be designed for emission in the mid-infrared and may provide a cheaper alternative to III-V and II-VI based mid-infrared laser sources for applications such as environmental sensing and process diagnostics. The other targeted energy range for Si-based intersubband emitters is the terahertz region, which is important for gas tracing and security surveillance, e.g., at airports. Combined with sensitive and fast photodetectors, applications of Si-based quantum cascade lasers to on-chip signal processing as well as chip-to-chip data communications may also be considered. The operating wavelength of such systems would certainly not reach the 1.5-μm telecommunication band because of the fairly low band offsets, but since for chip-to-chip communications free-space transmission rather than fiber coupling is required, the restriction to the mid-infrared (5 to 10 μm) is of less importance. As we will show, the bottleneck and thus a mandatory prerequisite for all these developments will be further progress in the deposition techniques of strained SiGe layers, in particular using *molecular beam epitaxy* (MBE) on virtual substrates. The understanding of the band structure, and in particular the complex multibands of the valence band, however, is now well developed.

Intersubband spectroscopy is used routinely for the characterization of the epilayer quality. For example, studies are conducted to verify the abruptness and thermal stability of Si/SiGe heterointerfaces as well as to measure band structure parameters such as the strain-dependent heterojunction band offsets. Pioneering studies were devoted to the investigation of two-dimensional electron and hole gases in Si inversion and accumulation layers in the late 1970s and 1980s [1]. During this period, many of the fundamental effects and theoretical considerations concerning the intersubband resonances, such as many-body effects, interface scattering, and coupling to plasmons, were demonstrated. This is remarkable as these studies were undertaken long before intersubband transitions were explored in group III-V materials, such as the lattice-matched GaAlAs [2] and InGa/AlAs systems. In common with the quantum Hall effect, discovered in Si inversion layers but more widely explored in GaAs/AlGaAs heterostructures, the rapid development of intersubband devices in the late 1980s and 1990s was mainly realized with III-V quantum structures [3, 4]. However, judging from the substantial progress made recently in the epitaxy of Si/SiGe heterostructures deposited on strain-compensating relaxed buffer substrates, similar device developments can be expected in the near future with Si-based intersubband systems. The high standard of today's Si/Ge epitaxy is demonstrated by the *high-resolution transmission electron microscopy* (HRTEM) study, shown in Fig. 9.1. The depicted Si/SiGe multi-quantum-well structure has been grown on relaxed buffer substrate by low-temperature MBE. Uniform and abrupt interfaces were obtained, and intersubband absorption studies revealed transitions with narrow line width (~14 meV), as shown below. The

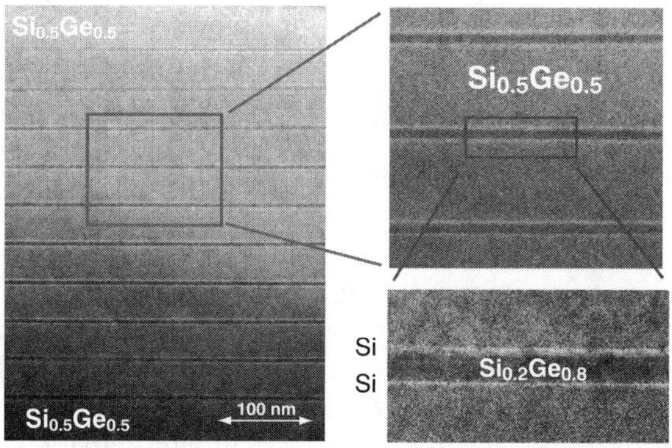

Figure 9.1 High-resolution cross-sectional TEM image of a 10-period $Si/Si_{0.2}Ge_{0.8}$ multiple-quantum-well structure grown on a relaxed SiGe buffer layer at 300°C. The structure is strain-symmertrized as the average Ge concentration of the 12-Å-thick Si barriers and 40-Å-thick SiGe wells adds up to that of the relaxed buffer (50%). (Source: E. Müller, PSI, Switzerland.)

ultimate goal of this area of research—and the ultimate proof of the progress made in the epitaxial growth and band structure engineering of group IV materials—is the development of Si-based quantum cascade lasers. How far we are from this milestone remains an open question. This chapter aims to look into this question and give an overview of the state of the research.

We review the main progress that has been made in the past few years in intersubband spectroscopy and device development with Si-based quantum structures. Specifically we emphasize those aspects of intersubband physics that are most important for Si/SiGe systems, such as the polarization selection rules of intersubband transitions in the multivalley conduction bands and strongly mixed valence bands. Sections 9.2 and 9.3 are devoted to the theoretical background, experimental methods, and results of various intersubband spectroscopy studies performed on Si-based *quantum wells* (QWs) and *quantum dots* (QDs). Section 9.4 then discusses the main device applications of these systems, namely, photodetectors and quantum cascade emitters.

9.2 Theoretical Background of Intersubband Transitions in Si/SiGe Quantum Wells

Depending on the substrate material and the strain configuration, various options are available to realize an intersubband system with Si/SiGe QWs. For example, when these layers are grown on a Si substrate, most of the band offset will occur in the valence bands, and only the holes will be effectively confined in the SiGe layers, while the Si layers form the barriers. By growing on top of a thick relaxed buffer layer of SiGe, valley splitting in the strained Si (or Ge) layers enables sufficient confinement for the electrons to be achieved as well. The most commonly used layer sequences, for n-type and p-type intersubband systems, are schematically shown in Fig. 9.2.

From the numerous investigations published since the 1990s, our knowledge of intersubband absorption in Si/SiGe QWs has reached a high level of maturity. The driving force for these studies has been the prospect of developing infrared detectors that can be integrated with large-scale Si circuits. Consequently, the main effort has been to elucidate the dependence of the intersubband absorption coefficient on the direction of light polarization. In particular, the possibility of absorption at normal incidence to the sample surface (defined in the following as the z direction), which is important for the realization of focal plane arrays, is of great interest. For the description of valence band states and intersubband dipolar matrix elements in strained Si/SiGe multiple QWs, envelope wave function approaches using $\mathbf{k}\cdot\mathbf{p}$ models of various complexity have been applied [5–17].

Experiments and theory have shown that normal-incidence intersubband absorption (i.e., with light polarized parallel to the QWs, or in the xy plane) is possible in both n-type and p-type Si/SiGe QWs. For the

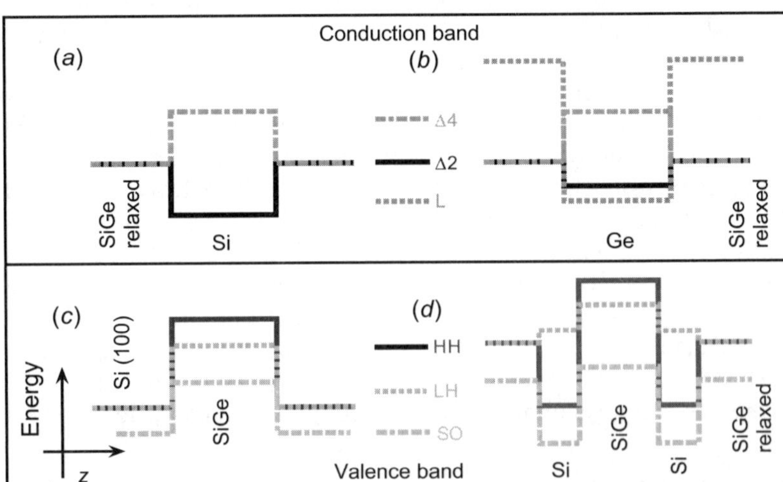

Figure 9.2 Schematic representation of the band structure of frequently explored SiGe intersubband systems. (*a*) Si QW grown on top of a SiGe relaxed buffer. Electrons in the ground state belong to the two-fold conduction band minima Δ2. (*b*) Ge QW on a (100) relaxed SiGe substrate; electrons occupy the three-fold minima in the L direction, i.e., along [111]. Due to the tilt of the constant energy ellipsoids relative to the (100) growth direction, intersubband transitions couple to the xy as well as to the z polarization components. (*c*) Valence band SiGe QW coherently strained to a Si (001) substrate. The ground state belongs to the heavy-hole (HH) band. At the Γ point of the Brillouin zone, i.e., at $k_{\text{par}} = 0$, this HH band is decoupled from the light-hole (LH) and split-off (SO) bands. (*d*) Strain-symmetrized SiGe QW grown on a (100) relaxed SiGe buffer layer: the compressively strained SiGe QW is embedded by tensile-strained Si barrier layers. The HH levels are much more strongly confined than the LH/SO levels.

case of transitions between conduction subbands, the coupling to the xy polarization component is provided by the ellipsoid shape of the constant-energy surfaces, but only when the z (i.e., growth) direction does not match any of the main axes of the ellipsoids [5, 6]. This requirement is a drawback for n-type Si QWs (type A, Fig. 9.2a), because the sixfold degenerate Si conduction band minima at the Δ point are oriented along the [100] direction, which is the preferred growth direction in Si microelectronics. Better compatibility to Si electronics is thus obtained from Ge QWs (type B, Fig. 9.2b) as their minima are at the L point of the Brillouin zone and thus are oriented along the [111] directions [5].

For hole intersubband transitions, the main difference from the conduction band QW states is the presence of the nearly degenerate (on the scale of typical confinement energies) *heavy-hole* (HH), *light-hole* (LH), and *spin-orbit split-off* (SO) bands at the Γ point of the valence band. Due to the reduced crystal symmetry caused by the QW potential and the strain in the QW layers, the valence band QW states are mixtures between these bands. The coupling of xy-polarized light to intersubband transitions is a direct consequence of this multiband structure of the valence band. As demonstrated by Fromherz et al. [7], transitions between states belonging to the same subspace (for example, HH-to-HH transitions) couple exclusively to the z polarization for vanishing in-plane momentum k_{par}. However, transitions between the HH and LH/SO subspace couple also to the xy polarization. In both cases, the strength of these transitions is determined by the dipole matrix element between the respective initial and final QW states. For finite in-plane wave vector, these selection rules are relaxed, as transitions between the subspaces of the QW states with nonvanishing overlap integral become allowed for both directions of polarization [7]. As valence band mixing (e.g., the admixture of LH/SO into the HH band) depends on strain and confinement and is increasing with increasing k_{par}, the actual coupling strength for the two polarizations depends on sample parameters such as temperature, doping concentration, and Ge concentration in the QW. In practice, for weakly to moderately doped systems with Fermi wave vector $k_F \sim 0.03$ Å^{-1} [which corresponds to a *two-dimensional* (2D) carrier density $n_{2\mathrm{D}} \sim 10^{12}$ cm^{-2}], this mixing is weak, particularly for the HH ground state. Excited HH levels may show mixing with the LH/SO bands, particularly in case of near degeneracy of the levels, as shown below. At the low carrier densities, energy level calculations based on a 6×6 $\mathbf{k}\cdot\mathbf{p}$ Luttinger-Kohn hamiltonian offer accurate results. However, for $k_F \geq 0.05$ Å^{-1} the contributions from the remote conduction bands at the Γ point should be included to higher than second order, as shown in a comparison between the 6×6 and the 14×14 $\mathbf{k}\cdot\mathbf{p}$ hamiltonians by El Kurdi et al.

[8]. Note, however, that some of the thus required **k·p** coupling parameters for the higher conduction bands are not accurately known, and their determination by intersubband experiments is difficult because of the broad intersubband resonances measured, particularly at high carrier densities.

A comparison between simulations based on the 6 × 6 hamiltonian and experiments is shown in Fig. 9.3 for valence band QWs of type C [7] (as in Fig. 9.2*c*) and D [18] (as in Fig. 9.2*d*). Good agreement is obtained in both cases by using the same model with the same band parameters, taken from Refs. 19 and 20. In addition, the theory is found to give an accurate description of the resonant line shapes and explains in detail the rich substructures appearing in the tails of the absorption peaks of QWs of type D [21]. In this latter work, the Ge concentration dependence of the Luttinger parameters as given in Ref. 22 has been used. From the good agreement with the experiment it is concluded that the control of the growth as well as the modeling seems highly accurate and thus should allow the realization and simulation of highly complex intersubband systems, such as those needed for the development of efficient quantum cascade emitters.

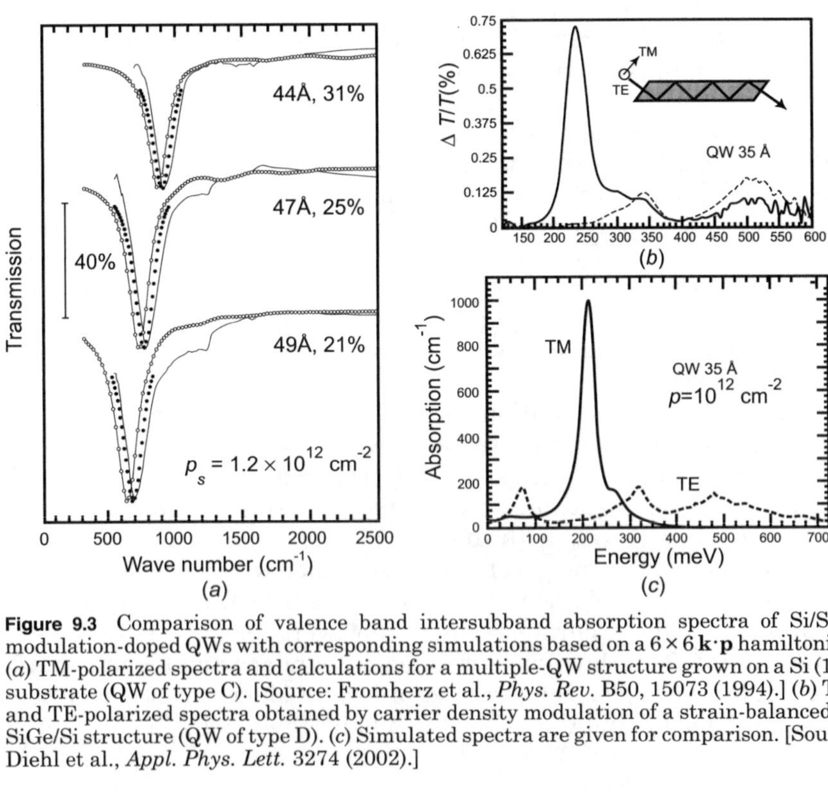

Figure 9.3 Comparison of valence band intersubband absorption spectra of Si/SiGe modulation-doped QWs with corresponding simulations based on a 6 × 6 **k·p** hamiltonian. (*a*) TM-polarized spectra and calculations for a multiple-QW structure grown on a Si (100) substrate (QW of type C). [Source: Fromherz et al., *Phys. Rev.* B50, 15073 (1994).] (*b*) TM- and TE-polarized spectra obtained by carrier density modulation of a strain-balanced Si/SiGe/Si structure (QW of type D). (*c*) Simulated spectra are given for comparison. [Source: Diehl et al., *Appl. Phys. Lett.* 3274 (2002).]

Another important aspect, particularly for the development of quantum cascade devices, is knowledge of the intersubband energy relaxation processes, since lasing requires the lifetime of the upper state to be longer than that of the lower state. The first attempt to design efficient intersubband emitters in the SiGe material system can be found in earlier work by Sun and Friedman [23]. Later, Reimann, Kaindl, and Woerner [24] developed a detailed picture of deformation-potential scattering of holes in QW structures. By symmetry, the deformation-potential scattering occurs predominantly between subbands of different types, e.g., between HH and LH subbands. The corresponding scattering rates were calculated for a Si/$Si_{0.5}Ge_{0.5}$ QW pseudomorphically grown on Si (100). Distinct scattering onsets were found at well widths between 5.5 and 4.5 nm, where the separation between the HH2 and the LH1 levels becomes equal to the optical phonon vibrational energies of the Ge–Ge, Si–Ge, and Si–Si bonds. This is illustrated in Fig. 9.4. For the deformation-potential scattering, it appears that the scattering rate increases monotonically with

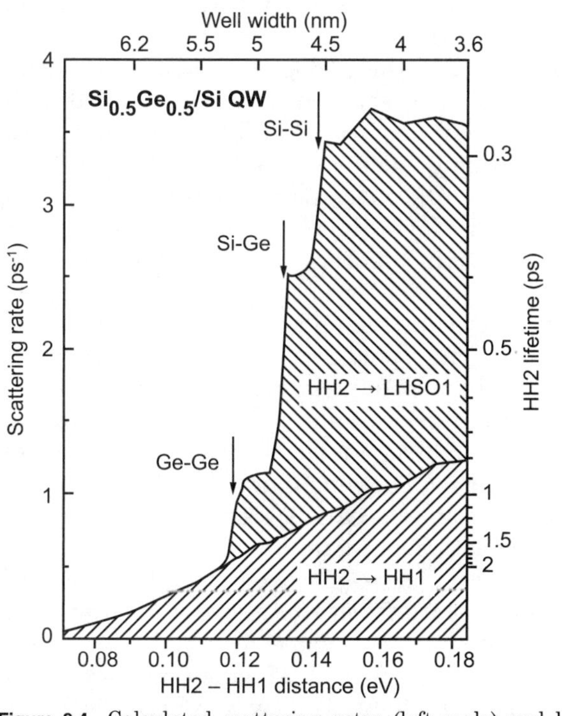

Figure 9.4 Calculated scattering rates (left scale) and lifetimes (right scale) at zero temperature from optical-phonon deformation scattering in $Si_{0.5}Ge_{0.5}$/Si multiple QWs as a function of the HH2-HH1 energy separation (varied by changing the well width, see upper scale). The arrows show the threshold for scattering by Ge-Ge, Si-Ge, and Si-Si type optical phonons. [Source: Reimann et al., *Phys. Rev.* B65, 045302 (2001).]

intersubband separation energy, as more relaxation channels become available, due to scattering to higher-k_{par} states. This is fundamentally different from the case of polar-optical-phonon scattering, where the scattering rate is largest for momentum transfers close to zero. These considerations must be incorporated into the design of Si-based quantum cascade lasers to get efficient depopulation of the lower state, while keeping the upper state sufficiently long lived. The resulting design strategies depend on the desired emission wavelength as described in greater detail below.

9.3 Intersubband Spectroscopy

9.3.1 Experimental techniques

The experimental techniques used to study intersubband absorption in SiGe QWs are very similar to those applied for the study of III-V intersubband systems. In most cases, the samples contain multiple QWs and, to increase the intersubband absorption signal, are investigated in the waveguide geometry shown in Fig. 9.5*a*. Infrared light is coupled in and out of the wedged facets of the parallelogram-shaped sample and undergoes multiple passes through the undoped substrate and doped QW layers. The input polarization is set by a wire grid polarizer to be aligned normal (TM polarization) or parallel (TE polarization) to the deposited layers. The absorption in such layers is strongly reduced for TE-polarized light if a metallic film is evaporated on the sample top surface. Conversly, a metallic layer increases the absorption for TM light [7]. Due to the oblique incidence, this coupling scheme does not provide light coupling in pure z polarization. In the TM configuration z and xy polarization is mixed, while in the TE configuration only the xy polarization is applied. The z polarization coupling without admixture of xy polarization could be obtained by attenuated total reflection geometry via a germanium prism tightly pressed on the sample surface [25]. The transmitted light is analyzed with a *Fourier transform infrared* (FTIR) spectrometer. The sensitivity of such experiments is strongly enhanced by modulating the carrier density with an electric field, applied between a Schottky front gate and the electrically contacted QW layers (c.f. Fig. 9.5*b*). Phase-sensitive detection of the modulated signal while stepping the interferometer allows resolution of the intersubband absorption spectrum of a single QW [26].

Carrier density modulation can also be obtained by photoexcitation of the samples with above-bandgap light, as shown in Fig. 9.5*c*. Due to the long electron-hole recombination lifetimes in indirect bandgap semiconductors such as Si and Ge, carrier densities exceeding 10^{18} cm^{-3} in bulk samples and 10^{11} cm^{-2} in QWs are obtained at moderate light

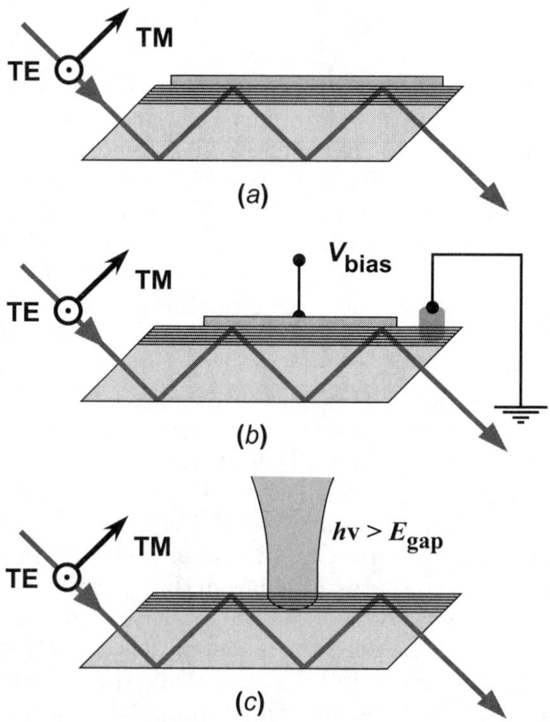

Figure 9.5 Frequently used sample configurations for intersubband transmission experiments in waveguide geometry for transverse magnetic and transverse electric polarization. (*a*) Samples with a top metallic layer offer good overlap of the TM mode with the multiple-QW layers near the surface. TE transmission provides the reference signal. (*b*) Carrier density modulation with a Schottky gate gives improved signal-to-noise performance, enabling detection of intersubband absorption of single wells, and gives the means to study the intersubband absorption under the influence of an E field. (*c*) Carriers are provided by above-bandgap optical excitation. Because of the long interband carrier lifetime in Si at low temperatures (> microseconds), substantial carrier densities are obtained at moderate excitation intensities.

intensities of 1 W/cm^2 [27]. However, Fabry-Perot interference in the multiple-QW stack due to refractive index variations and from the Drude free-carrier absorption by photo-generated carriers is found to complicate the interpretation of the spectra [28]. In doped SiGe QWs, the photoexcited intersubband absorption is found to differ from the direct absorption, which has been attributed to the fact that this method is sensitive to excess carriers located above the Fermi energy E_F and away from the Brillouin zone center [29]. Thus far, only hole intersubband absorption has been observed with this technique. This can be attributed in part to weak confinement in the conduction band, at least in systems that are pseudomorphically strained to Si. For systems grown on relaxed buffers with larger conduction band offsets, the measurement of electron intersubband absorption in the presence of photoexcitation should be possible as well.

In addition to these linear spectroscopy techniques used to determine the intersubband energy separations, nonlinear and time-resolved measurements have been performed to explore the dynamics of intersubband transitions in SiGe QWs. Early pump-probe experiments were carried out with 3-nm-thick *p*-type $Si_{0.8}Ge_{0.2}$ QWs using a free-electron laser source, and they revealed intersubband relaxation lifetimes of between 400 and 700 fs [30]. By using 150-fs-long mid-infrared pulses generated by parametric amplification in a GaSe crystal, intersubband relaxation processes with a time constant of 250 fs were subsequently resolved in 4.4-nm-thick, *p*-type $Si_{0.5}Ge_{0.5}$ multiple QWs [31]. Longer relaxation times were observed in wider wells [32], where the intersubband energy separations are small compared to the optical phonon energies (c.f. Fig. 9.4). This makes these systems convenient for studies with free-electron laser sources, as demonstrated by Murzyn et al. [33] and Kelsall et al. [34].

As an alternative to pump-probe techniques, the upper-state lifetime in quantum cascade structures can be determined from the quantitative analysis of their emission efficiency. Note, however, that this method can produce reliable results only if the detector response, the collection efficiency η_{opt} of the experimental setup, and the optical matrix element of the intersubband transition are known, and if an appropriate injector is built into the structure to provide efficient carrier injection into the upper state. The technique is based on the expression for the quantum efficiency η_{QC} (i.e., the number N_P of emitted photons per period divided by the number of carriers in the excited state N_X) in terms of the radiative and nonradiative lifetimes

$$\eta_{QC} = \frac{N_P}{N_X} = \frac{P}{\eta_{opt} Nh\nu}\frac{e}{J} = \frac{\tau_{nrad}}{\tau_{nrad} + \tau_{rad}}$$

where $h\nu$ is the photon energy, J is the current, N is the number of periods in the cascade structure, P is the detected power spectrally integrated over the resonance, and unit injection efficiency, that is, $N_X \sim J/e$, is assumed. Since τ_{rad} can be calculated from the dipole matrix element, τ_{nrad} can be obtained from the equation above. Regardless of its restrictions, this method has been found to be very useful to determine the nonradiative lifetime of the upper state [35, 36] and to estimate the achievable gain in SiGe quantum cascade structures [37].

9.3.2 SiGe and SiGeC quantum wells

As mentioned above, intersubband spectroscopy is an effective tool to investigate the interface quality of the layers forming the QWs and thus is of great use to characterize and optimize layer deposition techniques. Initial intersubband spectra of Si-based heterostructures were from *p*-type Si/SiGe multiple QWs grown by MBE [38–40]. Since these samples were doped in the wells, the spectra were very broad, independent of the actual QW dimensions, and only the fundamental HH resonance could be identified. Subsequently, normal-incidence absorption was demonstrated [41], and the intersubband spectra were simulated [42].

Better-resolved intersubband transitions were measured in *n*-type Si wells grown by MBE on a relaxed (100) $Si_{0.7}Ge_{0.3}$ buffer layer [43]. The barriers were formed by $Si_{0.5}Ge_{0.5}$ layers doped with Sb by secondary ion implantation techniques. The intersubband resonances were found to shift from 7 meV to approximately 30 meV as the well width was varied from 100 to 50 Å. Through simulations, this shift was found to be compatible with a discontinuity of 310 meV between the doubly-degenerate Δ2 conduction band minima in the symmetrically strained Si and $Si_{0.5}Ge_{0.5}$ layers. The intersubband line width and mobility of this sample were about 3 meV and 15 $m^2/(V \cdot s)$, respectively, depending on the doping concentration and well width.

Electron intersubband absorption has been investigated in 50-Å-thick well-doped SiGe/Si/SiGe QWs grown by MBE on relaxed (110) SiGe substrate layers [42]. At the doping densities of $\sim 10^{20}$ cm^{-3} used in these samples, exchange and depolarization energy shifts were found to contribute considerably to the measured intersubband transition energy of ~200 meV. As expected for (110) oriented substrates, these QWs showed intersubband absorption of a similar strength for both the in-plane (TE) and the out-of-plane (TM) polarization [44]. Comparable electron intersubband absorption of TE-polarized light at similar wavelengths (near 7 μm, that is, ~200 meV) was also obtained in delta-doped Ge QWs grown on relaxed (100) SiGe substrate layers [45]. The coupling to the TE polarization component is provided in this case by the tilted energy ellipsoids of the Ge *L* valleys [5].

In addition, intersubband studies have been undertaken to investigate the interfacial smearing in narrow Si/SiGe multiple QWs grown by MBE using Sb to control the interface abruptness by suppressing the Ge surface segregation [46]. From the close agreement between the observed and calculated transition energies versus QW thickness (c.f. Fig. 9.6) the authors concluded that the Ge segregation was suppressed to less than 0.4 nm. Incidentally, in these samples the exchange and depolarization shifts were found to give a relatively small (< 10 meV) contribution to the intersubband transition energy, consistent with the chosen carrier concentration of $\sim 2 \times 10^{18}$ cm^{-3}. Similar conclusions regarding the interface abruptness were obtained for Si/SiGe QWs grown using hydrogen as a segregant-assist in Si gas-source MBE [47].

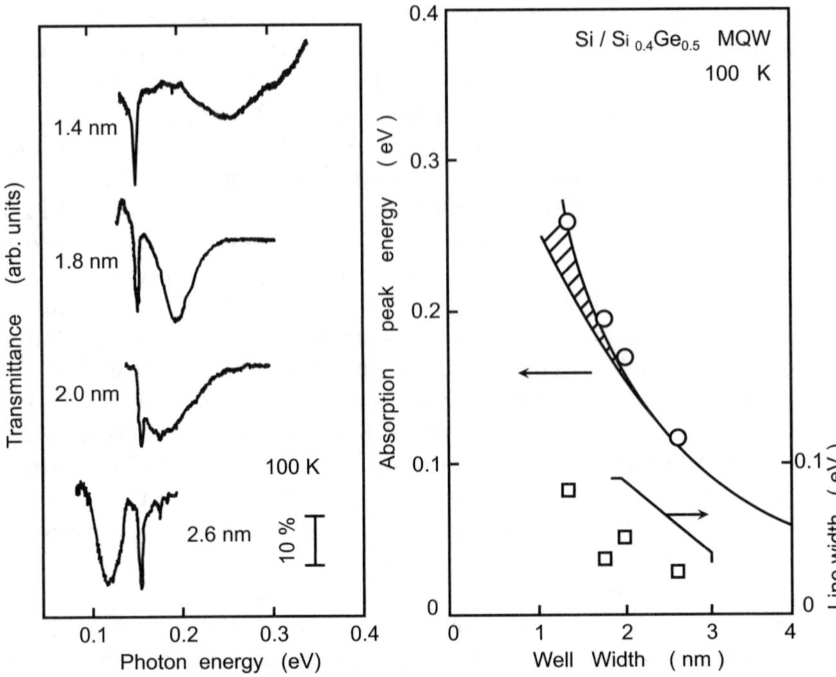

Figure 9.6 (*a*) Intersubband study of the heterointerface quality in Si/$Si_{0.4}Ge_{0.6}$ multiple QWs prepared by segregant-assisted growth with Sb. (*b*) Well width dependence of the peak intersubband absorption energy (circles) and resonance line width (squares). The solid line shows the intersubband transition energy calculated with a Kronig-Penney model. The shadow marks the formation of minibands in the multiple QWs. (Reprinted with permission from [46]. Copyright 1992, American Institute of Physics.)

Regarding intersubband transitions in the valence bands, a detailed study was carried out in pseudomorphically strained SiGe QWs with well thickness and Ge concentration ranging from 26 to 65 Å and from 20% to 50%, respectively [7]. The spectra measured with TM-polarized light could be accurately described by an envelope function band structure calculation and subsequent dielectric simulations of the multilayer stack. In this work, a line width of ~20 meV was observed for HH transitions in QWs with well-confined excited states, which represents the narrowest value ever published for *p*-type QW structures pseudomorphically grown on Si. The sensitivity of the measurement was not sufficient to resolve the TE component of the intersubband absorption. In a later work by the same group with a Si/SiGe multiple-QW stack capped by a 1.2-μm-thick Si overlayer, TE absorption was observed and found to be in qualitative and quantitative agreement with theory [46]. In the same work, the difference in polarization dependence between intersubband absorption and photoconductivity was addressed. The latter process showed a stronger response to TE than TM polarization, in contrast to the case of absorption, indicating the importance of the subsequent transport processes for the photocurrent [48]. To our knowledge, the thus far smallest line width value for intersubband transitions in *p*-type SiGe is ~14 meV and is realized in 60-Å-wide QWs of 75% Ge concentration grown on 50% relaxed buffer substrate [21].

Intersubband absorption spectra for both TE and TM polarization have also been investigated in *p*-type $Si_{1-x}Ge_x$ ($x = 0.2$) QWs grown by ultrahigh-vacuum *chemical vapor deposition* (CVD) [49]. Using the phenomenological Drude model, the contributions from free-carrier absorption and intersubband absorption were separately identified in these measurements. The resulting line width of the fundamental HH resonance was approximately 30 meV. This larger value compared to the MBE-grown QWs of Ref. 7 is likely related not to the growth technique but rather to the weaker confinement of the QWs used in this work. This leads to a broadening of the intersubband absorption peak due to the mixing of the HH2 subband with the continuum. The observed pronounced contribution from free-carrier absorption may also be related to the use of shallow QWs. The growth of SiGeC QWs by CVD has been studied in a comparison of the photoinduced intersubband absorption of 3-nm-thick $Si_{1-x-y}Ge_xC_y$ ($x = 17\%$, $y = 1\%$) QWs and carbon-free reference QWs [50]. Similar spectra were obtained from these two systems, indicating effective substitutional incorporation of the carbon atoms and high-quality epitaxial growth. However, in both systems high photopumping levels (> 100 W/cm^{-2}) were required to discriminate intersubband absorption from free-carrier absorption, and the resonance peaks were rather broad.

In addition, polarization-resolved intersubband absorption has been used in combination with photoluminescence [51] and Raman spectroscopy [52] to study the thermal stability of pseudomorphically grown Si/SiGe multiple QWs. In general, the Ge and Si atoms in these systems undergo some interdiffusion, which blurs the rectangular shape of the QWs. In turn, this reduces the confinement strength and changes the Ge distribution and thus the strain profile of the heterostructure. As this effect acts differently on the HH, LH, and SO subbands, the thermal relaxation process due to interdiffusion can be observed in the ratio of the TM and TE polarized spectra, which are sensitive to transitions between different pairs of such valence subbands. This behavior has been experimentally observed [51], and was later interpreted with $\mathbf{k}\cdot\mathbf{p}$ simulations [52]. From Raman and infared studies, the activation process during annealing of pseudomorphic Si/SiGe QWs was proposed to occur in two steps, first by strain relaxation (starting at ~750°C) and then by Ge/Si interdiffusion (from ~ 900°C on) [53]. However, interdiffusion in Si/SiGe multilayers on relaxed SiGe pseudosubstrate observed by means of in situ X-ray reflectivity has been found to occur at $T > 790°C$ [54], in agreement with Refs. 51 and 52.

The magnitude of the intersubband absorption signal in the presence of above-bandgap photoexcitation—and its dependence on the photoexcitation intensity and modulation frequency—have also been investigated to study the recombination kinetics in pseudomorphic *p*-type Si/SiGe QWs [27]. This method provides a direct measure of the excess carrier density because the intersubband absorption strength depends on the density of confined holes. From this experiment, an interband decay rate of 2.5 μs at 80 K was obtained. Furthermore, the contributions from linear (monomolecular) and quadratic (bimolecular) recombination mechanisms were identified from the modulation frequency dependence of the intersubband absorption signal. The latter mechanism is mainly due to nonradiative recombination of the (thermally excited) holes with the electrons distributed in the wells. This description accurately reproduces the experimental dependence of the photoexcitation efficiency on temperature.

The need for thick (~ micrometer) multilayer QW structures to achieve quantum cascade structures which provide high gain and good optical mode overlap in infrared waveguides has triggered the interest of the community in strain-compensated growth on relaxed SiGe buffer layers. Strain compensation is obtained by balancing the number of tensile layers (e.g., the Si layers) with compressive layers (the SiGe layers). Fortunately, the source material (and time) demanding growth of the relaxed SiGe buffer can be avoided these days, as epitaxy-ready pseudosubstrates consisting of Si (100) covered by multimicrometer-thick relaxed buffer layers have recently become available. Such growth templates have been developed mainly for the realization of highly conducting electron channels in strained Si for high-speed applications in mainstream electronics [55, 56]. The relaxed buffer layers are usually deposited by a fast chemical vapor deposition process on large-size Si (100) substrates. After the epitaxy, planarization by chemical mechanical polishing is sometimes performed.

To verify band parameters, such as the band offsets, the Luttinger effective mass parameters, and the deformation potential, in strain-compensated SiGe systems, and to prove the accuracy of state-of-the-art band structure calculations as well as the growth quality on Ge-rich pseudosubstrates, detailed intersubband absorption studies and intersubband modeling of *p*-type Si/ $Si_{1-x}Ge_x$ QWs (x varied from 70% to 85%) grown by MBE on $Si_{0.5}Ge_{0.5}$ buffer layers were performed [18, 22]. Ge content and layer thicknesses were accurately determined by X-ray measurements [22]. By using the Schottky gate modulation technique, the intersubband absorption signal was observed up to room temperature for both TM and TE polarizations. The large HH band offset (> 0.5 eV) results in the observation of well-resolved resonances with transition energies in excess of 300 meV. As mentioned previously, the energy of the fundamental HH transition, as well as the main characteristics of the TM and TE spectra, was found to be in good agreement with model calculations (c.f. Fig. 9.3). This indicates that the growth and modeling of strain-compensated SiGe QWs are well under control, opening new possibilities for the realization of elaborate quantum structures such as quantum cascade emitters, as described in the section entitled "Mid-Infrared Injection Devices."

The importance of resonant intravalence band mixing has been addressed in a recent study of the polarization dependence of the fundamental intersubband resonance in similar strain-balanced *p*-type SiGe QWs [57]. The mixing is found to become strongest in samples where the HH2 level develops into degeneration with the LH/SO ground state. This situation occurs for $Si_{0.2}Ge_{0.8}$ QWs approximately 6 nm wide, grown on 50% relaxed buffer substrate. Under these conditions, strong anticrossing occurs, repulsing the levels by as much as 30 meV, which goes along with the transfer of oscillator strength from the TM to TE polarizations, and vice versa. Experimentally, the mixing is observed by the growing intensity of the TE absorption of the HH1-to-HH2 transition as well as apparent TM coupling of the HH1-to-LH1 transition, as shown in Fig. 9.7.

The Schottky gate modulation technique not only offers high sensitivity by discriminating carrier-induced absorption from the background, but also allows the study of intersubband transitions under an applied electric field. By using this technique, the coupling between confined states in a double QW consisting of symmetrized Si/SiGe layers has been investigated [26]. This structure was grown by MBE on a relaxed buffer substrate of 50% Ge and is shown in Fig. 9.8*b*. By tuning the voltage between the *two-dimensional* (2D) hole gas and the gate between +1 and –0.5 V, oscillator strength is transferred from the direct to the diagonal transition while the transition energies show anticrossing (c.f. Fig. 9.8*a*). The experimental data could be accurately reproduced by calculations based on a simple one-band model, assuming depletion from the Ti gate corresponding to an electric field of 130 kV/cm. In particular, the observed coupling strength of 40 meV, obtained from the minimal energy difference between the two resonances (c.f. Fig. 9.8*c*), is in agreement with the calculated value, assuming a 10.9-Å-thick square Si barrier. The nominal thickness of this layer was 11 Å, which again demonstrates the high level of accuracy provided by today's Si MBE technology.

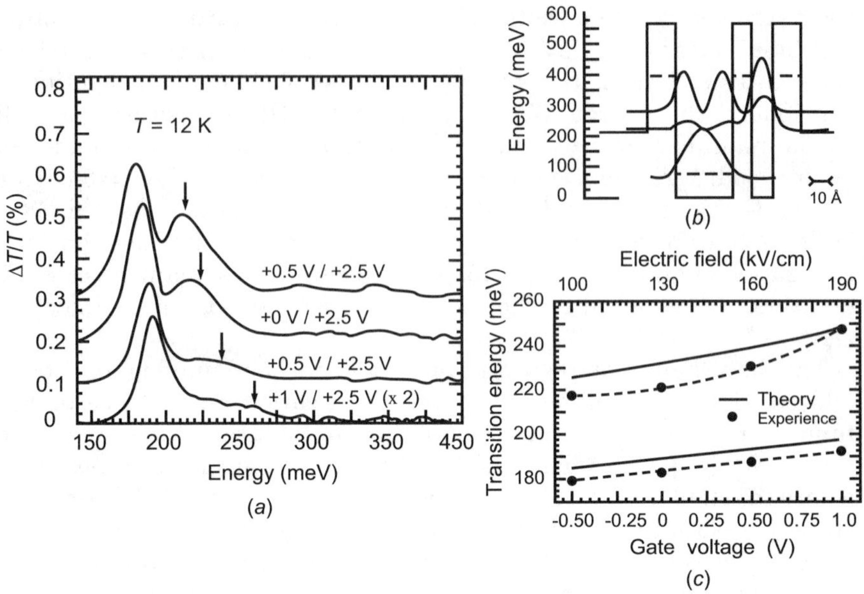

Figure 9.7 (*a*) Schottky bias-dependent absorption spectra of a $Si_{0.2}Ge_{0.8}$/Si/ $Si_{0.2}Ge_{0.8}$ double-QW structure. The spectra are referenced to the spectrum taken at +2.5-V bias when the QW is depleted. (*b*) Band structure of the double-QW structure under flat band condition. (*c*) Comparison of experimental and calculated transition energies. In the calculations the electric field (upper scale) is taken to consist of the external field determined by the gate voltage and the constant depletion field assumed to be 130 kV/cm. [Source: Diehl et al., *Appl. Phys. Lett.* 84, 2498 (2004).]

9.3.3 Quantum dots

Intersubband spectroscopy of self-assembled semiconductor *quantum dots* (QDs) is a subject of interest both for fundamental studies and for the development of infrared photodetectors and lasers. Compared to (two-dimensional) QWs, the use of (zero-dimensional) QDs for intersubband infrared devices has some advantages, such as their sharp deltalike density of states, reduced intersubband relaxation rates, favorable polarization properties, as well as lower dark current and hence lower noise in photodetectors. These features, combined with the relatively high band offsets and suitability for monolithic integration with Si signal processing and read-out electronics, make Ge islands on Si particularly attractive. For a comprehensive review of Si/Ge-based QDs and nanostructures, see Ref. 58.

Such QDs are formed by the deposition of several (five to nine) *monolayers* (MLs) of germanium onto a Si (100) substrate. The dot dimensions and densities depend on the growth parameters, such as the substrate temperature [59]. In the lateral direction, typical diameters are between 20 and 80 nm, while heights span from 1.5 to

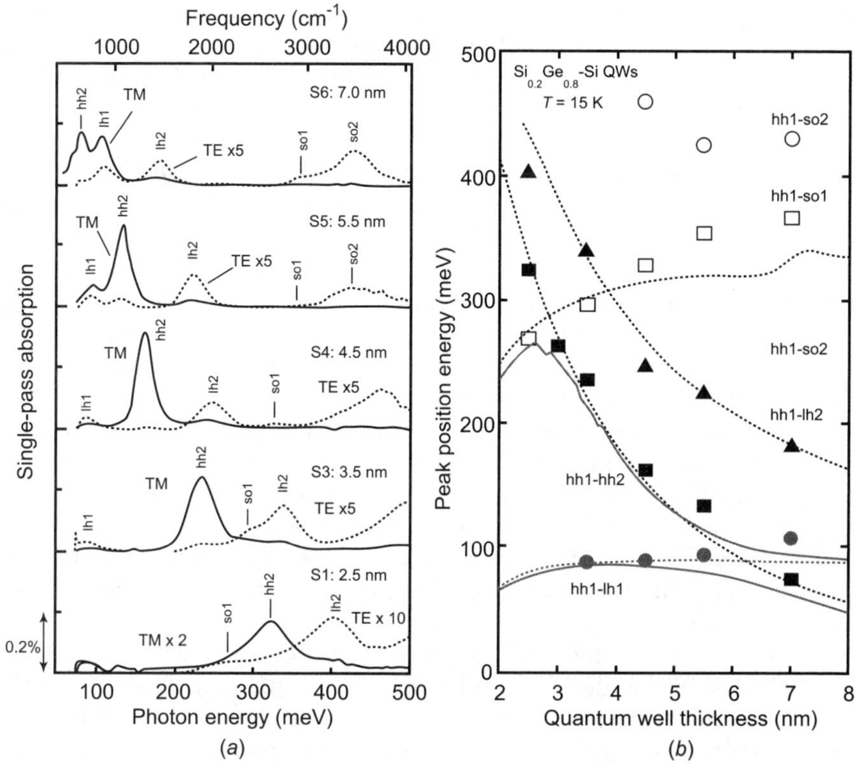

Figure 9.8 (*a*) TM- and TE-polarized intersubband absorption spectra of strain-compensated $Si_{0.2}Ge_{0.8}$/Si QWs measured at 15 K for well width thicknesses between 2.5 and 7.0 nm. (*b*) Dependence of peak positions on QW thickness. The solid curve shows transition energies calculated by a self-consistent six-band $\mathbf{k\cdot p}$ model. The dotted curves show the transition energies calculated at the band edge ($k_{par} = 0$) where the mixing between HH and LH/SO states is absent. At resonance, when the LH1 level is approaching the HH2 level, the transitions from the HH1 ground state to the HH2 and LH1 appear at equal strength, as is found for sample S7 in TM and in sample S6 in TE. [Source: Tsujino et al., *Phys. Rev.* B72, 153315 (2005).]

10 nm. Therefore, quantum confinement is strongest along the growth direction. Dot densities exceeding 10^{11} cm^{-2} have been obtained [60]. As is typical for systems grown in the Stranski-Krastanov mode, a wetting layer of approximately 3 ML of Ge develops before the dots are formed. Due to strain fields, ordered stacking is obtained for adjacent dot layers. During overgrowth with Si, the Ge islands flatten and their Ge concentration tends to decrease due to intermixing. When low-temperature capping at about 300°C is used, segregation and intermixing of Ge are reduced, as shown by a red-shifted photoluminescence [61, 62].

The first intersubband absorption spectra from boron-doped multiple Ge QDs grown by MBE showed broad resonances (~100 meV) centered at 200 to 250 meV [63]. The absorption signal was significantly stronger for TM-polarized input light compared to TE polarization. This behavior is similar to the case of transitions between HH states in a QW, which suggests that the dots were flattened as expected because of the rather high growth temperature of 650°C used in this study. In a subsequent work with self-assembled *p*-type Si/Ge QDs grown by MBE at 525°C, the fundamental TM resonance was found at a similar energy (about 250 meV) but with one-half the spectral width [64]. Through the application of a bias in the lateral direction, using TiAu electrodes evaporated as interdigitated fingers with a 60-μm spacing, the normal-incidence photoconductivity was also investigated in the same samples. Compared to the absorption, the spectral weight was found to be shifted to higher energies (400 meV), which is explained by bound-to-continuum transitions. At a bias of 1 V, the bound-to-bound HH transition was also observed. This photoconductivity response at normal incidence, i.e., for the "forbidden" *xy* polarization, is evidence of the relaxation of the QW selection rules caused by the three-dimensional hole confinement in the QDs. The strong measured photocurrent due to bound-to-continuum transitions is promising for applications in large-area photodetectors [65].

Si/Ge self-assembled QDs grown by CVD at 550°C have been investigated by Boucaud et al. [66]. Their photo-induced infrared intersubband absorption spectra showed a distinct resonance at approximately 300 meV. Since this resonance appeared to be active for *xy* polarization, and its energy closely matched the experimental energy difference between the Si bandgap (1.155 meV) and the QD photoluminescence energy (845 meV), it was attributed to the transition between the QD ground states and the continuum states. Furthermore, the photo-induced infrared absorption in these samples was found to saturate at a pump intensity of 10 mW/cm^2, which was attributed to filling of the QD ground states, i.e., the creation of two holes per QD. From the estimated dot density of 2×10^9 cm^{-2}, the observed absorption modulation ($\Delta T/T \sim 0.3\%$) was calculated to correspond to an interband cross section of $\sigma \sim 2 \times 10^{-13}$ cm^2, which is three orders of magnitude larger than the experimental value found by the same group for InAs/GaAs QDs [67]. However, no physical interpretation was given for this apparent giant cross section. An underestimation of the dot density may in part explain the unexpected result.

In Fig. 9.9 is shown the saturation of the photo-induced infrared absorption observed in a dense array of Ge QDs, however at the much higher pump intensity of 110 mW/cm^2 [60]. In this study, 10 Ge layers were deposited by MBE at a low growth temperature of 300°C. The Si

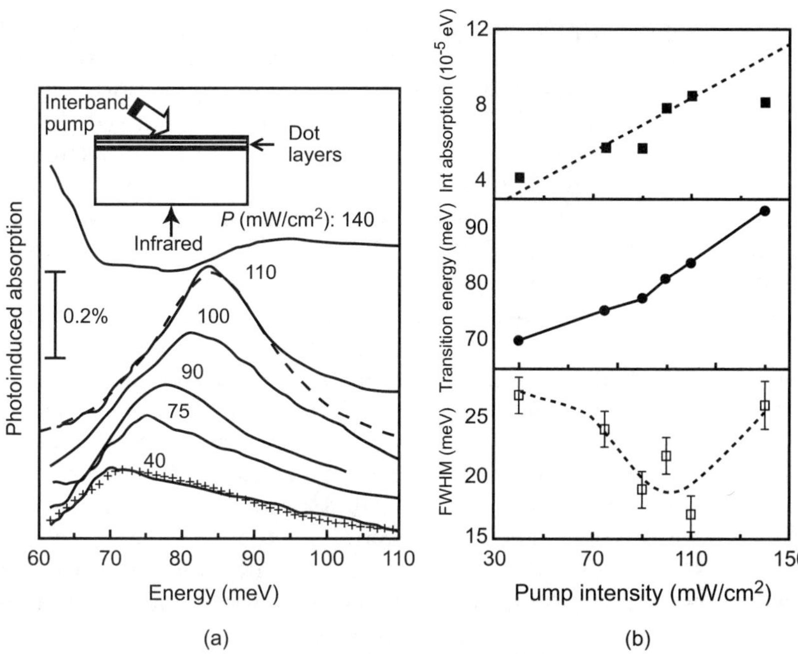

Figure 9.9 (*a*) Photo-induced absorption spectra of a dense array of Ge self-assembled QDs measured at quasi-normal incidence for different values of the pump power. The crosses show a calculated inhomogeneously broadened absorption line, assuming 20% size inhomogeneity. The experimental data at the incident pump power of 110 mW/cm^2 are fitted with a lorentzian (dashed line). (*b*) From top to bottom: dependence on pump intensity of integrated absorption, transition energy, and absorption line full width at half maximum (FWHM). [Source: Yakimov et al., *Phys. Rev.* B62, 9939 (2000).]

capping layers were deposited while the temperature was ramped up to 500°C, leading to good preservation of the island shape and large QD density. The saturation was again attributed to the occupation of two holes per QD. With the assumption that essentially all the photo-generated holes were trapped in the QDs, due to the extremely long electron-hole recombination lifetime in Si, an interband absorption cross section σ on the order of 10^{16} cm^2 (as in InAs/GaAs QDs) was correspondingly estimated. Furthermore in this study, the photo-induced infrared resonance was found to become narrow and to shift from 70 to 85 meV as the pump intensity was increased from 40 to 110 mW/cm^2, as shown in Fig. 9.9*b*. This effect was attributed to dynamic screening by the in-plane polarized collective excitation [60].

Such description reminds one of the results obtained by numerical simulations of intersubband absorption due to disorder localized electrons in a center-doped QW [68, 69] and by interlevel electromagnetic response calculations for a rectangular lattice of individual spherical quantum dots [70]. For these systems, similar shifts and line narrowing with increasing electron density have been predicted. Note, however, that in a perfect planar situation, the in-plane polarized dipole interaction would result rather in a red-shifted collective absorption peak [68]. The apparent blue shift may thus be related to the fact that in Yakimov's experiment [60] with stacked quantum dot arrays the dynamic dipole interaction is dominated by interlayer coupling. In analogy to the situation of the z dipole momentum interaction of in-plane distributed intersubband excitations, the dipole coupling then is repulsive and the collective resonance thus blue-shifted.

A more recent study considered the infrared absorption properties of 20 layers of self-assembled Si/Ge islands embedded in a p-i-n structure [71]. This active region was set in a mid-infrared waveguide formed by 3-μm-thick SiGe layers with 2% Ge concentration. Both active region and waveguide were grown by CVD at a total vapor pressure and substrate temperature of approximately 140 mbar and 700°C, respectively. Under a positive bias, carriers in this structure are injected into the QDs, giving rise to absorption in the mid-infrared, as well as enhanced interband transmission at near-infrared wavelengths related to band filling in the Ge islands. A long-wavelength resonance at 185 meV was observed and assigned to the bound-to-continuum transition between HH bound levels and the wetting layer ground state. This resonance was found to saturate at a current density of 20 A/cm^2. By comparing the observed HH absorption strength with the calculated absorption cross section and mode filling factor, the carrier density in the active region at saturation was found to be $3 \times 10^{18} cm^{-3}$. At such high carrier concentrations, Auger recombination becomes important, and an estimate for the Auger recombination rate was obtained, slightly enhanced (by a factor of 3 to 4) with respect to bulk Si.

Except for Ref. 60, interlevel spectroscopy thus far has been performed on capped Ge QD systems grown at relatively high temperatures $T > 500$ K. The Si overgrowth at these temperatures leads to interdiffusion of Si into the dots, which reduces the confinement energies and makes the system similar to QWs. Consequently, future investigations should be conducted on better-defined QDs with high Ge concentration. Issues of interest include depolarization effects for xy polarized excitations and the study of dot size and carrier density dependencies. The latter may be investigated by applying the Schottky gate technique (using transparent gates) or optical modulation.

9.4 Toward Applications

9.4.1 Infrared photodetectors

Infrared photodetectors are undisputedly one of the most important potential applications of intersubband transitions in Si-based materials. Therefore, as described in the previous section, several investigations of intersubband absorption in these QW and QD systems have addressed detector-related issues such as normal-incidence response and wavelength tuning. It turns out that among the investigated systems the most promising candidates for device applications are *p*-type SiGe QWs and Ge QDs. These systems can be grown pseudomorphically on Si (100), and thus they are compatible with Si microelectronics. Moreover, they offer a large band offset for wavelength tuning and bandgap engineering.

Quantum well infrared photodetectors (QWIPs) in Si/Ge systems commonly utilize intersubband transitions from the HH ground state to excited states that are close to or above the Si barriers. Thus, as proposed originally [72], the photo-excited holes can give rise to a photocurrent which is collected under the appropriate bias conditions by the doped contact layers above and underneath the multiple-QW structure. For a recent review of the fundamental concept and performance of QWIPs (mainly in III-V semiconductors), see Ref. 73. Typical performance values for SiGe QWIPs, based on a 10-period stack of 30-Å-thick $Si/Si_{1-x}Ge_x$ ($x = 35\%$) QWs [74], are peak responsivities of 50 mA/W at 5 μm under normal incidence. The 50-nm-thick Si barrier layers were used in this structure to suppress the dark current due to tunneling, yielding a 300 K background limited performance at T_{BLIB} = 85 K, similar to QWIPs made of *p*-type GaAs/AlGaAs QWs. However, at present, *n*-type GaAs-based devices with appropriate coupling schemes for quasi-normal incidence perform much better than any *p*-type QWIP [73].

Recent development efforts of Si/Ge infrared detectors have focused on dark current suppression schemes, multiwavelength detection, laterally biased photoconductivity, and the use of multiple QD layers. For instance, with the introduction of a *p*-doped, 30-nm-wide SiGe emitter layer in a $Si/Si_{1-x}Ge_x$ ($x = 23\%$) QWIP structure, the dark current was reduced considerably even at high bias voltages of up to 5 V [75]. Such an emitter layer establishes an ohmic contact to the top p^{++} Si contact layer but forms a Schottky-type contact to the SiGe multiple-QW region. Probably because of the reduced barrier height in this structure, the peak responsivity was found to be shifted to a lower energy (~200 meV) and a lower value (7 mA/W). The TM response (as measured in the 45 degree waveguide geometry) was found to exceed the TE response at wavelengths around 8 μm. (~150 meV). The TE component was strongest between 3 and 4.5 μm (400 and 250 meV).

Wide-range wavelength tuning of the peak responsivity has been demonstrated by utilizing charge transfer from a deep to a shallower well in an externally applied electric field [76] (c.f. Fig. 9.10). This QWIP structure consists of several repetitions of a fundamental unit containing five coupled QWs. Depending on the bias, the carriers populate the ground state of the first or the fifth QW in each unit. As these QWs differ in Ge concentration, the spectral onsets of their photocurrent response, which are related to transitions from their ground states to continuum states, differ as well. When the ground states of the five QWs in each unit become energetically aligned (miniband formation), at a bias of around −1 V, a photocurrent starts to flow in the direction of carrier escape. In the wavelength range from 5.9 μm (at +5 V) to 3.3 μm (at −5 V), a peak responsivity of order 1 mW/A and T_{BLIP} = 80 K resulting in a $D^* \sim 10^9\,\mathrm{cm}\sqrt{\mathrm{Hz}}/\mathrm{W}$ were measured in this device.

Infrared photodetectors based on QDs (QDIPs) are particularly promising because dots are intrinsically suited to normal incidence, their band offsets can generally be larger than in QWs of the same materials, and, finally, they can provide reduced dark currents and

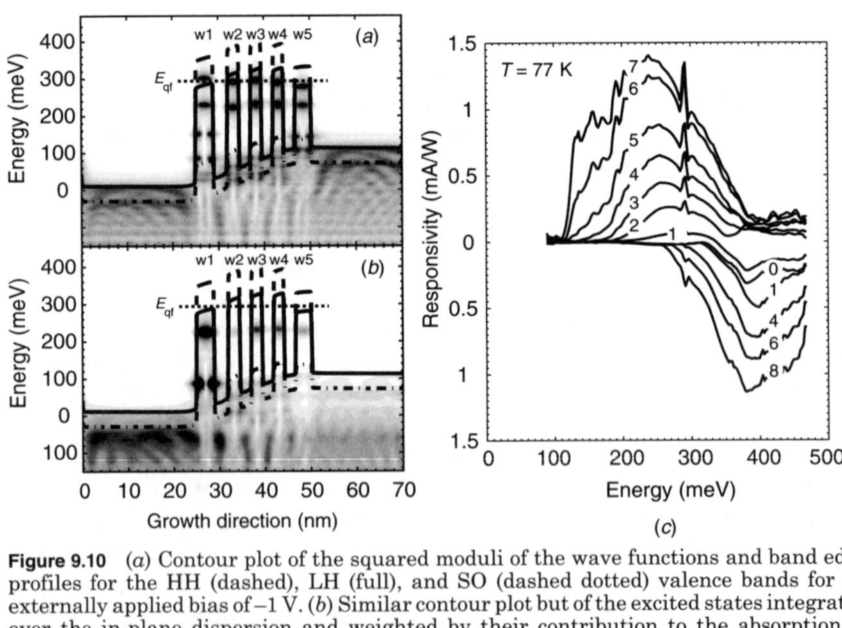

Figure 9.10 (*a*) Contour plot of the squared moduli of the wave functions and band edge profiles for the HH (dashed), LH (full), and SO (dashed dotted) valence bands for an externally applied bias of −1 V. (*b*) Similar contour plot but of the excited states integrated over the in-plane dispersion and weighted by their contribution to the absorption of normal-incident radiation. (*c*) Responsivity spectra of a cascade injector QWIP measured at T = 77 K and at voltages corresponding to the indicated labels. [Source: Rauter et al., *Appl. Phys. Lett.* 83, 3879 (2003).]

higher internal gains. Because the carrier confinement is effective in all three spatial dimensions, QD layers can be used to fabricate devices with photocurrent flowing in the vertical direction, similar to QWIPs, but also in the in-plane (lateral) direction. A comparison of mid-infrared photocurrent generation in Ge QDs embedded in vertical and lateral detector structures has been reported in Ref. 77. Apart from a 40-meV red shift (to 284 meV) of the peak response in the lateral detectors, the photoresponse was found to be very similar in the two types of devices. The responsivity of all devices was found to extend to above 600 meV but showed a steep edge at about 250 meV. In the lateral devices, when a 0.5-V bias was applied over the 60-μm spaced interdigitated contacts, peak responsivities of 10 mA/W were obtained. The vertical devices showed one-half this responsivity at the same bias, which of course corresponds to a much higher electric field.

In a subsequent paper [78], the same group presented the intraband response of devices where the QD layers were grown adjacent to $Si_{1-x}Ge_x$ (x = 25%) QWs. The in-plane photoresponse was found to increase with decreasing separation d between the QD and QW layers. For d = 10 nm, the responsivity at 20 K increased to 90 mA/W. Above 20 K, however, the responsivity decreased rapidly, and at 40 K it had already dropped by a factor of 10. This loss of sensitivity was attributed to thermal excitation of the holes to the Si valence band continuum, which decreases the hole lifetime and thus the photoresponse.

The photoconductivity of the previously mentioned high-density (3×10^{11} cm^{-2}) SiGe dot arrays [60] has also been investigated [79]. To study their photocurrent response, the dot arrays were embedded in p^{++} contact layers, with doping provided by boron delta sheets separated from the dot layers by 10-nm-thick intrinsic Si spacers. The photoresponse was measured under normal incidence in a single pass at room temperature. Two peaks at around 65 and 135 meV were observed and ascribed to transitions from the ground states to excited bound states. In spite of the low measured quantum efficiency of ~0.1%, the detectivity D^* at the resonance energies was found to be in the 10^8 cm$\sqrt{\text{Hz}}/\text{W}$ range, which is comparable to uncooled thermal pyroelectric detectors, demonstrating the potential of this type of Ge dot array detectors for room temperature applications.

State-of-the-art D^* figures for QD systems are on the order of 10^{10} cm$\sqrt{\text{Hz}}/\text{W}$, as measured in photoconductor devices based on InAs QDs operating above liquid nitrogen temperature [80, 81]. To obtain similar D^* figures, the Ge dot devices need to be operated at much lower temperatures. This was confirmed in a study of QDIPs containing large dots of 20-nm height and 200-nm diameter, where the significance of the Si barrier thickness to reduce the dark current and the importance of the bias to tune the wavelength response were investigated [82].

Using thick Si barriers under low bias, operation from 0.15 to 0.6 eV (9 to 2 μm) was demonstrated with D^*_{max} near 2×10^{10} cm$\sqrt{Hz}$/W, but these values were obtained at $T = 30$ K. Under a bias of 1 V, the response could also be tuned to the 20-μm regime (0.06 eV), with a D^* of about 10^9 cm$\sqrt{Hz}$/W. Similarly, the need for low-temperature operation to reduce the dark current was found in systems consisting of stacks of self-assembled QD layers grown by CVD [83]. In these devices, the 300 K background limited operation required a temperature of 25 K [84], which again is exceedingly low for many practical applications.

To summarize, the limitations of Si-based QWIPs concerning dark current and spectral response can be partly solved by using QDs. The main remaining problem is the very low operation temperature. Systematic studies to investigate the temperature-dependent properties of Si/Ge QDIPs would thus be highly valuable. What is interesting to note at this point is the well-behaved response at 300 K for both mid-infrared and longer-wavelength devices.

9.4.2 Quantum cascade lasers

The demonstration of a Si-based *quantum cascade* (QC) laser would represent an important technological achievement from the point of view of both fundamental and applied semiconductor research. Such a device would combine the speed and computing power of microprocessors with the enormous communication bandwidth of optics, and therefore it could have a huge impact on the Si microelectronics industry. In addition, mid-infrared and terahertz QC lasers are technologically important for many applications including chemical sensing (e.g., in explosive detection), free-space atmospheric optical communications, and medical analysis. Si would be cheaper than anything else because of the mature, cost-effective, and large scale Si- processing. Thus Si QC lasers may successfully compete with the present III-V devices. Moreover, SiGe QC lasers are particularly interesting for the wavelength region around 30 μm (~40 meV), which is essentially inaccessible to III-V semiconductors because of their *reststrrahlen* band in this spectral range.

So far QC lasers have only been realized with III-V semiconductors, and they utilize electron transitions between conduction subbands. In these devices, depopulation of the lower laser states is controlled by resonant phonon-assisted energy relaxation, which is particularly fast due to the polar nature of III-V compounds. On the other hand, Si and Ge are nonpolar materials, and furthermore quantum confinement of conduction band electrons in most types of SiGe QWs is limited due to the small band offsets and the large effective masses ($m^* \sim m_0$) for vertical transport. Thus, the challenge in developing a Si/SiGe QC laser is to design a suitable active material based on the quasi-degenerate HH, LH, and SO subbands, and to find an alternative mechanism for the depopulation of the lower laser states. Additionally, because of the large effective masses of the SiGe valence bands ($m^*_{\mathrm{SiGe}} \sim 0.2m_0$ versus $m^*_{\mathrm{GaAs}} \sim 0.07m_0$ for electrons), the energy levels and the electrical transport are more sensitive to thickness fluctuations, because the well and barrier layers need to be considerably thinner than in III-V based QC structures. Finally, the optical losses due to free-carrier absorption in the doped contact and QW layers are significant.

QC active layers are among the most complex structures ever prepared by epitaxial techniques. This is especially true in the case of the Si/SiGe system because of a number of characteristic features of these materials, namely, the strong affinity of Si and Ge with oxygen, the strong tendency of Ge atoms to segregate and diffuse at the interface, and the possibility of island formation driven by the large strain fields. As a result, low growth temperatures are required for the epitaxy of Si/SiGe QC structures. The challenge is to realize uniform and abrupt interfaces while maintaining the electrical and optical high quality of the crystal. Many groups all over the world are currently working toward this goal, and after the first demonstration of intersubband emission in a Si-based cascade structure [35], further progress is been reported.

For clarity, in the following the development of QC emitters in Si is presented in two parts. The section "Mid-Infrared Injection Devices" is based on bound-to-bound or bound-to-continuum intersubband transitions in active stages similar to those of the more traditional III-V mid-infrared QC lasers. The section "Long-Wavelength Emitters" describes terahertz emitters employing transitions between closely spaced subbands within a well or between neighboring wells in simple multiple-QW structures. As the spacing between the relevant energy levels in this case is less than the optical phonon energies (for example, 37 meV for the Ge-Ge optical vibrational modes), the intersubband scattering lifetimes can become very long, which is favorable to achieve high emission efficiency in the terahertz range and, eventually, population inversion.

Long-wavelength emitters. Si-based QC structures are considered to be particularly well suited for the realization of terahertz radiation sources [85]. This follows from the experimental observation of strong electroluminescence at far-infrared wavelengths and of long upper-state lifetimes; furthermore, theoretical studies have predicted that large population inversions can be obtained in these structures.

The first demonstration of terahertz emission with Si-based QC devices was carried out with a simple structure consisting of 30 periods of 5-nm-wide Si barriers and 8-nm-wide $Si_{1-x}Ge_x$ ($x = 28\%$) wells, grown by CVD on a relaxed buffer substrate with 23% Ge concentration [86]. Band structure calculations were utilized to interpret the measured spectral features. An emission peak at around 100 μm (12.4 meV, 2.9 THz) was observed and assigned to vertical transitions between the first-excited LH and the HH ground state. In a similar structure [consisting of 100 periods of 2.2-nm-thick Si barriers and 4.4-nm-thick $Si_{1-x}Ge_x$ ($x = 28\%$) wells on a relaxed buffer substrate with 21% Ge concentration], the observed emission features were attributed to diagonal interwell transitions [87]; c.f. Fig. 9.11*a*. In both of these structures, the analysis of the emission spectra is complicated by the appearance of impurity lines, which, by their spectral position, were unambiguously assigned to boron (B) acceptors in silicon [87]. Given the low Si content and large strain of the active layers in these samples, such impurity lines probably originate from the substrate, and consequently they cannot be explained by direct carrier injection because no current flows through the substrate. Instead, the impurity signals seem to be related to blackbody-type of emission. In any case, these lines were found to disappear at $T > 40$ K where thermal ionization sets in. A systematic analysis of the emission spectrum may be required for comparsion of the knowledge of the absorption spectrum. Such data are, however, not available for these structures, possibly due to experimental difficulties related to the wide frequency spread of the oscillator strength.

Despite these complications, the polarization-resolved features in the complex emission spectra of the second structure were assigned to transitions from HH1′ to LH1 (at 8 meV), LH1′ to LH1 combined with HH1′ to HH1 (at 25 meV), and LH1′ to HH1 (at 40 meV); c.f. inset of Fig. 9.11*a* (here the primes denote subbands in the QW upstream). In the experiment, the lowest-energy resonance was found to shift from 5 to 8 meV as the bias voltage was varied from 3.8 to 6.5 V [87], as shown in Fig. 9.11*b*. This shift represents only a fraction of the applied voltage divided by the period number (10 mV per 1-V bias), which calls into question the assumptions made regarding the current flow and/or field distribution. The stark shift may also be reduced due to contact resistances.

Calculations of the energy levels, scattering processes, and vertical transport in these SiGe terahertz emitters have been carried out by Ikonic et al. [88–90]. The relative importance of non-polar-optical and acoustic phonon scattering relative to alloy scattering has been discussed in relation to temperature and intersubband transition energies [88]. It was found that alloy scattering is the dominant mechanism, particularly for low-energy transitions at low temperature. Under these conditions, upper-state population lifetimes on the order

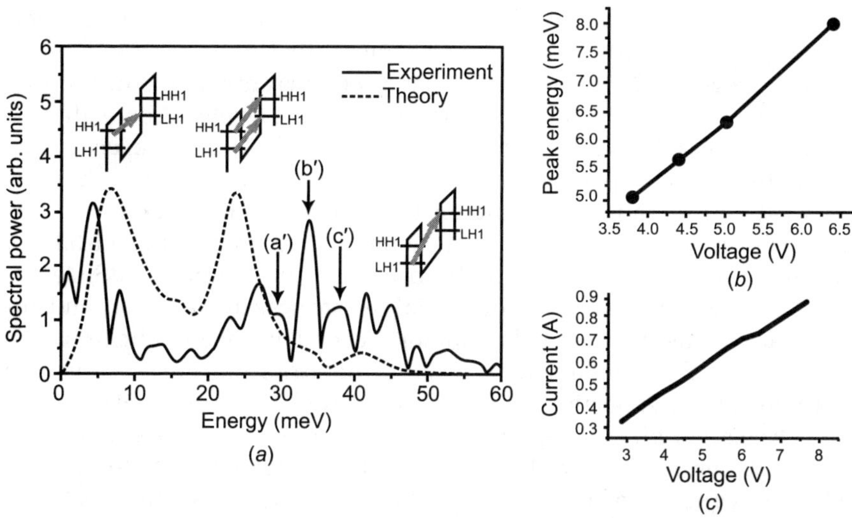

Figure 9.11 (*a*) Long-wavelength edge emission from a 100-period Si/SiGe multiple-QW cascade structure measured under an applied bias of 6.4 V (10% duty cycle) at 4.2 K (point dashed line) and theoretical predictions (solid line). The spectral features are assigned to interwell intersubband transitions from the HH1 and LH1 states in each well to the HH1 and LH1 states in the QW downstream; c.f. inset. Spectral features labeled by (*a′*) through (*c′*) are identified as impurity transitions involving B acceptors in bulk Si. (*b*) Peak energy versus applied voltage for the low-energy resonance. (*c*) Current-voltage characteristics of the device. [Source: Bates et al., *Appl. Phys. Lett.* 83, 4092 (2003).]

of 3 to 10 ps were predicted. To include carrier dynamics and transport in this theoretical description, the same authors have employed the method of the self-consistent energy balance [89] and *Monte Carlo* (MC) simulations [90]. For both methods, the assumption is made that the electric field is uniform across the active region, and hole-hole scattering, interface roughness, and scattering into the next-nearest neighboring wells have all been neglected thus far in the modeling. From their analysis of the bias-dependent subband occupation probabilities in structures with a single QW per period (similar to the design that was investigated experimentally in Ref. 87), population inversion between HH1′ and LH1 at high electric fields (~80 keV/cm) was predicted [89]. In fact, the calculation predicts that independent of the bias, the population of HH1 is higher than in LH1. According to the authors, such a stable population relation between HH1 and LH1 is plausible because of the better confinement of the HH1 states—due to the higher band offset of HHs compared to LHs—which leads to a longer dwell time of carriers in HH1. In theory, the performance of these structures is found to be much better than that of structures based on the inverted mass feature of the LH1 in-plane dispersion [91]. Unfortunately, in an experiment where only spontaneous emission is measured [87, 88], no conclusive statement concerning population inversion can be made, and thus experimentally this prediction has not yet been verified. Moreover, the calculated currents are found to disagree with experiment. When the HH1′ and LH1 levels become aligned (at approximately 42 keV/cm^2), current densities of ~5kA/cm^2 are predicated, while experimentally the current flowing through the 240-μm × 240-μm mesa at 4.2 K and for a 5-V bias was 0.55 A (or 0.65 kA/cm^2) (c.f. Fig. 9.11*c*). The current is monotonically increasing with bias, different from the case of Si MBE-grown strain-balanced Si/SiGe superlattices, where negative differential resistances due to resonant tunneling as well as the sign of domain formation have been indicated [92].

It appears that model predictions of the state lifetimes are in fair agreement with experiments on unbiased QW structures. In a recent work, Kelsall et al. [34] reported a degenerate pump-probe experiment carried out on multi-QW structures made of alternating 2.2-nm-thick Si barriers and SiGe wells of 6.6-nm (4.4-nm) thickness and 34% (25%) Ge concentration. In contrast to earlier measurements with multiple-QW structures where the QWs were isolated from one another by SiGe spacer layers [33], in this case the decay was well fitted by a single exponential. This confirmed the hypothesis [33] that the previously observed dual exponential decay arose from scattering into intermediate states in the spacer layers. The measured absorption recovery times were found to lie between 2 and 25 ps, with the time increasing with decreasing LH1-to-HH1 separation [34]. This observation confirms the importance of the deformation potential scattering which gets suppressed for reduced subband spacing below the energy of the Ge-Ge optical phonon mode. The temperature dependence was weak, which is explained by alloy scattering dominating phonon scattering.

Finally, it should be mentioned that the realization of SiGe terahertz lasers will require the development of appropriate waveguide structures [93]. Toward this goal, the use of buried tungsten silicide layers for confinement of the optical mode has been investigated [94].

Mid-infrared injection devices. Intersubband electroluminescence from Si-based QC emitters was first observed in pseudomorphically grown Si/SiGe short-period superlattices consisting of five QWs per period [35]. These superlattices were designed to achieve a strong overlap of the lowest HH miniband states in the active wells at an electric field of 50 keV/cm (c.f. Fig. 9.12, left panel). Consistent with this design, the emission spectrum, also shown in the figure, was found to be centered at the calculated HH1-HH2 transition energy of 130 meV; furthermore, the light output was purely TM polarized, as expected for transitions between HH states. The corresponding full width at half maximum was

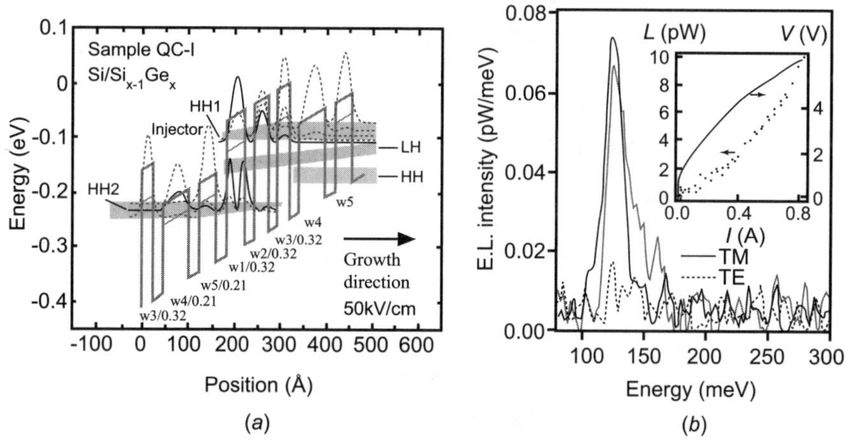

Figure 9.12 (a) Schematics of the valence band potential energy profile of a Si/SiGe QC emitter consisting of five wells per period at the electric field of 50 kV/cm. The emission involves transitions between the HH1 and HH2 states, whose wave functions squared are shown by the solid lines. The wave functions of HH injector miniband states are given by the dotted lines. The shadowed areas denote the energy levels of the injector miniband as well as the intermediate LH states. (b) Electroluminescence spectra measured at T = 50 K and I = 250 mA with 50% duty cycle (dark solid line). The polarization dependence is investigated at 760 mA and T = 85 K. Practically no signal is obtained with TE-polarized light (dotted curve), giving strong evidence of transitions between the HH2 and HH1 states. [Source: Dehlinger et al., *Science* 290, 2278 (2000).]

only 22 meV, close to the best values obtained for pseudomorphic p-type SiGe QWs in linear absorption [7]. The emission efficiency of these structures was measured, using an FTIR setup that was calibrated for absolute intensities with an InGaAs/AlInAs intersubband LED of well-known radiative and nonradiative lifetimes. The radiative emission efficiency and the nonradiative lifetime were found to depend strongly on the design of the active QW. Design issues include the blocking of the carrier escape to the continuum and the actual nonradiative decay time from the upper to the lower state. This issue has been addressed in a subsequent study where the emission efficiencies of cascade structures with 45-Å-wide $Si_{1-x}Ge_x$ active QWs of x = 32% and 41% (QC-1 and QC-2) were compared to those of a cascade structure with 39-Å-wide (x = 42%) QWs (QC-3) [95]. The quantum efficiency of QC-1 and QC-3 was found to be relatively low, while that of QC-2 was reasonably high. The deduced nonradiative lifetime was on the order of 500 fs [35].

To explain these observations, it is necessary to consider the main nonradiative decay paths in these structures, illustrated schematically in Fig. 9.13. The direct relaxation from the injector to the HH ground state (path 1 in Fig. 9.13) primarily depends on the barrier thickness, and thus should not differ much among the three structures. The escape to the continuum (path 2 in Fig. 9.13) should be similar for QC-2 and

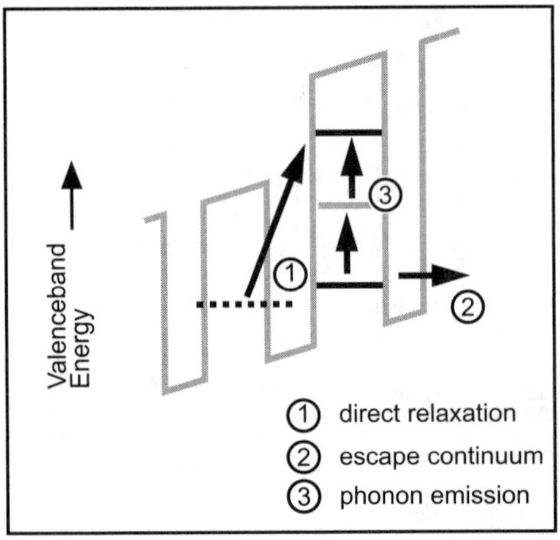

Figure 9.13 Schematic representation of the possible energy relaxation paths in SiGe QC structures (only the two main QWs of each period are shown, i.e., the adjacent injector and emitter wells). (1) Nonradiative relaxation from the injector to the emitter ground state; (2) escape to the continuum; and (3) relaxation via an intermediate LH level by optical-phonon emission.

QC-3, given their similar Ge content in the active wells, and thus is also not likely related to the higher efficiency of QC-2. Therefore, the process that appears to give the dominant contribution to the nonradiative relaxation time is the emission of optical phonons via the deformation potential. As predicted in Ref. 24, this scattering mechanism is most efficient between subbands of different symmetry, such as HH-to-LH subbands. In QC-1 and QC-3 the energy separation between subbands HH2 and LH1 is larger than, and relatively close to, the energy of the Si-Si phonon mode (~60 meV). Thus, phonon emission is very efficient in these structures. And vice versa, in QC-2 the separation between HH2 and LH1 is about 40 meV, so that phonon emission is essentially disallowed except for the lowest but also weakest Ge-Ge mode. From Fig. 9.4, taken from Ref. 24, we find that the lifetime in systems where only the lowest phonon mode (Ge-Ge) is available for emission is about 0.5 to 1 ps, consistent with the measured value in QC-2.

The current- and temperature-dependent emission characteristics of similar SiGe cascade structures but with seven QWs per period have been measured and compared to the results of a **k·p** calculation by Bormann et al. [96]. The theory was found to give an appropriate description of the observed polarization selection rules, confinement energies, and low-energy emission line broadening at high currents. The emission was again ascribed to HH transitions. For this structure a lifetime of 400 fs was calculated, but no comparison with experiments was possible, since the setup used was not calibrated to measure absolute efficiencies.

The question of the upper-state lifetime was addressed by the same group in a subsequent study where the emission efficiency associated with diagonal transitions was determined as a function of barrier width and corresponding oscillator strength [36]. For several QC structures, these authors computed the ratio of the measured emission efficiency divided by the calculated oscillator strength, from which they obtained the upper-state lifetime to within a multiplicative constant. This result is plotted in Fig. 9.14, compared with calculated values. A significant increase of scattering time with increasing barrier width is observed, consistent with calculations. Compared to the case of vertical transitions, the scattering time in this system can be increased by more than one order of magnitude. Unfortunately, the oscillator strength—and hence the optical gain—are correspondingly reduced by a similar factor. The authors thus concluded that diagonal transitions may enable intersubband population inversion, but at the price of a reduced optical matrix element. A further drawback to the use of diagonal transitions in SiGe-based QC structures is the considerable line width increase, e.g., from 26 to 39 meV [36], due to interface roughness scattering. This is analogous to the case of intersubband absorption in coupled wells reported in Ref. 26, where the line widths of vertical and diagonal transitions were 21 and 36 meV, respectively.

Recent research on Si-based QC emitters has focused on the mid-infrared electroluminescence from strain-compensated structures with high Ge concentration, prepared on relaxed buffer substrates. This step

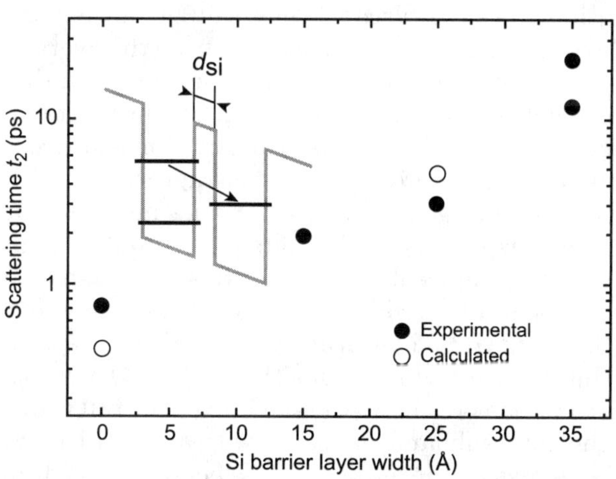

Figure 9.14 Comparison between calculated and experimental nonradiative scattering times as derived from electroluminescence intensities in Si/SiGe cascade emitters with optically active diagonal transitions across Si barriers of variable width. The experimental points (relative values) have been scaled to match the calculated result for the structure with a 15-Å-wide Si barrier. [Source: Bormann et al., *Appl. Phys. Lett.* 83, 5371 (2002).]

was needed to overcome the strain limitations concerning the number of QWs per period (five to seven) and the number of periods (10 to 15) in the active layer. In principle, the resulting increase in design freedom allows one to fabricate more complex cascade structures with improved carrier extraction from the lower states.

The first mid-infrared emitters using the strain symmetrization concept were realized by Diehl et al. [97] with $Si/Si_{1-x}Ge_x$ ($x = 80\%$) cascade structures grown on 50% SiGe relaxed buffer substrates. As can be recognized from the band structure shown in Fig. 9.15*a*, these QC emitters are based on the bound-to-continuum design; i.e., the excited state is confined in the active well, but the lower state extends over several QWs and belongs to a broad miniband. The main advantage of this design is that the various relaxation paths within the miniband allow for fast depopulation of the lower state. In the device of Fig. 9.15, the Ge concentration averaged over one cascade period corresponds to that of the relaxed buffer substrate (50%), so that the accumulated strain in each period is almost compensated. It is not exactly compensated because Vegard's law does not apply for the elastic constants, as demonstrated for the case of the AlGaAs system [98]. Within the period, the strain averaged over neighboring Si and SiGe layers is changing gradually, as shown in Fig. 9.15*b*, but the strain forces stay well below the limit imposed by strain relaxation. For example, the maximal thickness of the Si layers in the above structure (25 Å) is much less than the critical thickness for a Si layer on a 50% relaxed buffer layer to form misfit dislocations, which is 10 nm. Similarly, on the length scale of a cascade period (40.4 nm for the structure above) the Ge concentration does not deviate from the average by more than 2.5%, as shown in Fig. 9.15*b*. On the other hand, on this length scale SiGe layers with even a 20% mismatch would be stable against the spontaneous formation of misfit dislocations. Emission from this structure was found to occur at 176 meV, in good agreement with the calculations. The current voltage *(I-V)* characteristics measured at 80 K show a clear onset, corresponding to the voltage required to align the minibands leading to a noticeable current flow through the structure. This onset occurs at a lower bias than the design value, but was found to scale well among three identical samples with 3, 15, and 30 periods, indicating that the current is flowing through the QW states and not through the dislocations present in the structure. The full width at half maximum of the intersubband emission was about 45 meV, measured close to the emission onset, and has been ascribed to scattering due to interface roughness.

Using interface roughness parameters estimated from X-ray reflectivity, from high-resolution cross-sectional transmission electron microscopy, and from the measured relation between thickness and

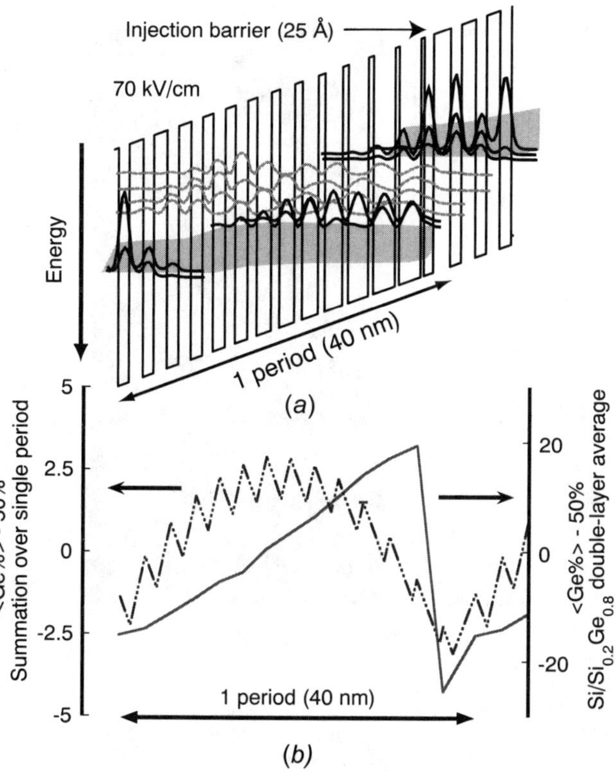

Figure 9.15 (*a*) Schematic diagram of a Si/SiGe QC structure of the bound-to-continuum type, which is strain-compensated to a relaxed buffer with 50% germanium content. The moduli squared of the wave functions of the HH states relevant for the optical transition and of some of the interdispersed LH states are given by the solid and dashed lines, respectively. Note that the energy axis is upside down. (*b*) Ge concentration averaged over one cascade period, given in relation to the 50% Ge concentration of the relaxed buffer substrate. The peak values of the Ge concentration averaged over one period (dashed line, left-hand scale) and averaged over adjacent Si and $Si_{0.2}Ge_{0.8}$ layers are non-critical, i.e., the respective length scales—period length (40 nm) and layer thicknesses (4 Å $< t <$ 25 Å)—are considerably smaller than the critical layer thicknesses for which, at corresponding Ge concentrations, dislocations would be formed. This makes it possible to grow many periods of this structure. [Source: Diehl et al., *Physica* E16, 315 (2003).]

low-temperature Hall mobility in Si/SiGe QWs [99], Tsujino et al. calculated the electroluminescence line width of the emitters as a function of the correlation length of the interface roughness in the vertical (i.e., growth) direction [100]. As shown in Fig. 9.16, the measured line width could be reproduced, assuming a vertical correlation length in the range of 1.4 to 2.2 nm. This implies that partial cancellation of the interface roughness does occur in the investigated cascade structures, as the active QWs are separated by less than this amount.

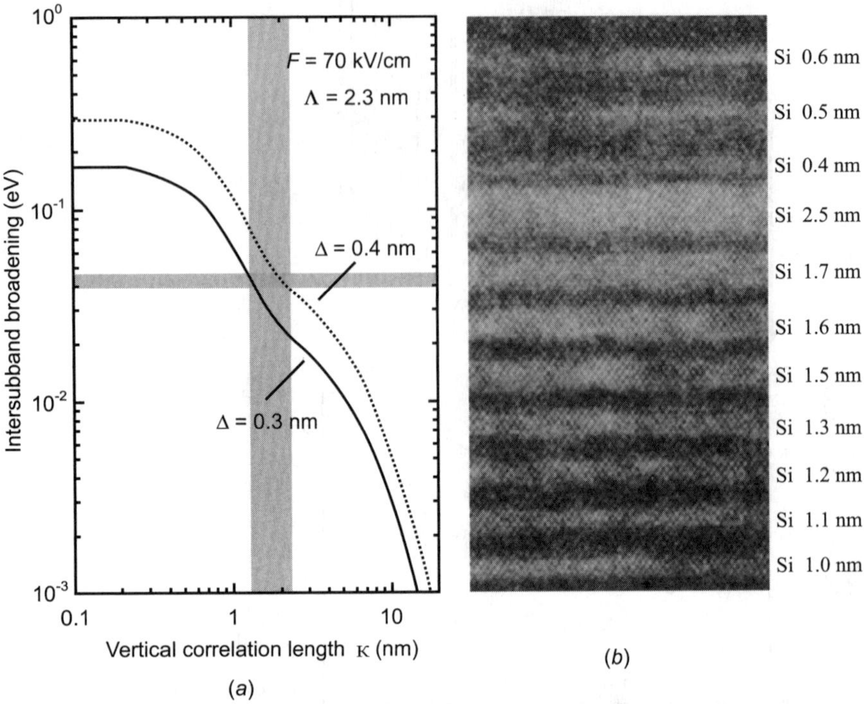

Figure 9.16 (*a*) Calculated line width of the intersubband electroluminescence spectrum of the SiGe/Si QC structure shown on Fig. 9.15*a*, assumed to be limited by interface roughness scattering, plotted as a function of the vertical correlation length κ. The values for the lateral roughness and lateral correlation length used in this plot, 0.3 nm < Δ < 0.4 nm and Λ = 2.3 nm, respectively, were taken from Hall measurements and transmission electron microscopy (TEM). The observed emission line broadening of ~45 meV is consistent with interfaces that are correlated at separations up to 1.8 nm ± 0.4 nm. (*b*) TEM cross section showing the Si barriers (bright) and the $Si_{0.2}Ge_{0.8}$ wells (dark). [Source: Tsujino et al., *Appl. Phys. Lett.* 86, 062,113 (2005).]

A detailed analysis of the lifetime performance and current limitations of mid-infrared Si-based QC structures grown on relaxed buffer layers and based on the bound-to-continuum design has been presented recently by Borak et al. [37]. From the measured emission efficiencies of 10^{-5} to 10^{-6} such devices and calculated radiative lifetime of ~95 ns, an upper-state lifetime between 10 and 100 fs was deduced, assuming a 100% injection efficiency. The emission efficiency started to level off at current densities of ~6.5 kA/cm^2. From these data, an upper bound for the gain coefficient G per period was estimated at ~1 to 2 cm^{-1}. This is a very low number, particularly in relation with the *free-carrier absorption* (FCA) coefficient of the active layer itself. By using the effective mass $m^* = 0.2m_0$ and a momentum relaxation time of ~10 fs [corresponding to a mobility of ~100 cm^2/(V · s)] the FCA coefficient was estimated to be ~15 cm^{-1} at the device emission photon energy of ~200 meV and carrier concentration of 5 × 10^{17} cm^{-3}. Therefore, a substantial performance improvement is required to surmount the free-carrier losses as well as the additional losses in a laser waveguide device. Varying the doping density may not be an effective strategy in this respect, since for a parabolic dispersion the ratio *G*/FCA is approximately independent of carrier concentration. Rather, a substantial increase in upper-state lifetime and/or a reduction of the line width, e.g., by reducing the effect of nonparabolicity by working closer to $k_{par} = 0$, will likely be required to improve the gain and hence achieve lasing.

A possible approach to achieve the required performance improvement is to design structures in which the overlap of the upper emission state with the LH/SO states is decreased and/or the energy separation between these states is reduced to below the optical phonon energies. This turns out to be difficult, because the LH/SO states tend to disperse when the HH states align and form a miniband. Further improvements may be obtained by using nonvertical transitions to extend the upper-state lifetime, as in the work by Bormann et al. [36]. The results presented by Tsujino et al. [100], showing the effect on the emission line width of vertical correlations in interface roughness, also provide important guidelines in the design of improved Si-based QC structures. Aside from these considerations relating to the active layer, several aspects of the device physics also have to be addressed and optimized before intersubband laser action can be demonstrated in Si/SiGe QWs. These include the design of the optical waveguide to obtain high mode filling, uniform carrier injection, and minimal optical losses from the electrical contacts. A detailed discussion of these aspects is, however, beyond the scope of this chapter.

9.5 Summary and Outlook

Progress over the last 15 years of intersubband-based spectroscopy and applications of the Si/SiGe system is reviewed. Technical details of various intersubband experiments covering linear absorption, emission, modulation by electric and light fields, and time-solved spectroscopy are introduced. Most of the investigations discussed involve *p*-type intersubband systems because of the favorable (compared to the conduction band systems) large band offsets and low confinement masses in the valence band. Corresponding polarization-dependent selection rules of intersubband and intralevel transition within and between the multivalence band are summarized, and experiments are presented that demonstrate level mixing and crossing. Measurement configurations are elucidated for which normal-incidence absorption is taking place.

Advanced Si epitaxy, in particular for growth on relaxed buffer substrate, is demonstrated by narrow intersubband line width. Excellent agreement is found for band structure calculation exposed to detailed intersubband studies of *p*-type Si/SiGe QWs grown strain-compensated on relaxed buffer substrate.

Advances in Si-based infrared detection are illuminated by giving examples from pseudomorphic QW systems to QD systems with two-dimensional hole channels to improve responsivity. It is concluded that the application potential of Si-based infared photodetectors is still valid, provided that reduction of dark currents at temperatures beyond 30 K can be realized. A detailed summary is given of recent efforts to develop a Si-based intersubband laser. Systems presented include, for the terahertz regime, single-QW superlattices and, for mid-infrared emission, multiperiod cascade structures, consisting of tunneling injector, active QWs for vertical or diagonal emission, and continuum-type depopulation sections. Presently, a gain deficit due to interface-roughness-broadened emission and short upper-state lifetimes, due to immersed LH/SO states, combined with strong free-carrier absorption is hindering the latter system from successfully lasing. It is argued that formation of electric field domains is occurring in the biased superlattices, in contradiction to the implications made for modeling of the high-field-current transport.

Research opportunities are suggested for the study of depolarization and many-body effects in the intersubband absorption of QWs and interlevel transitions in QDs. The latter systems resemble a strongly disordered two-dimensional system, for which phase locking of the distributed dipoles by the electromagnetic field is predicted to generate large depolarization shifts: toward the blue for perpendicular (to the QW) polarization and toward the red for parallel polarization. However, interlevel absorption performed on Ge island arrays shows a blue shift for increasing level population. Exploring light modulation techniques to study the intersubband resonances of two- and three-dimensional strain-field ordered arrays of Ge QD islands—obtained by low-temperature Si capping—may allow clarification of this matter by experiment.

9.6 Acknowledgments

The author wishes to thank A. Borak, P. Bouceaud, T. Fromherz, R. Paiella, and D. Paul for critical reading of and suggestions on essential parts of the manuscript. I would like also to acknowledge the many collaborators working with me over the last 10 years on the topics presented here. For frequent discussion of the presented material, I wish to thank S. Tsujino. Finally, I would like to thank

G. Abstreiter, T. Fromherz, D. Paul, K. Reimann, and A. Yakimov for permission to use various figures presented in this chapter.

References

1. For a review see T. Ando, A. B. Fowler, and F. Stern, *Rev. Modern Phys.* 54, 437 (1982).
2. L. C. West and S. J. Eglash, *Appl. Phys. Lett.* 46, 1156 (1985).
3. For a review see "Intersubband Transitions in Quantum Wells," E. Rosencher, B. Vinter, and B. Levine, eds., *Proceedings of a NATO Advanced Workshop CY,* Sept. 9–14, 1991, CL Cargèse, France, 1992.
4. M. Helm, in *Intersubband Transitions in Quantum Wells: Physics and Device Applications I,* H. C. Liu and F. Capasso, eds., *Semiconductors and Semimetals* 62, 99 (2000).
5. S. K. Chun, and K. L. Wang, *Phys. Rev.* B46, 7682 (1992).
6. W. I. Xu, Y. Fu, M. Willander, and S. C. Shen, *Phys. Rev.* B49, 13760 (1994).
7. T. Fromherz, E. Koppensteiner, M. Helm, G. Bauer, J. F. Nutzel, and G. Abstreiter, *Phys. Rev.* B50, 15073 (1994).
8. M. El Kurdi, G. Fishman, S. Sauvage, and P. Boucaud, *Phys. Rev.* B68, 165333 (2003).
9. Cho, S. M. Lee, H. H., Journ. Appl. Phys. 73, 1918 (1993).
10. G. Hionis, M. Tsetseri, A. Zora, and G. P. Triberis, *Superlattices and Microstructures* 28, 151 (2000).
11. W. L. Xu, Y. Fu, M. Willander, and S. C. Shen, *Phys. Rev.* B48, 11477 (1993).
12. A. Kahan, M. Chi, and L. Friedman, *J. Appl. Phys.* 75, 8012 (1994).
13. T.-S. Liou, T. Wang, and C.-Y. Chang, *J. of Appl. Phys.* 77, 6646 (1995).
14. S. Ridene, K. Boujdaria, H. Bouchriha, and G. Fishman, *Phys. Rev.* B64, 1 (2001).
15. K. Boujdaria, S. Ridene, S. B. Radhia, O. Zitouni, H. Bouchriha, and G. Fishman, *J. Appl. Phys.* 92, 2586 (2002).
16. D. E. Weeks, S. H. Yang, M. R. Gregg, S. J. Novotny, K. D. Greene, and R. L. Hengehold, *Phys. Rev.* B65, 195314 (2002).
17. Y.-Y. Lin, and J. Singh, *J. Appl. Phys.* 96, 1059 (2004).
18. L. Diehl, H. Sigg, G. Dehlinger, D. Grützmacher, E. Müller, U. Gennser, I. Sagnes, T. Fromherz, Y. Campidelli, O. Kermarrec, D. Bensahel, and J. Faist, *Appl. Phys. Lett.* 80, 3274 (2002).
19. C. G. Van der Walle *Phys. Rev.* B39, 1871 (1989).
20. M. M. Rieger and P. Vogl, *Phys. Rev.* B48, 14276 (1993).

21. T. Fromherz, M. Meduna, G. Bauer, A. Borak, C. V. Falub, S. Tsujino, H. Sigg, and D. Grützmacher, *J. Appl. Phys.* 78, 044501 (2005).
22. R. Winkler, M. Merkler, T. Darnhofer, and U. Rössler, *Phys. Rev.* B53, 10858 (1993).
23. G. Sun, and L. Friedman, *Phys. Rev.* B53, 3966 (1996).
24. K. Reimann, R. A. Kaindl, and M. Woerner, *Phys. Rev.* B65, 045302 (2002).
25. H. Sigg, M. H. Kwakernaak, B. Margot, D. Erni, P. van Son and K. Köhler, *Appl. Phys. Lett.* 67, 2827 (1995).
26. L. Diehl, A. Borak, S. Mentese, D. Grützmacher, H. Sigg, U. Gennser, I. Sagnes, Y. Campidelli, O. Kermarrec, D. Bensahel, and J. Faist, *Appl. Phys. Lett.* 84, 2497 (2004).
27. E. Dekel, E. Ehrenfreund, D. Gershoni, P. Boucaud, I. Sagnes, and Y. Campidelli, *Phys. Rev.* B56, 15734 (1997).
28. L. Wu, P. Boucaud, J.-M. Lourtioz, F. H. Julien, I. Sagnes, Y. Campidelli, and P.-A. Badoz, *Appl. Phys. Lett.* 67, 3462 (1995).
29. P. Boucaud, L. Gao, Z. Moussa, F. Visocekas, F. H. Julien, J.-M. Lourtioz, I. Sagnes, Y. Campidelli, and P.-A. Badoz, *Appl. Phys. Lett.* 67, 2948 (1995).
30. P. Boucaud, F. H. Julien, R. Prazeres, J.-M. Ortega, I. Sagnes, and Y. Campidelli, *Appl. Phys. Lett.* 69, 3069 (1996).
31. R. A. Kaindl, M. Wurm, K. Reimann, M. Woerner, T. Elsaesser, C. Miesner, K. Brunner, and G. Abstreiter, *Phys. Rev. Lett.* 86, 1122 (2001).
32. W. Heiss, E. Gornik, H. Hertle, B. Murdin, G. M. H. Knippels, C. J. G. M. Langerak, F. Schaffler, and C. R. Pidgeon, *Appl. Phys. Lett.* 66, 3313 (1995).
33. P. Murzyn, C. R. Pidgeon, J.-P. R. Wells, I. V. Bradley, Z. Ikonic, R. W. Kelsall, P. Harrison, S. A. Lynch, D. J. Paul, D. D. Arnone, D. J. Robbins, D. Norris, and A. G. Cullis, *Appl. Phys. Lett.* 80, 1456 (2002).
34. R. W. Kelsall, Z. Ikonic, P. Murzyn, C. R. Pidgeon, P. J. Phillips, D. Carder, P. Harrison, S. A. Lynch, P. Townsend, D. J. Paul, S. L. Liew, D. J. Norris, and A. G. Cullis, *Phys. Rev.* B71, 115326 (2005).
35. G. Dehlinger, L. Diehl, U. Gennser, H. Sigg, J. Faist, K. Ensslin, D. Grützmacher, and E. Müller, *Science* 290, 2277 (2000).
36. I. Bormann, K. Brunner, S. Hackenbuchner, G. Abstreiter, S. Schmult, and W. Wegscheider, *Appl. Phys. Lett.* 83, 5371 (2003).
37. A. Borak, S. Tsujino, C. V. Falub, M. Scheinert, L. Diehl, E. Müller, H. Sigg, D. Grützmacher, U. Gennser, I. Sagnes, S. Blunier, T. Fromherz, Y. Campidelli, O. Kermarrec, D. Bensahel, and J. Faist , *MRS Symp. Proc.* 832, F4.2.1 (2005).

38. R. P. G. Karunasiri, J. S. Park, K. L. Wang, and Li-Jen Cheng, *Appl. Phys. Lett.* 56, 1342 (1990).
39. R. P. G. Karunasiri, J. S. Park, Y. J. Mii, and K. L. Wang, *Appl. Phys. Lett.* 57, 2585 (1990).
40. J. S. Park, R. P. G. Karunasiri, Y. J. Mii, and K. L. Wang, *Appl. Phys. Lett.* 58, 1083 (1991).
41. J. S. Park, R. P. G. Karnasiri, and K. L. Wang, *Appl. Phys. Lett.* 61, 681 (1992).
42. S. K. Chun, D. S. Pan, and K. L. Wang, *Phys. Rev.* B47, 15638 (1993).
43. H. Hertle, G. Schuberth, E. Gornik, G. Abstreiter, and F. Schaffler, *Appl. Phys. Lett.* 59, 2977 (1991).
44. C. Lee, and K. L.Wang, *Appl. Phys. Lett.* 60, 2264 (1992).
45. C. Lee, and K. L. Wang, *Appl. Phys. Lett.* 64, 1256 (1994).
46. K. Fujita, S. Fukatsu, Y. Shiraki, H. Yaguchi, and R. Ito, *Appl. Phys. Lett.* 61, 210 (1992).
47. D. K. Nayak, N. Usami, S. Fukatsu, and Y. Shiraki, *Appl. Phys. Lett.* 63, 3509 (1993).
48. T. Fromherz, P. Kruck, M. Helm, G. Bauer, J. F. Nutzel, and G. Abstreiter, *Appl. Phys. Lett.* 68, 3611 (1996).
49. J. M. Berroir, S. Zanier, Y. Guldner, J. P. Vieren, I. Sagnes, Y. Campidelli, P. A. Badoz, and B. Gil, and R.-L. Aulombard, *Proceedings of Semiconductor Heteroepitaxy: Growth Characterization and Device Applictions CY* July 4–7 1995, CL Montpellier, France, P.314, 1995.
50. P. Boucaud, J.-M. Lourtioz, F. H. Julien, P. Warren, and M. Dutoit, *Appl. Phys. Lett.* 69, 1734 (1996).
51. P. Boucaud, L. Wu, C. Guedj, F. H. Julien, I. Sajnes, Y. Campidelli, and L. Garchery, *J. Appl. Phys.* 80, 1414 (1996).
52. R. Oszwaldowski, and G. Fishman, *Molecular Physics Reports CT National Symposium on Crystalline Microstructures CY* July 2–4 1997, CL Poznan, Poland, vol. 21, p. 99, 1998.
53. B. Adoram, D. Krapf, J. Shappir, A. Sa'ar, M. Levy, R. Beserman, S. G. Thomas, and K. L. Wang, *Appl. Phys. Lett.* 75, 2232 (1999).
54. M. Meduna, J. Novak, C. V. Falub, G. Chen, G. Bauer, S. Tsujino, D. Grutzmacher, E. Muller, Y. Campidelli, O. Kermarrec, D. Bensahel, and N. Schell, *J. Phys. D. Appl. Phys.* 38, A121 (2005).
55. M. Y. A. Yousif, O. Nur, M .Willander, C. J. Patel, C. Hernandez, Y. Campidelli, D. Bensahel, and R. N. Kyutt, *Solid-State Electronics* 45, 1869 (2001).
56. G. Isella, D. Chrastina, B. Rossner, T. Hackbarth, H. Herzog, U. Konig, and H. von Känel, *Solid-State Electronics* 48, 1317 (2004).
57. S. Tsujino, A. Borak, C. V. Falub, T. Fromherz, L. Diehl, H. Sigg, and D. Grützmacher, *Phys. Rev.* B72, 153315 (2005).

58. K. Brunner, *Reports on Progress in Physics* 65, 27 (2002).
59. Y.–W. Mo, D. E. Savage, B. S. Swartzentruber, and M. G. Lagally, *Phys. Rev. Lett.* 65, 1020 (1990).
60. A. I. Yakimov, A. V. Dvurechenskii, N. P. Stepina, and A. I. Nikiforov, *Phys. Rev.* B62, 9939 (2000).
61. U. Denker, M. Stoffel, O. G. Schmidt, and H. Sigg, *Appl. Phys. Lett.* 82, 454 (2003).
62. M. Stoffel, U. Denker, G. S. Kar, H. Sigg, and O. G. Schmidt, *Appl. Phys. Lett.* 83, 2910 (2003).
63. J. L. Liu, W. G. Wu, A. Balandin, G. L. Jin, and K. L. Wang, *Appl. Phys. Lett.* 74, 185 (1999).
64. T. Fromherz, W. Mac, A. Hesse, G. Bauer, C. Miesner, K. Brunner, and G. Abstreiter, *Appl. Phys. Lett.* 80, 2093 (2002).
65. C. Miesner, O. Rothig, K. Brunner, and G. Abstreiter, *Appl. Phys. Lett.* 76, 1027 (2000).
66. P. Boucaud, V. Le Thanh, S. Sauvage, D. Debarre, and D. Bouchier, *Appl. Phys. Lett.* 74, 401 (1999).
67. S. Sauvage, P. Boucaud, J.-M. Gerard, and V. Thierry-Mieg, *Phys. Rev.* B58, 10562 (1998).
68. C. Metzner, and G. H. Döhler, *Physica* E7, 718 (2000).
69. R. N. Riemann, C. Metzner, and G. H. Döhler, *Phys. Rev.* B65, 115304 (2002).
70. V. Bondarenko, M. Zaluzny, and Y. Zhao, *Phys. Rev.* B71, 115304 (2005).
71. M. El Kurdi, P. Boucaud, S. Sauvage, F. Aniel, G. Fishman, O. Kermarrec, Y. Campidelli, D. Bensahel, G. Saint-Girons, I. Sagnes, and G. Patriarche, *J. Appl. Phys.* 97, 83525 (2005).
72. B.F. Levine, *J. Appl. Phys.* 74, R1 (1993).
73. H.C. Liu, in *Intersubband Transitions in Quantum Wells: Physics and Device Applications*, H. C. Liu, and F. Capasso, eds., *Semiconductors and Semimetals* 62, 129 (2000).
74. P. Kruck, M. Helm, T. Fromherz, G. Bauer, J. F. Nutzel, and G. Abstreiter, *Appl. Phys. Lett.* 69, 3372 (1996).
75. D. Krapf, B. Adoram, J. Shappir, A. Sa'ar, S. G. Thomas, J. L. Liu, and K. L. Wang, *Appl. Phys. Lett.* 78, 495 (2001).
76. P. Rauter, T. Fromherz, G. Bauer, L. Diehl, G. Dehlinger, H. Sigg, D. Grützmacher, and H. Schneider, *Appl. Phys. Lett.* 83, 3879 (2003).
77. C. Miesner, K. Brunner, and G. Abstreiter, *Physica Status Solidi B, Basic Research* 224, 605 (2001).
78. D. Bougeard, K. Brunner, and G. Abstreiter, *Physica E-, Low-Dimensional Systems & Nanostructures* 16, 609 (2003).
79. A. I. Yakimov, A. V. Dvurechenskii, A. I. Nikiforov, and Y. Y. Proskuryakov, *J. Appl. Phys.* 89, 5676 (2001).

80. J. Jiang, S. Tsao, T. O'Sullivan, W. Zhang, H. Lim, T. Sills, K. Mi, M. Razeghi, G. J. Brown, and M. Z. Tidrow, *Appl. Phys. Lett.* 84, 2166 (2004).

81. S. Chakrabarti, A. D. Stiff-Roberts, P. Bhattacharya, S. Gunapala, S. Bandara, S. B. Rafol, and S. W. Kennerly, *IEEE Photonics Technol. Lett.* 16, 2004 (2004).

82. Y. H. Peng, C. C. Chen, C. H. Kuan, and H. H. Cheng, *Solid-State Electron.* 47, 1775 (2003).

83. N. Rappaport, E. Finkman, T. Brunhes, P. Boucaud, S. Sauvage, N. Yam, V. Le Thanh, and D. Bouchier, *Appl. Phys. Lett.* 77, 3224 (2000).

84. N. Rappaport, E. Finkman, P. Boucaud, S. Sauvage, T. Brunhes, V. Le Thanh, D. Bouchier, and S. E. Schacham, *Infrared Phys. & Technol.* 44, 513 (2003).

85. A. Borak, *Science* 308, 638 (2005).

86. S. A. Lynch, R. Bates, D. J. Paul, D. J. Norris, A. G. Cullis, Z. Ikonic, R. W. Kelsall, P. Harrison, D. D. Arnone, and C. R. Pidgeon, *Appl. Phys. Lett.* 81, 1543 (2002).

87. R. Bates, S. A. Lynch, D. J. Paul, Z. Ikonic, R. W. Kelsall, P. Harrison, S. L. Liew, D. J. Norris, A. G. Cullis, W. R. Tribe, and D. D. Arnone, *Appl. Phys. Lett.* 83, 4092 (2003).

88. Z. Ikonic, P. Harrison, and R. W. Kelsall, *Phys. Rev.* B64, 245311 (2001).

89. Z. Ikonic, P. Harrison, and R. W. Kelsall, *J. Appl. Phys.* 96, 2004 (2004).

90. Z. Ikonic, R. W. Kelsall, and P. Harrison, *Phys. Rev.* B69, 235308 (2004).

91. L. Friedman, G. Sun, and R. A. Soref, *Appl. Phys. Lett.* 78, 401 (2001).

92. S. Tsujino, S. Mentese, L. Diehl, E. Müller, B. Hass, D. Bächle, S. Stutz, D. Grützmacher, Y. Campidelli, O. Kermarrec, and D. Bensahel, *Appl. Surf. Science* 224, 337 (2004).

93. Z. Ikonic, R. W. Kelsall, and P. Harrison, *Semiconductor Sci. and Technol.* 19, 76 (2004).

94. R. W. Kelsall, Z. Ikonic, P. Harrison, S. A. Lynch, P. Townsend, D. J. Paul, D. J. Norris, S. L. Liew, A. G. Cullis, X. Li, J. Zhang, M. Bain, and H. S. Gamble, *Opt. Materials* 27, 851 (2005).

95. H. Sigg, G. Dehlinger, L. Diehl, U. Gennser, S. Stutz, F. Faist, D. Grützmacher, K. Ensslin, and E. Müller, *Physica E CT Rutherford Memorial Workshop on Semiconductor Nanostructures CY* Feb. 5–9, 2001 CL, Queenstown, New Zealand, vol. 11, no. 2–3, pp. 240 (2001).

96. I. Bormann, K. Brunner, S. Hackenbuchner, G. Zandler, G. Abstreiter, S. Schmult, and W. Wegscheider, *Appl. Phys. Lett.* 80, 2260 (2002).

97. L. Diehl, S. Mentese, E. Müller, D. Grützmacher, H. Sigg, U. Gennser, I. Sagnes, Y. Campidelli, O. Kermarrec, D. Bensahel, and J. Faist, *Appl. Phys. Lett.* 81, 4700 (2002).

98. S. Gehrsitz, H. Sigg, N. Herres, K. Bachem, K. Köhler and F. K. Reinhart, *Phys. Rev.* B60, 11601 (1999).

99. S. Tsujino, C. V. Falub, E. Müller, M. Scheinert, L. Diehl, U. Gennser, T. Fromherz, A. Borak, H. Sigg, D. Grützmacher, Y. Campidelli, O. Kermarrec, and D. Bensahel, *Appl. Phys. Lett.* 84, 2829 (2004).

100. S. Tsujino, A. Borak, E. Müller, M. Scheinert, C. V. Falub, H. Sigg, D. Grützmacher, M. Giovannini, and J. Faist, *Appl. Phys. Lett.* 86, 62113 (2005).

Chapter

10

All-Optical Modulation and Switching in the Communication Wavelength Regime Using Intersubband Transitions in InGaAs/AlAsSb Heterostructures

Arup Neogi

Department of Physics and Materials Engineering
University of North Texas
Denton, Texas

10.1 Introduction

The *International Technology Roadmap for Semiconductors* has reported the presence of a potential "red brick wall" or "100-nm wall" that by 2010 could block further scaling of integrated electronic circuits, as predicted by Moore's law [1]. A novel photonic technology that goes beyond the diffraction limit is thus likely to be needed to meet the demands of the future semiconductor industry. At the same time, communication networks currently operating in the 10 to 40 Gb/s range are expected to exceed 10 Tb/s by 2015, creating new challenges in transmission, switching, and networking [1]. This will necessitate ultrafast and subwavelength-scale photonic integrated circuits including devices such as switches, sources, and interconnects, e.g., to accommodate 1000 × 1000 channels on a single substrate, and to process each channel at much higher data rates than allowed by conventional electronics [2, 3].

To address the issue of physical scaling, novel functionalities are being explored based on local electromagnetic interactions between nanometric elements and an optical near-field wave, which are not limited by the principles of conventional wave optics such as interference and diffraction. Devices with fundamentally different physical principles, e.g., based on the concepts of surface elementary excitations and nanofabrication technology, are being developed that can overcome fundamental constraints in transmission or emission of light from nanoscale structures.

At the same time, there is a strong tendency to replace "slow" electronic devices with "fast" photonic ones (Fig. 10.1) [4]. This drive is especially evident in the areas of computing and communications. It requires the development of an optical analog of an electronic integrated circuit capable of routing, controlling, and processing optical signals. The use of optical and optoelectronic devices for ultrafast switching and their integration in photonic circuits in turn require novel approaches to light manipulation. In particular, it will require (active) nonlinear photonic elements capable of performing, in the optical domain, logic operations analogous to their electronic counterparts without optical-to-electrical and electrical-to-optical signal conversion (Figs. 10.1*b* and 10.2).

The fundamental understanding and exploitation of quantum and nonlinear optical effects for the development of such optical switches and modulators promise to revolutionize the field of nanophotonic devices and systems. A particularly interesting example, discussed in this chapter, is nonlinear devices based on intersubband transitions in semiconductor quantum wells, whose development integrates in a novel way aspects of quantum physics, optoelectronics, material science, and telecommunication engineering.

10.2 Femtosecond All-Optical Switches: The Challenge and Present Status

Electronic circuits conventionally used for data switching are limited in their speed due to either the carrier's transit times and lifetimes or *RC* time constants. Switching devices for ultrafast operation in the picosecond and femtosecond regimes must be all-optical, so that any limitation of the electronic nature can be avoided. Such ultrafast all-optical switches are a prerequisite for future *optical time-division multiplexing* (OTDM) systems. Also, in the *wavelength-division multiplexing* (WDM) architecture, the development of high-speed all optical switches is of great importance to realize networks with wavelength transparency. At the optical nodes of such networks, various optical signal switching and routing elements are required to carry out functions such as *add/drop multiplexing* (ADM) and cross-connect. Thus, different types of switching devices suitable for

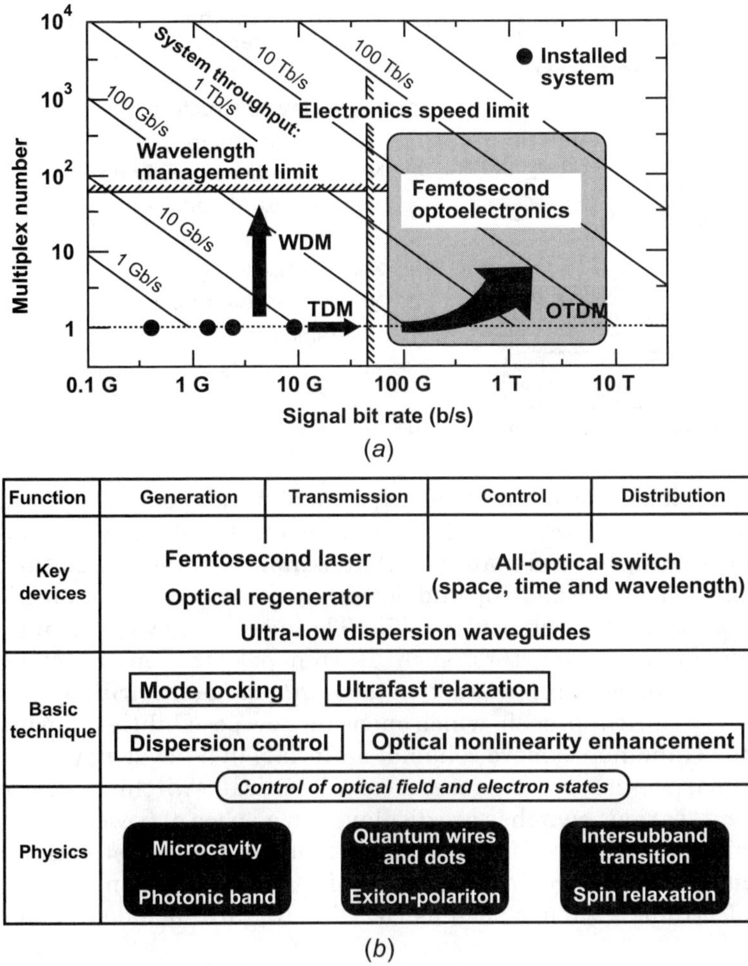

Figure 10.1 (*a*) Different approaches to improve system throughput shown in a space of channel multiplex number as a function of signal bit rate per channel. (*b*) Key femtosecond devices required for future ultrahigh-throughput system operation, elementary functions, and basic techniques as well as novel material and semiconductor physics, which can be used in the development of these devices.

operation in the time, space, and wavelength divisions need to be developed.

The essential features required for all these devices include an ultrafast response time, high repetition rate, low switching energy, high extinction ratio, and low insertion loss. A few different approaches can be used to develop practical, ultrafast all-optical switches with the above-mentioned characteristics. One approach is to develop new device structures that enable ultrafast operation without relying on the

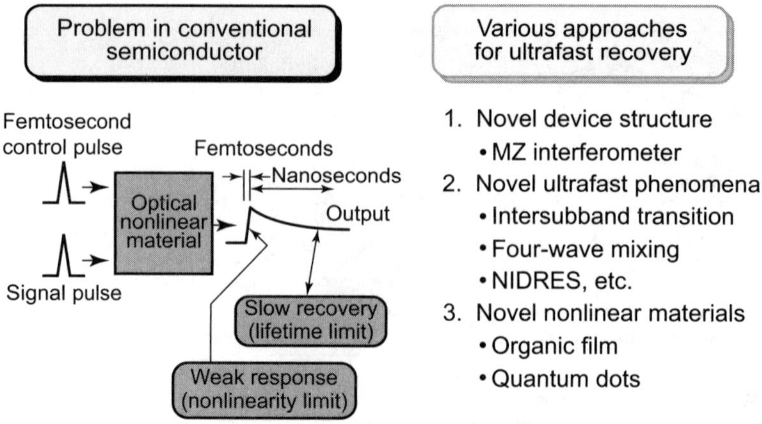

Figure 10.2 Overview of ultrafast optical switching.

material's intrinsic speed but at the expense of device complexity, i.e., *symmetric Mach-Zehnder* (SMZ) interferometers [5]. The second approach is to utilize novel materials that exhibit ultrafast relaxation, such as organic materials [6] and low-temperature-grown *multiple quantum well* (MQW) structures [7]. The use of novel ultrafast phenomena in semiconductors, such as spin polarization switching due to the D'yakonov-Perel interaction [8] or *intersubband* (ISB) optical switching, is a third approach, which opens up new possibilities of using more mature (and hence more reliable) materials in a novel way.

To summarize the present status of all-optical switching device development, various switches are displayed in a space of power per bit versus switching time in Fig. 10.3 [4, 9]. A large number of existing devices, including electronic transistors, Josephson junction devices, and optoelectronic switches such as SEEDs [10], are included in this

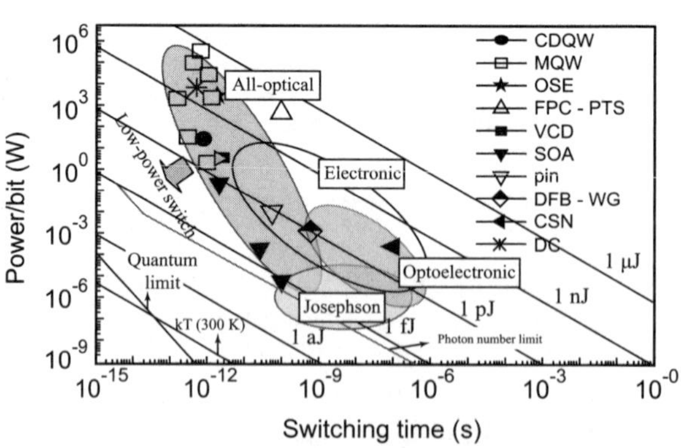

Figure 10.3 A plot of switching power per bit versus switching time for a variety of devices including all-optical, electronic, optoelectronic, and Josephson junction devices.

figure for comparison. All-optical devices include symmetric Mach-Zehnder switches [11], *polarization-discriminating SMZ* (PZ-SMZ) switches [12], *low-temperature-grown surface reflection all-optical switches* (LOTOS) [13], spin polarization switches based on MQWs (SPS-MQW) [14], bistable devices based on *semiconductor optical amplifiers* (SOAs) [15], conventional MQW etalons [16], *optical Stark effect* (OSE) devices operating at low temperature [17], and *current-induced directional coupler* (CIDC) switches [18]. In addition, recent studies using *ISB transitions* (ISBTs) in *coupled double quantum wells* (CDQWs) [19, 20] and MQWs [21] have demonstrated very promising results of switching operation in the femtosecond regime and with switching energy on the order of 10 pJ or less.

While impressive results have already been achieved in the development of these all-optical switching devices, further improvements mainly directed toward the reduction of the switching energy are needed. These may be obtained through the enhancement of the optical nonlinearities utilized in the switching process, by the introduction of novel physics and materials, and by novel device structures and system architectures. In general, intrinsic material nonlinearities tend to be weak, but can in some cases be strongly enhanced through careful material engineering. Specifically, to maximize the nonlinear optical susceptibilities, one needs to maximize the products of the dipole moments of the relevant optical transitions, and to minimize the detunings between the energies of the photons participating in the nonlinear interaction and the transition energies [22, 23].

Compared to optical transitions in discrete atomic or molecular-level systems, band-to-band transitions in semiconductors are generally off-resonant due to the electron and hole dispersions within the parabolic conduction and valence bands. As a result, semiconductor materials exhibit relatively weak optical nonlinearities; however, these can be substantially improved by using discrete states in low-dimensional structures such as *quantum wells* (QWs) and *quantum dots* (QDs). In particular, ISBTs in these systems have been recognized early on as very promising for nonlinear optical devices due to their remarkable design flexibility.

To optimize the efficiency and speed of all-optical switching, we combine two approaches to enhance optical nonlinearities. One is resonance enhancement via eigenfunction engineering: We design a structure whose geometric properties enhance the ISB nonlinear interaction in the communication wavelength regime, in particular using the wide-conduction-band-offset InGaAs/AlAsSb QW system. The second approach is to engineer the ISB scattering rate in a multilevel quantum confined system by manipulating the doping level, which changes the switching speed as well as the on/off contrast ratio of the switching process. Nonlinear Kerr coefficients 12 orders of magnitude larger than in bulk AlGaAs have been measured in such systems [24].

10.3 Interband and Intersubband Transitions in Wide-Conduction-Band-Offset Materials

Recently investigated microphotonic techniques for all-optical communications utilize optical nonlinearities based on carrier-induced ISBTs in semiconductor QWs. Intersubband transitions have several advantages for device applications compared to the more conventional band-to-band transitions originating from both exciton recombination and free-carrier processes. These advantages include the ability to control the ISBT energies independently of the material bandgap and to obtain resonance enhancement of ISB optical nonlinearities, as well as the giant ISB dipole moment matrix elements. In addition, ISBTs are characterized by ultrafast (< 1-ps) lifetimes due to *longitudinal optical* (LO) phonon scattering. In the past few years, several groups including ours have successfully demonstrated ultrafast ISB carrier relaxation at 1.55 μm, an important milestone for all-optical modulation and switching in QWs [25–29].

Motivated by these advantageous features of ISBTs, current activities to extend the transmission capacity of fiber-optic networks beyond 1 Tb/s include the development of bandgap-engineered ISB systems [30] with transition wavelengths in the near infrared. These can be realized with QWs and QDs based on wide-conduction-band-offset semiconductor systems, such as III nitrides [31], II-VI compounds (for example, CdS/ZnSe/BeTe) [26], and antimonides [32]. In conventional semiconductor optoelectronic devices, band-to-band transitions (*interband transitions,* or IBTs) are used, whose response time is limited by the electron-hole recombination lifetime (~nanoseconds) [34]. For ultrafast all-optical switching in future OTDM systems, the use of ISBTs in QWs [34–38] or QDs [39] is advantageous, primarily because of the very fast ISB relaxation times. These lifetimes are determined by LO-phonon scattering and are typically in the range of ~50 fs to 3 ps (depending on the electron effective mass), that is, two to three orders of magnitude faster than the interband carrier lifetime.

Following several investigations of ISB all-optical switching, recently the operation of digital (NOR) gates for ultrafast all-optical data processing has also been demonstrated, based on a combination of ISB and IB transitions in QWs [40]. All-optical NOR gating had already been demonstrated using optical fibers [41] or interband SOAs [42]. However, these devices suffer from a low efficiency due to the small nonlinearities in silica fibers, or a low operation speed due to long carrier lifetimes in SOAs. Both of these drawbacks can be overcome by using the large nonlinearities and subpicosecond absorption recovery lifetimes in strongly coupled QWs.

The fundamental requirements for a semiconductor heterostructure (QW or QD) appropriate for ultrafast all-optical modulation or switching in the communication wavelength regime are as follows:

1. A large conduction band offset between the quantum well and the barrier materials, suitable for ISBT energies exceeding 0.8 eV (1.55 μm)
2. Ultrafast ISB relaxation rate governed by LO-phonon scattering and dependent upon the overlap of the electron wave functions of the conduction subband states
3. Favorable optical properties including transparency at the relevant wavelength, narrow ISB absorption peaks, and a large third-order ISB optical nonlinearity [$\chi^{(3)}$] for low-threshold operation

There are also a few accepted limitations to the performance of devices based on ISBTs between conduction subbands. The foremost limitation is the need for the incident radiation to be linearly polarized along the growth direction of the QWs. This is related to the spherically symmetric (s-like) nature of the conduction subband electronic states. Techniques such as grating coupling, textured surfaces, or end firing into a waveguide can be used to deal with this limitation for most device applications. There have also been some efforts to achieve normal-incidence ISBTs, i.e., using indirect-gap semiconductors, transitions between quantized valence band states, or QDs. In general, the possibility of normal-incidence operation would add versatility to all ISB devices including all-optical modulators and switches.

10.3.1 Material systems

The AlGaAs/GaAs system has been one of the most extensively studied material systems for the development of mid-infrared and far-infrared (terahertz) ISB devices since the first observation of ISBTs in 1985 [43]. The negligible mismatch between AlAs and GaAs (~0.1%) allows for practically unstrained epitaxial growth on GaAs substrates. However, the conduction band offset between GaAs and AlAs is insufficient to support ISBTs in the optical communication wavelength region (~0.8 eV).

Presently there are four different material systems that are being extensively studied for all-optical ISB devices at communication wavelengths: InGaAs/AlAsSb QWs lattice-matched on InP [44], InGaAs/AlAs QWs [45], AlGaN/GaN QWs [46] and QDs [47], and II-VI systems such as CdS/ZnSe/BeTe QWs [26].

Optoelectronic devices based on InP substrates have been gaining prominence in the past several years due to the availability of lattice-matched ternary and quaternary alloys useful for the visible and near-infrared wavelength regions. This is illustrated in Fig. 10.4. For InGaAs QWs (lattice-matched to InP substrates) and AlAs barriers, large conduction band and valence band offsets (~1.3 and ~1.1 eV, respectively) can be obtained. Also, because the light-hole (LH) mass in InGaAs is smaller than in AlAs, whereas the heavy-hole (HH) masses are comparable, the energy separation between the first HH and LH subbands is relatively large. This is further enhanced by a built-in splitting caused by strain. This enhanced separation between HH and LH states is useful for avoiding LH conduction band transitions in case of interband light modulation. However, the InGaAs/AlAs/InP system is seriously limited by the large lattice mismatch between the AlAs barrier and the InP substrate. There are also segregation effects observed in the presence of high indium content, which becomes significant in the narrow wells essential for communication wavelength devices. For such narrow wells (~2 nm) there are reports of intervalley transfer of electrons from the Γ valley in the well to the X valley in the barrier [49]. The other promising InP-based material system consists of lattice-matched InGaAs/AlAsSb QWs, which are described in detail in the next section.

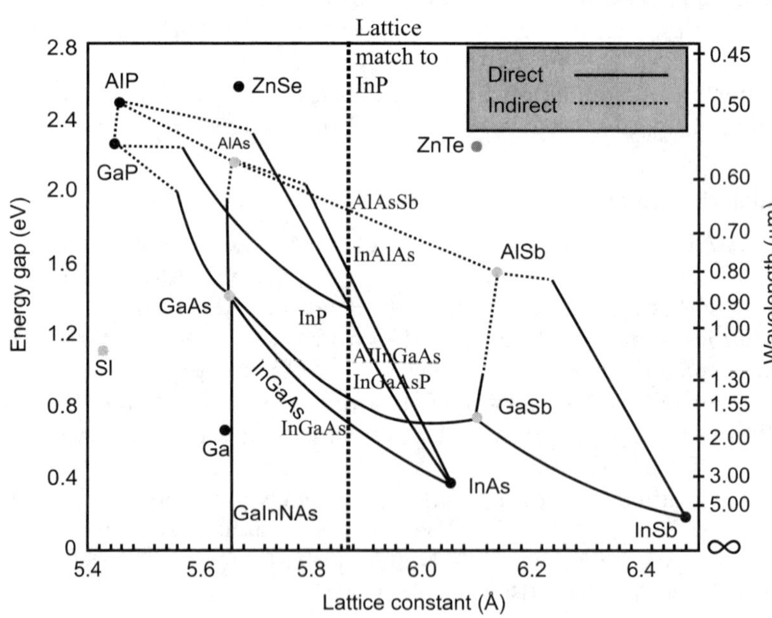

Figure 10.4 InP-based material system.

The other relevant III-V material system is the nitrogen-based AlGaN/GaN system grown on sapphire or SiC substrates, which yields a type I heterostructure with a large conduction band offset [48]. However, the growth of nitride semiconductors is still not mature due to the lack of native substrates such as GaN. Thus, similar to the case of InGaAs/AlAs heterostructures, Al(Ga)N/GaN QWs are highly strained, which restricts the total number of wells that can be grown pseudomorphically, limiting the effective ISB absorption within the active medium [49, 50]. In addition, while in InGaAs/AlAs QWs near-infrared ISBTs are restricted by In segregation and carrier leakage [51], in Al(Ga)N-GaN QWs they are strongly influenced by the intrinsic electric fields due to spontaneous and piezoelectric polarizations [52].

Finally, recent reports by Akimoto et al. have suggested the possibility of using II-VI QW systems such as CdS/ZnSe QWs grown on BeTe substrates [26] for realizing near-infrared ISBTs with subpicosecond scattering rates.

10.3.2 Properties of antimonide quantum wells

InGaAs-AlAsSb QWs lattice-matched to InP have been found to be ideally suited for near-infrared ISB optoelectronic devices due to their large conduction band offset [54], large [$\chi^{(3)}$] [55], and subpicosecond optical response [56, 57]. In these QWs the well and barrier layers consist of $In_{0.53}Ga_{0.47}As$ and $AlAs_{0.56}Sb_{0.44}$, respectively, both of which are lattice-matched to the InP substrate. Compared to AlAs, AlAsSb provides a larger conduction band offset with InGaAs, which is favorable for ISBTs at the optical communication wavelengths of 1.3 and 1.55 μm. On the other hand, interband transitions in these QWs are relatively weak due to the staggered (type II) lineup with the valence band states in the InGaAs layers being unconfined. In general, interband transitions in Sb-based systems can be both direct and indirect. For the ideal case of perfectly smooth interfaces, the InGaAs/AlAsSb system can be converted to a type I QW either by reducing the Ga fraction from 47% to 25% or by reducing the Sb fraction from 44% to 30%. The conduction band offset in the resulting strained QWs is still sufficiently large to yield ISBTs around 1.55 μm, with a relatively strong and fast interband absorption around 1.3 μm compared to the lattice-matched structure. This type I InGaAs/AlAsSb system is also less strained than InGaAs/AlAs or AlGaN/GaN QWs operating in the near-infrared region, and its ISB absorption is not limited by carrier leakage from the conduction band states due to intervalley transitions.

However, the immiscibility gap of the Sb material system makes it very difficult to form abrupt interfaces, which affects the ISB absorption characteristics. In general, interface abruptness is an important issue in devices based on ISBTs, as the transition energy is a very sensitive function of the well width. To achieve ISBTs at communication wavelengths, the well width needs to be very narrow, usually 6 to 9 *monolayers* (MLs), so that even a single-ML fluctuation in well width causes a significant shift in the transition energy. By using dimeric group V sources and with As termination at the heterojunction planes, the interfaces become more abrupt and the band alignment of InGaAs/AlAsSb QWs is shown to be type I with a conduction band offset of 1.6 eV.

One of the unique features of the InGaAs/AlAsSb material system is the effect of doping or n-type carrier density on the interface properties and band offset. High doping in the well or barrier is necessary for devices based on ISBTs, and it can lead to shifts in the transition energies, especially in narrow QWs. In the case of InGaAs-AlAsSb QWs with a uniform doping concentration of $\sim 1 \times 10^{19}$ cm^{-3} in the barriers, an extremely high level of out-diffusion of In and Ga atoms from the wells to the barriers, and of Al atoms from the barriers to the wells (especially from the surface side AlAsSb barrier layer), has been observed [86].

Normally, at the usual growth temperatures of MBE, bulk diffusion is not operative due to the lack of vacancies, and the atomic arrangements are determined during growth by surface or near-surface processes. The surface mobility of atoms is necessary to obtain a high-quality material, but can simultaneously lead to an exchange between substrate atoms and impinging atoms. For arsenides, the direction and extent of the segregation process follow the order In > Ga > Al. Indium and gallium show different diffusion behavior: In tends to diffuse toward the surface side, while Ga diffuses toward the substrate side. In doped InGaAs-AlAsSb QWs, indium and gallium diffusion could be caused by charged vacancies produced in AlAsSb by high doping with Si. Incorporation of Sb into the InGaAs wells is also clearly observed. The diffusion rates of In, Ga, and Al atoms in undoped systems are significantly lower.

The basic material parameters relevant for all-optical ISB processes are summarized in Table 10.1. It is observed that GaN-based QWs have the largest scattering rate and saturation intensity due to their comparatively higher LO-phonon energy. Furthermore, the InGaAs system has the lightest effective mass, leading to more efficient electron transport compared to the GaN and ZnSe systems. In addition, the state of material synthesis of nitride or II-VI compounds is still immature.

Thus, the remainder of this chapter focuses on ISB modulation and switching in InP-based antimonide QWs.

TABLE 10.1 Material System for Near-Infrared ISBT

Parameters	InGaAs	GaN	ZnSe
Effective mass m_e^*/m_0	0.042	0.2	2.7
LO-phonon energy $\hbar\omega_{LO}$ (meV)	36	88	31
Dielectric const.			
ε_s	14.1	9.5	8.1
ε_∞	11.6	5.35	5.75
Basic scattering rate W_0 (ps^{-1})	6.7	121	40
Dephasing time T_2	Slow	Fast	Fast
Saturation intensity l_s	Low	High	High

InGaAs QW: InGaAs/AlAs/Inp (MIT); InGaAs/AlAs/GaAs (Kyoto Univ., Stanford Univ.); InGaAs/AlAs/Inp (FESTA, NTT).

Nitride QW: GaN/AlGaN (Toshiba, Lucent Tech., Sophia Univ., Univ. Tokyo).

II-VI QW: CdS/ZnSe/BeTe (AIST).

10.3.3 Interband transitions in $In_{0.53}Ga_{0.47}As/AlAs_{0.56}Sb_{0.44}$ quantum wells

Despite the potentialities of InGaAs/AlAsSb QWs, there have been very few reports of their optical properties until recently, due to the difficulties involved in growing the AlAsSb material which includes two group V elements with a large miscibility gap [54]. The composition control of these alloys becomes problematic, because of the different incorporation rates of the two group V elements. In particular, the incorporation of Sb and As into $AlAs_ySb_{1-y}$ is reported to depend critically on the substrate temperature and on the Sb and As fluxes [58].

The valence band maximum of the AlAsSb barrier is located about 70 meV above the valence band maximum of InGaAs, whereas the lowest conduction band minimum resides in the InGaAs layers [59]. The conduction band offset varies depending on the growth condition in the range of 1.6 to 1.7 eV at the Γ point and therefore provides excellent electronic confinement. This type II (staggered) band alignment leads to a spatial separation of electrons and holes, and it allows for interband optical emmision at a photon energy smaller than the bandgap energy of both materials forming the heterostructure. The structural quality of the interfaces is more important in such type II systems as compared to type I heterostructures such as GaAs/AlGaAs, because of the carrier localization at the interfaces where the spatially indirect optical transitions take place.

A schematic band diagram of the InGaAs/AlAsSb heterostructure is given in Fig. 10.5. In this system electrons (holes) are confined in the

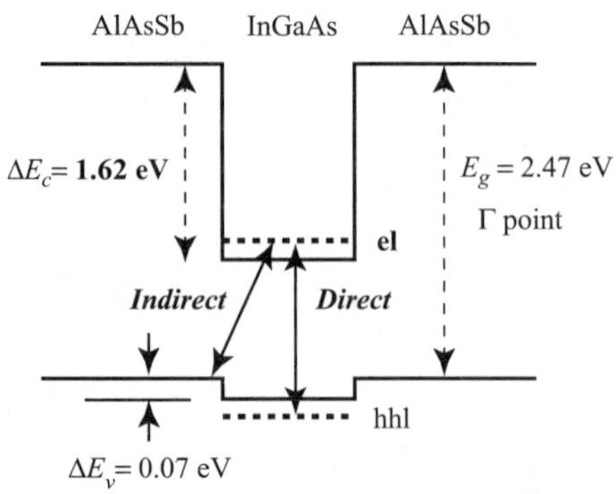

Figure 10.5 Band diagram of a type II InGaAs/AlAsSb heterostructure.

InGaAs (AlAsSb) layers. Thus, the lowest-energy interband transition is spatially indirect (type II). Another possible optical transition in this structure is a direct (type I) transition inside the InGaAs layer between a confined electron state and a quasi-bound hole state. It is observed that the nature of the interband transitions in these InGaAs/AlAsSb heterostructures also depends on the width of the InGaAs layers.

Photoluminescence (PL) measurements of interband transitions in InGaAs/AlAsSb QWs show both type I and type II transitions [59]. This is clearly illustrated in Fig. 10.6*a* for the case of 20-nm InGaAs layers in a low-excitation regime.

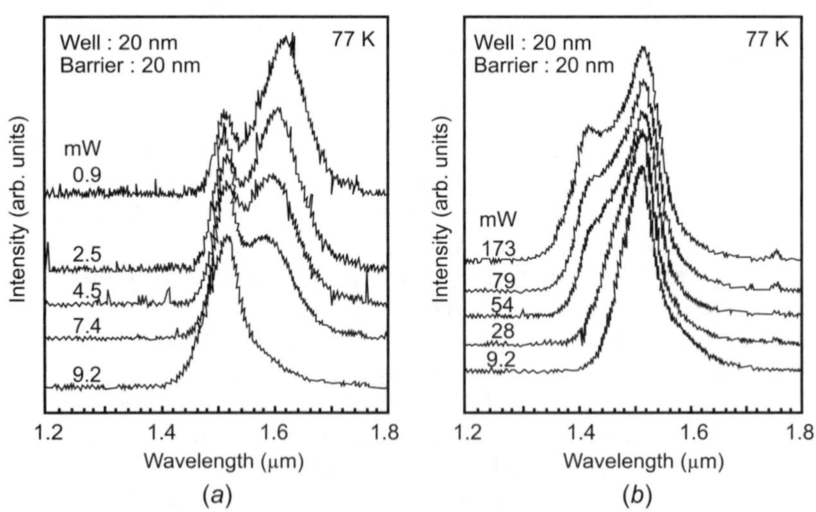

Figure 10.6 Excitation power dependence of the low-temperature PL spectrum of a 20-nm unintentionally doped InGaAs/AlAsSb QW. Results obtained (*a*) at low excitation powers and (*b*) at high excitation powers.

1. The lower energy peaks in this figure correspond to type II transitions between electrons confined in the InGaAs well and holes confined in the AlAsSb barriers.
2. The higher-energy peaks correspond to direct type I transitions associated with the formation of novel excitons involving a confined electron and a quasi-bound hole, both in the InGaAs layer.

A transition caused by the recombination of confined electrons in the first excited subband and quasi-bound holes, both in the InGaAs layer, is also observed in this 20-nm InGaAs heterostructure in the higher excitation power regime (Fig.10.6*b*).

CdSe/ZnTe [60] and ZnSe/BeTe [61] heterostructures also exhibit a direct type I transitions in type II QWs. A comparison of the band structures of such systems explains these optical transitions involving a quasi-bound state in a single barrier layer. In material systems with large band offsets such as CdSe/ZnTe and ZnSe/BeTe, continuum or above-barrier states exist that are strongly confined in the barrier layers. They can be explained by simply considering the interference of electron waves reflected at the interfaces (planes of potential discontinuity). However, the small valence band offset of 70 meV in InGaAs/AlAsSb QWs does not provide sufficiently strong reflections to effectively confine the above-barrier states within the barrier layers. This is similar to the behavior of below-barrier states in a shallow well, where significant wave function spreading into neighboring layers occurs.

Another possible mechanism that can effectively confine the above-barrier states is the Coulomb attraction between electrons and holes. The presence of spatially separated InGaAs and AlAsSb layers with high densities of electrons and holes, respectively, produces an electric field which in turn gives rise to strong bending of the conduction and valence bands [62]. The resulting local potential minima in the InGaAs layers can confine the majority of photo-generated holes and thus enhance the intensity of the direct (type I) transitions in structures with wide InGaAs layers. On the other hand, as the InGaAs layer thickness is decreased, the intensity of these direct transitions becomes weak. This explains the observed well width dependence of the PL spectra shown in Fig. 10.7.

Quantum confinement in InGaAs-AlAsSb QWs was first observed by measuring the PL due to interband indirect transitions as a function of well width in narrow InGaAs layers. This is shown in Fig. 10.8, where the PL spectra are normalized to their respective peak intensities. In these measurements, the excitation power was 2 mW, and all the QWs were uniformly doped to 10^{19} cm^{-3}. As expected, the PL peak position undergoes a blue shift as the QW width is reduced. These PL spectra

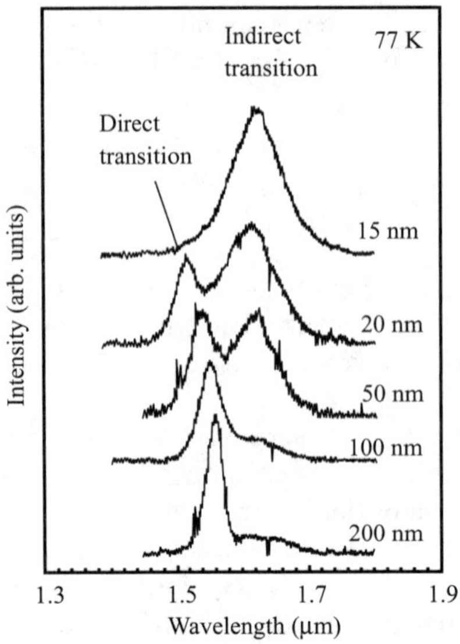

Figure 10.7 Photoluminescence spectra from 15-, 20-, 50-, 100-, and 200-nm InGaAs samples. The AlAsSb layers are 200 nm thick. Both direct (type I) and indirect (type II) transitions are observed.

are relatively broad (80 to 100 meV) compared, e.g., to typical AlGaAs/GaAs QWs, because of the indirect nature of these transitions in type II structures, and also likely due to the interface steps characteristic of the InGaAs/AlAsSb heterojunction. As discussed earlier, the diffusion of group III atoms is significant, especially in the presence of high-Si

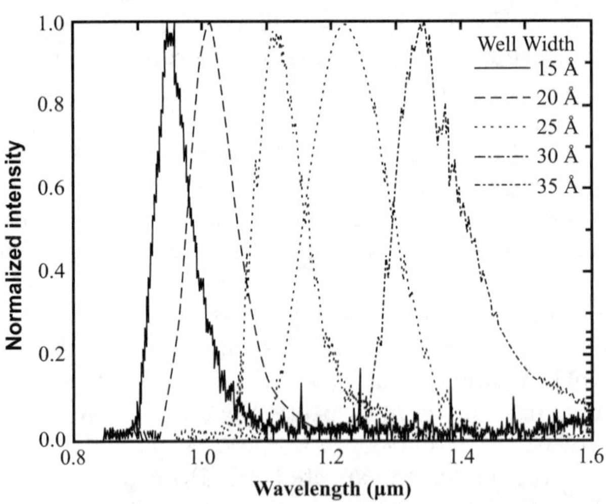

Figure 10.8 Well width dependence of the low-temperature PL spectrum of InGaAs/AlAsSb QWs doped to 10^{19} cm^{-3}, measured with an excitation power of 2 mW.

doping concentrations, and can cause changes in the band alignment of the system. In particular, band offset calculations predict a change in band alignment from type I to type II as a result of the formation of a GaAsSb layer at the interface.

As a first step toward the use of ISBTs for the modulation of interband light absorption in these QWs, the spontaneous interband carrier relaxation rate without ISB excitation was studied with time-resolved PL measurements [56]. Experimental results from a 20-Å InGaAs/AlAsSb QW sample with a Si doping density of 10^{19} cm^{-3} are shown in Fig. 10.9. The sample was pumped with optical pulses from a frequency-doubled mode-locked Ti : Sapphire laser at 100 MHz, and the PL signal was measured by using a streak camera. The pump wavelength was 385 nm, well above the absorption edge of the AlAsSb barriers. Figure 10.9*a* shows PL spectra from this sample for an excitation power of 10 mW measured by integrating between a time delay of 0.2 and 0.8 ns. A broad PL line width of 55 meV is observed at 10 K, with a major contribution due to the intermixing at the interface leading to well width fluctuations.

Figure 10.9*b* shows the time-resolved PL decay observed at various wavelengths for an excitation power of 5 mW. In type I QWs, the PL lifetime is usually shorter than in bulk materials, because of the stronger overlap of the electron and hole wave functions in quantum confined structures. Similarly, one should expect a faster PL decay involving confined QW states compared to the decay from unbound barrier states. The oscillator strength of a type II transitions is also expected to be only a few percent of that of type I transitions. These expectations are confirmed by transmission spectra measurements. The interband carrier lifetime in the sample of Fig. 10.9*b* exhibits two components: a fast component related to intraband transitions from higher-k states, mainly due to scattering induced by interface roughness; and a slower component corresponding to the interband transitions. The faster carrier relaxation is observed at energies slightly higher than the resonant PL peak energy. A bi-exponential decay is clearly observed for the luminescence at 980 and 1000 nm. For instance, the luminescence at 980 nm exhibits a fast component with a decay time of 450 ps, followed by a slower relaxation of 2.1 ns. This relaxation rate is comparable to or slightly longer than the carrier lifetime measured in InGaAs/InAlAs or InGaAs/InP type I QWs [63, 64], and is much faster than in other type II InGaAs QWs which exhibit a transition (Γ-X and X-X) lifetime of 15 ns [65]. This indicates that the doped InGaAs/AlAsSb system used in this work has either a type I structure or a type II structure with an extremely shallow valence band offset. It can also be observed from Fig. 10.9*b* that the relaxation at lower energies (< 1.24 eV) is dominated by the slower component, and

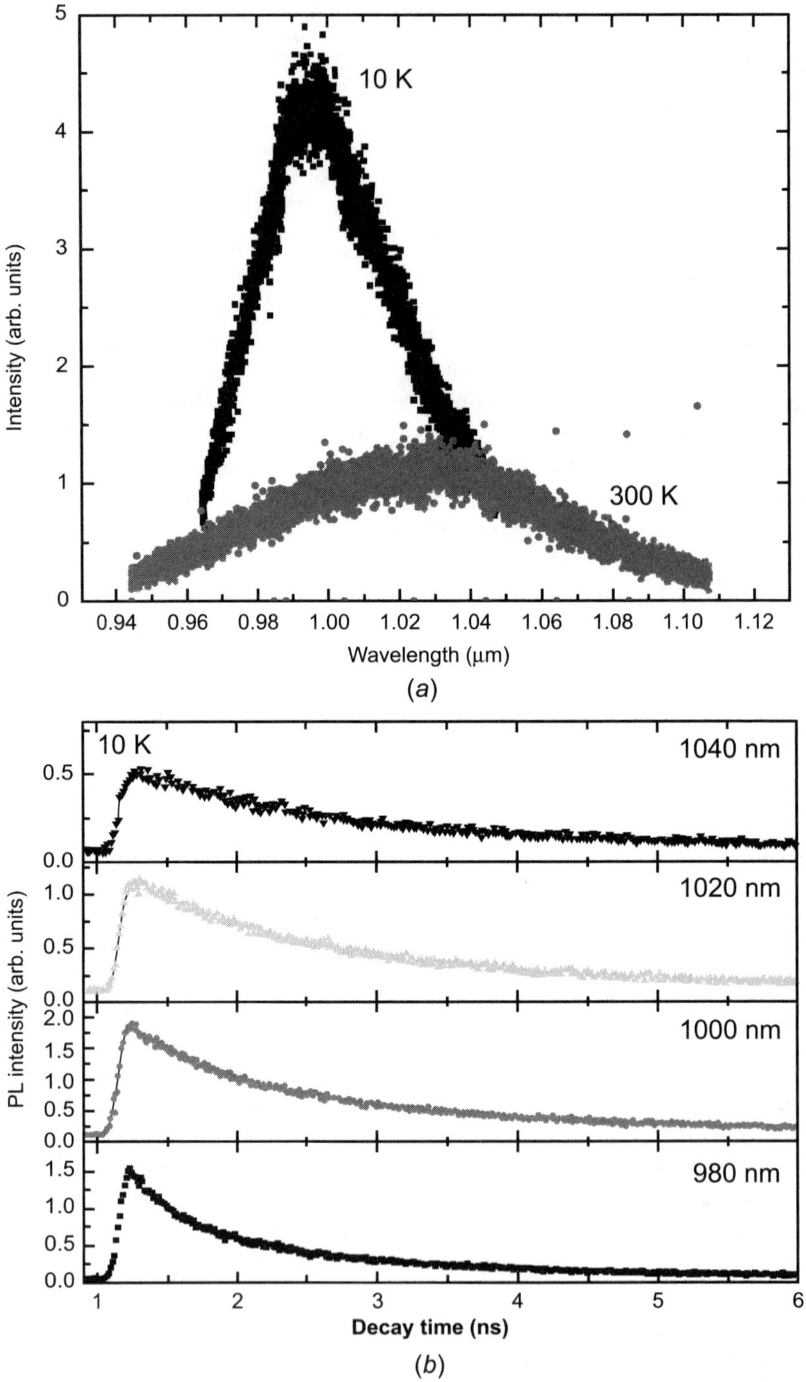

Figure 10.9 (*a*) Photoluminescence spectra of a 20-Å InGaAs/AlAsSb QW sample at 10 and 300 K. (*b*) PL decay at a pumping level of 5 mW for various luminescence wavelengths.

the decay process essentially reduces to a single exponential with a lifetime of ~2 to 3 ns.

10.3.4 Near-infrared intersubband transitions in $In_{0.53}Ga_{0.47}As/AlAs_{0.56}Sb_{0.44}$ quantum wells

Single-quantum-well systems. Intersubband transitions in InGaAs/AlAsSb QWs depend critically on the conduction band offset, which is strongly influenced by the Si doping level. The larger the doping concentration (above 10^{18} cm^{-3}), the lower the interface quality and the smaller the conduction band offset. To study ISB absorption in these heterostructures, near-infrared polarization-resolved transmission spectra were measured at room temperature with a BRUKER IFS 66 vs FTIR, using a near-infrared white-light source and a cooled InSb detector. Figure 10.10 shows ISB absorption spectra measured in these experiments [53]. It is observed that at high carrier concentrations ($\rho \sim 10^{19}$ cm^{-3}), the ISBT wavelength decreases with decreasing well width but tends to saturate around 1.9 to 2.0 μm for wells narrower than 2 nm. The shortest transition wavelength achieved with normal growth conditions was 1.9 μm for a 1.5-nm well. At the same time, the ISB absorption was found to be considerably higher compared to strained QWs; for example, in the case of a 3.0-nm well, the peak ISBT absorption coefficient was measured to be about 5000 cm^{-1} at 2.4 μm. The observed transition wavelength is red-shifted compared to theoretical estimates based on a simple envelope wave function approximation. This deviation increases for narrower wells, as the interface gradient of 2 to 3 monolayers/interface becomes comparable to the intended well thickness of 5 to 6 monolayers. The interface quality and the well width uniformity are significantly improved by reducing the doping level. For example, the ISBT energy can be reduced to 1.72 μm by lowering the carrier concentration to 10^{17} cm^{-3}.

The AlAsSb-InGaAs heterointerface is likely to form a terracelike stepped structure due to the intermixing of group V atoms [66]. This asymmetric QW structure results in an increase of the ground subband energy and a lowering of the upper conduction subband on reducing the well width, leading to a saturation of the ISBT wavelength. The reduction in ISB absorption strength with well width observed in Fig. 10.10 is also associated with the decrease in the active layer thickness (and number of carriers in the QWs), since the well number has been maintained constant. The conduction band offset at the InGaAs/AlAsSb heterojunction is insufficient to yield ISBTs at 1.5 μm, especially for heavily doped QWs. By reducing the doping level to ~3×10^{17} cm^3 it is

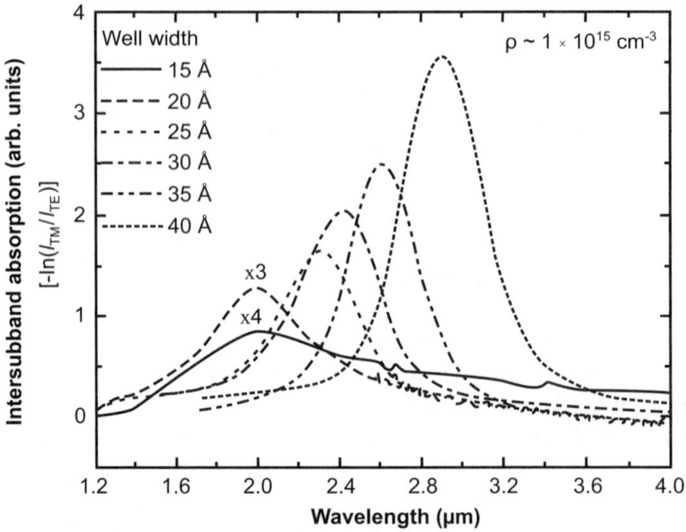

Figure 10.10 Well width dependence of the ISB absorption spectrum of uncoupled InGaAs/AlAsSb QWs, where the absorption is estimated from the logarithmic ratio of TM and TE transmission spectra. Each sample consists of a repetition of 88 QWs uniformly doped to 10^{19} cm^{-3}.

possible to shorten the ISBT wavelength to the communication wavelength regime, but at the expense of a significant drop in ISB absorption strength. A more effective approach to achieve ISBTs at 1.3 and 1.55 μm with appreciable absorption suitable for device applications is to use strongly coupled QW system.

Coupled quantum well systems. *Coupled double quantum well* (C-DQW) structures provide unique opportunities for engineering novel semiconductor systems with large optical nonlinearities in the infrared [67]. This approach has inherent advantages for ultrafast optical switching and modulation due to its flexibility in tailoring quantized energy levels and carrier relaxation processes [68]. In C-DQWs, the ISBT energies can be controlled through the splitting of quantized electron states that occurs when the barrier thickness is reduced and the coupling between the wells becomes sufficiently strong (Fig. 10.11*a*). As a result, larger ISB energy separations can be achieved with wider well widths compared to single QWs [69]. Also note that the ISB transition wavelength and absorption strength in these structures depend on the doping level and temperature due to the carrier-induced space-charge electric field [70].

ISB absorption spectra calculated for a doped ($n = 1 \times 10^{18}$ cm^{-3}) InGaAs-AlAsSb C-DQW system using Bastard's three-band model are

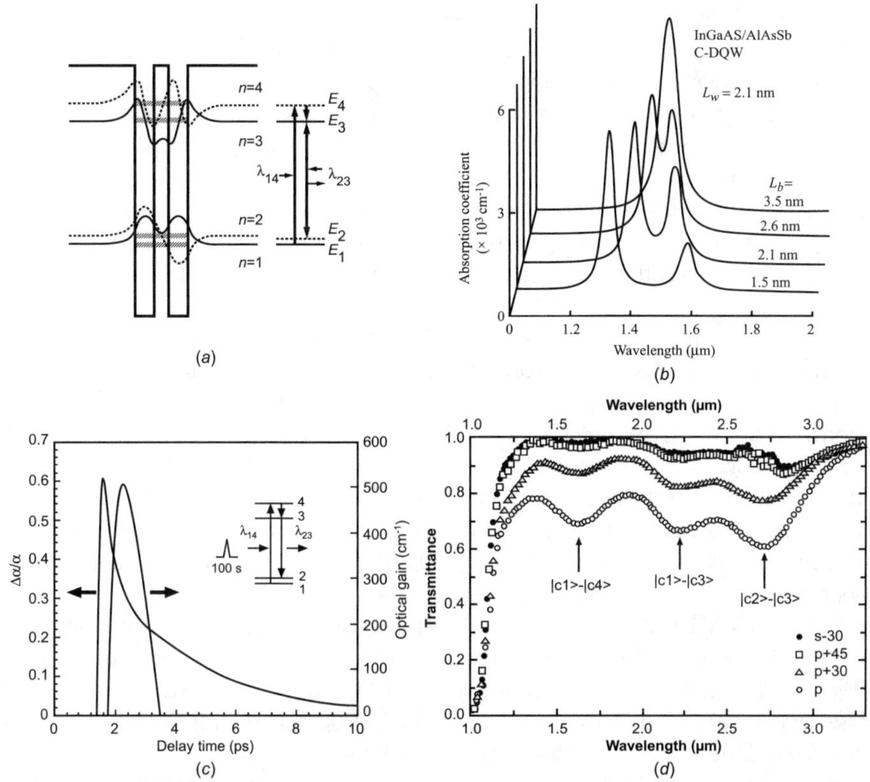

Figure 10.11 (*a*) Schematic conduction band diagram of a coupled double QW system. (*b*) Calculated dependence of the ISB absorption spectra on the central barrier width in InGaAs/AlAsSb C-DQWs. (*c*) Calculated transient response of the 1–4 absorption and the 3–2 optical gain of a C-DQW designed for optical switching between 1.3 and 1.55 μm, after irradiation of a control beam resonant to the 1–4 transition. (*d*) Measured transmission spectra of an InGaAs/AlAsSb C-DQW structure with an AlAs central barrier.

shown in Fig. 10.11*b* for different widths of the center barrier. As shown in the figure, when this barrier is sufficiently narrow, two separate absorption peaks corresponding to the 1–4 and 2–3 transitions are observed. All-optical absorption modulation can be achieved based on the interaction between these two ISBTs. Specifically, the population difference between the second and third subbands can be reduced by irradiating a light beam (control light) resonant to the 1–4 transition, assuming that the 4–3 relaxation is sufficiently fast. The latter condition can be realized by designing the 4–3 energy difference to be larger than and close to the LO-phonon energy. As a result, the absorption of the 2–3 transition decreases in the presence of the control wave, and modulation of a signal beam resonant to this transition is achieved. Although the absorption coefficient slightly decreases as the barrier thickness is reduced, it remains larger than 10^3 cm^{-1}, which is an order of magnitude larger than that of the 1–4 transition in conventional uncoupled QWs, and sufficiently large for device operation.

The design of the subband structure also has considerable influence on the response speed (Fig. 10.11*c*). For example, rate equation calculations have indicated that the ISB absorption recovery time can be reduced to 300 fs in an optimized triple-coupled QW structure [71]. At the same time, the optical switching energy can be significantly reduced compared to single QWs due to enhanced nonlinear optical coefficients.

Intersubband transitions in InGaAs/AlAsSb C-DQWs are significantly influenced by the strong intermixing of In, Sb, and As in the well and barrier layers, which causes substantial interfacial roughness. To deal with this problem, the use of an AlAs (instead of AlAsSb) central barrier layer has been introduced to reduce intermixing and thus improve the quality and uniformity of the QWs [55, 56]. Shown in Fig. 10.11*d* are the measured polarization-resolved near-infrared transmission spectra of such a C-DQW structure designed for ISB absorption at 1.55 μm, consisting of two nominally identical wells of ~2.7 nm (9 ML) width. At room temperature the interband absorption edge of these QWs occurs around 1.2 to 1.3 μm. The spectra of Fig. 10.11*d* reveal an asymmetry in the C-DQW structure induced by width fluctuations of about 0.5 to 1.0 ML in both the well and the central barrier. Due to this asymmetry an absorption dip is observed near 2.2 μm corresponding to the |c1>–|c3> transition, which is forbidden in symmetric structure due to parity selection rules. The other two dominant transitions in the near-infrared regime are due to the |c1>–|c4> transition around 1.55 μm and the |c2>–|c3> transition around 2.7 μm, which also indicates that at 300 K the Fermi level lies above the |c2> state in this particular structure. The magnitude and spectral broadening are similar for the different ISBTs in this structure, but in general depend critically on the actual sheet density. The absorption coefficient $\alpha_{|c1>-|c4>}$ for the |c1>–|c4> transition is 850 cm^{-1}.

A schematic description of all-optical switching in a waveguide based on InGaAs-AlAsSb QWs is shown in Fig. 10.12. An active (modulating) layer consisting of the C-DQW structure just described can be employed to provide the desired large and ultrafast optical nonlinearity. In this device, a control wave resonant to the 1–4 transition can be used to modulate the transmission of an input signal resonant to the 2–3 transition, through modification of the subband carrier densities via ISB absorption. As a result, only those pulses of the input signal that are coupled in the waveguide (nearly) simultaneously with a control pulse are transmitted out.

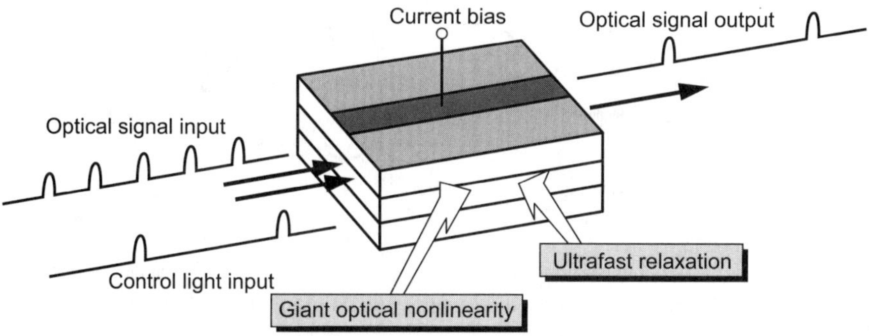

Figure 10.12 Schematic description of all-optical modulation or switching in a waveguide device.

10.4 All-Optical Modulation of Interband Resonant Light via Intersubband Resonant Light

Absorption of interband resonant light can be modified in the presence of an ISB resonant control wave. The ultrafast modulation of a *continuous wave* (cw) interband signal based on this approach was first demonstrated by Noda et al. using a resonantly excited free-carrier induced nonlinearity in doped QWs [72]. The response time of this all-optical modulation scheme is on the order of 1 ps, determined by the ISB energy relaxation rate [73, 74]. Its efficiency depends on the ISB excitation intensity. To achieve high nonlinearities in the femtosecond regime, we have proposed the use of ultrashort light pulses both for the ISB control wave and for the IB signal [34]. In a three-level undoped QW system, the manipulation of interband resonant light absorption by ISB resonant coupling light is analogous to the phenomenon of electromagnetically induced transparency in atomic systems [75]. In doped QWs the all-optical switching is due to a real carrier density modulation.

To describe this all-optical modulation process, we consider a three-level QW system as shown in Fig. 10.13. A weak probe (or signal) wave is tuned resonantly to the interband transition (from the heavy-hole subband $|hh1>$ to the first conduction subband $|c1>$ at $k_{II} = 0$), while a control wave consisting of a train of ultrashort pulses is tuned to the ISBT between the conduction subbands $|c1>$ and $|c2>$. The one-electron density matrix equation of motion under the rotating-wave approximation can be used to adequately describe the population distribution and transition probabilities among these various subbands. The polarizations of the interband signal wave and ISB control wave are assumed to be parallel (TE) and perpendicular (TM)

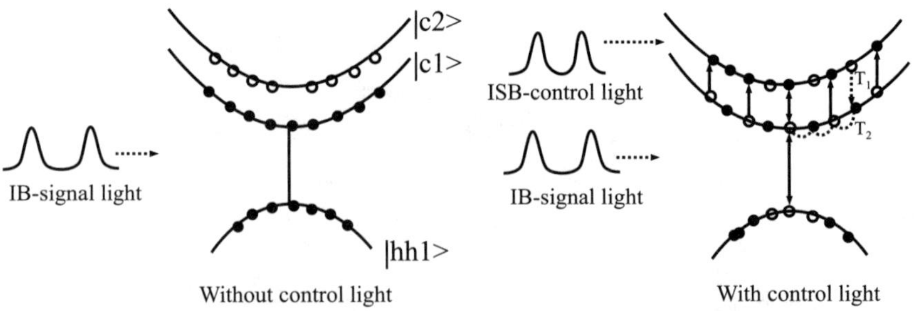

Figure 10.13 Schematic description of all-optical modulation of interband resonant light via ISB-Ts.

to the QW layer, respectively, so as to maximize the relevant transition dipole moments.

The optical fields of the signal or probe wave and of the control wave can be written as

$$\mathcal{F}_p(t) = F_p(t)\exp\left(-i\omega_p t\right) + c.c. \tag{10.1}$$

and

$$\mathcal{F}_c(t) = F_c(t)\exp\left(-i\omega_c t\right) + c.c. \tag{10.2}$$

respectively. The total electronic polarization P in the QWs arising from the simultaneous presence of these interband probing field and ISB coupling field is

$$\begin{aligned} P(t) &= \varepsilon_0\,\chi(t)\left[F_p(t) + F_c(t)\right] \\ &= \frac{2}{V}\sum_k\left[\mu_{ab}\rho_{ba}(t) + \mu_{ba}\rho_{ab}(t) + \mu_{bc}\rho_{cb}(t) + \mu_{cb}\rho_{bc}(t)\right] \end{aligned} \tag{10.3}$$

where $\chi(t)$ is the complex susceptibility; the subscripts a, b, and c refer to the three subbands $|hh1>$, $|c1>$, and $|c2>$, respectively; the summation extends over all values of the in-plane wave vector; V is the sample volume; and the factor of 2 arises from the spin degeneracy. The probe absorption coefficient α can be obtained from the imaginary part Im $[\chi_{ba}]$ of the linear susceptibility of the medium at the probe frequency, by assuming that the polarization due to the probe field varies slowly on the scale of the optical period.

The generalized density matrix equations for the diagonal and off-diagonal elements are obtained from a time-dependent perturbative analysis as follows [38, 44]:

$$\frac{\partial \rho_{jj}(t)}{\partial t} = \left[(\partial \rho\ /\partial t)_{\text{relax}}\right]_{jj} + i\left[\Omega_{ji},\ f_{c,p}(t)\rho_{ij}(t)\right.$$
$$- \Omega_{lj} f_{c,p}(t)\rho_{ji}(t) + \Omega_{jk} f_{c,p}(t)\rho_{kj}(t)$$
$$\left.- \Omega_{kj} f_{c,p}(t)\rho_{jk}(t)\right]$$

and (10.4)

$$\frac{\partial \rho_{jk}(t)}{\partial t} = -\left(i\Delta_{jk} + \Gamma_{jk}\right)\rho_{jk}(t) + i\{\Omega_{jk} f_{c,p}(t)\left[\rho_{kk}(t)\right.$$
$$\left.- \rho_{jj}(t)\right] + \Omega_{ji} f_{c,p}(t)\rho_{ik}(t)$$
$$- \Omega_{lk} f_{c,p}(t)\rho_{ji}(t)\}$$

where Ω_{ji} is the Rabi frequency of the optical field tuned to the transition between states $|j>$ and $|i>$; $f_c(t)$ and $f_p(t)$ are the pulse shapes of the control and signal waves, respectively; Γ_{ji} is the dephasing rate between levels j and i; and $\hbar\Delta_{jk} = E_j - E_k - \hbar\omega_{c,p}$ is the detuning of the relevant optical field from its resonant transition energy. The relaxation hamiltonian matrix for the QW system includes a phenomenological description of the many-body effects leading to transverse phase relaxation in the off-diagonal elements, and to energy relaxation of the carriers in the diagonal elements [76]. LO-phonon scattering is considered to be the dominant dephasing process in these structures.

The effective steady-state interband absorption coefficient can be expressed as

$$\begin{aligned}\alpha\left(\omega_p\right) &= \alpha^{(1)}\left(\omega_p\right) + \alpha^{(3)}\left(\omega_p\right)\\ &= \alpha_1 - \chi^{(3)}\left(\omega_p,\ -\omega_p,\ \omega_p\right)\cdot I_p\\ &\quad + \chi^{(3)}\left(\omega_p, -\omega_c, \omega_c\right)\cdot I_c\end{aligned} \tag{10.5}$$

where $I_p = |F_p|^2$, $I_c = |F_c|^2$, and $\chi^{(3)}$ $(\omega_p, -\omega_p, \omega_p)$ and $\chi^{(3)}$ $(\omega_p, -\omega_c, \omega_c)$ are the nonlinear absorption coefficients due to interband and ISB resonant light, respectively. The latter is the dominant third-order nonlinear optical susceptibility and is several orders of magnitude larger than the interband transitions related term $\chi^{(3)}$ $(\omega_p, -\omega_p, \omega_p)$.

As the origin of the all-optical modulation or switching lies in the carrier density modulation by the control light pulses, the carrier concentration in the conduction subband $|c1>$ effectively determines the change in the absorption coefficient of the signal wave due to the ISB transitions. If we neglect coherent nonlinear effects, the modulation depth can be estimated from the difference in interband absorption with the $|c1\rangle$ states partly occupied due to doping (in the absence of control light) and with the $|c1\rangle$ states empty (in the presence of control light). Theoretical spectra of this interband absorption coefficient, computed with the model just described, are plotted in Fig. 10.14, in the presence (solid lines) and absence (dotted lines) of the control wave, for three different values of the QW doping density. As shown in the figure, the maximum change in absorption occurs near the band edge of the QW, where the resonant third-order nonlinear susceptibility is largest. Furthermore, the effect of the ISB control light is enhanced at higher carrier concentrations, where the lower k states are completely occupied, resulting in a very small interband absorption

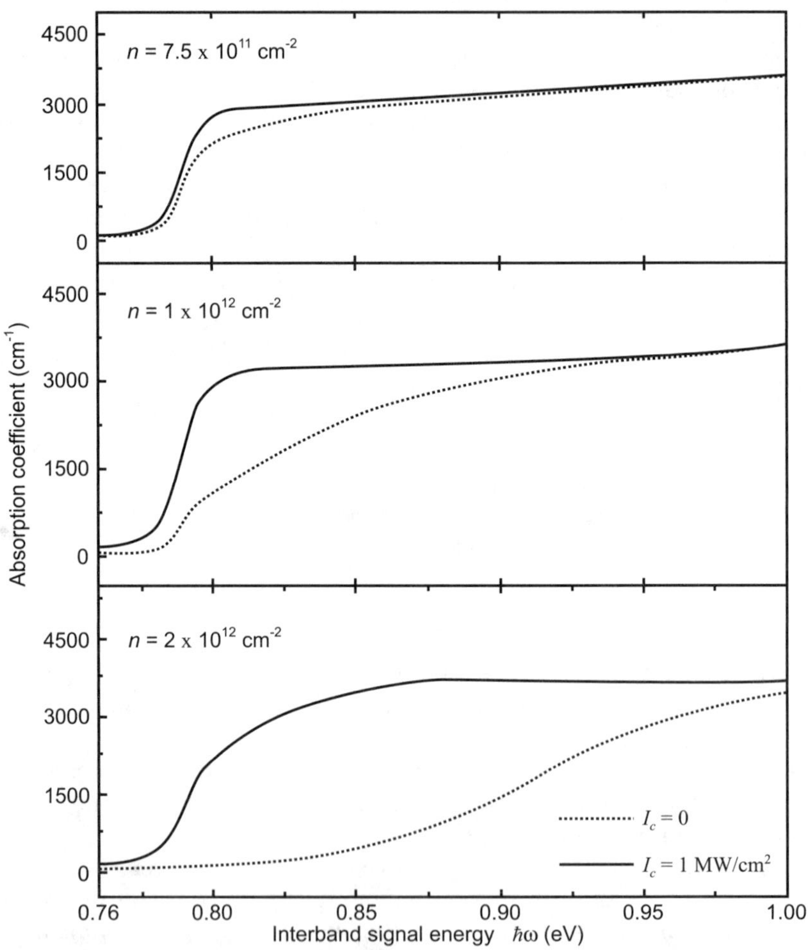

Figure 10.14 Steady-state absorption coefficient of interband resonant (signal) light in the three-level system of Fig. 10.13, for different values of the QW doping concentration. Solid (dotted) lines represent the absorption in presence (absence) of the ISB control beam.

in the absence of control light. The ISB control wave effectively induces a red shift in the interband absorption edge, which is proportional to the ISB energy separation in case of highly doped QWs. However, many-body effects will influence the modulation process and limit the maximum achievable modulation depth.

10.4.1 All-optical modulation or absorption recovery in InGaAs/AlAsSb single quantum wells

Figure 10.15 shows a pump-probe measurement system used to evaluate the absorption recovery dynamics associated with ISBTs. Pump pulses with a width of 150 fs and a wavelength of 1.95 μm are used to excite ISBTs in InGaAs/AlAsSb QWs, while probe pulses with a spectrum broadened by a sapphire plate (about 1μm) are used to monitor the resulting interband absorption changes, as indicated in the inset of the figure. Figure 10.16 shows examplary pump-probe data measured with an 80-period InGaAs/AlAsSb MQW sample with 2-nm wells doped to 1×10^{19} cm^{-3}. From this plot, which shows the interband (probe) absorption change as a function of time delay after the ISB excitation, a relaxation time of 2.1 ps is determined [77].

Initially, in the absence of ISB pump light, the conduction subband states below the Fermi level are populated, and interband absorption involving these states does not occur. When the ISB resonant light is coupled to the sample, carriers from the bottom of the first conduction subband $|c1\rangle$ are pumped to the second subband on a time scale of ~200 fs governed by the width of the pump pulses [78]. The resulting empty states in $|c1\rangle$ lead to an enhancement of the interband absorption rate from $|hh1\rangle$ to $|c1\rangle$, or equivalently to a red shift in the interband absorption edge [79]. The states in $|c1\rangle$ are subsequently filled through two simultaneous processes – the excitation of carriers from the valence

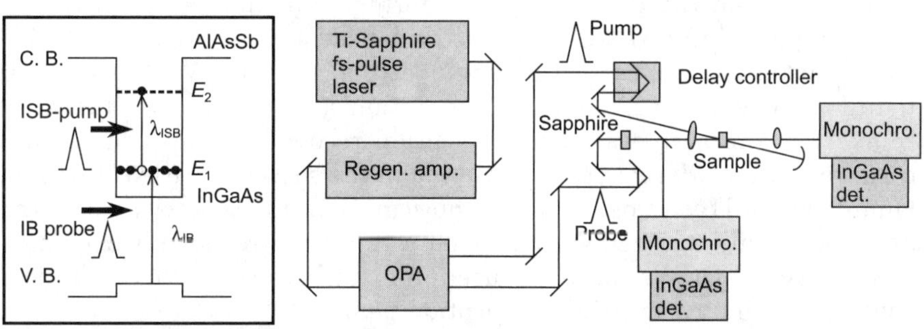

Figure 10.15 Pump-probe measurement system used to study ISB relaxation dynamics in InGaAs/AlAsSb QWs. The inset indicates the functions of the pump and probe beams. The optical parametric oscillator (OPA) generates an output beam near 1.9 μm and an idler beam whose spectrum is broadened by a sapphire crystal around 1 μm.

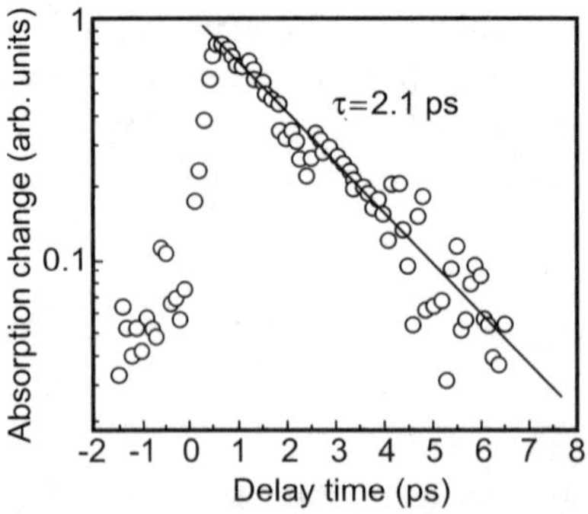

Figure 10.16 Temporal change of interband absorption coefficient in an InGaAs/AlAsSb MQW sample in the presence of ISB excitation.

subbands by interband transitions and the relaxation of carriers from excited conduction subband states by ISB and intrasubband transitions. Such relaxation mainly involves LO-phonon emission and hot carrier effects and occurs on a time scale of a few picoseconds. Intrasubband relaxation is dominant at lower probe photon energies, as the empty states created near the bottom of $|c1\rangle$ are immediately filled up through intrasubband relaxation from higher k states. This explains the particularly fast absorption recovery observed for lower-energy (longer wavelength) interband resonant light, as compared to signals probing higher-energy states, which are filled only through ISB relaxation process. The optical switching or modulation process has also been investigated for slightly off-resonant ISB pump pulses.

10.4.2 Ultrafast all-optical modulation in coupled quantum well structures

To gate OTDM signals beyond 1 Tb/s, all-optical switching with response time faster than 1 ps is required. However, the ISB relaxation time for near-infrared wavelengths shorter than 2 μm is slower than 2 ps in single InGaAs/AlAsSb QWs, as evidenced by the data of Fig. 10.16. Using optimized wave function engineering, coupled QWs can be used to tailor the electron-LO phonon interaction so as to overcome this slow relaxation and accelerate the absorption recovery time to the subpicosecond range. Moreover, as discussed in the section entitled "Coupled Quantum Well Systems," ISBTs at optical communication wavelengths are more easily realized with coupled well structures than with single QWs. Also C-DQWs may be used for multiwavelength all-optical switching utilizing multiple absorption peaks.

To demonstrate ultrafast all-optical modulation at speeds exceeding 1 Tb/s, a C-DQW was fabricated consisting of 2.8-nm-thick $In_{0.8}Ga_{0.2}As$

wells coupled by 4-monolayer AlAs central barriers, and separated by thick AlAsSb layers with a thin AlAs termination [57, 71]. The QWs were doped with Si at 1.5×10^{19} cm^{-3}. As shown in Fig. 10.17*a*, these samples exhibit two ISB absorption peaks, one due to the $|c1\rangle - |c4\rangle$ transition at ~1.6 μm and the other due to $|c2\rangle - |c3\rangle$ transition at ~2.2 μm. The absorption coefficient at 1.6 μm was estimated to be 1200 cm^{-1}. The transient interband absorption change induced by ISB resonant pulses at 1.62 μm, as depicted in Fig. 10.17*b*, is presented in Fig. 10.17*c*. For probe wavelengths in the range 1.1 to 1.25 μm, the induced interband absorption shows ultrafast decay (500 to 800 fs), considerably faster than in the case of single QWs (2 to 3 ps) or even C-DQWs with lower doping density. The transient spectra near the interband absorption edge were integrated in wavelength to estimate the net absorption recovery time: A value of about 600 fs was obtained from the initial fast decay of this integrated interband absorption change. Such fast decay is mainly due to the accelerated recovery of electrons in the first subband due to ultrafast relaxation of electrons from the second subband.

More recently, Naruse et al. utilized the optical modulation of interband resonant light by ISB resonant light in InGaAs/AlAsSb C-DQWs to demonstrate optical gating operation at communication wavelengths [40]. First, they pointed out that an all-optical NOT gate can be realized by using the ISB resonant light as the input signal while simultaneously irradiating the device with interband resonant light. The output signal is the modulated interband resonant wave. In the presence of the input signal, the incident interband resonant beam is absorbed so that the output signal is a logic 0. And vice versa, when there is no input ISB resonant light, the device is transparent for the interband resonant wave and the output signal is a logic 1. The NOT truth table can therefore be implemented all-optically by using this device.

NOR operation was then proposed and demonstrated in a proof-of-principle experiment. In this scheme, two input signals (input *A* and input *B*) acting as the ISB resonant light are added incoherently to each other, and they are simultaneously coupled to the device together with an interband resonant wave. The output signal, which is the modulated interband resonant wave, is low when at least one input signal is nonzero and is high otherwise; the NOR gate truth table is therefore achieved. In practice, this is accomplished by properly selecting the signal intensities, since ISBTs cause absorption saturation. Incidentally, the number of inputs may be more than two. This NOR gate could be used for detecting a specific element among distinct channels, e.g., to find an empty time slot for OTDM add/drop [80] or an empty buffer among channels, NOR-based matching [85], and so forth.

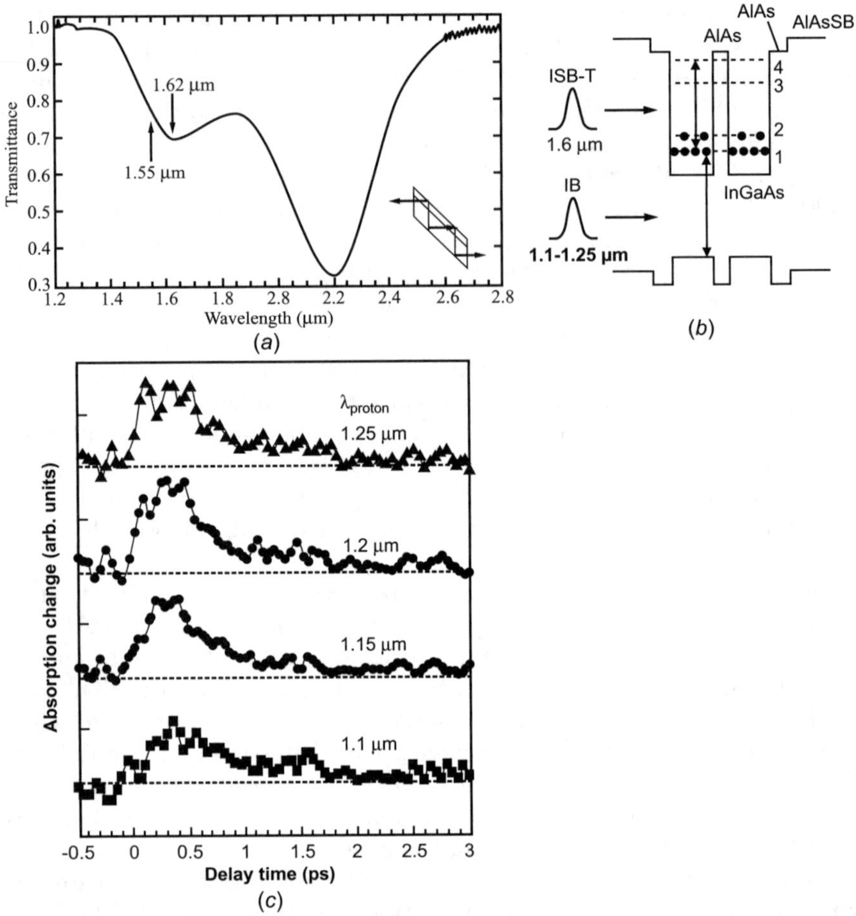

Figure 10.17 (*a*) Near-infrared *p*-polarized transmission spectrum of an InGaAs/AlAsSb C-DQW structure with AlAs central barrier and spacer layers. (*b*) Schematics of ISB pump and interband probe experiment in the C-DQW structure shown. (*c*) Measured interband absorption recovery at different wavelengths in the range from 1.1 to 1.5 μm, in the presence of an ISB pump wave at 1.6 μm resonant to the 1–4 transition of the C-DQW system shown in (*b*). (Figure kindly provided by H. Yoshida, T. Simoyama, A. V. Gopal, J. Kasai, T. Mozume, and H. Ishikawa.)

10.5 Intersubband All-Optical Switching in Coupled InGaAs/AlAsSb Quantum Wells

In coupled QWs (Fig. 10.11*a*), the relaxation time between closely spaced subbands |*c*1>–|*c*2> or |*c*3>–|*c*4> can be less than 200 to 300 fs, which is one order of magnitude shorter than the relaxation time between widely separated subbands such as |*c*2>–|*c*3> [68]. Theoretical calculations based on the rate equation model predict that the (|*c*1>–|*c*4>) absorption recovery time can be shortened not only by utilizing C-DQW structures but also by increasing the electron density in the QWs. When the Fermi level is below the first conduction subband ($\rho < 10^{18}$ cm^{-3}), the absorption recovery time is weakly dependent on the electron density. And vice versa, for carrier densities $> 10^{19}$ cm^{-3}, with the Fermi level lying between the closely spaced subbands |*c*1>–|*c*2>, the recovery time is strongly dependent on the carrier density and is on the order of 600 to 700 ps, which is about 3 times faster than in a single-QW structure. The carrier relaxation process and absorption recovery time also depend on the optical excitation intensity. Although the recovery becomes slower as the pulse energy is increased, it remains faster than 700 fs for energies upto 1 pJ/μm^2.

The ISB carrier relaxation lifetime and absorption recovery time of the 1–4 transition have been measured in a pump-probe experiment [71]. A C-DQW ridge waveguide was fabricated with ISB absorption at the optical communication wavelength of 1.55 μm. Both the pump and probe pulses were generated by using an optical parametric oscillator (300-fs pulse width, 80-MHz repetition rate) and polarized along the TM direction for ISB absorption. The waveguide consisted of an 80-period C-DQW structure sandwiched between InAlAs cladding layers, as shown in Fig. 10.18*a*. The ISB absorption spectrum of this C-DQW structure, measured at various temperatures, is shown in Fig. 10.18*b*. In the pump-probe setup, the transmitted probe pulses were selectively measured by obtaining a beat signal with a reference shift of 1 MHz. An ultrafast ISB decay time of 690 fs was observed, and the corresponding absorption recovery time was estimated to be 730 fs. Akiyama and colleagues then used this device to demonstrate ultrafast demultiplexing (DEMUX) of a series of four pulses at an effective rate of 1 THz, by changing the relative timing of the control pulse (Fig. 10.18*c*) [84]. This all-optical switching was performed with a switching energy of 27 pJ, and extensive effort is underway in the FESTA Laboratories to optimize the QW structure for a lower operating threshold. In addition, the feasibility of this device for operation with high-repetition-rate control pulses has also been investigated [71]. Although there are relatively slow decay components (such as due to the |*c*2>–|*c*3> relaxation path), 0.3- to 1-THz operation without significant pattern effects has been shown to be possible in an appropriate device configuration, by increasing the control pulse energy.

10.6 Enhancement of the Figure of Merit of Sb-Based Optical Modulators and Switches

The efficiency of ISBT-based Tb/s DEMUX and 2R (pulse *retiming* and *reshaping*) regeneration is defined in terms of the switching

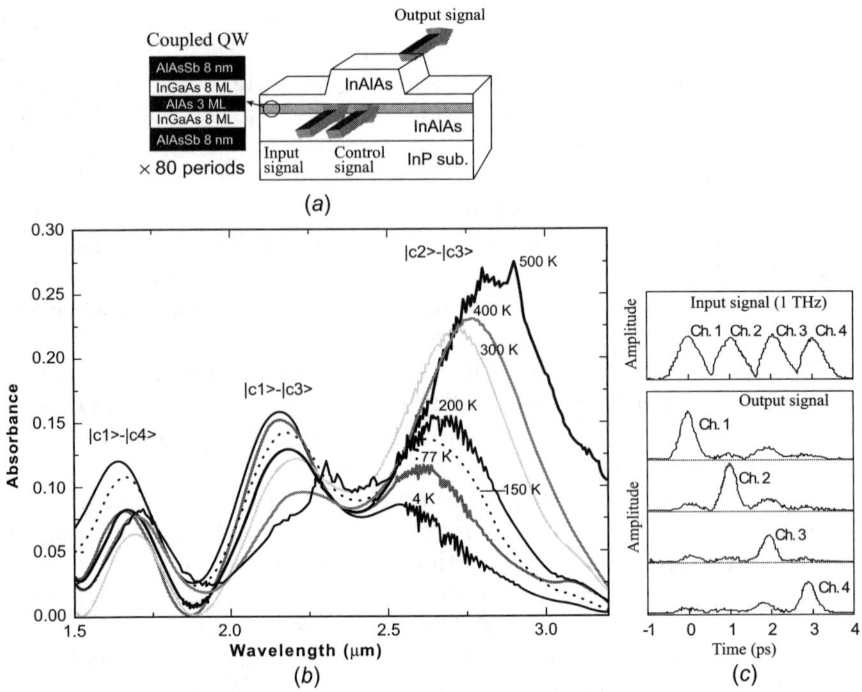

Figure 10.18 (*a*) Ridge waveguide all-optical switch incorporating an InGaAs/AlAsSb C-DQW system. (*b*) ISB absorbance spectra of the C-DQW structure at $L_w \sim 2.7$ nm, $L_b = 1.0$ nm, and $\rho = 10^{19}$ cm^{-3}, measured at various temperatures. (*c*) Input and output traces showing demultiplexing operation with the waveguide of (*a*) at an equivalent repetition rate of 1 THz.

response time and the switching energy. A commonly used *figure of merit* (FoM) is defined by the ratio of the third-order susceptibility $\{\chi^{(3)}\}$ and the product of the linear absorption coefficient α_0 and the relaxation time τ. The switching energy, related to the ISB absorption saturation intensity I_s, is proportional to the inverse of this FoM. Efficient ISB all-optical switching therefore requires a large $\chi^{(3)}$ and/or weak absorption and fast relaxation. Due to the interdependence of these three parameters on the dipole moment μ, optimizing the FoM in single QWs is nontrivial. Furthermore, interfacial roughness and interfacial phonon modes in QWs based on a same material system can lead to a wide variety of scattering processes, resulting in a large variation in saturation intensity. These processes further complicate the optimization of I_s. However, general guidelines can be drawn from the following expression, which is valid for a two-level system such as single QWs, under steady-state excitation and with homogeneous broadening:

$$I_s \approx \frac{cn\varepsilon_0\hbar^2}{2\mu^2\,\tau\,T_2} \tag{10.6}$$

where T_2 is the dephasing time, n is the refractive index, c is the velocity of light in vacuum, $\hbar$ is Planck's constant, and ε_0 is the vacuum dielectric constant.

To measure I_s, a tunable *optical parametric amplifier* (OPA) pumped by the output of a regenerative amplifier at 800 nm was used, with pulse width of 150 to 200 fs, a repetition rate of 1 kHz, wavelength in the range from 1.7 to 2.5 μm, and intensities up to 10^9 to 10^{10} W·cm^{-2} [55]. The sample transmission was measured as a function of the incident light intensity over a dynamic range spanning more than three decades using attenuation filters. A comparison of the interband and ISB transition energies in typical InGaAs/AlAsSb QWs reveals that the latter is close to the interband two-photon transition energy, leading to the possibility of interband transitions at high optical powers due to multiphoton processes. In narrower wells with larger conduction subband energy separation (as required for ISBTs at communication wavelengths) in the presence of optical pumping by ISB resonant light, *two-photon absorption* (TPA) in the InGaAs buffer layer or InP substrate is also likely to occur.

Figure 10.19*a* depicts the measured change in transmittance as a function of OPA input power at 2.4 μm through a 200-period MQW sample with a well width of 30 Å and peak ISB absorption at 2.4 μm. In the weak field limit there is very little transmission of the incident light due to the strong ISB absorption; saturation of this absorption is then observed with increasing input power; finally, if the OPA (average) power is increased beyond a few hundred microwatts, strong absorption is again observed. This is due to interband TPA, which is allowed when the ground state conduction subband is depopulated by optical pumping to the upper state via ISBTs. The interband gap occurs at 1.0 eV, which is nearly at the two-photon resonance of the ISB pump excitation with a photon energy of ~0.5 eV. The transmission of s-polarized light is shown in the inset of Fig. 10.19*a*. Since s-polarized light does not excite ISBTs, it does not undergo any interband TPA as the first conduction subband is highly occupied in these doped QWs. TPA at this particular wavelength occurs at a much higher intensity compared to I_s, and therefore it would not affect the performance of ISB optical switches operating at lower input power levels. In narrower QWs, as the resonant ISBT energy increases ($\hbar\omega_{\mathrm{ISB}} \sim 0.65$ eV) and the TPA energy ($2\hbar\omega_{\mathrm{ISB}}$) exceeds the interband transition energy, the threshold power for TPA is expected to decrease, thereby affecting the ISBT. This is a

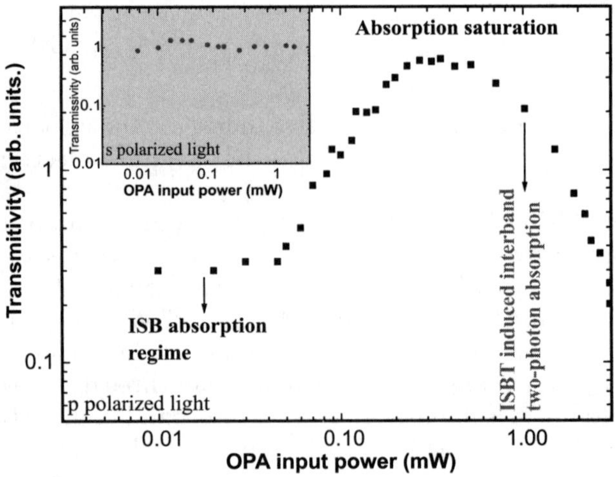

(*a*)

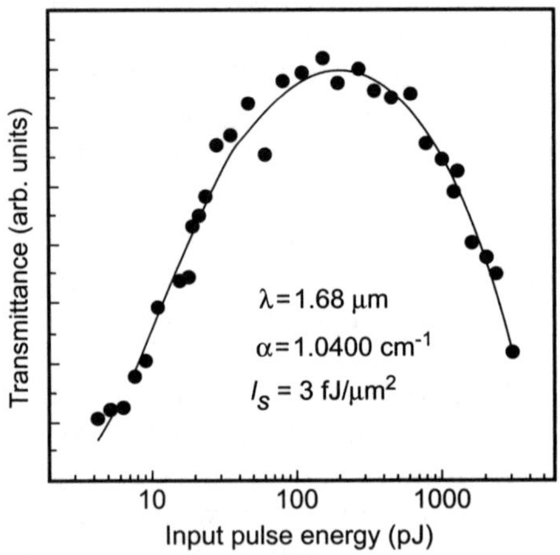

(*b*)

Figure 10.19 (*a*) Absorption saturation of ISB resonant light at 2.4 μm in 3.0-nm-wide lattice-matched single QWs. (*b*) Absorption saturation data measured with 90-fs pulses at 1.68 μm in strained QWs with AlAs spacer layers at the interfaces. This material shows a very low saturation energy of about 3 fJ/μm^2.

unique feature of this narrow-bandgap material system with large conduction band offset [82].

The saturation intensity can be estimated from the above data by fitting them to an expression of the form $\alpha(I) = \alpha(0)/(1 + I/I_s) + \beta I$, where $\alpha(0)$ is the linear absorption coefficient, I is the incident intensity, and β is the two-photon absorption coefficient. The saturation intensity depends on the wavelength of the optical excitation. In 2.0-nm InGaAs uncoupled QWs with ISB absorption peak at 1.88 μm, $\alpha(0)$ and I_s in the presence of excitation at 1.72 and 2.2 μm are found to be, respectively, 781 and 719 cm^{-1}, and 4.2 and 5400 MW/cm^2. The nonlinear susceptibility $\chi^{(3)}$ at these wavelengths, that is, 1.72 μm, 1.88 and 2.2 μm, is 1.5×10^{-14}, 6.6×10^{-15}, and 1.2×10^{-17} m^2/V^2, respectively [55].

As discussed in the preceding sections, μ and τ can be optimized by quantum design in coupled QW systems. At the same time, such optimization also involves choosing the right materials. In particular, to achieve long dephasing times and large peak value of $\chi^{(3)}$ for ISBTs near 1.55 μm, InGaAs with a large indium composition of about 72% (and hence relatively small bandgap) is more efficient. Improvements in the interface quality and the density of confined carriers in such $In_{0.72}Ga_{0.28}As$/AlAsSb QWs can be achieved by introducing a 1.5-nm-thick strain-compensating AlAs layer at the well-barrier interface. With this material choice, a reduction in I_s has been obtained and attributed to increased doping density, reduced inhomogeneities, and reduced effective masses m^* leading to lower carrier-carrier scattering rates [83]. In fact, at 1.68 μm these strained QWs with high In content show a very low saturation energy of about 3 fJ/μm^2 (Fig. 10.19*b*). Incidentally it should be mentioned that while one would expect the reduced effective mass of these QWs to play a role, calculations do not show any large dependence of dephasing time on m^* [84]. In any case, by using Eq. 10.6, which is valid for a strictly homogeneously broadened system, it is predicted that the saturation intensity can be lowered by a factor of about 45 compared to lattice-matched QWs. Further improvements are expected from a reduction in layer-width inhomogeneities.

10.7 Conclusions

All-optical modulation and switching in the terabits per second regime can be achieved using ISBTs in InGaAs/AlAsSb QWs. This material system is lattice-matched to InP substrates and is ideally suited for the communication wavelength region. A waveguide structure has been used to achieve ultrafast (equivalent to 1 THz) optical-time-division demultiplexing operation at 1.55 μm, with control pulse energy as low as 27 pJ. The nonlinear optical susceptibility, response time, and saturation intensity (which is directly related to the switching energy) can be optimized by using strain-engineered and coupled QW system.

References

1. O. Wada, 2000, *Opt. and Quantum Electron.* **32**, 453.
2. A. Y. Cho, D. L. Sivco, H. M. Ng, C. Gmachl, A. Tredicucci, A. L. Hutchinson, S. N. G. Chu, and F. Capasso, 2001, *J. Cryst. Growth*, **227–228**, 1.

3. T. Kawazoe, K. Kobayashi, S. Sangu, and M. Ohtsu, 2003, *Appl. Phys. Lett.*, **82**, 2957.
4. T. Kamiya, 1999, *Femtosecond Technology—Basic Research to Application Prospects*, T. Kamiya, F. Saito, H. Yajima, and O. Wada, eds., Berlin, Springer, chap. 1, pp. 1–11.
5. S. Nakamura, Y. Ueno, and K. Tajima, 1998, *IEEE Photonics Tech. Lett.* **10**, 1575.
6. M. Kajita, M. Shimizu, A. Neogi, and T. Ishihara, 2001, *Int. J. Mod. Phys.*, **B15**, 3741.
7. R. Takahashi, Y., Kawamura, and H. Iwamura, 1996, *Appl. Phys. Lett.*, **68**, 153.
8. A. Tackeuchi, O. Wada, and Y. Nishikawa, 1997, *Appl. Phys. Lett.*, **70,** 1131.
9. O. Wada, 2004, *New J. Phys.* **6**, 183.
10. P. W. Smith, 1972, *Bell System Tech. J.* **61**, 1975; and 1987, *IEEE Circuits and Devices Mag.* **3**, 9.
11. K. Tajima, 1993, *Jpn. J. Appl. Phys.*, **32**, L1746.
12. S. Nakamura, Y. Ueno, K. Tajima, T. Sugimoto, T. Kato, T. Shimoda, M. Itoh, H. Hatakeyama, T. Tamanuki, and T. Sasaki, 2000, *IEEE Photonics Tech. Lett.*, **12,** 425.
13. A. Hirano, H. Kobayashi, H. Tsuda, R. Takahashi, M. Asobe, K. Sato, and K. Hagimoto, 1998, *IEE Electron. Lett.* **34,** 198.
14. Y. Nishikawa, A. Tackeuchi, M. Yamaguchi, S. Muto, and O. Wada, 1996, *IEEE J. Select. Top. Quant. Electron.* **2,** 661.
15. H. Nobuhara, K. Kondo, S. Yamakoshi, and K. Wakao, 1989, *IEE Electron. Lett.* **25**, 1485.
16. A. Migus, A. Antonetti, D. Hulin, A. Mysyrowicz, H. M. Gibbs, N. Peyghambarian, and J. L. Jewell, 1985, *Appl. Phys. Lett.* **46,** 70.
17. D. Hulin, A. Mysyrowicz, A. Antonetti, A. Migus, W. T. Masselink, H. Morkoc, H. M. Gibbs, and N. Peyghambarian, 1986, *Appl. Phys. Lett.* **46**, 749.
18. R. Jin, J. P. Sokolo, P. A. Harten, C. L. Chuang, S. G. Lee, M. Warren, H. M. Gibbs, N. Peyghambarian, J. N. Pllky, and G. A. Pubanz, 1990, *Appl. Phys. Lett.* **56**, 993.
19. A. Neogi, H. Yoshida, T. Mozume, N. Georgiev, and O. Wada, 2001, *IEEE J. Select. Top. Quant. Electron.* **7**, 710.
20. A. V. Gopal, H. Yoshida, T. Simoyama, N. Georgiev, T. Mozume, and H. Ishikawa, 2003, *IEEE J. Quant. Electron.* **39**, 299.
21. N. Iizuka, K. Kaneko, N. Suzuki, T. Asano, S. Noda, and O. Wada, 2000, *Appl. Phys. Lett.* **77**, 64.
22. Y. R. Shen, *Principles of Nonlinear Optics*, John Wiley, New York, 1984.
23. A. Yariv, *Quantum Electronics*, John Wiley, New York, 1988, chap. 15.

24. S. Lan, A. V. Gopal, K. Kanamoto, and H. Ishikawa, 2004, *Appl. Phys. Lett.* **84,** 5124.

25. T. Mozume, H. Yoshida, A. Neogi, and M. Kudo, 1999, *Jpn. J. Appl. Phys.* **38,** 1286.

26. R. Akimoto, K. Akita, F. Sasaki, and S. Kobayashi, 2002, *Appl. Phys. Lett.* **80**, 2433.

27. C. Gmachl, H. M. Ng, S.-N. G. Chu, and A. Y. Cho, 2000, *Appl. Phys. Lett.* **77,** 3722.

28. Kh. Moumanis, A. Helman, F. Fossard, M. Tchernycheva, A. Lusson, F. H. Julien, B. Damilano, N. Grandjean, and J. Massies, 2003, *Appl. Phys. Lett*, **82**, 868.

29. T. Mozume, J. Kasai, N. Georgiev, T. Simoyama, A. V. Gopal, and H. Yoshida, *Jpn. J. Appl. Phys.*, to be published.

30. J. Faist, F. Capasso, D. L. Sivco, C. Sirtori, A. L. Huchinson, and A. Y. Cho, 1994, *Science* **264**, 553.

31. N. Suzuki and N. Iizuka, 1998, *Jpn. J. Appl. Phys.* **37**, L369.

32. T. Akiyama, N. Georgiev, T. Mozume, H. Yoshida, A. V. Gopal, and O. Wada, 2001, *IEE Electron. Lett.* **37,** 129.

33. O. Wada, A. Tackeuchi, Y. Nishikawa, and T. Nishimura, "Physics and Simulation of Optoelectronic Devices VI," Paper 3283-66, presented at Photonics West, Feb. 1998, San Jose, CA, *Proc. SPIE.*

34. A. Neogi, O. Wada, Y. Takahashi, and H. Kawaguchi, 1998, *Opt. Lett.* **23,** 1212.

35. E. Rosencher, A. Fiore, B. Vinter, V. Berger, Ph. Bois, and J. Nagle, 1996, *Science* **271**, 168.

36. S. Noda, M. Ohya, Y. Muromoto, T. Asano, and A. Sasaki, 1995, *IEEE J. Quant. Electron.* **31**, 1683.

37. M. C. Nuss, W. H. Knox, and U. Koren, 1996, *IEE Electron. Lett.* **32,** 1311.

38. A. Neogi, Y. Takahashi, and H. Kawaguchi, 1996, *IEEE Quant. Electron.* **32**, 701.

39. P. Bhattacharya, D. Klotzkin, O. Qasaimeh, W. Zhou, S. Krishna, and D. Zhu, 2000, *IEEE J. Select. Top. Quant. Electron.* **6,** 426.

40. M. Naruse, H. Yoshida, T. Miyazaki, F. Kubota, and H. Ishikawa, 2005,*IEEE Photon. Tech. Lett.* 17, 1701.

41. M. N. Islam, 1990, *Opt. Lett.* **15**, 417.

42. A. Sharaiha, H. W. Li, F. Marchese, and J. Le Bihan, 1997, *IEE Electron. Lett.* **33,** 323.

43. L. C. West, and S. J. Eglash, 1985, *Appl. Phys. Lett*, **46**, 1156.

44. A. Neogi., H. Yoshida, T. Mozume, and O. Wada, 1999, *J. Appl. Phys.* **85**, 3352.

45. Y. Hirayama, J. H. Smet, L. H. Peng, C. G. Fonstad, and E. P. Ippen, 1994, *Jpn. J. Appl. Phys.* **33**, 890.

46. C. Gmachl, S. V. Frolov, H. M. Ng, S.-N. G. Chu, and A. Y. Cho, 2001, *IEE Electron. Lett*, **37,** 378.
47. I. Waki, C. Kumtornkittikul, Y. Shimogaki, and Y. Nakano, 2003, *Appl. Phys. Lett.* **82,** 4465.
48. N. Iizuka, K. Kaneko, N. Suzuki, T. Asano, S. Noda, and O. Wada, 2000, *Appl. Phys. Lett.* **77**, 648.
49. T. Asano, and S. Noda, 1998, *Jpn. J. Appl. Phys.* **37**, 6020.
50. B. Sung, H. C. Chui, M. M. Fejer, and J. S. Harris, Jr., 1997, *IEE Electron. Lett.* **33**, 818.
51. T. Asano, S. Noda, T. Abe, and A. Sasaki, 1996, *Jpn. J. Appl. Phys.* **35**, 1285.
52. N. Suzuki and N. Iizuka, 1997, *Jpn. J. Appl. Phys.* **36**, L1006.
53. A. Neogi, H. Yoshida, T. Mozume, N. Georgiev, T. Akiyama, and O. Wada, 2000, *Physica E*, **7**, 183.
54. Y. Nakata et al., 1999, *Proc. Mater. Res. Soc. Symp.* **198,** 289.
55. A. V. Gopal, H. Yoshida, A. Neogi, T. Mozume, N. Georgiev, O. Wada, and H. Ishikawa, 2001, *Jpn. J. Appl. Phys.* **40**, L1015.
56. A. Neogi, H. Yoshida, T. Mozume, N. Georgiev, T. Akiyama, A. Takeuchi, and O. Wada, 2001, *Opti. and Quant. Electron.* **33**, 975.
57. A. Neogi, H. Yoshida, T. Mozume, N. Georgiev, and O. Wada, 2001, *IEEE. J. Select. Top. Quant. Electron.* **7**, 710.
58. S. Nemeth, B. Grietens, H. Bender, and G. Borghs, 1997, *Jpn. J. Appl. Phys.* **36,** 3426.
59. T. Mozume, N. Georgiev, H. Yoshida, A. Neogi, and T. Nishimura, 2000, *J. Vac. Sci. Tech.* **B18**, 1586.
60. H. Luo, W. C. Chou, N. Samarth, A. Petrou, and J. K. Furdyana, 1993, *Solid State Comm.* **85**, 691.
61. A. V. Platonov, D. R. Yakovlev, U. Zehnder, V. P. Kochereshko, W. Ossau, F. Fischer, Th. Litz, A. Waag, and G. Landwehr, 1998, *J. Cryst. Growth* **184–185**, 801.
62. S. V. Zaitsev, V. D. Kulakovskii, A. A. Maksimov, D. A. Pronin, I. I. Tartakovskii, N. A. Gippius, M. Th. Litz, F. Fisher, A. Waag, D. R. Yakovlev, W. Ossau, and G. Landwehr, 1997, *JETP Lett.* **66,** 376.
63. S. Gupta, P. K. Bhattacharya, J. Pamulapati, and G. Mourou, 1991, *J. Appl. Phys.* **69**, 3219.
64. U. Cebulla, G. Bacher, A. Forcel, G. Mayer, and W. T. Tsang, 1989, *Phys. Rev.* **B39,** 6257.
65. C. Lugand, T. Benyattou, G. Guillot, T. Venet, M. Gendry, G. Hollinger, and B. Sermage, 1997, *Appl. Phys. Lett.* **70**, 3257.
66. C. Yang, 1999, International Conference on Mid-infrared Optoelectronics Materials and Devices, Paper I1, Institute of Electrical Engineers, London.

67. F. Capasso, C. Sirtori, and A. Y. Cho, 1994, *IEEE J. Quant. Electron.* **30**, 1313.

68. H. Yoshida, T. Mozume, T. Nishimura, and O. Wada, 1998, *IEE Electron. Lett.* **34**, 913.

69. H. Yoshida, T. Mozume, A. Neogi, and O. Wada, 1999, *IEE Electron. Lett.*, **35**, 1103.

70. Y. Lavon, A. Saar, J. Wang, J.-P. Leburton, F. H. Julien, and R. Planel, 1996, *Appl. Phys. Lett.* **69**, 197.

71. T. Akiyama, N. Georgiev, T. Mozume, H. Yoshida, A. V. Gopal, and O. Wada, 2002, *IEEE Photon. Technol. Lett.* **14**, 495.

72. S. Noda, T. Uemura, T. Yamashita, and A. Sasaki, 1992, *IEEE J. Quant. Electron.* **28**, 493.

73. S. Noda, M. Ohya, T. Sakamoto, and A. Sasaki, 1996, *IEEE J. Quant. Electron.* **32**, 448.

74. A. Neogi, Y. Takahashi, and H. Kawaguchi, 1997, *IEEE J. Quant. Electron.* **33**, 2060.

75. D. S. Lee, and K. J. Malloy, 1994, *IEEE J. Quant. Electron.* **30**, 85.

76. C. Weisbuch and B. Vinter, *Quantum Semiconductor Structure Fundamentals and Applications*, Academic Press, Boston, 1991.

77. O. Wada et al., 1999, *Femtosecond Technology—From Basic Research to Application Prospects,* T. Kamiya, F. Saito, H. Yajima, and O. Wada, eds., Springer, Berlin, chap. III-1, pp. 59–78.

78. A. Neogi, H. Yoshia, T. Mozume, N. Georgiev, and O. Wada, 2001, *Jap. J. Appl. Phys., Lett.* **240**, L558.

79. A. Neogi, 1997, *Opt. Commun.*, 1997, **133**, 479.

80. S. Diez, R. Ludwig, and H. G. Weber, 1999, *IEEE Photon. Technol. Lett.* **11**, 60.

81. H. Miyatake, M. Tanaka, and Y. Mori, 2001, *IEEE J. Solid State Circuits* **36**, 956.

82. G. Hendorfer, M. Seto, H. Ruckser, J. Jantsch, M. Helm, G. Brunthaler, W. Jost, H. Obloh, K. Köhler, and D. J. As, 1993, *Phys. Rev.* **B48**, 2328.

83. A. Neogi, H. Yoshida, and O. Wada, 2001, *Phys. Rev.* **B63**, 23530.

84. T. Simoyama, T. Akiyama, T. Mozume, H. Yoshida, A. V. Gopal, and H. Ishikawa, 2002, *Proc. Eur. Conf. Optical Communications (ECOC2002),* Copenhagen, Denmark, Sept., Paper 7.3.1.

85. H. Miyatake, M. Tanaka, and Y. Mori, 2001, "A design for high-speed low-power CMOS fully parallel content-addressable memory macros," *IEEE J. Solid-State Circuits* **36,** 956.

86. T. Mozume et al., 2001, *J. Cryst. Growth* **227–228**, 577.

Index

ABOUT THE EDITOR

Roberto Paiella is an Assistant Professor in the Department of Electrical and Computer Engineering at Boston University. He has extensive research experience in photonics and semiconductor device physics, and has designed and developed several novel devices in the areas of all-optical switching, intersubband optoelectronics, photonic integrated circuits, and high-speed diode lasers. Dr. Paiella holds a B.S. in Electrical Engineering from Columbia University and a Ph.D. in Applied Physics from the California Institute of Technology. Prior to joining Boston University, he worked at Bell Laboratories, Lucent Technologies, and in the Optoelectronics Division of Agere Systems.

ABOUT THE CD-ROM